D1922343

Edited by
Andreas Lendlein and
Adam Sisson

Handbook of
Biodegradable Polymers

Further Reading

Loos, K. (Ed.)

Biocatalysis in Polymer Chemistry

2011
Hardcover
ISBN: 978-3-527-32618-1

Mathers, R. T., Maier, M. A. R. (Eds.)

Green Polymerization Methods

Renewable Starting Materials, Catalysis and Waste Reduction

2011
Hardcover
ISBN: 978-3-527-32625-9

Yu, L.

Biodegradable Polymer Blends and Composites from Renewable Resources

2009
Hardcover
ISBN: 978-0-470-14683-5

Elias, H.-G.

Macromolecules

2009
Hardcover
ISBN: 978-3-527-31171-2

Matyjaszewski, K., Müller, A. H. E. (Eds.)

Controlled and Living Polymerizations

From Mechanisms to Applications

2009
ISBN: 978-3-527-32492-7

Matyjaszewski, K., Gnanou, Y., Leibler, L. (Eds.)

Macromolecular Engineering

Precise Synthesis, Materials Properties, Applications

2007
Hardcover
ISBN: 978-3-527-31446-1

Fessner, W.-D., Anthonsen, T. (Eds.)

Modern Biocatalysis

Stereoselective and Environmentally Friendly Reactions

2009
ISBN: 978-3-527-32071-4

Janssen, L., Moscicki, L. (Eds.)

Thermoplastic Starch

A Green Material for Various Industries

2009
Hardcover
ISBN: 978-3-527-32528-3

Edited by Andreas Lendlein and Adam Sisson

Handbook of Biodegradable Polymers

Synthesis, Characterization and Applications

WILEY-
VCH

WILEY-VCH Verlag GmbH & Co. KGaA

The Editors

Prof. Andreas Lendlein
GKSS Forschungszentrum
Inst. für Chemie
Kantstr. 55
14513 Teltow
Germany

Dr. Adam Sisson
GKSS Forschungszentrum
Zentrum f. Biomaterialentw.
Kantstraße 55
14513 Teltow
Germany

All books published by **Wiley-VCH** are carefully produced. Nevertheless, authors, editors, and publisher do not warrant the information contained in these books, including this book, to be free of errors. Readers are advised to keep in mind that statements, data, illustrations, procedural details or other items may inadvertently be inaccurate.

Library of Congress Card No.: applied for

British Library Cataloguing-in-Publication Data
A catalogue record for this book is available from the British Library.

Bibliographic information published by the Deutsche Nationalbibliothek
The Deutsche Nationalbibliothek lists this publication in the Deutsche Nationalbibliografie; detailed bibliographic data are available on the Internet at < http://dnb.d-nb.de>.

© 2011 Wiley-VCH Verlag & Co. KGaA, Boschstr. 12, 69469 Weinheim, Germany

All rights reserved (including those of translation into other languages). No part of this book may be reproduced in any form – by photoprinting, microfilm, or any other means – nor transmitted or translated into a machine language without written permission from the publishers. Registered names, trademarks, etc. used in this book, even when not specifically marked as such, are not to be considered unprotected by law.

Cover Design Grafik-Design Schulz, Fußgönheim
Typesetting Toppan Best-set Premedia Limited, Hong Kong
Printing and Binding Fabulous Printers Pte Ltd, Singapore

Printed in Singapore
Printed on acid-free paper

ISBN: 978-3-527-32441-5
ePDF ISBN: 978-3-527-63583-2
ePub ISBN: 978-3-527-63582-5
Mobi ISBN: 978-3-527-63584-9
oBook ISBN: 978-3-527-63581-8

Contents

Preface

Degradable polyesters with valuable material properties were pioneered by Carothers at DuPont by utilizing ring-opening polymerization approaches for achieving high molecular weight aliphatic poly(lactic acid)s in the 1930s. As a result of various oil crises, biotechnologically produced poly(hydroxy alkanoates) were keenly investigated as greener, non-fossil fuel based alternatives to petrochemical based commodity plastics from the 1960s onwards. Shortly afterwards, the first copolyesters were utilized as slowly drug releasing matrices and surgical sutures in the medical field. In the latter half of the 20th century, biodegradable polymers developed into a core field involving different scientific disciplines such that these materials are now an integral part of our everyday lives. This field still remains a hotbed of innovation today. There is a burning interest in the use of biodegradable materials in clinical settings. Perusal of the literature will quickly reveal that such materials are the backbone of modern, biomaterial-based approaches in regenerative medicine. Equally, this technology is central to current drug delivery research through biodegradable nanocarriers, microparticles, and erodible implants, which enable sophisticated controlled drug release and targeting. Due to the long historic legacy of polymer research, this field has been able to develop to a point where material compositions and properties can be refined to meet desired, complex requirements. This enables the creation of a highly versatile set of materials as a key component of new technologies. This collected series of texts, written by experts, has been put together to showcase the state of the art in this ever-evolving area of science.

The chapters have been divided into three groups with different themes. Chapters 1–8 introduce specific materials and cover the major classes of polymers that are currently explored or utilized. Chapters 9–14 describe applications of biodegradable polymers, emphasizing the exciting potential of these materials. In the final chapters, 15–16, characterization methods and modelling techniques of biodegradation processes are depicted.

Materials: Lendlein *et al.*, then Ienczak and Aragão, start with up-to-date reviews of the seminal polyesters and biotechnologically produced polyesters, respectively. Other chapters concern polymers with different scission moieties and behaviors. Domb *et al.* provide a comprehensive review of polyanhydrides, which is followed by an excellent overview of poly(ortho esters) contributed by Heller. Amino acid- based materials and degradable polyurethanes make up the subject of the next two chapters by Katsarava and Gomurashvili, then Puiggali *et al.*, respectively. Synthetic polysaccharides, which are related to many naturally occurring biopolymers, are then described at length by Dumitriu, Dräger *et al.* To conclude the individual polymer-class section, biodegradable polyolefins, which are degraded oxidatively, and are intended as degradable commodity plastics, are covered by Wiles *et al.*

Applications: The two chapters by Ikada and Shakesheff give a critical update on the status of biodegradable materials applied in regenerative therapy and then in drug delivery systems. From there, further exciting applications are described; shape-memory polymers and their potential as implant materials in minimally invasive surgery are discussed by Lendlein *et al.*; Huh *et al.* highlight the importance of biodegradable hydrogels for tissue expander applications; Franke *et al.* cover how implants can be used to aid regenerative treatment of mucosal defects in surgery; Khandare and Kumar review the relevance of biodegradable dendrimers and dendritic polymers to the medical field.

Methods: Van der Zee gives a description of the methods used to quantify biodegradability and the implications of biodegradability as a whole; Watanabe and Kawai go on to explain methods used to explore degradation through modelling and simulations.

The aim of this handbook is to provide a reference guide for anyone practising in the exploration or use of biodegradable materials. At the same time, each chapter can be regarded as a stand alone work, which should be of great benefit to readers interested in each specific field. Synthetic considerations, physical properties, and erosion behaviours for each of the major classes of materials are discussed. Likewise, the most up to date innovations and applications are covered in depth. It is possible upon delving into the provided information to really gain a comprehensive understanding of the importance and development of this field into what it is today and what it can become in the future.

We wish to thank all of the participating authors for their excellent contributions towards such a comprehensive work. We would particularly like to pay tribute to two very special authors who sadly passed away during the production time of this handbook. Jorge Heller was a giant in the biomaterials field and pioneered the field of poly(ortho esters). Severian Dimitriu is well known for his series of books on biodegradable materials, which served to inspire and educate countless scientists in this area. Our sincerest thanks go to Gloria Heller and Daniela Dumitriu for their cooperation in completing these chapters. We also acknowledge the untiring administrative support of Karolin Schmälzlin, Sabine Benner and Michael Schroeter, and the expert cooperation from the publishers at Wiley, especially Elke Maase and Heike Nöthe.

Teltow, September 2010

Andreas Lendlein
Adam Sisson

List of Contributors

Gláucia Maria Falcão de Aragão
Federal University of Santa Catarina
Chemical and Food Engineering Department
Florianópolis, SC 88040-900
Brazil

Marc Behl
Center for Biomaterial Development,
Institute of Polymer Research
Helmholtz-Zentrum Geesthacht
Kantstr. 55
14513 Teltow
Germany

Emo Chiellini
University of Pisa
Department of Chemistry and Industrial Chemistry
via Risorgimento 35
Pisa 56126
Italy

Andrea Corti
University of Pisa
Department of Chemistry and Industrial Chemistry
via Risorgimento 35
Pisa 56126
Italy

Salvatore D'Antone
University of Pisa
Department of Chemistry and Industrial Chemistry
via Risorgimento 35
Pisa 56126
Italy

Avi Domb
Hebrew University
School of Pharmacy
Department of Medicinal Chemistry
Jerusalem 91120
Israel

Gerald Dräger
Gottfried Wilhelm Leibniz Universität Hannover
Institut für Organische Chemie
Schneiderberg 1B
30167 Hannover
Germany

Severian Dumitriu[†]
University of Sherbrooke
Department of Chemical Engineering
2400 Boulevard de l'Université
Sherbrooke, Quebec J1K 2R1
Canada

Lourdes Franco
Universitat Politècnica de Catalunya
Departament d'Enginyeria Química
Av. Diagonal 647
08028 Barcelona
Spain

Ralf-Peter Franke
Centre for Biomaterial Development and Berlin-Brandenburg Centre for Regenerative Therapies (BCRT)
Institute of Polymer Research
Helmholtz-Zentrum Geesthacht GmbH
Kantstr. 55
14513 Teltow
Germany
and
University of Ulm
Central Institute for Biomedical Engineering
Department of Biomaterials
89069 Ulm
Germany

Yourong Fu
Akina, Inc.
West Lafayette, IN 47906
USA

Rosemarie Fuhrmann
University of Ulm
Central Institute for Biomedical Engineering
Department of Biomaterials
89069 Ulm
Germany

John Garner
Akina, Inc.
West Lafayette, IN 47906
USA

Zaza Gomurashvili
PEA Technologies
709 Mockingbird Cr.
Escondido, CA 92025
USA

Jorge Heller[†]
PO Box 3519, Ashland, OR 97520
USA

Bernhard Hiebl
Centre for Biomaterial Development and Berlin-Brandenburg Centre for Regenerative Therapies (BCRT)
Institute of Polymer Research
Helmholtz-Zentrum Geesthacht
Kantstr. 55
14513 Teltow
Germany

Kang Moo Huh
Chungnam National University
Department of Polymer Science and Engineering
Daejeon 305-764
South Korea

Jaciane Lutz Ienczak
Federal University of Santa Catarina
Chemical and Food Engineering Department
Florianópolis, SC 88040-900
Brazil

Yoshito Ikada
Nara Medical University
Shijo-cho 840
Kashihara-shi
Nara 634-8521
Japan

Jay Prakash Jain
National Institute of Pharmaceutical
Education and Research (NIPER)
Department of Pharmaceutics
Sector 67
S.A.S. Nagar (Mohali) 160062
India

Friedrich Jung
Centre for Biomaterial Development
and Berlin-Brandenburg Centre for
Regenerative Therapies (BCRT)
Institute of Polymer Research
Helmholtz-Zentrum Geesthacht
Kantstr. 55
14513 Teltow
Germany

Ramaz Katsarava
Iv. Javakhishvili Tbilisi State
University
Institute of Medical Polymers and
Materials
1, Chavchavadze ave.
Tbilisi 0179
Georgia
and
Georgian Technical University
Centre for Medical Polymers and
Biomaterials
77, Kostava str.
Tbilisi 75
Georgia

Fusako Kawai
Kyoto Institute of Technology
Center for Nanomaterials and Devices
Matsugasaki
Sakyo-ku, Kyoto 606-8585
Japan

Jayant Khandare
Piramal Life Sciences Ltd.
Polymer Chem. Grp
1 Nirlon Complex
Off Western Express Highway
Goregaon (E), Mumbai 400063
India

Andreas Krause
Gottfried Wilhelm Leibniz Universität
Hannover
Institut für Organische Chemie
Schneiderberg 1B
30167 Hannover
Germany

Neeraj Kumar
National Institute of Pharmaceutical
Education and Research (NIPER)
Department of Pharmaceutics
Sector 67
S.A.S. Nagar (Mohali) 160062
India

Sanjay Kumar
Piramal Life Sciences Ltd.
Polymer Chem. Grp
1 Nirlon Complex
Off Western Express Highway
Goregaon (E), Mumbai 400063
India

Andreas Lendlein
Center for Biomaterial Development
and Berlin-Brandenburg Center for
Regenerative Therapies, Institute of
Polymer Research
Helmholtz-Zemtrum Geesthacht
Kantstr. 55
14513 Teltow
Germany

Lena Möller
Gottfried Wilhelm Leibniz Universität Hannover
Institut für Organische Chemie
Schneiderberg 1B
30167 Hannover
Germany

Kinam Park
Purdue University
Department of Biomedical Engineering and Pharmaceutics
West Lafayette, IN 47907-2032
USA

Jordi Puiggalí
Universitat Politècnica de Catalunya
Departament d'Enginyeria Química
Av. Diagonal 647
08028 Barcelona
Spain

Dorothee Rickert
Marienhospital Stuttgart
Böheimstrasse 37
70199 Stuttgart
Germany

Alfonso Rodríguez-Galán
Universitat Politècnica de Catalunya
Departament d'Enginyeria Química
Av. Diagonal 647
08028 Barcelona
Spain

Michael Schroeter
Center for Biomaterial Development
Institute of Polymer Research
Helmholtz-Zentrum Geesthacht
Kantstr. 55
14513 Teltow
Germany

Kevin M. Shakesheff
The University of Nottingham
School of Pharmacy, STEM
NG 7 2RD
UK

Adam L. Sisson
Center for Biomaterial Development and Berlin-Brandenburg Center for Regenerative Therapies, Institute of Polymer Research
Helmholtz-Zentrum Geesthacht
Kantstr. 55
14513 Teltow
Germany

Thanh Huyen Tran
Chungnam National University
Department of Polymer Science and Engineering
Daejeon 305-764
South Korea

Masaji Watanabe
Okayama University
Graduate School of Environmental Science
1-1, Naka 3-chome
Tsushima, Okayama 700-8530
Japan

David Mckeen Wiles
Plastichem Consulting
Victoria, BC V8N 5W9
Canada

Maarten van der Zee
Wageningen UR
Food & Biobased Research
P.O. Box 17
6700 AA Wageningen
The Netherlands

Jörg Zotzmann
Center for Biomaterial Development
Institute of Polymer Research
Helmholtz-Zentrum Geesthacht
Kantstr. 55
14513 Teltow
Germany

1
Polyesters

Adam L. Sisson, Michael Schroeter, and Andreas Lendlein

1.1
Historical Background

1.1.1
Biomedical Applications

Biomaterials are defined as any materials intended to interface with biological systems to analyze, treat, or replace any tissue, organ, or function of the body [1]. The current trend in biomaterial development is shifted toward the use of biodegradable materials that have definite advantages in the fields of tissue engineering [2] and drug delivery [3]. The general principle is to use a material that achieves a specific therapeutic task and is subsequently, over time, degraded and removed harmlessly from the body. As an increasingly relevant part of the medical device and controlled release industry, biodegradable polymers are used to fabricate temporary scaffolds for tissue regeneration, medical sutures, and nano- or micro-scale drug delivery vehicles [4–6].

The important properties that are required for biodegradable biomaterials can be summarized as follows:

- Nontoxic and endotoxin-free, aiming to minimalize unwanted foreign body responses upon implantation.
- Degradation time should be matched to the regeneration or required therapy time.
- Mechanical properties must be suited to the required task.
- Degradation products should be nontoxic and readily cleared from the body.
- Material must be easily processed to allow tailoring for the required task.

Although natural polymers such as collagen have been used in medical applications throughout history, synthetic polymers are valuable also, as they allow us to tailor properties such as mechanical strength and erosion behavior. Naturally

Handbook of Biodegradable Polymers: Synthesis, Characterization and Applications, First Edition. Edited by Andreas Lendlein, Adam Sisson.
© 2011 Wiley-VCH Verlag GmbH & Co. KGaA. Published 2011 by Wiley-VCH Verlag GmbH & Co. KGaA.

occurring biopolymers are typically degraded by enzymatic means at a rate that may be difficult to predict clinically. Furthermore, natural polymers may have unwanted side effects arising from inherent biological activity. This has led to the widespread use of biodegradable synthetic polymers in therapeutic applications. Of this class, biodegradable aliphatic polyesters, which are degraded hydrolytically, are by far the most employed.

1.1.2
Poly(Hydroxycarboxylic Acids)

All polyesters are, in principle, hydrolytically degradable. However, only (co)polyesters with short aliphatic chains between ester bonds typically degrade over the time frame required for biomedical applications. The major group of this material are the poly(hydroxycarboxylic acids), which are prepared via ring-opening polymerization of lactones or cyclic diesters. Indeed, the first biodegradable polyester used as a medical suture in the 1960s was based on the polyglycolide. Scheme 1.1 shows the most common monomers and the polymers they produce. These can be summarized as diglycolide, stereogenic dilactides, lactones such as ε-caprolactone and stereogenic β-butyrolactone, the cyclic trimethylene carbonate, and *p*-dioxanone. As the polymerization methods of these monomers are broadly applicable to each, copolymers such as poly(lactide-*co*-glycolide) are readily produced.

Another source of poly(hydroxycarboxylic acids) is from bacteria, which store polyesters as their energy source [7]. These polymers are known as polyhydroxyalkanoates (PHAs) in the literature. The most common polymer derived from bacteria is poly(3-hydroxybutyrate), which has the same structure as the polymer which can be obtained from optically active β-butyrolactone [8]. Poly(3-hydroxybutyrate) formed in this way is strictly stereoregular, showing the (*R*) configuration. Biotechnologically produced polymers are discussed in more details in Chapter 2 of this handbook.

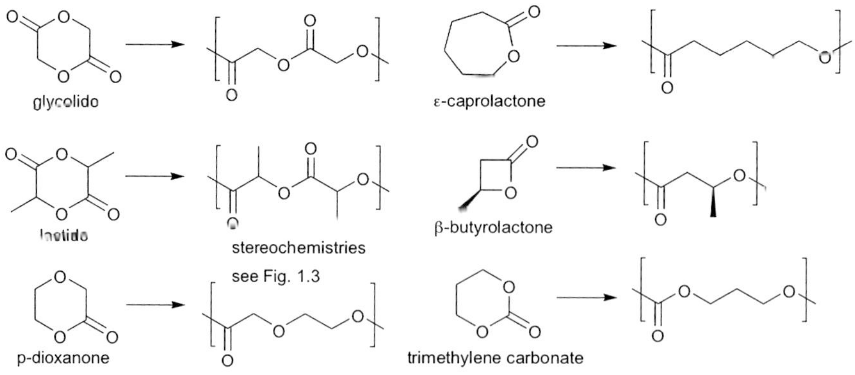

Scheme 1.1 Common cyclic monomers for the preparation of polyester derivatives.

1.2 Preparative Methods

1.2.1 Poly(Hydroxycarboxylic Acid) Syntheses

Polyesters can be synthesized via the direct condensation of alcohols and acids. This may take the form of condensing dialcohols and diacids, for example, AA + BB systems, or the direct condensation of hydroxycaboxylic acid monomers, for example, AB systems. Various catalysts and coupling reagents may be used but typically the polyesters formed in this manner have low and uncontrolled molecular weight and are not suitable for biomedical applications. The majority of cases where a high degree of polymerization was obtained came via ring-opening polymerizations of cyclic monomers of the type shown in Scheme 1.1 [9]. The cyclic dilactones are prepared from the corresponding hydroxycarboxylic acid by elimination of water in the presence of antimony catalysts such as Sb_2O_3 [10]. These dimers have to be purified rigorously if high degrees of polymerization are sought, as impurities such as water and residual hydroxycarboxylic acids can hinder polymerization. Enantiomerically pure lactic acids are typically produced by fermentation.

Ring-opening polymerizations may be initiated by nucleophiles, anionically, cationically, or in the presence of coordinative catalysts. Representative mechanisms are shown in Scheme 1.2. However, precise mechanisms may vary from case to case and are an ongoing important area of study [11, 12]. As a testament to the popularity of the ring-opening polymerization approach, over 100 catalysts were identified for the preparation of polylactide [13].

The typical complex used for the industrial preparation of polyglycolide derivatives is tin(II)-bis-(2-ethylhexanoate), also termed tin(II)octanoate. It is commercially available, easy to handle, and soluble in common organic solvents and in melt monomers. High molecular weight polymers up to 10^6 Da and with narrow polydispersities are obtained in a few hours in bulk at 140–220 °C. Approximately 0.02–0.05 wt% of catalyst is required. Care must be taken when polymerizing dilactides, if stereochemistry is to be preserved. This means that milder conditions are to be selected relative to the homopolymerization of diglycolide.

For the copolymerization of dilactide and diglycolide catalyzed with tin(II) octoate, different reactivities are observed. A chain with a growing glycolide end will add a further diglycolide with a preference of 3:1. With a terminal lactide unit, the preference for diglycolide is 5:1. Due to this, glycolide blocks tend to form, separated by single dilactides. One possibility to improve the homogeneity of the composition of the obtained polyesters is the online control of the monomer ratio by addition of further monomer. However, this method is technically complicated.

The mechanism is a nonionic coordinated insertion mechanism, which is less prone to the side reactions commonly found in ionic polymerizations, such as transesterification or racemization [14, 15]. It has been found that the addition of alcohols to the reaction mixture increases the efficiency of the tin catalyst albeit

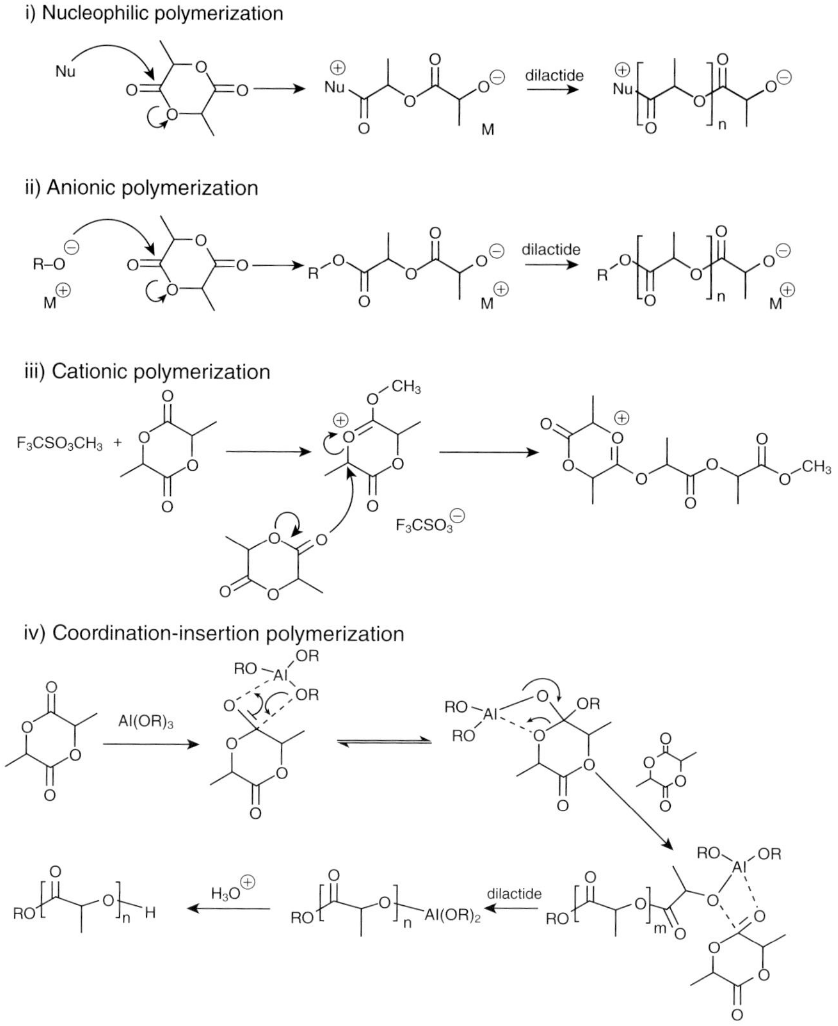

Scheme 1.2 Overview of various mechanisms relevant to polylactide synthesis.

by a disputed mechanism [16]. Although tin(II)octoate has been accepted as a food additive by the U.S. FDA, there are still concerns of using tin catalysts in biomedical applications.

Aluminum alkoxides have been investigated as replacement catalysts. The most commonly used is aluminum isopropoxide, which has been largely used for mechanistic studies [17]. However, these are significantly less active than tin catalysts requiring prolonged reaction times (several hours to days) and affording polymers with molecular weights generally below 10^5 Da. There are also suspected links between aluminum ions and Alzheimer's disease. Zinc complexes, especially

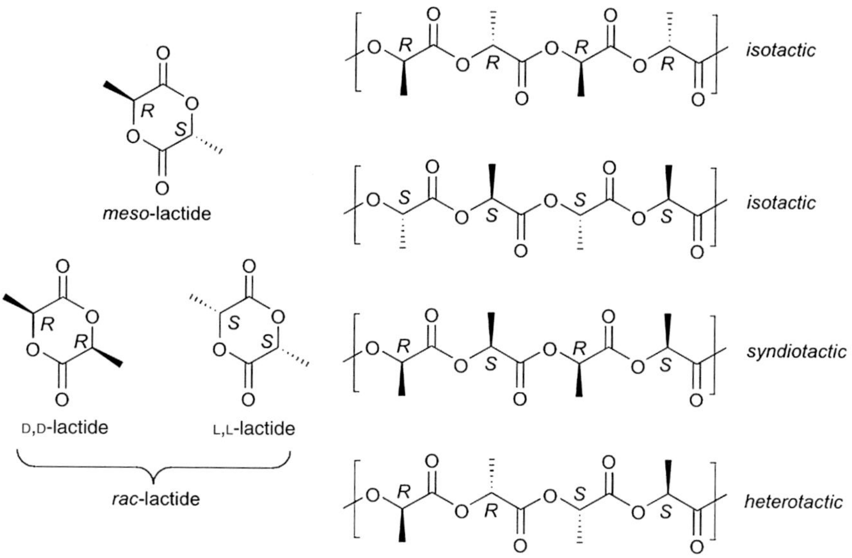

Scheme 1.3 Stereochemical possibilities observed with polylactide synthesis.

zinc(II)lactate, are a plausible replacement with low toxicity and activities of the same order as aluminum complexes [18]. Zinc powder may also be used but then the active species has been identified as zinc(II)lactate in the preparations of polylactide [19]. Iron salts and particularly iron(II)lactate show comparable activity, but prolonged reaction times mean that some racemization occurs in the synthesis of high molecular weight (50,000 Da) poly(L-lactide) [20].

As can be seen in Scheme 1.3, polylactides can exist in isotactic, syndiotactic, and heterotactic blocks. The configuration has obvious consequences for the material properties of the final polymer. While the ring-opening polymerization of L,L-dilactide or D,D-dilactide leads to isotactic polymers, the polymers of the *rac*-dilactide should consist mainly of isotactic diads. This is due to the fact that *rac*-dilactide is commonly used as a mixture of D,D- and L,L-dilactide with very little *meso*-dilactide content. The formation of syndiotactic diads is expected in the case of *meso*-dilactide polymerization, but the longer range sequence structure of such polymers is typically atactic. Due to the expense of producing stereopure dilactides, a kinetic resolution procedure was developed whereby chiral SALEN (salicylimine) ligands in combination with aluminum isopropoxide catalyst produced isotactic polylactides from *rac*-dilactide. Optical purities were high at 50% conversion; kinetics show that the catalyst system has a 28:1 preference toward one isomer. By choosing the appropriate SALEN ligand enantiomer, selective polymerization of either L,L- or D,D-dilactide could be achieved [21]. Similar approaches using SALEN ligands have been employed to produce syndiotactic and heterotactic polylactides [22].

Tin(II)octoate is also the most common catalyst used for the polymerization of cyclic lactone monomers such as ε-caprolactone; although the mechanism of polymerization may differ [23]. In addition, rare-earth metal complexes have been shown to work as effective catalysts leading to high molecular weight polylactones with low polydispersity [24, 25]. An efficient cationic ring-opening polymerization of lactones has been developed using scandium trifluoromethanesulfonate as catalyst. Poly(ε-caprolactone) with narrow polydispersity and a molecular weight in the order of 10^4 Da was produced in quantitative yield after 33 h at room temperature in toluene. Only 0.16 mol% of catalyst was required. Similar results were obtained for poly(δ-valerolactone). Notably, the reaction was relatively tolerant to the presence of moisture and other contaminants [26].

1.2.2
Metal-Free Synthetic Processes

The use of low molecular weight organic molecules to catalyze ring-opening polymerization is a rapidly expanding field [27]. Organocatalytic routes toward polyesters typically involve a nucleophilic polymerization mechanism; problems with side reactions are minimal and products with high molecular weights and very narrow polydispersities can be formed. In addition, organocatalysts may be significantly more abundant and less toxic than metal containing catalysts. These factors point toward future large-scale synthesis applications. Many examples are based around traditional acyl substitution catalysts such as phosphines [28], and pyridine-derivatived nucleophiles [29]. In more recent developments, *N*-heterocyclic carbenes have shown great promise, allowing the synthesis of polylactides with molecular weights up to 10^4 Da [30]. Supramolecular catalysts are known, which can stabilize transition-states in a noncovalent fashion and thus can exert a great effect on reaction rate and mechanism. To this end, thiourea-containing supramolecular catalysts have shown excellent promise [31].

The enzyme-catalyzed synthesis of polyesters is another technique that is being developed as a very ecologically friendly process with several benefits over conventional chemical polymerization [32]. Enzymatic reactions are often extremely regio- and stereospecific, so unwanted side reactions can be largely eliminated. Lipase-catalyzed ring-opening polymerization has been applied to many substrates including a wide range of lactones and lactides. As an example, poly(D,L-lactide) with molecular weights up to 10^5 Da could be synthesized in bulk, but recovery yields were relatively low [33]. Among the problems associated with enzymatic polymerization are the high cost of enzymes, long reaction times, and relatively low molecular weight products. These challenges are to be met before enzymes can be used for industrial scale synthesis.

1.2.3
Polyanhydrides

Polyanhydrides are an important class of biodegradable polymers which are closely related to the polyesters [34, 35]. Monomers used are commonly hydrophobic long chain fatty-acid-derived diacids or aromatic group containing diacids such as shown in Scheme 1.4a. Polyanhydrides are treated in detail in Chapter 3. They are

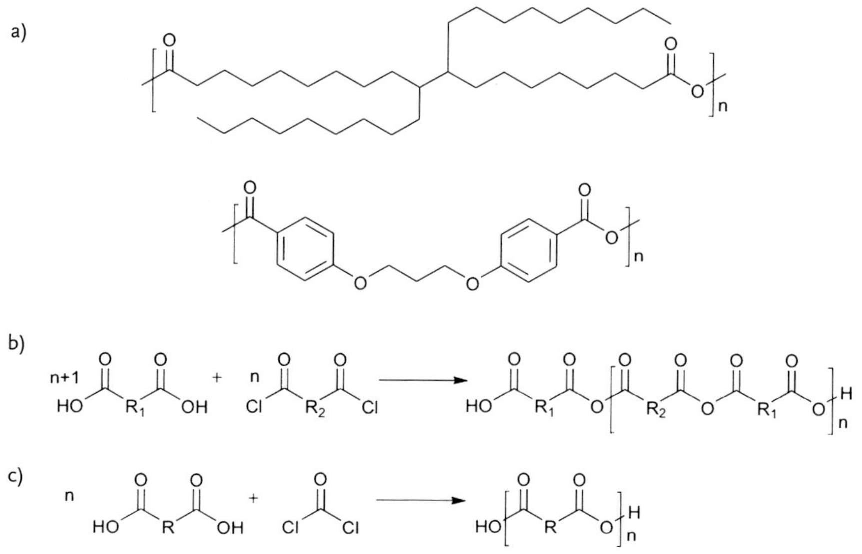

Scheme 1.4 Typical polyanhydrides and synthetic methods.

mentioned here for comparative reasons, as they typically degrade in an alternative manner to poly(hydroxycarboxlic acids) due to their hydrophobic nature (Section 1.4).

Aliphatic diacids can be polycondensated to polyanhydrides by reaction with acetic acid anhydride. The reaction proceeds in two steps. First of all, oligomeric polyanhydrides with terminal acetate groups are received, further reacting to high molecular weight products at elevated temperatures and under vacuum. Using catalysts like cadmium acetate in the second step, average molecular weights of 10^5 Da are reached. Under comparable conditions, glutaric acid and succinic acid form cyclic monomers in contrast to sebacic acid. The reaction of dicarboxylic acids and diacidchlorides results in poor molecular weight products (Scheme 1.4b). A method to gain high molecular weight products even at low temperatures is the use of phosgene as condensation agent (Scheme 1.4c). Formed hydrochloric acid in the reaction is sequestered and removed from the growing polymer by use of insoluble proton scavengers.

1.3 Physical Properties

1.3.1 Crystallinity and Thermal Transition Temperatures

As shown in Table 1.1, high molecular weight polyglycolides, polylactides, and copolymers thereof are typically strong, stiff materials with high modulus (E) and

Table 1.1 Material properties of various poly(lactide-*co*-glycolides) [5].

Comonomer proportion (mol%)			Polymer properties				
Diglycolide	L,L-dilactide	*rac*-Dilactide	T_g (°C)	T_m (°C)	E (GPa)	σ_B (MPa)	ε_B (%)
0	100	0	57	174	3.6	58	2.1
0	84	16	55	124			
0	75	25	60	–	3.4	46	1.6
0	50	50	60	–	3.3	46	3.2
0	25	75	59	–	2.8	41	2.9
0	0	100	59	–	3.2	48	8.7
100	0	0	36	228	7.0		15.0
90	10	0	37	200			
75	25	0	44	–			
50	50	0	44	–			
25	75	0	52	–			
75	0	25	43	–			
25	0	75	54	–			

tensile strength (σ_B). These properties are of similar magnitude to those found within human hard tissues (bones, ligaments, tendons) [36], and are useful for the biomaterial applications mentioned in Section 1.1.1.

Polyglycolide is of high crystallinity, 40–55%, and has a relatively high melting point of 228 °C. The glass transition temperature is 36 °C. Polyglycolide is insoluble in most organic solvents with the exception of highly fluorinated solvents, which must be taken into account when processing materials. Upon copolymerization with dilactides, amorphous materials are produced if the diglycolide content is less than 25%. The glass transition temperature rises from 36 °C to 54 °C as the amount of dilactide monomers are incorporated into the polymer. Poly(L-lactide) has slightly lower crystallinity of 37%; $T_g = 57$ °C and $T_m = 174$ °C. Incorporation of *rac*-dilactide as a comonomer gradually decreases the crystallinity and at 25% *rac*-dilactide content amorphous polymers result. Poly(D,L-lactide) with a glass transition temperature of 65 °C is completely amorphous. The bacterially produced poly(3-hydroxybutyrate) is highly crystalline at 60–80% and has $T_g = 10$ °C and T_m of 179 °C. The modulus is 3.5 GPa. As shown in Table 1.2, incorporation of 3-hydroxyvaleric acid as comonomer leads to a softer and more elastomeric material [37].

Materials comprised of the other major groups of poly(hydroxycarboxylic acid) are considerably softer and more elastic. The polyetherester polydioxanone has a melting point of 115 °C and a glass transition temperature in the range of −10–0 °C. The crystallinity is approximately 55%. Polydioxanone has a lower modulus (1.5 GPa) than the polylactide materials, and loses mechanical strength at a higher rate during hydrolytic degradation. Poly(ε-caprolactone) is a semicrystalline

Table 1.2 Physical properties of various poly([3-hydroxybutyrate]-*co*-[3-hydroxyvalerate])s [5].

Amount of 3-hydroxyvaleric acid (mol%)	T_g (°C)	T_m (°C)	*E* (GPa)	ε_B (%)
0	10	179	3.5	6
9	6	162	1.9	
20	−1	145	1.2	
25	−6	137	0.7	

polymer with a melting point in the range of 59–64 °C. The glass transition temperature is −60 °C. Poly(ε-caprolactone) has a relatively very low modulus (0.4 GPa) but an extremely high elongation at breakage of over 700%. Poly(trimethylene carbonate) is an elastomeric polyester with high flexibility but limited mechanical strength, and is the most commonly employed in copolymers to increase elasticity [38].

1.3.2
Improving Elasticity by Preparing Multiblock Copolymers

While the degradation rate and the degradation behavior of the polymers described previously are adjustable, the mechanical properties of these materials are only of restricted variability. The homopolymers of the α-hydroxycarboxylic acids are highly crystalline, brittle materials. The elongations at break are relatively low compared to polymers such as polyethylene terephthalate ($\varepsilon_R = 100\%$) and polypropylene ($\varepsilon_R = 400\%$) [37]. The mechanical properties are sufficient for the production of fibers [39]. However, in addition to the described polymer systems, elastic, tough materials are desirable. A concept to realize this requirement is the preparation of phase-segregated block copolymers. One segment should be crystallizable and act as crosslinking unit to give the material the desired strength. The second segment should be amorphous, with a low glass transition temperature that is responsible for the elasticity. This principle is shown in Figure 1.1.

One method to generate high molecular weight multiblock copolymers is to the co-condensation of two bifunctional linear prepolymers, known as telechelic polymers [40]. A group of copolyesterurethanes can be used as a case in point [41]. Poly([3-*R*-hydroxybutyrate]-*co*-[3-*R*-hydroxyvalerate])-diol is used as crystallizable segment. It is prepared by transesterification of high molecular weight bacterially produced polyester with a low molecular weight diol. The number average molecular weight, M_n, of this telechelic polymer ranges from 2100 to 2500 g mol^{-1}. The soft segments are telechelic copolyester diols (M_n = 500–3000 g mol^{-1}), prepared by ring-opening polymerization of lactones with a low molecular weight diol. A low molecular weight diisoyanate was used to link the polymer blocks through urethane linkages. In this way, the final polymers may be considered as poly(ester urethane)s. With the correct conditions, products with an average molecular

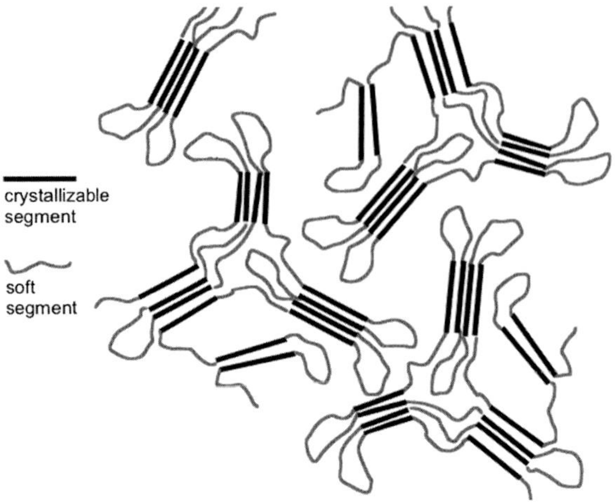

Figure 1.1 Schematic diagram of the morphology of multiblock copolymers.

Table 1.3 Influence of the weight content of hard segments in multiblock copolymers on material properties [41].

Hard segment content (wt%)	E (MPa)	ε_R (%)	T_g (°C)	T_m (°C)
31	60	1250	−24	132
46	186	1130	−18	134
61	400	270	−10	136

weight of more than 10^5 Da were obtained. As can be seen in Table 1.3, lowering the weight percentage of hard segments lowers the material modulus and increases the elongation at break. These polymers have low glass transition temperatures, which prevents them from forming brittle materials at body temperature.

Routes to prepare polyester-based block copolymers have been widely studied [42]. The most direct route is via sequential addition of monomers to systems polymerizing under living conditions. However, this is not broadly applicable to polyester synthesis as the monomers must have comparable reactivities under one set of conditions. This approach is more effective for the copolymerization of similarly functionalized lactones, although the resultant blocks generally have similar physical properties, so phase segregation is not realized [43]. The large difference in reactivity ratio between dilactides and lactones makes it difficult to synthesize such block copolymers. However, an elegant approach using a cyclic tin oxide catalyst has been developed to produce ABA-type triblocks [44, 45]. The ring-expansion mechanism is outlined in Scheme 1.5. This process forms telechelic polymers that can be crosslinked directly with diisocyanates or activated diesters.

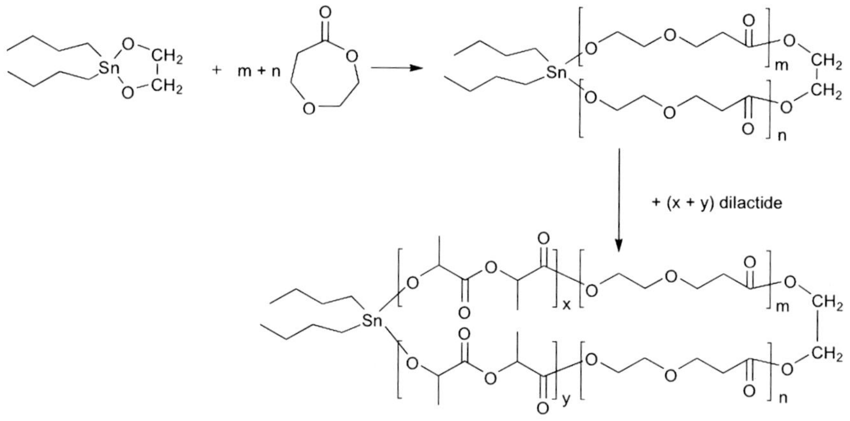

Scheme 1.5 Ring-expansion mechanism for the preparation of ABA triblock polyesters.

1.3.3
Covalently Crosslinked Polyesters

A further method to produce elastomeric polyester-based materials is the preparation of crosslinked amorphous polyesters. In this case, crystalline regions are absent but mechanical strength is given by the inherent rigidity of the network. Such materials are often described as "cured" polymers [46]. Due to the absence of crystalline regions, erosion occurs more homogeneously, and properties can be tailored by composition. Such elastomers were prepared by the photopolymerization of methacrylate functionalized star-shaped poly([ε-caprolactone]-*co*-[*rac*-lactide]) (see Section 1.5.2). Networks suitable for implant materials were obtained, with physical properties adjustable by selecting the molecular weight of the precured polymers [47]. Poly(diol citrates) were synthesized by reacting citric acid with various diols to form a covalent crosslinked network via polycondensation [48]. The physical properties and degradation characteristics could be controlled by choosing different diols and by controlling the crosslink density of the polyester network. Biocompatible materials with elongations at break as high as 500% could be obtained. Other common crosslinkers used for curing polymers are multifunctional isocyanates and acid chlorides.

1.3.4
Networks with Shape-Memory Capability

Polymer networks can be designed in a way that they become capable of a shape-memory effect [49]. Such materials possess the ability to memorize a permanent shape, which can substantially differ from their temporary shape. The transition from the temporary to the permanent shape could be initiated by an external stimulus such as a temperature increase above a characteristic switching temperature of the polymer (thermally induced shape-memory effect). Exemplary shape-memory biodegradable polyesters have been prepared as networks of star-shaped polymers crosslinked by diisocyanates [50], or as photocrosslinkable macrodimethacrylates [51]. Biodegradable shape-memory polymers will be covered in more detail in Chapter 8.

1.4 Degradation Mechanisms

In general, two different mechanisms for the biodegradation of polyesters are discussed in literature: bulk degradation and surface erosion [52]. In the bulk degradation process, water diffuses into the polymer matrix faster than the polymer is degraded. The hydrolyzable bonds in the whole polymer matrix are cleaved homogeneously. Therefore, the average molecular weight of the polymer decreases homogeneously. In the case of surface erosion, the diffusion rate of water into the polymer matrix is slower than the degradation rate of the macromolecules. The degradation only takes place in the thin surface layer while the molecular weight of the polymer in the bulk remains unchanged. Surface erosion is a heterogeneous process, with a rate strongly dependent on the shape of the test sample (e.g., size of the surface) [53].

The majority of polyester materials undergo bulk hydrolysis, as will be explained in the following section. Polyanhydride materials differ from the common polyesters by the fact that they undergo linear mass loss by surface erosion mechanisms [54]. The hydrophobic chains preclude water penetration into the bulk of the material, thus negating bulk erosion mechanisms.

1.4.1 Determining Erosion Kinetics

Erosion rates can be determined *in vitro* and *in vivo* [55]. For *in vitro* experiments, the polymers are exposed to an aqueous solution, in which ionic strength, pH-value, and temperature can be varied. The degradation products of the polymer can be isolated from the aqueous solution and characterized. The addition of enzymes is also possible. Furthermore, the polymers can be exposed to cell- and tissue cultures. By suitable selection and systematic variation of the *in vitro* test conditions, the influence of single parameters on the degradation behavior of the polymer can be determined. Accelerated degradation tests at elevated temperature (usually 70 °C) serve as preliminary experiments for planning the 37 °C experiments and to give reference to the extended degradation behavior of the materials. Another method to accelerate hydrolysis tests is the elevation of the pH-value of the degradation medium to the alkaline region (as a rule 0.01 or 0.1 M NaOH solution). The reaction of cell- and tissue cultures in contact with the material gives information on the compatibility of the partially degraded polymer samples and their degradation products.

In vivo experiments are performed with different species such as dogs, monkeys, rats, mice, and sheep [56]. To investigate the degradation behavior, the implants are typically placed subcutaneously or intramuscularly. The tissue compatibility of the polymer can be determined by histological investigations. There are several characteristics to follow in implant material degradation, for example, height, weight, and mechanical properties of the test device. An additional method for *in vivo* tests is marking of the implant with ^{14}C or by fluorescent chromophores. In this case, it can be observed where the fragments of the degraded polymer and the degradation products in the test animal remain. Furthermore, the change in thermal properties, crystallinity, and the surface properties (wettability, roughness), depending on the degradation- and implantation time duration, can be determined.

1.4.2
Factors Affecting Erosion Kinetics

Poly(hydroxycarboxylic acid)s degrade via the bulk process [57]. The degradation process can be divided into three parts. In the first step, water is absorbed and the polymer swells. Several ester bonds are cleaved already, but there is no mass loss. In the second step, the average weight is significantly reduced. As ester bonds are cleaved, carboxylic groups are formed, which autocatalyze the hydrolysis. During this period, the polymer loses mechanical strength. The third step is characterized by mass loss of the test sample and an increase in degradation rate. The degradation of an implanted material is completed when oligomeric and low-molecular-weight fragments are dissolved in the surrounding medium. The dissolved polymer fragments are then hydrolyzed to the free hydroxycarboxylic acids. Degradation products, many of which occur naturally within the metabolic cycle (e.g., lactic acid), are typically removed from the body without toxic effect. To some extents, smaller crystalline segments may remain, which are eliminated from the body by phagocytosis [58].

The degradation times of several poly(hydroxycarboxylic acids) are summarized in Table 1.4. The differences in degradation rates may be rationalized mainly by the ability of water to permeate the polymers (crystallinity and hydrophobicity), and in the case of polylactides the presence of an α-methyl group which hinders hydrolysis on steric grounds. Copolymers, due to the greater prevalence of amorphous regions, are generally degraded faster than homopolymers [59].

The higher degradation rate of poly(*rac*-lactide) compared to poly(L-lactide) is due to the higher crystallinity of the isotactic poly(L-lactide). Polyglycolide, being less hindered at the scission site, is degraded relatively quickly. The degradation rate of poly(lactide-*co*-glycolide) can be finely tuned by varying the monomer content. Polydioxanone and poly(ε-caprolactone) are also less sterically hindered but increased hydrophobicity hinders erosion. Poly(ε-caprolactone), with a pentylene ($-C_5H_{10}-$) chain, erodes an order of magnitude slower than polyglycolide.

Due to the high crystallinity, poly(3-*R*-hydroxybutyrate) is relatively slowly degraded. In this case, a certain amount of surface erosion takes place first. With

Table 1.4 Comparative degradation times of different polyesters in physiological conditions [5].

Polymer	Degradation time (months)
Poly(L-lactide)	18–24
Poly(*rac*-lactide)	12–16
Polyglycolide	2–4
Poly(3-hydroxybutyrate)	≫36
Polydioxanone	6–12
Poly(ε-caprolactone)	>24
Poly([L-lactide]-*co*-glycolide) 50:50	2
Poly([*rac*-lactide]-*co*-glycolide) 85:15	5
Poly([*rac*-lactide]-*co*-[ε-caprolactone]) 90:10	2

proceeding degradation, a loss of weight and increasing porosity lead to a bulk degradation. Poly([3-*R*-hydroxybutyrate]-*co*-[3-*R*-hydroxyvalerate]) is hydrolyzed more rapidly due to lowered crystallinity. The degradation of the poly(3-*R*-hydroxybutyrate) can be accelerated by microorganisms and enzymes [60]. Bacterially produced poly(3-*R*-hydroxybutyrate) is degraded faster than synthetic poly(3-*R,S*-hydroxybutyrate), while poly(3-*S*-hydroxybutyrate) is not hydrolyzed at all [61]. Enzymatic digestion proceeds at the surface of the sample as proved by scanning electron microscopy. The enzyme, because of high molecular weight, is too large to diffuse into the bulk of the material. Another important factor to take into account when considering the processing of these polymers is that above 200 °C thermal degradation can occur [62].

1.5 Beyond Classical Poly(Hydroxycarboxylic Acids)

1.5.1 Alternate Systems

In addition to the broadly studied groups described previously, there are some notable examples to include when discussing biodegradable polyesters. Scheme 1.6 shows a selection of such alternate polyester structures. Various amino acids were converted to the corresponding α-hydroxycarboxylic acid and then polymerized to give stereogenic α-substituted polyglycolide analogs [63, 64]. This approach could give access to polymers with more tailored properties and sophisticated material properties. Polydepsipeptides are an interesting class of poly(ester amide) which are copolymers of α-hydroxycarboxylic acids and α-amino acids [65]. They are synthesized from morpholine diones (analogs of diglycolide, in which one lactone is replaced by a lactam group) by many of the same procedures outlined for polyglycolide synthesis. Polydepsipeptides degrade through the ester bonds, whereas the amide linkages remain intact under physiological conditions [66].

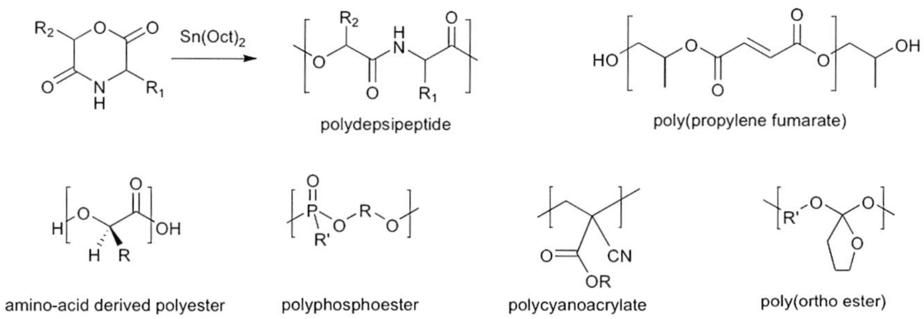

Scheme 1.6 Alternative polyesters and closely related polymers.

There is the possibility to select the amino acid component to incorporate functional groups into the chain in a facile manner [67].

Poly(propylene fumarates) are a bulk eroding class of polyester which are synthesized typically via transesterification [68, 69]. Although molecular weights are generally low, the unsaturated polymer backbone can be photochemically crosslinked to provide polymer networks with desired properties for implant materials. The general concept of crosslinking by photopolymerization has been extended to the other polyester classes, typically by methacrylate functional groups, which have been appended to polymeric precursors [70]. Supramolecular polyesters have been developed by attaching self-recognizing binding units on either end of telechelic polylactones [71]. These polymers are shown to self-assemble via noncovalent means into long strands composed of multiple individual blocks. Such systems show a high level of sensitivity to environmental conditions, and as such may be considered as "smart" polyester systems.

Although not strictly polyesters, some interesting analogs remain. Poly(ortho esters) are materials which are studied mainly as drug delivery vehicles [72]. Like polyanhydrides, they have very labile bonds but are hydrophobic in nature. Water is precluded from the bulk polymer, hence retarding degradation. As such, polyorthoesters have fairly linear surface erosion behavior, ideal for controlled drug release. Polyphosphoesters are hydrolytically degraded polymers that have an extra degree of versatility due to the pentavalent nature of the phosphorous atom [73]. They are highly hydrophilic and show good biocompatibility. Copolymers with other esters such as lactides have been prepared and studied for the range of applications from tissue engineering to drug delivery. Poly(alkyl cyanoacrylates) are rapidly degrading polymers which have widespread medical applications [74]. Although they only contain ester units in the side chains, they are degraded not only at the carbonyl position but at the carbon–carbon sigma bond of the polymer backbone chain. This behavior separates this class of material markedly from the polyesters.

1.5.2
Complex Architectures

While the properties of linear polyesters have been widely studied, branched polyesters are becoming the subject of increasing study. The search for more complex architectures is stimulated mainly by the desire to lower crystallinity, increase the amount of endgroups for further functionalization, or to further control degradation behavior.

Star-shaped poly(caprolactones) with molecular weights in the 10^4 Da range were prepared starting with triol or tetraol multifunctional initiators [75]. These multivalent products were further reacted with glycolide derivatives to give a star-shaped block copolymer of poly([ε-caprolactone]-block-[lactide-*alt*-glycolide]). This globular architecture has a soft segment on the interior with a more crystalline outer sphere. Hyperbranched polyesters of $M_n \approx 10^4$ Da were prepared by copolymerizing polylactide with an alcohol-containing lactone as a latent AB_2 monomer [76]. This dramatically changed the material properties of the polymer with reduced crystallinity and lowered glass transition temperature due to the branching. Poly(ether ester) dendrimers were synthesized from lactic acid and glycerol [77]. These structures have perfect branching radiating from a single point, and make an interesting class of polyvalent biocompatible material.

Polysaccharides have been used as a multivalent scaffold, from which to graft polyesters. Polylactide grafted onto pullulan showed a relatively increased degradation rate which was attributed to the branching effect and the increased hydrophilicity of the pullulan core [78]. Comb polymers were prepared by grafting multiple polylactides to a linear polymethacrylate chain [79]. Highly crystalline regions were observed at the interdigitating comb regions, leading to novel material properties. Interesting and complex structures were formed from polylactides attached to ligands which are able to aggregate into defined supramolecular complexes with Cu(I) ions [80]. Size defined nanoparticles resulted.

1.5.3
Nanofabrication

The previous analyses of polyester materials were given largely in the context of mechanical properties required for tissue engineering or medical implant applications. Polyesters with nanoscale dimensions, however, are widely studied as nanovehicular delivery agents for drugs and biodiagnostic molecules [81, 82]. For the "top-down" approach, submicron polyester spheres can be prepared by a range of molding and lithography techniques [83]. In addition, electrospinning techniques are emerging as a powerful tool for the preparation of nanoscale fibers [84]. Biodegradable polyesters may be formulated with inorganic nanomaterials to provide nanobiocomposites which have relatively controllable physical properties [85]. Polyester/clay biocomposites have been the subject of considerable study.

"Bottom-up" approaches allow the preparation of nanoscale materials without the need for further processing. Various nanospheres of polylactides and polylactones were prepared in the size range of 80–200 nm using a miniemulsion technique. Endocytic, cellular uptake of fluorescently labeled particles was observed, with kinetics revealing that polyesters are endocytozed much faster than polystyrene particles [86]. Polylactide nanoparticles presenting mannose residues at their surface were prepared with dimensions of 200–300 nm by a nanoprecipitation technique [87]. Biochemical assays were used to quantify their recognition of lectin proteins. Such nanoparticles are designed to be specifically recognized by mannose receptors, which are highly expressed in cells of the immune system. Future applications as vaccine delivery agents are anticipated. Thiol-capped polylactides were used to coat photoluminescent quantum dots, which are designed as drug delivery nanovehicles with both diagnostic imaging properties and controlled drug release properties [88]. Another route to nanoscale particles is via micelle formation of amphiphilic block copolymers [89]. Polyethylene glycol-block-lactides were shown to form polymeric micelles, into which paclitaxel as a model drug was encapsulated. Biodistribution and drug release behaviors were studied.

References

1 Williams, D.F. (2009) On the nature of biomaterials. *Biomaterials*, **30**, 5897–5909.
2 Langer, R.S. and Vacanti, J.P. (1993) Tissue engineering. *Science*, **260**, 920–926.
3 Sokolsky-Papkov, M., Agashi, K., Olaye, A., Shakesheff, K., and Domb, A.J. (2007) Polymer carriers for drug delivery and tissue engineering. *Adv. Drug Deliv. Rev.*, **59**, 187–206.
4 Nair, L.S. and Laurencin, C.T. (2007) Biodegradable polymers as biomaterials. *Prog. Polym. Sci.*, **32**, 762–798.
5 Lendlein, A. (1999) Polymere als Implantwerkstoffe. *Chem. Unserer Zeit*, **33**, 279–295.
6 Vert, M. (2009) Degradable and bioresorbable polymers in surgery and in pharmacology: beliefs and facts. *J. Mater. Sci. Mater. Med.*, **20**, 437–446.
7 Lenz, R.W. and Marchessault, R.H. (2005) Bacterial polyesters: biosynthesis, biodegradable plastics and biotechnology. *Biomacromolecules*, **6**, 1–8.
8 Shelton, J.R., Lando, J.B., and Agostini, D.E. (1971) Synthesis and characterization of poly(β-hydroxybutyrate). *J. Polym. Sci. Polym. Lett. B*, **9**, 173–178.
9 Okada, M. (2002) Chemical synthesis of biodegradable polymers. *Prog. Polym. Sci.*, **27**, 87 –133.
10 Gilding, D.K. and Reed, A.M. (1979) Biodegradable polymer for use in surgery – polyglycolic–polylacticacid homopolymer and copolymers.1. *Polymer*, **20**, 1459–1464.
11 Dechy-Cabaret, O., Martin-Vaca, B., and Bourissou, D. (2004) Controlled ring-opening polymerization of lactide and glycolide. *Chem. Rev.*, **104**, 6147–6176.
12 Albertsson, A.-C. and Varma, I.K. (2003) Recent developments in ring opening polymerization of lactones for biomedical applications. *Biomacromolecules*, **4**, 1466–1486.
13 Gupta, A.P. and Kumar, V. (2007) New emerging trends in synthetic biodegradable polymers – polylactide: a critique. *Eur. Polym. J.*, **43**, 4053–4074.
14 Leenslag, J.W. and Pennings, A.J. (1987) Synthesis of high-molecular-weight poly(L-lactide) initiated with tin 2-ethylhexanoate. *Macromol. Chem. Phys.*, **188**, 1809–1814.
15 Kricheldorf, H.R. and Dunsing, R. (1986) Polylactones, 8. Mechanism of the catalytic polymerization of L,L-dilactide. *Macromol. Chem. Phys.*, **187**, 1611–1625.
16 Kowalski, A., Duda, A., and Penczek, S. (2000) Kinetics and mechansim of cyclic esters polymerization initiated with tin(II) octoate.3. Polymerization of L,L-dilactide. *Macromolecules*, **33**, 7359–7370.
17 Degée, P., Dubois, P., Jerôme, R., Jacobsen, S., and Fritz, H.G. (1999) New catalysis for fast bulk ring-opening polymerization of lactide monomers. *Macromol. Symp.*, **144**, 289–302.
18 Kricheldorf, H.R. and Damrau, D.-O. (1997) Polylactones, 37. Polymerizations of L-lactides initiated with Zn(II)L-lactate

and other resorbable Zn salts. *Macromol. Chem. Phys.*, **198**, 1753–1766.
19 Schwach, G., Coudane, J., Engel, R., and Vert, M. (1999) Ring-opening polymerization of D,L-lactide in the presence of zinc metal and zinc lactate. *Polym. Int.*, **46**, 177–182.
20 Kricheldorf, H.R. and Damrau, D.-O. (1997) Polylactones, 38. Polymerization of L-lactide with Fe(II)lactate and other resorbable Fe(II) salts. *Macromol. Chem. Phys.*, **198**, 1767–1774.
21 Majerska, K. and Duda, A. (2004) Stereocontrolled polymerization of racemic lactide with chiral initiator: combining stereoselection and chiral ligand-exchange mechanism. *J. Am. Chem. Soc.*, **126**, 1026–1027.
22 Ovitt, T.M. and Coates, G.W. (2002) Stereochemistry of lactide polymerization with chiral catalysts: new opportunities for stereocontrol using polymer exchange mechanisms. *J. Am. Chem. Soc.*, **124**, 1316–1326.
23 Kowalski, A., Duda, A., and Penczek, S. (1998) Kinetics and mechanism of cyclic esters polymerization initiated with tin(II) octoate, 1. Polymerization of ε-caprolactone. *Macromol. Rapid Commun.*, **19**, 567–572.
24 Yamashita, M., Takamoto, Y., Ihara, E., and Yasuda, H. (1996) Organolanthanide-initiated living polymerizations of ε-caprolactone, δ-valerolactone and β-propiolactone. *Macromolecules*, **29**, 1798–1806.
25 Shen, Y., Shen, Z., Zhang, Y., and Yao, K. (1996) Novel rare earth catalysts for the living polymerization and block copolymerization of ε-caprolactone. *Macromolecules*, **29**, 8289–8295.
26 Nomura, N., Taira, A., Tomioka, T., and Okada, M. (2000) A catalytic approach for cationic living polymerization: $Sc(OTf)_3$-catalyzed ring-opening polymerization of lactones. *Macromolecules*, **33**, 1497–1499.
27 Kamber, N.E., Jeong, W., Waymouth, R.M., Pratt, R.C., Lohmeijer, B.G.G., and Hedrick, J.L. (2007) Organocatalytic ring-opening polymerization. *Chem. Rev.*, **107**, 5813–5840.
28 Myers, M., Connor, E.F., Glauser, T., Möck, A., Nyce, G., and Hedrick, J.L. (2002) Phosphines: nucleophilic organic catalysts for the controlled ring-opening polymerization of lactides. *J. Polym. Sci. A Polym. Chem.*, **40**, 844–857.
29 Nederberg, F., Connor, E.F., Möller, M., Glauser, T., and Hedrick, J.L. (2001) New paradigms for organic catalysts: the first organocatalytic living polymerization. *Angew. Chem. Int. Ed.*, **40**, 2712–2715.
30 Connor, E.F., Nyce, G.W., Myers, M., Möck, A., and Hedrick, J.L. (2002) First example of *N*-heterocyclic carbenes as catalysts for living polymerization: organocatalytic ring-opening polymerization of cyclic esters. *J. Am. Chem. Soc.*, **124**, 914–915.
31 Dove, A.P., Pratt, R.C., Lohmeijer, B.G.G., Waymouth, R.M., and Hedrick, J.L. (2005) Thiourea-based biofunctional organocatalysis: supramolecular recognition for living polymerization. *J. Am. Chem. Soc.*, **127**, 13798–13799.
32 Varma, I.K., Albertsson, A.-C., Rajkhowa, R., and Srivastava, R.K. (2005) Enzyme catalyzed synthesis of polyesters. *Prog. Polym. Sci.*, **30**, 949–981.
33 Matsumura, S., Mabuchi, K., and Toshima, K. (1997) Lipase-catalyzed ring-opening polymerization of lactide. *Macromol. Rapid Commun.*, **18**, 477–482.
34 Tamada, J. and Langer, R. (1992) The development of polyanhydrides for drug delivery applications. *J. Biomater. Sci. Polym. Ed.*, **3**, 315–353.
35 Kumar, N., Langer, R.S., and Domb, A.J. (2002) Polyanhydrides: an overview. *Adv. Drug Deliv. Rev.*, **54**, 889–910.
36 Sabir, M.I., Xu, X., and Li, L. (2009) A review on biodegradable polymeric materials for bone tissue engineering applications. *J. Mater. Sci.*, **44**, 5713–5724.
37 Poirier, Y., Dennis, D.E., Nawrath, C., and Somerville, C. (1993) Progress toward biologically produced biodegradable thermoplastics. *Adv. Mater.*, **5**, 30–36.
38 Pêgo, A.P., Poot, A.A., Grijpma, D.W., and Feijen, J. (2003) Biodegradable elastomeric scaffolds for soft tissue engineering. *J. Control. Rel.*, **87**, 69–79.
39 Leenslag, J.W., Gogolewski, S., and Pennings, A.J. (1984) Resorbable materials of poly(L-lactide). V. Influence of secondary structure on the mechanical

properties and hydrolyzability of poly(L-lactide) fibers produced by a dry-spinning method. *J. Appl. Polym. Sci.*, **29**, 2829–2842.

40 Hiltunen, K., Härkönen, M., Seppälä, J.V., and Väänänen, T. (1996) Synthesis and characterization of lactic acid based telechelic prepolymers. *Macromolecules*, **29**, 8677–8682.

41 Lendlein, A., Neuenschwander, P., and Suter, U.W. (1998) Tissue-compatible multiblock copolymers for medical applications, controllable in degradation rate and mechanical properties. *Macromol. Chem. Phys.*, **199**, 2785–2796.

42 Castillo, R.V. and Müller, A.J. (2009) Crystallization and morphology of biodegradable single and double crystalline block copolymers. *Prog. Polym. Sci.*, **34**, 516–560.

43 Löfgren, A., Albertsson, A.-C., Dubois, P., Jérôme, R., and Teyssié, P. (1994) Synthesis and characterization of biodegradable homopolymers and block copolymers based on 1,5-dioxepan-2-one. *Macromolecules*, **27**, 5556–5562.

44 Pospiech, D., Komber, H., Jehnichen, D., Häussler, L., Eckstein, K., Scheibner, H., Janke, A., Kricheldorf, H.R., and Petermann, O. (2005) Multiblock copolymers of L-lactide and trimethylene carbonate. *Biomacromolecules*, **6**, 439–446.

45 Stridsberg, K. and Albertsson, A.C. (2000) Controlled ring-opening polymerization of L-lactide and 1,5-dioxepan-2-one forming a triblock copolymer. *J. Polym. Sci. A*, **38**, 1774–1778.

46 Amsden, B. (2007) Curable, biodegradable elastomers: emerging biomaterials for drug delivery and tissue engineering. *Soft Matter*, **3**, 1335–1348.

47 Amsden, B.G., Misra, G., Gu, F., and Younes, H.M. (2004) Synthesis and characterization of a photo-cross-linked biodegradable elastomer. *Biomacromolecules*, **5**, 2479–2486.

48 Yang, J., Webb, A.R., Pickerill, S.J., Hageman, G., and Ameer, G.A. (2006) Synthesis and evaluation of poly(diol citrate) biodegradable elastomers. *Biomaterials*, **27**, 1889–1898.

49 Lendlein, A. and Langer, R. (2002) Biodegradable, elastic, shape-memory polymers for potential biomedical applications. *Science*, **296**, 1673–1676.

50 Lendlein, A., Zotzmann, J., Feng, Y., Alteheld, A., and Kelch, S. (2009) Controlling the switching temperature of biodegradable, amorphous, shape-memory poly(*rac*-lactide)urethane networks by incorporation of different comonomers. *Biomacromolecules*, **10**, 975–982.

51 Choi, N.-Y. and Lendlein, A. (2007) Degradable shape-memory polymer networks from oligo([L-lactide]*ran*-glycolide)dimethacrylates. *Soft Matter*, **3**, 901–909.

52 von Burkersroda, F., Schedl, L., and Göpferich, A. (2002) Why degradable polymers undergo surface erosion or bulk erosion. *Biomaterials*, **23**, 4221–4231.

53 Wu, X.S. (1995) *Encyclopedic Handbook of Biomaterials and Bioengineering* (eds D.L. Wise, D.J. Trantolo, D.E. Altobelli, M.J. Yaszemski, J.D. Gresser, and E.R. Schwarz), Marcel Dekker, New York, pp. 1015–1054.

54 Göpferich, A. and Tessmar, J. (2002) Polyanhydride degradation and erosion. *Adv. Drug Deliv. Rev.*, **54**, 911–931.

55 Zhang, Z., Kuijer, R., Bulstra, S.K., Grijpma, D.W., and Fiejen, J. (2006) The *in vivo* and *in vitro* degradation behavior of poly(trimethylene carbonate). *Biomaterials*, **27**, 1741–1748.

56 Lan, C.X.F., Hutmacher, D.W., Schantz, J.-T., Woodruff, M.A., and Teoh, S.H. (2009) Evaluation of polycaprolactone scaffold degradation for 6 months *in vivo* and *in vitro*. *J. Biomed. Mater. Res. A*, **90A**, 906–919.

57 Göpferich, A. (1997) Polymer bulk erosion. *Macromolecules*, **30**, 2598–2604.

58 Woodward, S.C., Brewer, P.S., Moatemed, F., Schindler, A., and Pitt, C.G. (1985) The intracellular degradation of poly(ε-caprolactone). *J. Biomed. Mater. Res.*, **19**, 437–444.

59 Miller, R.A., Brady, J.M., and Cutright, D.E. (1977) Degradation rates of oral resorbable implants (polylactides and polyglycolates): rate modification with changes in PLA/PGA copolymer ratios. *J. Biomed. Mater. Res.*, **11**, 711–719.

60 Pouton, C.W. and Akhtar, S. (1996) Biosynthetic polyhydroxyalkanoates and their potential in drug delivery. *Adv. Drug Deliv. Rev.*, **18**, 133–162.

61 Timmins, M.R. and Lenz, R.W. (1996) Effect of tacticity on enzyme degradability of poly(β-hydroxybutyrate). *Macromol. Chem. Phys.*, **197**, 1193–1215.

62 Carrasco, C., Dionisi, D., Martinelli, A., and Majone, M. (2006) Thermal stability of polyhydroxyalkanoates. *J. Appl. Polym. Sci.*, **100**, 2111–2121.

63 Kolitz, M., Cohen-Arazi, N., Hagag, I., Katzhendler, J., and Domb, A.J. (2009) Biodegradable polyesters derived from amino acids. *Macromolecules*, **42**, 4520–4530.

64 Cohen-Arazi, N., Katzhendler, J., Kolitz, M., and Domb, A.J. (2008) Preparation of new α-hydroxy acids derived from amino acids and their corresponding polyesters. *Macromolecules*, **41**, 7259–7263.

65 Feng, Y. and Guo, J. (2009) Biodegradable polydepsipeptides. *Int. J. Mol. Sci.*, **10**, 589–615.

66 Ohya, Y., Toyohara, M., Sasakawa, M., Arimura, H., and Ouchi, T. (2005) Thermosensitive biodegradable polydepsipeptide. *Macromol. Biosci.*, **5**, 273–276.

67 Barrera, D.A., Zylstra, E., Lansbury, P.T., Jr., and Langer, R. (1993) *J. Am. Chem. Soc.*, **115**, 11010–11011.

68 Yaszemski, M.J., Payne, R.G., Hayes, W.C., Langer, R., and Mikos, A.G. (1996) *In vitro* degradation of a poly(propylene fumarate)-based composite material. *Biomaterials*, **17**, 2127–2130.

69 He, S., Yaszemski, M.J., Yasko, A.W., Engel, P.S., and Mikos, A.G. (2000) Injectable biodegradable polymer composites based on poly(propylene fumarate) cross-linked with poly(ethylene glycol)-dimethacrylate. *Biomaterials*, **21**, 2389–2394.

70 Ifkovitz, J.L. and Burdick, J.A. (2007) Review. Photopolymerizable and degradable biomaterials for tissue engineering applications. *Tissue Eng.*, **13**, 2369–2385.

71 van Beek, D.J.M., Gillissen, M.A.J., van As, B.A.C., Palmans, A.R.A., and Sijbesma, R.P. (2007) Supramolecular copolyesters with tunable properties. *Macromolecules*, **40**, 6340–6348.

72 Heller, J. and Barr, J. (2004) Poly(ortho esters) – from concept to reality. *Biomacromolecules*, **5**, 1625–1632.

73 Zhao, Z., Wang, J., Mao, H., and Leong, K.W. (2002) Polyphosphoesters in drug and gene delivery. *Adv. Drug Deliv. Rev.*, **55**, 483–499.

74 Vauthier, C., Dubernet, C., Fattal, E., Pinto-Alphandary, H., and Couvreur, P. (2003) Poly(alkylcyanoacrylates) as biodegradable materials for biomedical applications. *Adv. Drug Deliv. Rev.*, **55**, 519–548.

75 Dong, C.-M., Qiu, K.-Y., Gu, Z.-W., and Feng, X.-D. (2001) Synthesis of star-shaped poly(ε-caprolactone)-*b*-poly(DL-lactic acid-*alt*-glycolic acid) with multifunctional initiator and stannous octoate catalyst. *Macromolecules*, **34**, 4691–4696.

76 Tasaka, F., Ohya, Y., and Ouchi, T. (2001) One-pot synthesis of novel branched polylactide through the copolymerization of lactide with mevalonolactone. *Macromol. Rapid. Commun.*, **22**, 820–824.

77 Carnahan, M.A. and Grinstaff, M.W. (2001) Synthesis and characterization of polyether–ester dendrimers from glycerol and lactic acid. *J. Am. Chem. Soc.*, **123**, 2905–2906.

78 Ohya, Y., Maruhashi, S., and Ouchi, T. (1998) Graft polymerization of L-lactide on pollulan through the trimethylsilyl protection method and degradation of the graft copolymers. *Macromolecules*, **31**, 4662–4665.

79 Zhao, C., Wu, D., Huang, N., and Zhao, H. (2008) Crystallization and thermal properties of PLLA comb polymer. *J. Polym. Sci. B Polym. Phys.*, **46**, 589–598.

80 Hoogenboom, R., Wouters, D., and Schubert, U.S. (2003) L-lactide polymerization utilizing a hydroxy-functionalized 3,6-bis(2-pyridyl)pyridazine as supramolecular (Co)initiator: construction of polymeric (2 × 2) grids. *Macromolecules*, **36**, 4743–4749.

81 Mohamed, F. and van der Walle, C.F. (2008) Engineering biodegradable polyester particles with specific drug targeting and drug release properties. *J. Pharm. Sci.*, **97**, 71–87.

82 Soppimath, K.S., Aminabhavi, T.M., Kulkarni, A.R., and Rudzinski, W.E. (2001) Biodegradable polymeric nanoparticles as drug delivery devices. *J. Control. Rel.*, **70**, 1–20.

83 Lu, Y. and Chen, S.C. (2004) Micro and nano-fabrication of biodegradable polymers for drug delivery. *Adv. Drug Deliv. Rev.*, **56**, 1621–1633.

84 Greiner, A. and Wendorff, J.H. (2007) Electrospinning: a fascinating method for the preparation of ultrathin fibers. *Angew Chem. Int. Ed.*, **46**, 5670–5703.

85 Bordes, P., Pollet, E., and Avérous, L. (2009) Nano-biocomposites: biodegradable polyester/nanoclay systems. *Prog. Polym. Sci.*, **34**, 125–155.

86 Musyanovych, A., Schmitz-Wienke, J., Mailänder, V., Walther, P., and Landfester, K. (2008) Preparation of biodegradable polymer nanoparticles by miniemulsion technique and their cell interactions. *Macromol. Biosci.*, **8**, 127–139.

87 Rieger, J., Freichels, H., Imberty, A., Putaux, J.-L., Delair, T., Jérôme, C., and Auzély-Velty, R. (2009) Polyester nanoparticles presenting mannose residues: toward the development of new vaccine delivery systems combining biodegradability and targeting properties. *Biomacromolecules*, **10**, 651–657.

88 Hou, X., Li, Q., Jia, L., Li, Y., Zhu, Y., and Cao, A. (2009) New preparation of structurally symmetric, biodegradable poly(L-lactide) disulfides and PLLA-stabilized, photoluminescent CdSe quantum dots. *Macromol. Biosci.*, **9**, 551–562.

89 Liggins, R.T. and Burt, H.M. (2002) Polyether-polyester diblock copolymers for the preparation of paclitaxel loaded polymeric micelle formulations. *Adv. Drug Deliv. Rev.*, **54**, 191–202.

2
Biotechnologically Produced Biodegradable Polyesters

Jaciane Lutz Ienczak and Gláucia Maria Falcão de Aragão

2.1
Introduction

Polyhydroxyalkanoates (PHAs) are polyesters synthesized by many microorganisms as a carbon and energy storage material [1].

The interest in establishing PHA as an alternative plastic to conventional petrochemical-based plastics was first motivated because it can be produced from renewable carbon sources and since they are biodegradable. Fuel-based polymers are extensively used due to their easy manufacturing and low cost of production. Unfortunately, these same qualities can transform them into an important environmental problem because they are cheap and disposable. The great demand for this kind of polymer production generates pollution and problems related with the disposal in landfills because these materials are resistant to degradation [2]. In response to rising public concern regarding the effects of fuel-based materials in the environment, biopolymers are a reality that can minimize these problems. Biopolymers are polymeric materials structurally classified as polysaccharides [3, 4], polyesters [5–7], or polyamides [8]. The main raw material for manufacturing them is a renewable carbon source, usually carbohydrates such as sugar cane, corn, potato, wheat, beet, or a vegetable oil extracted from soybean, sunflower, palm, or other plants. Currently, biopolymers of interest include thermoplastic starch [9], polylactides (PLA) [10], xanthan [3], polyamides cyanophycin, and the PHA class which includes the most studied biopolymer, poly(3-hydroxybutyric) (P[3HB]) and its copolymer poly(3-hydroxybutyrate-*co*-3-hydroxyvalerate) (P[3HB-*co*-3HV]) [7, 11, 12].

PHAs are able to replace synthetic polymers because they have very similar properties with the advantage of being completely degradable to water and carbon dioxide in aerobic conditions [13]. Depending on the monomer composition, the properties of PHA polymers can range from thermoplastics to elastomeric. P(3HB) shows thermoplastic and mechanical properties similar to those of polypropylene [13, 14]. Despite the possibility of PHA applications in medicine, pharmaceutical, food, and chemical industries as an alternative to conventional plastic

Handbook of Biodegradable Polymers: Synthesis, Characterization and Applications, First Edition. Edited by Andreas Lendlein, Adam Sisson.
© 2011 Wiley-VCH Verlag GmbH & Co. KGaA. Published 2011 by Wiley-VCH Verlag GmbH & Co. KGaA.

[15], biodegradable plastics still have minimal participation in the market because of the high cost compared to fuel-based polymers. Therefore, many research groups are conducting studies to reduce the production costs of biopolyesters by using low-cost substrates [11, 16–19], large-scale fermentation methods [19–22], and metabolic engineering to develop strains with higher productivity and capable of assimilating renewable carbon sources. In addition, the PHA granule has various protein-based functions and has attracted interest due to the utilization of these bionanoparticles in medical and biotechnological applications [23, 24].

2.2 History

Beijerinck [25] observed granules in a microscope inside *Rhizobium* cells. Such granules were present in the "bacteroides" isolated from nodules and were described as being extremely refractile globules. Another microbiologist, Lemoigne [26], noticed that, when cultures of *Bacillus subtilis* were followed by autolysis in distilled water, the pH value decreased because of the formation of an unknown acid. This acid was subsequently identified as monomer of poly-β-hydroxybutyric acid [27]. In the same period, Stapp [28], analyzing the results of other researchers, suggested that *Azotobacter chroococcum* inclusions could be easily extracted with chloroform, and identified this structure as poly-β-hydroxybutyrate. In 1958, the functional P(3HB) pathway was proposed by Macrae and Wilkinson [29]. They observed that *Bacillus megaterium* stored the polymer especially when the ratio glucose/nitrogen in the medium was high, and that the subsequent degradation occurred quickly with the absence of the carbon source. PHA's potential usefulness has been recognized since the first half of the 1960s through patents related to P(3HB) production process [30]; extraction from the producing biomass [31]; plasticization with additives [32]; the use unextracted as a polymer mixed with other cell material [33]; and pure for absorbable prosthetic devices [34]. In a review about the regulatory role and energy resource microorganisms, published in 1973 by Dawes and Senior [35], P(3HB) was found to be a microbial resource material storage as starch and glycogen. In the period between 1974 [36] and 1989 [37], other hydroxyalkanoates (HAs) have been identified besides 3HB, such as 4-hydroxybutyrate (4HB), 3-hydroxyhexanoate (3HH$_x$), 3-hydroxyoctanoate (3HO), 3-hydroxyvalerate (3HV), among others. The identification of copolymer poly(3-hydroxybutyrate-*co*-3-hydroxyvalerate) (P[3HB-*co*-3HV]) has led to a positive impact on research and commercial interest because the homopolymer (P[3HB]) is brittle and has a low extension break. This lack of flexibility limits its range of application in relation to the copolymer, which has a much lower melting point and is less crystalline [38]. The industrial production of these polymers began in 1980 by the UK chemical group Imperial Chemical Industry (ICI) [39]; after that, it started to be produced by others industries (Table 2.1).

In the past few decades, PHA researchers have been living a period of interest for metabolic engineering [40], and site-directed mutagenesis of the enzymes

Table 2.1 Summary of industrial PHA production: past to present.

Industry	History	Product
Chimie Linz	Currently Biomer	P(3HB)
Biomer	Production from sucrose	P(3HB)
Mitsubishi	Production from methanol, with the name Biogreen.	P(3HB)
PHB Industrial	Production from sugar cane sucrose, with the name Biocycle.	P(3HB-*co*-3HV)
Zeneca Bioproducts (previously ICI).	A development program has been carried out (1980–1990) for PHA commercialization (Biopol). Biopol is a copolymer family P(3HB-*co*-3HV) produced by microorganisms using glucose and propionic acid as substrates. Zeneca has begun P(3HB) and P(3HB-*co*-3HV transgenic plant production.	P(3HB-*co*-3HV)
Monsanto	Zeneca was incorporated by Monsanto in 1996 and has continued the development of this polymer. However, in 2001 Monsanto stopped their work.	P(3HB-*co*-3HV)
Metabolix	*Spin off* Industry. Nowadays possesses an agreement with Archer Daniels Midland Co (ADM) for copolymer production in industrial scale.	P(3HB-*co*-3HV)
Procter & Gamble	Nodax is a copolymer family produced from glucose and vegetable oils, above all palm oil.	P(3HB-*co*-3HHx)
Kaneka Co	Licensed by Procter & Gamble for Nodax production in industrial scale.	P(3HB-*co*-3HHx)

Adapted from "Biopolímeros e Intermediários Químicos" [40].

involved in PHA biosynthesis will most likely result in new polyesters [41]. In the 1980s, the first study for PHA production from recombinant microorganisms involving cloning of PHA biosynthetic genes [42, 43] was realized, and in the 1990s it was also possible to study transgenic plants as potential producers of PHA in the future [44–47]. Many studies have been made to determine tertiary and quaternary structures of PHA synthesis, which would allow researchers to understand the catalytic mechanisms, the substrate specificities of this group of enzymes, and probably also the factors that determine the molecular weight of produced PHA [11, 48–50]. Interest in granule-associated protein, large-scale production, and high productivity has also been noted, besides new methods of PHA recovery. In this history, it can be observed a scientific PHA evolution in eight decades of research and process development for biodegradable polymer production, initially recognized by lipophilic inclusions and currently being studied at molecular level by protein and metabolic engineering.

2.3 Polyhydroxyalkanoates – Granules Morphology

PHA occurs as an insoluble inclusion in the cytoplasm. Several structural models of granules have been proposed with the purpose of answering the questions related to granule formation, structure, size, and composition. In this context, the first structure, shown by Ellar [51], was a fibrilar PHA structure of granules with 10–15 nm in length and enclosed by a membrane approximately 2–4 nm thick. The same author proposed a granule composition of approximately 98% PHA and 2% proteins. In another study, Lundgren [52] suggested that the composition of the granule has a membrane with approximately 0.5% and 2% of lipid and protein, respectively, and not only proteins.

Dunlop and Robards [53] investigated the structure of P(3HB) granules in *Bacillus cereus* using freeze-etching methods and found that the granule has a central core amounting to 50% of granule volume and an outer coat with different densities. Ballard and coworkers [54] reported many responses in relation to the size and number of the granules of P(3HB) in *Alcaligenes eutrophus* (nowadays, *Cupriavidus necator*) by freeze-fracture and by using a cylindrical cell model to interpret the results. These authors proposed a granule diameter from 0.24 to 0.5 μm and the average number of granules per cell remained constant at 12.7 ± 1.0 and 8.6 ± 0.6 for two different scale experiments. Another important consideration by these authors was that the number of granules is fixed at the earliest stages of polymer accumulation and polymer accumulation ceases when a P(3HB) content of about 80% is attained, although PHA synthase activity remains high.

Initial studies about PHA characteristics inside the granule were developed in the 1960s through X-ray studies of solid P(3HB) and the results led initially to the conclusion that P(3HB) granules *in vivo* were crystalline [55, 56]. Nevertheless, Kawaguchi and Doi [57] have examined by X-ray diffraction the structure of native P(3HB) granules of *A. eutrophus* and concluded that the polymer presented an amorphous state. These authors concluded that the treatment of granules with alkaline hypochlorite, sodium hydroxide, aqueous acetone, or lipase initiated crystallization of P(3HB) by removing lipid components. Hence, the initial studies of the PHA granule structure proposed a noncrystalline structure, with the presence of fluid polyesters and a small amount of phospholipids and proteins, but there was no knowledge about the protein type and granule formation.

In a previous study about depolymerase, Foster *et al.* [58] observed serine residues in the active site of a functional depolymerase (PhaZ, structural gene *phaZ*) associated with isolated poly(3-hydroxyoctanoate) (P[3HO]) granules, which is a copolymer obtained by feeding *Pseudomonas oleovorans* with *n*-octanoic acid. Following this investigation, Foster *et al.* [59] studied with more details the presence of depolymerase in P(3HO) granules. The results revealed that (i) the *P. oleovorans* depolymerase remains active in isolated P(3HO) inclusion bodies; (ii) this enzymatic activity occurs in association with the organized protein lattice that encompasses the stored P(3HO) polymer; and (iii) depolymerase activity of isolated native P(3HO) granules showed a maximum degradation rate of $1.17\,mg\,h^{-1}$ at an optimum pH of 9.

Phasins (PhaP, structural gene *pha*P) are defined as a protein class that has a similar role as oleosins of triacylglycerol inclusions in seeds and pollen of plants [60]. These proteins have been identified from *A. eutrophus* (nowadays, *C. necator*) [61] and *Rhodococcus ruber* [62], and have been shown to influence the size of intracellular PHA granules. Phasins have been suggested to have a role as amphiphilic proteins (substance readily soluble in polar as well as in nonpolar solvents) in the interphase between the hydrophilic cytoplasm and the hydrophobic PHA molecule, and may also act as an anchor for the binding of other proteins such as PHA synthase [62]. Inside this group, the GA13 protein, studied by Schembri and coworkers [63], can be noted. These authors studied *Acinetobacter* RA3123, RA3849, RA3757, RA3762, and *Escherichia coli* DH5α to identify the 13-kDa PHA (GA13) granule-associated protein as the protein encoded by a structural gene located within the *Acinetobacter pha locus*. When the P(3HB) granule samples were examined, a protein of approximately 13 kDa (GA13) was identified in all four *Acinetobacter* P(3HB)-positive strains, shown to be the product of the $phaP_{AC}$ (gene encoded PhaP protein by *Acinetobacter*) gene in strain RA3849 and revealing the presence of two regions containing predominantly hydrophobic and amphiphilic amino acids. This may be involved in the anchoring of this protein into the phospholipid monolayer surrounding the PHA granule. *E. coli* showed a small amount of accumulated P(3HB), however, with large-sized granules. This fact may be related to the poor expression of GA13 protein in this strain, which is able to reduce PHA synthase (PhaC, structural gene *pha*C) activity.

Stuart *et al.* [64] reported that different microorganisms (*Ralstonia eutropha, Norcadia corallina, Azotobacter vinelandii,* and pseudomonads species) showed a different granule protein boundary in electron microscopy and SDS-PAGE. These results can be very interesting for biotechnologists since they indicate a natural "packaging" of polymer during biosynthesis. Maehara *et al.* [65] proposed a structural PHA granule model determining the distinct target DNA sequences for PhaR (a repressor protein which regulates PHA synthesis) binding and demonstrated that PhaR binds not only to DNA but also to PHA. These results confirm that PhaR has bifunctional characteristics, namely, binding abilities toward both PHA and DNA. PhaR is the first protein that interacts directly with PHA polymer. The recognition requirement for this interaction was relatively nonspecific, because PhaR bound to all forms of P(3HB)–crystalline, amorphous, and 3HB oligomers. PhaR recognizes and binds directly to the PHA polymer chains being synthesized, and then the expression of PhaP is initiated at the onset of dissociation of PhaR from an upstream element for *pha*P. During the elongation of PHA polymer chains, the PHA granules enlarge in size, and then the surfaces of PHA granules become covered by PhaP and other specific proteins before the other nonspecific proteins bind to the PHA granules. Under these conditions, the authors concluded that PhaR is a sensor for PHA synthesis in the cell.

According to the conventional classification of PHA granule-associated proteins proposed by Steinbüchel *et al.* [66], the following four distinct proteins can be defined functionally: class I comprises the PHA synthases, which catalyze the polymerization of the monomers of hydroxyacyl-CoA; class II comprises the PHA depolymerases, which are responsible for the intracellular degradation and mobilization of PHA; class III comprises the phasins (designated as PhaP), which probably form a protein layer at the surface of the PHA granule with phospholipids, lipids, and other proteins; and class IV comprises all other proteins. Figure 2.1 shows a likely model for the PHA granules.

Based on these observations, two models have been proposed for granule formation. The first one is the micelle model, in which the extended PHA chains covalently attached to the synthase aggregate initially into a micelle structure [56, 67]. The physical properties of the polymer are thus proposed to be the driving force for inclusion formation. The second model is the maturing model that was proposed by Stubbe and Tian [68], in which the hydrophobic synthase binds to the inner face of the plasma membrane, leading to a granule surface covered with a lipid monolayer. In this model, the biology of the system and the physical properties of the polymer are required for granule formation.

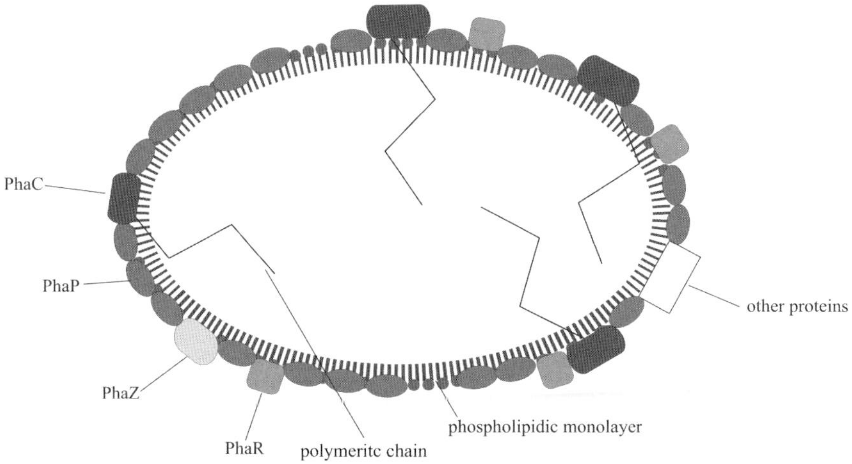

Figure 2.1 Model for the PHA granules. PhaC protein is a PHA synthase, PhaP protein is a phasin, PhaZ protein is a PHA depolymerase, PhaR protein is a sensor for PHA synthesis in the cell, the phospholipidic monolayer, and other proteins that surround the granule (based on [65], with modifications).

2.4 Biosynthesis and Biodegradability of Poly(3-Hydroxybutyrate) and Other Polyhydroxyalkanoates

2.4.1 Polyhydroxyalkanoates Biosynthesis on Microorganisms

Various microorganisms can accumulate a large amount of PHA inside their cells in response to the limitation of an essential nutrient. Many previous works were concerned with the control of the synthesis of PHA under unbalanced growth conditions [7].

Since the discovery of PHA-producing microorganisms by Lemoigne in 1925, there are over 300 types of microorganisms that accumulate PHA, belonging to the genus *Alcaligenes, Azobacter, Pseudomonads*, methylotrophs, and some recombinant microorganisms such as *E. coli* [13]. The Gram-negative bacteria *C. necator* has been the most widely used microorganism for the production of P(3HB). *C. necator* was previously categorized as *Hydrogenomonas eutropha, A. eutrophus, R. eutropha*, and *Wautersia eutropha* [69]. *C. necator* has also been used for the commercial production of P(3HB) by many industries [18, 34, 40].

Among the substrates required for PHA production, the carbon source has a primal significance in the case of P(3HB) production, since P(3HB) is composed only of C, H, and O atoms [70]. Microorganisms have the ability to produce PHA from various carbon sources including inexpensive and complex waste effluents. In the past years, our group has prepared works [11, 71–74] in order to reduce the production costs of P(3HB) and its copolymers by the use of renewable carbon sources.

Figure 2.2 shows P(3HB) (a PHA_{SCL}–short chain length PHA) production by *C. necator* in two phases: balanced growth and unbalanced growth (Pathways I and II, respectively); P(3HB-*co*-3HV) production from propionigenic substrates by *C. necator* (Pathways II and III); and PHA_{MCL} (medium chain length PHA) production from fatty acid *de novo* biosynthetic route and fatty acids β-oxidation (Pathways IV and V, respectively) according to the substrate.

P(3HB) production by *C. necator* occurs in two phases. The first phase comprises the exponential growth where all nutrients are present (balance growth–Pathway I in Figure 2.2) and the second phase shows a nutritional limitation of N, P, S, Mg, or O_2 in the presence of an excessive carbon source (unbalanced growth–Pathway II in Figure 2.2) [75]. Hence, the metabolism for the biomass production during balanced growth catabolizes carbohydrates via the Entner–Doudoroff pathway to pyruvate, which can be converted through dehydrogenation to acetyl-CoA. During reproductive growth (Pathway I), acetyl-CoA enters the tricarboxylic acid (TCA) cycle, releases CoASH, and is terminally oxidized to CO_2 generating energy in the form of ATP, reducing equivalents (NADH, NADPH, and $FADH_2$) and biosynthetic precursors (2-oxoglutarate, oxalacetate) [76]. Direct amination or transamination of the oxalocetate leads to the synthesis of amino

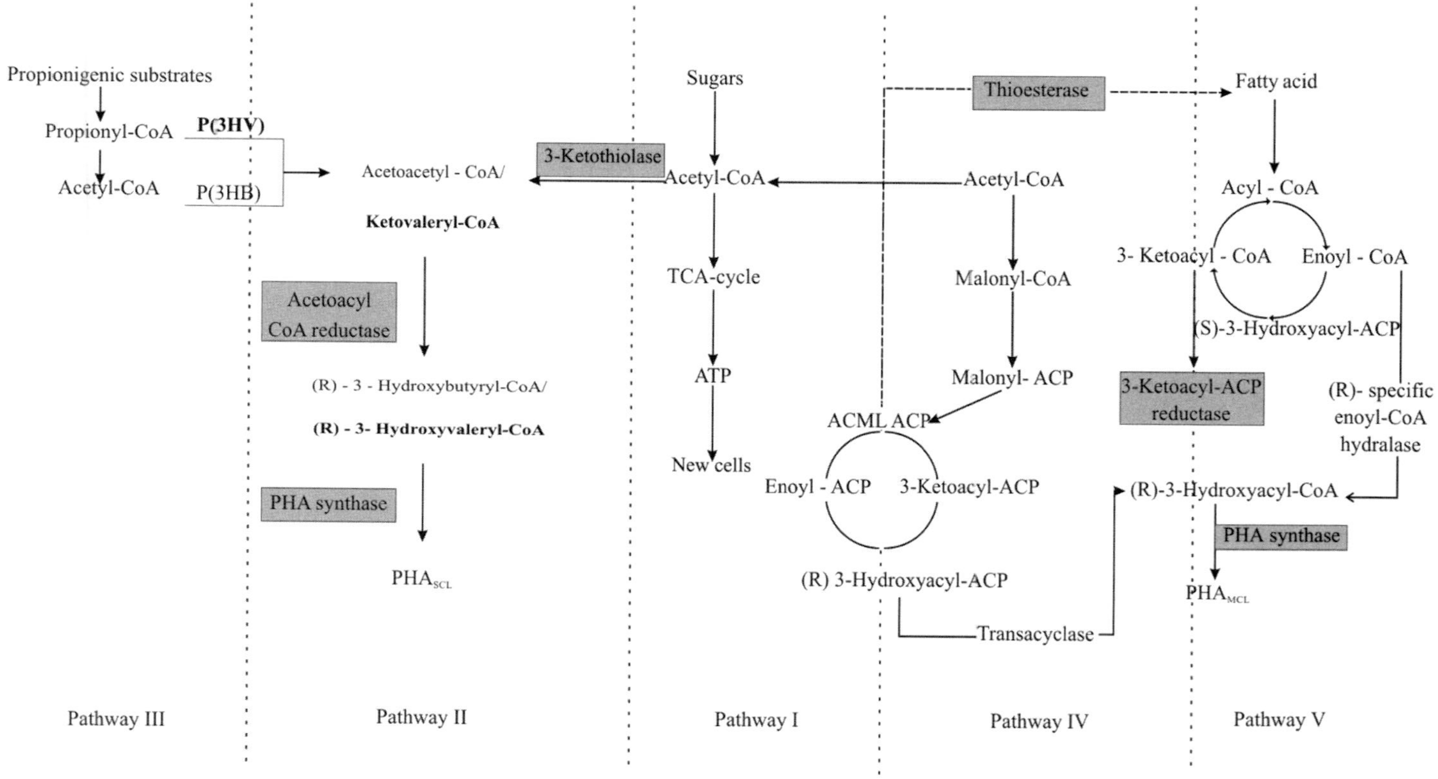

Figure 2.2 PHA production at pathways proposed for *C. necator*: (I) balanced growth for biomass production; (II) unbalanced growth for P(3HB) (a PHAS$_{SCL}$) production; (III) P(3HB-*co*-3HV) production from propionigenic substrates by *C. necator*, and proposed for pseudomonads; (IV) PHA$_{MCL}$ or PHAS$_{SCL}$ production from fatty acids *de novo* synthesis; and (V) PHA$_{MCL}$ production from fatty acids β-oxidation (based on [83] with modifications).

acids, which are incorporated into the polypeptide chains of nascent proteins. The rate of admission of acetyl-CoA into TCA cycle is dependent on the availability of sources of nitrogen, phosphorous, and other elements, as well as on the oxidative potential of the environment [6].

In Figure 2.2, Pathway II, limitations in nitrogen, phosphorous, oxygen [77, 78], magnesium, or sulfate [79] lead to P(3HB) production. This limitation causes cessation of protein synthesis leading to high concentrations of NADH and NADPH resulting in an inhibition of citrate synthase and isocitrate dehydrogenase and in a slowdown of the TCA cycle and the channeling of acetyl-CoA toward P(3HB) biosynthesis [35]. Acetyl-CoA no longer enters the TCA cycle at the same rate and instead is converted to acetoacetyl-CoA by 3-ketothiolase, the first enzyme of the P(3HB) biosynthetic pathway, which is inhibited by CoA.

According to Figure 2.2, Pathway II, three enzymes are involved in PHA_{SCL} production: 3-ketothioloase; acetoacyl-CoA reductase, and PHA synthase. The role of these enzymes is described below.

The first step for PHA formation is catalyzed by *3-ketothiolase*. Its mechanism includes two partial reactions which result in a condensation of two acetyl-CoA molecules to obtain acetoacyl-CoA. Two cystein residues are present in the active site of this enzyme and are responsible for the acetyl-CoA molecule ligament at the enzyme and for the activation of a second molecule of acetyl-CoA, hence entailing a condensation and formation of acetoacyl-CoA [1]. The enzyme catalyzes the reversible reaction shown in Eq. (2.1):

$$2\ \text{Acetyl-CoA} \leftrightarrow \text{acetoacyl-CoA} + \text{CoASH} \tag{2.1}$$

This 3-ketothiolase competes for acetyl-CoA with many other metabolic pathways, including acetate, citrate, and the fatty acids synthesis. This enzyme is inhibited by free CoASH molecules [80].

Acetoacyl-CoA reductase catalyzes the second step on PHA biosynthesis (Eq. (2.2)), converting acetoacyl-CoA into hydroxyacyl [1]:

$$\text{Acetoacyl-CoA} + \text{NADPH} + \text{H}^{+} \leftrightarrow \text{3-hydroxyacyl-CoA} + \text{NADP}^{+} \tag{2.2}$$

Two acetoacyl-CoA reductase types, with different specificities for substrates and coenzymes, were found in *C. necator*. The NADH-dependent enzyme is active in D(−) and L(+) substrates, while a NADPH-dependent one is stereospecific or active only at C4 to C6 D(−)3-hydroxyacyl-CoA substrates. During the P(3HB) synthesis, acetoacetyl-CoA is reduced to D(−)3- hydroxybutyryl-CoA, catalyzed by NADPH-dependent enzymes [5].

PHA synthase is the key enzyme for PHA biosynthesis. This enzyme catalyzes ester formation through the polymerization of D(−)3-hydroxyacyl-CoAs units, resulting in the polymer. The wide monomer variety that composes PHA is related to large substrate PHA synthase specificities. In this context, *C. necator* PHA synthase is able to polymerize 3-hydroxy, 4-hydroxy, and 5-hydroxyalkanoates from the 4 and 5 carbon hydroxyacyl-CoA, D-isomers [5, 12]. This enzyme is shown in two forms – a soluble form in the cytoplasm (balanced growth) and associated with P(3HB) granules (unbalanced growth) [5].

Based on the types of monomer incorporated into PHA, various metabolic pathways have been shown to be involved in the generation of these monomers [12, 81].

Biosynthesis of poly(3-hydroxybutyrate-*co*-3-hydroxyvalerate) P(3HB-*co*-3HV) requires, besides 3HB-CoA (3-hydroxybutyryl-CoA), also 3-hydroxyvaleryl-CoA (3HV-CoA). The latter is also required if other copolyesters containing 3HV or poly(3HV) homopolyester are synthesized. 3HV-CoA ([R]-3-hydroxyvaleryl-CoA) is obtained from the condensation of acetyl-CoA and propionyl-CoA into 3-ketovaleryl-CoA (Figure 2.2, Pathways II and III) and a subsequent reduction in the condensation product to 3HV-CoA. The specific substrates for P(3HB-*co*-3HV) production can be propionic acid [7, 11, 82], valeric acid, heptanoic acid, or nonanoic acid. Steinbüchel and Lütkte-Eversloh [12] cited many other sources for poly-(3HV) production. Among them, *n*-pentanol (metabolized from *Paracoccus denitrifricans*), valine, isoleucine, threonine, and methionine are considered precursor substrates for 3HV containing PHA.

Fatty acid *de novo* biosynthesis (Figure 2.2, Pathway IV) is the main route during growth on carbon sources, like gluconate, acetate, or ethanol, that are metabolized to acetyl-CoA, for the PHA_{MCL} synthesis by pseudomonads like *Pseudomonas putida, P. aeruginosa, P. aureofaciens, P. citronellolis,* and *P. mendocina* [81, 83, 84]. From the results of labeling studies, nuclear magnetic resonance spectroscopy, and gas chromatography mass spectroscopy ([85, 86], cited by [84]), authors concluded that the precursors of PHA_{MCL} biosynthesis from simple carbon sources are predominantly derived from (*R*)-3-hydroxyacyl-ACP intermediates occurring during the fatty acid *de novo* biosynthetic route. Since the constituents of P(3HB) and PHA represent the *R* configuration and since PHA_{SCL} and PHA_{MCL} synthases are highly homologous, the intermediates in fatty acid metabolism are presumably converted to (*R*)-3-hydroxyacyl-CoA before polymerization (Figure 2.2, Pathway V). Nevertheless, some other routes of PHA synthesis are also possible. Other conceivable alternatives are the release of free fatty acids by the activity of a thioesterase with a thiokinase, subsequently activating these fatty acids to the corresponding hydroxyacyl-CoA thioesters or chain elongation with 3-ketothiolase, or β-oxidation of synthesized fatty acids.

2.4.2
Plants as Polyhydroxyalkanoates Producers

Another important example for establishing PHA biosynthesis is the production by plants. With this strategy, the steps necessary to produce the substrates used in a fermentative process are no longer required, as naturally occurring carbon dioxide and sunlight serve as carbon and energy sources, respectively [6]. In the first investigations reported, the plant *Arabidopsis thaliana*, harboring the PHA genes of *C. necator,* was used to produce P(3HB). An endogenous plant 3-ketothiolase is present in the cytoplasm of this plant as part of the mevalonate pathway. Thus, creation of the P(3HB) biosynthetic pathway in the cytoplasm was theoretically simple, requiring only the expression of two additional enzymes – acetoacyl-CoA reductase and PHA synthase. Today, it is possible to produce P(3HB) and P(3HB-*co*-3HV) utilizing transgenic plants. However, further studies to enhance the PHA percentage and productivity in these plants are necessary [44].

2.4.3
Microbial Degradation of Polyhydroxyalkanoates

Biodegradation is a natural process, mediated by microorganisms that, through their enzymes, can hydrolyze the macromolecules, altering the structure of the material. In all processes of degradation, hydrolysis or hydrolytic degradation is the initial step of the process [87]. Products from the degradation should not be ecotoxic or harmful to the environment. When biodegradation occurs in an aerobic environment, the products formed are usually carbon dioxide, water, organic material, and biomass, while in an anaerobic one, methane and water are generated.

A remarkable characteristic of PHA is its biodegradability in various environments where a considerable number of microorganisms excrete PHA depolymerases to hydrolyze solid PHA into water-soluble oligomers and monomers, which are also utilized as nutrients for their own cells [88].

PHA is degraded in various environments such as soil, sewage, sea water, and lakes. Biodegradation depends on several factors such as microbial activity in the environment, humidity, temperature, pH, and molecular weight of the polymer [89]. The most important factors related to the PHA characteristics which influence its degradation are (i) stereospecificity, since only monomers in configuration (*R*) are hydrolyzed by depolymerases; (ii) crystallinity, since the degradation decreases with higher crystallinity; (iii) molecular weight, because the polymers of low molecular weight are generally degraded more rapidly than those with high molecular weight; and (iv) monomeric composition of PHA [90].

Schneider and coworkers [87] performed a study to determine the biodegradation of P(3HB) films produced from *C. necator* with different oleic acid concentrations (0, 0.3, 0.9, 1.5, and 3.0 $g\,L^{-1}$) as a nutritional supplement for the culture. The biodegradation process was carried out in 5×5 cm P(3HB) films, which were buried in beakers in 17-cm-deep soil for 0, 7, 14, and 21 days. The soil humidity was kept between 20% and 30%, based on the dry weight of soil. Figure 2.3 shows the biodegradation of P(3HB) films shown in the Schneider *et al.* [87] study.

According to Figure 2.3, the sample obtained without oleic acid was only slightly decomposed during the studied period (21 days), and the films with 0.3 and 0.9 $g\,L^{-1}$ concentrations showed advanced destruction after 14 days, and were completely destroyed a week later. The films synthesized with 1.5 and 3.0 $g\,L^{-1}$ of oleic acid showed an advanced degradation stage after only 7 days and, after 14 days, it was not possible to find fragments of these samples anymore. These results indicate that kinetic biodegradation of the films in soil was faster with the increase in oleic acid concentration.

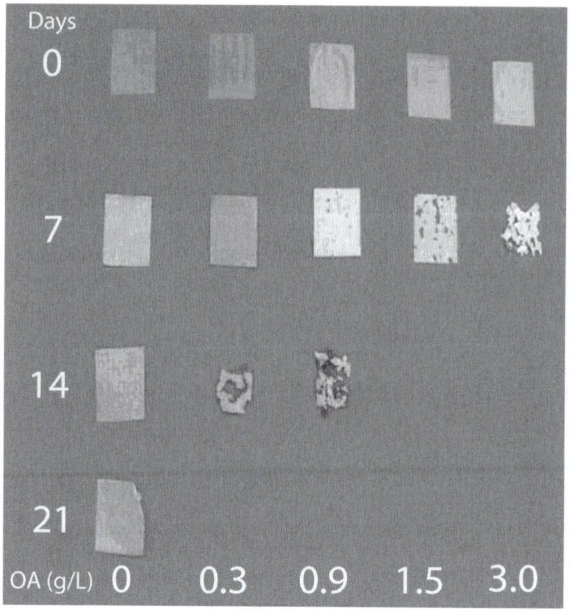

Figure 2.3 Visual analysis of P(3HB) with different oleic acid content after biodegradation in soil during 0, 7, 14, and 21 days(Ref. [88]).

2.5
Extraction and Recovery

It is estimated that PHA recovery can represent about 50% of the total production cost depending on variables such as the separation process and PHA content in the biomass (%) [91]. As PHA is an intracellular product, the methods adopted for its recovery focus either on its solubilization or on the solubilization of the non-PHA biomass (total biomass excluding PHA) [92]. Figure 2.4 shows methods for PHA extraction.

According to Figure 2.4, the pretreatment aims at the separation of the mixture biomass-fermented medium and can be realized by a mechanical process like centrifugation or a chemical process similar to flocculation either by acidification with sulfuric or phosphoric acids or by adding an alkalizing agent such as calcium hydroxide. One of the processes for PHA isolation is the cell lysis by chemical digestion, which occurs by the use of surfactant and chelate agents, such as anionic sodium dodecyl sulfate (SDS) [93] or synthetic palmitol carnitine [94], sodium hypochlorite, or sodium hypochlorite plus chloroform [95–97]. The enzymatic destruction of cell-wall structures is an effective method to obtain PHA and has been used to achieve lysis of Gram-negative bacteria [98]. Enzymatic destruction of cell-wall structures by less expensive enzymes like pepsin, trypsin, bromelain, papain, and other proteolytic enzymes in order to obtain PHA has been extensively

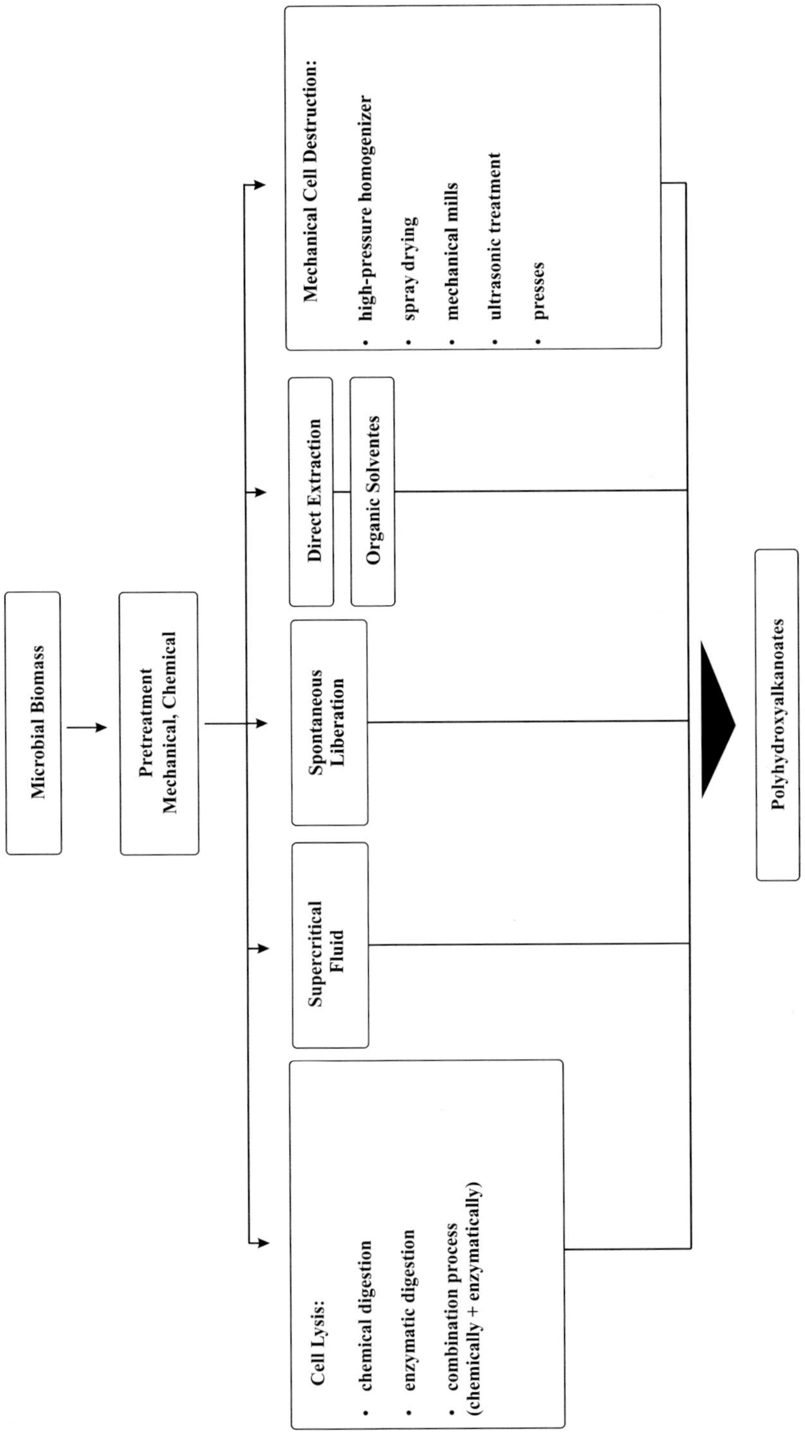

Figure 2.4 Methods for PHA extraction.

investigated [92, 99, 100]. The use of high pressure in continuously operating cell homogenizers is a usual downstream operation in biotechnology [101, 102]. Direct extractions with organic solvents, chloroform, have been realized by Dalcanton [73]. In this process, cells from *C. necator* cultures were centrifuged and after that they were submitted to extraction with chloroform for 2 h at 60 °C. The ratio of chloroform and biomass was 100 mL and 2 g containing 75% P(3HB), respectively. The results obtained for purity percentage and recovery percentage were 98 and 94, respectively. Other solvents can be used for PHA isolation, whether or not associated with another process, like dichloromethane, pyridine or dichloromethane/ethanol mixtures [31, 32], 1,2-propanodiol, glycerol formal [103], and propylene carbonate [104–106]. Nowadays, genetically modified cells, which have spontaneous liberation of intracellular PHA granules [107], are being studied.

In general, all processes for PHA isolation and purification are expensive, causing a large effluent generation, and it is necessary to develop a cleaner and less expensive process to allow PHA to be competitive in relation to fuel-based polymers.

2.6
Physical, Mechanical, and Thermal Properties of Polyhydroxyalkanoates

Solid-state P(3HB) is a compact right-handed helix with a twofold screw axis (i.e., two monomer units complete one turn of the helix) and a fiber repeat of 0.596 nm [108]. The stereoregularity of P(3HB) makes it a highly crystalline material. It is optically active, with a chiral carbon always in the *R* absolute configuration on biologically produced P(3HB).

The physical properties of PHA are influenced by the molecular weight of the polymer, which depends on the particular bacteria strain used and on the extraction process [109]. The molecular mass distribution of a polymer is a measure of the distribution of its individual molecules' molecular mass around the average molecular mass; a narrow distribution around a high average is usually desired [6].

In addition to being a function of the production organism and the strategy of production (fermentation duration, growth rate, carbon source concentration, etc.), the average molecular mass of PHA is affected by the method of extraction. Values until recently typically ranged from 2×10^5 to 2×10^6 Da [5].

Table 2.2 shows a comparison of properties between different PHA, polypropylene, and low-density polypropylene.

According to Table 2.2, P(3HB) and polypropylene show similar melting temperatures (T_m), although their chemical properties are completely different. P(3HB) has far lower solvent resistance, specially elongation to break, but it shows a better natural resistance to UV weathering. Physically, P(3HB) is stiffer and brittler than polypropylene [110]. However, when copolymer formation occurs with 3HB and 3HV monomer units, the properties of the material are altered as a consequence of decreased crystallinity and T_m. This results, in mechanical terms, in a decrease

Table 2.2 Comparison of PHA polymers' and common plastics' properties.

Sample	Melting temperature (°C)	Glass-transition temperature (°C)	Young's modulus (GPa)	Tensile strength (MPa)	Elongation to break (%)
P(3HB)	180	4	3.5	40	5
P(3HB-*co*-20 mol% 3HV)	145	−1	0.8	20	50
P(3HB-*co*-6 mol% 3HV)	133	−8	0.2	17	680
Polypropylene	176	−10	1.7	38	400
Low-density polyethylene	130	−30	0.2	10	620

Source: [81].

in stiffness (Young's modulus) and an increase in elongation to break, producing more desirable properties for commercial application [5].

It has been widely accepted that PHA_{MCL} shows amorphous and elastic properties with a lower melting point (T_m) and a lower degree of crystallinity compared with short-chain-length PHA (PHA_{SCL}) [110, 111]. Interestingly, studies demonstrated that the presence of over 30% long chain comonomer units such as HDD (hydroxydodecanoate−C_{12}) and HTD (hydroxytetradecanoate−C_{14}) in the PHA_{MCL} increased the T_m and the degree of crystallinity of PHA_{MCL}, leading to dramatic changes in PHA_{MCL} mechanical properties [111, 112] making them significantly different from those of the typical PHA_{MCL} [113].

2.7 Future Directions

PHA has quickly gained interest in both research and industry. Their structural versatility and characteristics have been investigated and new areas of exploitation are being discovered. The major drawback for extensive use of these polymers is their high production cost [114]. In this regard, research is continuing on their production from cheap raw materials [71, 72].

Decades of study have been dedicated to the composition and material properties of PHA in this crystallized form. A new perspective for industrial applications of PHA could be the use of native, isolated PHA granules as nano/microbeads in biotechnology and medicine. The PHA granules exhibit all important features of core–shell nanoparticles; their surface has been demonstrated to be easily modified and activity of surface-exposed proteins of interest could be shown [24, 83].

The recent Food and Drug Administration (FDA) approval for the clinical application of P(4HB) suggests a promising future for PHA [114]. The application of PHA in composites scaffolds showing potential for drug delivery has been developed for bone tissue engineering [115]. The emphasis is on the area of medical/biomedical applications including the development to regenerate bone tissue by combining PHA with bone marrow mesenchymal stem cells [116], neural stem cells nanofiber scaffolds [117], among others.

References

1 Madison, L.L. and Huisman, G.W. (1999) Metabolic engineering of poly(3-hydroxyalkanoates): from DNA to plastic. *Microbiol. Mol. Biol. Rev.*, **63** (**1**), 21–53.

2 Sotero, A.P. (2000) Plásticos biodegradáveis trazem melhoria ambiental. *Jornal de Plásticos*, **7**, 1017–1018.

3 Vauterin, L., Hoste, B., Kersteres, K., and Swings, J. (1995) Reclassification of *Xanthomonas*. *Int. Syst. Bacteriol.*, **45**, 472–489.

4 Garcia-Ochoa, F., Santos, V.E., Casas, J.A., and Gomez, E. (2000) Xanthan gum: production, recovery, and properties. *Biotechnol. Adv.*, **18**, 549–579.

5 Anderson, A. and Dawes, E.A. (1990) Occurrence, metabolism, metabolic role, and industrial uses of bacterial polyhydroxyalkanoates. *Microbiol. Rev.*, **54**, 450–472.

6 Braunegg, G., Lefebvre, G., and Genser, K.F. (1998) Polyhydroxyalkanoates, biopolyesters from renewable resources: physiological and engineering aspects. *J. Biotechnol.*, **65**, 127–161.

7 Byrom, D. (1992) Production of poly(β-hydroxybutyrate):poly(β-hydroxyvalerate) copolymers. *Microbiol. Rev.*, **103**, 247–250.

8 Elbahloul, Y., Krehenbrink, M., Reichelt, R., and Steinbüchel, A. (2005) Physiological conditions conducive to high cyanophycin content in biomass of *Acinetobacter calcoaceticus* strain ADP1. *Appl. Environ. Microbiol.*, **71**, 858–866.

9 Muller, C., Yamashita, F., and Laurindo, J. (2008) Evaluation of the effects of glycerol and sorbitol concentration and water activity on the water barrier properties of cassava starch films through a solubility approach. *Carbohydr. Polym.*, **72**, 82 87.

10 Jacobsen, S., Fritz, H.G., and Jérôme, R. (2004) Polylactide (PLA) – a new way of production. *Polym. Eng. Sci.*, **39** (**7**), 1311–1319.

11 Marangoni, C., Furigo, A., Jr., and Aragão, G.M.F. (2002) Production of poly(3-hydroxybutyrate-*co*-3-hydroxyvalerate) by *Ralstonia eutropha* in whey and inverted sugar with propionic acid feeding. *Proc. Biochem.*, **38**, 137–141.

12 Steinbüchel, A. and Lütke-Eversloh, T. (2003) Metabolic engineering and pathway construction for biotechnological production of relevant polyhydroxyalkanoates in microorganisms. *Biochem. Eng. J.*, **16**, 81–96.

13 Lee, S.Y. (1996) Bacterial polyhydroxyalkanoates. *Biotechnol. Bioeng.*, **49**, 1–14.

14 Ramsay, B.A. (1994) Physiological factor affecting PHA production, in: Physiology, kinetics, production and use of biopolymers. Proceedings of the Symposium on Physiology, Kinetics, Production and Use of Biopolymers, May 13–15, 1994, Schloss Seggau, Austria.

15 Khanna, S. and Srivastava, A.K. (2005) Statistical media optimization studies for growth and PHB production by *Ralstonia eutropha*. *Process Biochem.*, **40**, 2173–2182.

16 Cavalheiro, J.M.B., De Almeida, M.C.M.D., Grandfils, C., and da Fonseca, M.M.R. (2009) Poly(3-

hydroxybutyrate) production by *Cupriavidus necator* using waste glycerol. *Process Biochem.*, **44**, 509–515.

17 Mengmeng, C., Hong, C., Qingliang, Z., Shirley, S.N., and Jie, R. (2009) Optimal production of polyhydroxyalkanoates (PHA) in activated sludge fed by volatile fatty acids (VFAs) generated from alkaline excess sludge fermentation. *Bioresour. Technol.*, **100** (**3**), 1399–1405.

18 Copersucar (BR/SP)/Instituto de Pesquisas Tecnológicas do Estado de São Paulo S/A–IPT (BR/SP) (1991) Processo para produzir polihidroxialcanoatos a partir de açúcares extraídas da cana de açúcar. INPI PI9103116-8.

19 Rocha, R.C.S., Da Silva, L.F., Taciro, M.K., and Pradella, J.G.C. (2008) Production of poly(3-hyroxybutyrate-*co*-3-hyroxyvalerate) P(3HB-*co*-3HV) with a broad range of 3HV content at high yields by *Burkoholderia sacchari* IPT 189. *World J. Microbiol. Biotechnol.*, **24**, 427–431.

20 Diniz, S.C., Taciro, M.K., Gomez, J.G.C., and Pradella, J.G.C. (2004) High-cell-density cultivation of *Pseudomonas putida* IPT 046 and medium-chain-length polyhydroxyalkanote production from sugarcane carbohydrates. *Appl. Biochem. Biotechnol.*, **119**, 51–69.

21 Choi, J., Lee, S.Y., and Han, K. (1998) Cloning of the *Alcaligenes latus* polyhydroxyalkanoate biosynthesis genes and use of this genes for enhanced production of poly(3-hydroxybutyrate) in *Escherichia coli. Appl. Environ. Microbiol.*, **64**, 4879–4903.

22 Khanna, S. and Srivastava, A.K. (2006) Computer fed-batch cultivation for over production of PHB: a comparison of simultaneous an alternate feeding of carbon and nitrogen. *Biochem. Eng. J.*, **27**, 197–203.

23 Barnard, G.C., McCool, J.D., Wood, D.W., and Gerngross, T.U. (2005) Integrated recombinant protein expression and purification platform based on *Ralstonia eutropha. Appl. Environ. Microbiol.*, **71**, 5735–5742.

24 Grage, K. and Rehm, B.H.A. (2008) *In vivo* production of scFv-displaying biopolymer beads using a self-assembly promoting fusion partner. *Bioconj. Chem.*, **19**, 254–262.

25 Beijerinck, M.W. (1888) Cultur des *Bacillus radicola* aus den Knöllchen. *Bot. Ztg.*, **46**, 740–750.

26 Lemoigne, M. (1923) Production d'acide β-oxybutyrique par certaines bact'eries du groupe du *Bacillus subtilis. C. R. Hebd. Seances Acad. Sci.*, **176**, 1761–1765.

27 Lemoigne, M. (1926) Produit de déshydratation et de polymérisation de l'acide b-oxybutyrique. *Bull. Soc. Chim. Biol.*, **8**, 770–782.

28 Stapp, C. (1924) Zur Frage es Lebens un Wirksambeilsdauer der Knöllchenbakterien. *Angew. Bot.*, **4**, 152–159.

29 Macrae, R.M. and Wilkinson, J.R. (1958) Poly-β-hydroxybutyrate metabolism in washed suspensions of *Bacillus cereus* and *Bacillus megaterium. J. Gen. Microbiol.*, **19**, 210–222.

30 Baptist, J.N. (1963) Molded product containing poly-β-hydroxybutyric acid and process. US Patent 3, 107,172.

31 Baptist, J.N. (1962a) Process for preparing poly-β-hyroxybutyric acid. US Patent 3, 036,959.

32 Baptist, J.N. (1962b) Process for preparing poly-β-hyroxybutyric acid. US Patent 3, 044,942.

33 Baptist, J.N. (1965) Plasticize poly-β-hydroxybutyric acid and process. US Patent 3, 182,036.

34 Grace and Co (1963) Absorbable prosthetic devices and surgical sutures. British Patent Specification 1, 034,123.

35 Dawes, E.A. and Senior, P.J. (1973) The role and regulation of energy reserve polymers in microorganisms. *Adv. Microb. Physiol.*, **10**, 135–266.

36 Wallen, L.L. and Rohwedder, W.K. (1974) Poly-β-hydroxyalkanoate from activated sludge. *Environ. Sci. Technol.*, **8**, 576–585.

37 Huisman, G.W., Leeuwde, O., Eggink, G., and Wiltholt, B. (1989) Synthesis of poly-3-hydroxyalkanoates is a common feature of fluorescent pseudomonas. *Appl. Environ. Microbiol.*, **55**, 1949–1954.

38 Luzier, W.D. (1992) Material derived from biomass/biodegradable materials. *Proc. Natl. Acad. Sci.*, **89** (**3**), 835–838.

39 Holmes, P.A., Wright, L.F., and Collins, S.H. (1982) Copolyesters and process for their production. European Patent 69,497.

40 Pradella, J.G.C. Biopolímeros e Intermediários Químicos, in Relatório Técnico n° 84 396-205, http://www.anbio.org.br/pdf/2/tr06_biopolimeros.pdf (accessed March 2006), Centro de Tecnologia de Processos e Produtos, Laboratório de Biotecnologia Industrial-LBI/CTPP.

41 Steinbüchel, A. and Valentin, H.E. (1995) MiniReview: diversity of bacterial polyhydroxyalkanoic acids. *FEMS Microbiol. Lett.*, **128**, 219–228.

42 Slater, S.C., Voige, W.H., and Dennis, D.E. (1988) Cloning and expression in *Escherichia coli* of the *Alcaligenes eutrophus* H16 poly-beta-hydroxybutyrate biosynthetic pathway. *J. Bacteriol.*, **170**, 4431–4436.

43 Peoples, O.P. and Sinskey, A.J. (1989) Poly-beta-hydroxybutyrate biosynthesis in *Alcaligenes eutrophus* H16. Characterization of the genes encoding beta-ketothiolase and acetoacetyl-CoA reductase. *J. Biol. Chem.*, **264** (**26**), 15293–15298.

44 Poirier, Y., Dennis, D., Klomparens, K., Nawrath, C., and Somerville, C. (1992a) Perspectives on the production of polyhyroxyalkanoates in plants. *FEMS Microbiol. Rev.*, **103**, 237–246.

45 Poirier, Y., Nawrath, C., and Somerville, C. (1995) Production of polyhydroxy-alkanoates, a family of biodegradable plastics and elastomers in bacteria and plants. *Biotchnol.*, **13**, 142–150.

46 Rinehart, J.A., Petersen, M.W., and John, M.E. (1996) Tissue-specific and developmental regulation of cotton gene Fb12A-demonstration of promoter activity in transgenic plants. *Plant Physiol.*, **112**, 1331–1341.

47 Nawrath, C., Poirier, Y., and Somerville, C. (1994) Targeting of the polyhydroxybutyrate biosynthetic pathway to the plastics of *Arabiopsis thaliana* results in high levels of polymer accumulation. *Proc. Natl. Acad. Sci USA*, **91**, 12760–12764.

48 Lu, X., Zhang, W., Jian, J., Wu, Q., and Chen, G. (2005) Molecular cloning and functional analysis of two polyhydroxyalkanoate synthases from two strains of *Aeromonas hydrophila* spp. *FEMS Microbiol. Lett*, **243**, 149–155.

49 Rehm, B.H.A. and Steinbüchel, A. (1999) Biochemical and genetic analysis of PHA synthases an other proteins required for PHA synthesis. *Int. J. Biol. Macromol.*, **25**, 3–19.

50 Shin, H.D., Oh, D.H., Jung, Y.M., Ghim, S.Y., and Lee, Y.H. (2002) Comparison of *phbC* genes cloned from *Ralstonia eutropha* and *Alcaligenes latus* for utilization in metabolic engineering of polyhydroxyalkanoate biosynthesis. *Biotechnol. Lett.*, **24**, 539–545.

51 Ellar, D., Lundgren, D.G., Okamura, K., and Marchessault, R.H. (1968) Morphology of poly-beta-hydroxybutyrate granules. *J. Mol. Biol.*, **35** (**3**), 489–502.

52 Lundgren, D.G., Pfister, R.M., and Merrick, J.M. (1964) Structure of (poly-β-hydroxybutyrate acid) granules. *J. Gen. Microbiol.*, **34**, 441–446.

53 Dunlop, W.F. and Robards, A.W. (1973) Ultrastructural study of poly-β-hydroxybutyrate granules from *Bacillus cereus*. *J. Bacteriol.*, **114**, 1271–1280.

54 Ballard, D.G.H., Holmes, P.A., and Senior, P.J. (1987) Formation of polymers of β-hydroxybutyric acid in bacterial cells and a comparison of the morphology of the growth with the formation of polyethylene in the solid state, in *Recent Advances in Mechanistic and Synthetic Aspects of Polymerization* (eds I.M. Fontanille and A. Guyot), Reidel (Kluwer) Publishing, Lancaster, UK, pp. 293–314.

55 Alper, R., Lundgren, D.G., Marchessault, R.H., and Cote, W.A. (1963) Properties of poly-β-hydroxybutyrate. I. General considerations concerning the naturally occurring polymer. *Biopolymers*, **1**, 545–556.

56 Lundgren, D.G., Alper, R., Schnaitman, C., and Marchessault, R.H. (1965) Characterization of poly-β-hydroxybutyrate extracted from different bacteria. *J. Bacteriol.*, **89**, 245–251.

57 Kawaguch, Y. and Doi, Y. (1990) Structure of native poly(3-

hydroxybutyrate) granules characterized by X-ray diffraction. *FEMS Microbiol. Lett.*, **79**, 151–156.

58 Foster, L.J.R., Lenz, R.W., and Fuller, R.C. (1994) Quantitative determination of intracellular depolymerase activity in *Pseudomonas oleovorans* inclusions containing poly-3-hydroxyalkanoates with long alkyl substituents. *FEMS Microbiol. Lett.*, **133**, 279–282.

59 Foster, L.J.R., Stuart, E.S., Tehrani, A., Lenz, R.W., and Fuller, R.C. (1996) Intracellular depolymerase functionally and location in *Pseudomonas oleovorans* inclusions containing polyhydroxyoctanoate. *Int. J. Biol. Macromol.*, **19**, 177–183.

60 Murphy, D.J. (1993) Structure, function and biogenesis of storage lipid bodies and oleosins in plants. *Prog. Lipid Res.*, **32**, 246–280.

61 Wieczorek, R., Pries, A., Steinbüchel, A., and Mayer, F. (1995) Analysis of a 24-kDa protein associated with the polyhydroxyalkanoic acid granules in *Alcaligenes eutrophus*. *J. Bacteriol.*, **177**, 2425–2435.

62 Pieper-Fürst, U., Madkour, M.H., Mayer, F., and Steinbüchel, A. (1994) Purification and characterization of a 14-kilodalton protein that is bound to the surface of polyhydroxyalkanoic acid granules in *Rhodococcus ruber*. *J. Bacteriol.*, **176**, 4328–4337.

63 Schembri, M.A., Woods, A.A., Bayly, R.C., and Davies, J.K. (1995) Identification of a 13-kDa protein associated with the polyhydroxyalkanoic acid granules from *Acinetobacter* spp. *FEMS Microbiol. Lett.*, **133**, 277–283.

64 Stuart, E.S., Tehrani, A., Valentin, H.E., Dennis, D., Lenz, R.W., and Fuller, R.C. (1998) Protein organization on the PHA inclusion cytoplasm boundary. *J. Biotechnol.*, **64**, 137–144.

65 Maehara, A., Taguchi, S., Nishiyama, T., Yamane, T., and Doi, Y. (2002) A repressor protein, PhaR, regulates polyhydroxyalkanoate (PHA) synthesis via its direct interaction with PHA. *J. Bacteriol.*, **184 (14)**, 3992–4002.

66 Steinbüchel, A., Aerts, K., Babel, W., Föllner, C., Liebergesell, M., Madlour, M.H., Mayer, F., Pieper-Fürst, U., Pries, A., Valentin, H.E., and Wieczorek, R. (1995) Considerations on the structure and biochemistry of bacterial polyhydroxyalkanoic acid inclusions. *Can. J. Microbiol.*, **41**, 94–105.

67 Gerngross, T.U., Snell, K.D., Peoples, O.P., Sinskey, A.J., Cushai, E., Masamune, S., and Stubbe, J. (1994) Overexpression and purification of the soluble polyhydroxyalkanoate synthase from *Alcaligenes eutrophus*: evidence for a required post-transitional modification for catalytic activity. *Biochemistry*, **33**, 9311–9320.

68 Stubbe, J. and Tian, J. (2003) Polyhydroxyalkanoate (PHA) homeostasis: the role of PHA synthase. *Nat. Prod. Rep.*, **20**, 445–457.

69 Vanamme, P. and Coenye, T. (2004) Taxonomy of the genus *Cupriavidus*: a tale of lost and found. *Int. J. Syst. Evol. Microbiol.*, **54**, 2285–2289.

70 Shi, H., Shiraishi, M., and Simizu, K. (1997) Metabolic flux analysis for biosynthesis of poly(β-hydroxybutyric acid) in *Alcaligenes eutrophus* from various carbon sources. *J. Ferment. Bioeng.*, **84 (4)**, 579–587.

71 Aragão, G.M.F., Schmidell, W., Ienczak, J.L., and Fischer, S.A. (2009) Preparation of polyhydroxyalkanoates from a citric residue. PCT–WO2009/149529 A1.

72 Dalcanton, F., Ienczak, J.L., Fiorese, M.L., and Aragão, G.M.F. (2010) Produção de poli(3-hidroxibutirato) por *Cupriavidus necator* em meio hidrolisado de amido de arroz com suplementação de óleo de soja em diferentes temperaturas. *Quím. Nova*, **3 (33)**, 552–556.

73 Dalcanton, F. (2006) Produção, extração e caracterização de poli(3-hidroxibutirato) por Rasltonia eutropha em diferentes substratos. Federal University of Santa Catarina – Brazil. Dissertation.

74 Rodrigues, R.C. (2005) Condições de cultura para a produção de poli(3-hidroxibutirato) por Ralstonia eutropha a partir de resíduos de indústrias de alimentos. Federal University of Santa Catarina – Brazil. Dissertation.

75 Lee, S.Y., Kim, M.K., Chang, H.N., and Park, Y.H. (1995a) Regulation of poly-beta-hydroxytbutyrate biosynthesis by nicotinamide nucleotide in *Alcaligenes eutrophus*. *FEMS Microbiol. Lett.*, **131**, 35–39.
76 Pranamuda, H., Towkiwa, Y., and Tanaka, H. (1995) Microbial degradation of an aliphatic polyester with a high melting point, poly(tetramethylene succinate). *Appl. Environ. Microbiol.*, **61**, 1828–1832.
77 Schlegel, H.G., Gottschalk, G., and Von Bartha, R. (1961) Formation and utilization of poly-β-hydroxybutyric acid by knallgas bacteria (*Hyrogenomonas*). *Nature*, **191**, 463–465.
78 Schuster, E. and Schlegel, H.G. (1967) Chemolithotrophes wachstum von Hydrogenomonas H16 im chemostaten mit elektrolytischer Knallgaserzeugung. *Arch. Mikrobiol.*, **58**, 380–409.
79 Repaske, R. and Repaske, C. (1976) Quantitative requirements for exponential growth of *Alcaligenes eutrophus*. *Appl. Environ. Microbiol.*, **32**, 585–591.
80 Lee, S.Y. and Chang, H.N. (1995b) Production of poly(hydroxyalkanoic acid). *Adv. Biochem. Eng./Biotechnol.*, **52**, 27–85.
81 Sudesh, K., Abe, H., and Doi, Y. (2000) Synthesis, structure and properties of polyhydroxyalkanoates: biological polyesters. *Prog. Polym. Sci.*, **25**, 1503–1555.
82 Nonanto, R.V., Mantlatto, P.E., and Rossel, C.E.V. (2001) Integrate production of biodegradable plastic, sugar and ethanol. *Apll. Microbiol. Biotecnol.*, **57**, 1–5.
83 Grage, K., Peters, V., Palanisamy, R., and Rehm, B.H.A. (2009) Polyhydroxyalkanoates: from bacterial storage compound via renewable plastic to bio-bead, in *Microbial Production of Biopolymers and Polymer Precursor* (ed. B.H.A. Rehm), Caister Acaemic Press, Norfolk, UK, pp. 255–289.
84 Rehm, B.H.A., Krüger, N., and Steinbüchel, A. (1998) New metabolic link between fatty acid *de novo* synthesis and polyhydroxyalkanoic acid synthesis. *J. Biol. Chem.*, **273** (**37**), 24044–24051.
85 Eggink, G.W., De Waa, P., and Huijberts, G.N.M. (1992) The role of fatty acid biosynthesis and degradation in the supply of substrates of poly(3-hyroxyalkanoates) formation in *Pseudomonas putida*. *FEMS Microbiol. Rev.*, **105**, 759–764.
86 Huijberts, G.N.M., De Rijk, T., De Waard, P., and Eggink, G. (1994) ^{13}C nuclear magnetic resonance studies of *Pseudomonas putida* fatty acid metabolic routes involved in poly(3-hydroxyalkanoate) synthesis. *J. Bacteriol.*, **176** (**6**), 1661–1666.
87 Chiellini, E. and Solaro, R. (1996) Biodegradable polymeric materials. *Adv. Mat.*, **8** (**4**), 305–313.
88 Schneider, A.L.S., Silva, D., Formolo, M.C., Grigull, V.H., Mazur, L.P., Furlan, S.A., Aragão, G.M.F., and Pezzin, A.P.T. (2010) Biodegradation of poly(3-hydroxybutyrate) produced from *Cupriavidus necator* with different concentrations of oleic acid as nutritional supplement. *J. Pol. Environ.*, published online, DOI 10.1007/s10924-010-0184-1.
89 Khanna, S. and Srivastava, A.K. (2005) Computer simulated fed-batch cultivation for over production of PHB: a comparison of simultaneous and alternate feeding of carbon and nitrogen. *Biochem. Eng. J.*, **27**, 197–203.
90 Jendrossek, D., Schirmer, A., and Schlegel, H.G. (1996) Biodegradation of polyhydroxyalkanoic acids. *Appl. Microbiol. Biotechnol.*, **46**, 451–463.
91 Da Silva, L.F., Gomez, J.G.C., Rocha, R.C.S., Taciro, M.K., and Pradella, J.G.C. (2007) Produção biotecnológica de poli-hidroxialcanoatos para a geração de polímeros biodegradáveis no Brasil. *Quím. Nova*, **30** (**7**), 1732–1743.
92 Kapritchkoff, F., Viotti, A.P., Alli, R.C.P., Zuccolo, M., Pradella, J.G.C., Maiorano, A.E., Miranda, E.A., and Bonomi, A. (2006) Enzymatic recovery and purification of polyhydroxybutyrate produced by *Ralstonia eutropha*. *J. Biotechnol.*, **122**, 453–462.
93 Ramsay, J.A., Berger, E., Ramsay, B.A., and Chavarie, C. (1990) Recovery of poly-β-hydroxybutyric acid granules by a

surfactant-hypocholite treatment. *Biotechnol. Technol.*, **4**, 221–226.

94 Lee, K.M., Chang, H.N., Chang, Y.K., Kim, B.S., and Hahn, S.K. (1993) The lysis of Gram-negative *A. eutrophus* and *A. latus* by Palmitoyl carnitine. *Biotechnol. Technol.*, **7**, 295–300.

95 Berger, E., Ramsay, B.A., Ramsay, J.A., and Chavarie, C. (1989) PHB recovery by hypoclorite digestion of non-PHB biomass. *Biotechnol. Technol.*, **3**, 227–232.

96 Hahn, S.K., Chang, Y.K., Kim, B.S., and Chang, H.N. (1994) Optimization of microbial poly(3-hydroxybutyrate) recovery using dispersions of sodium hypochlorite solution and chloroform. *Biotechnol. Bioeng.*, **44**, 256–261.

97 Valappil, S.P., Misra, S.K., Boccaccini, A.R., Keshavarz, T., Bucke, C., and Roy, I. (2007) Large-scale production and efficient recovery of PHB with desirable material properties, from the newly characterized *Bacillus cereus* SPV. *J. Biotechnol.*, **132**, 251–258.

98 Repaske, R. (1956) Lysis of Gram-negative bacteria by lysozyme. *Biochim. Biophys. Acta*, **22** (**1**), 189–191.

99 Holmes, P.A. and Lim, G.B. (1984) Separation process. European Patent 145233.

100 De Koning, G.J.W. and Wiltholt, B. (1997) A process for the recovery of poly(hydroxyalkanoate) from part 1: solubilization. *Bioprocess. Eng.*, **17**, 7–13.

101 Tamer, M., Moo-Young, M., and Chisti, Y. (1998) Disruption of *Alcaligenes latus* for recovery of poly(β-hydroxybutyric acid): comparison of high-pressure homogenization, bead milling and chemically induced lysis. *Ind. Eng. Chem. Res.*, **37**, 1807–1814.

102 Ghatnekar, M.S., Pai, J.S., and Ganesh, M. (2002) Production and recovery of poly-3-hydroxybutyrate from *methylobacterium sp.* V49. *J. Chem. Technol. Biotechnol.*, **77**, 444–449.

103 Traussnig, H., Kloimstein, E., Kroath, H., and Estermann, R. (1990) Extracting agents for poly(D-(−)-3-hydroxybutyric acid). U.S. Patent 4, 968,611.

104 Lafferty, R.M. and Heinzle, E. (1977) Poly(β-hydroxybutiyric acid) solutions. The Patent Office London 1, 568,719.

105 Lafferty, R.M. and Heinzle, E. (1978) Cyclic carbonic acid esters as solvents for poly-(β-hydroxybutyric acid. U.S. Patent 4, 101,533.

106 Fiorese, M.L., Freitas, F., Pais, J., Ramos, A.M., Falcão de Aragão, G.M., and Reis, M.A.M. (2009) Recovery of P(3HB) from *Cupriavidus necator* biomass by solvent extraction with 1,2-propylene carbonate. *Eng. Life Sci.*, **9** (**6**), 454–461.

107 Jung, L., Phyo, K.H., Kim, K.C., Park, H.K., and Kim, I.G. (2005) Spontaneous liberation of intracellular polyhydroxybutyrate granules in *Escherichia coli. Res. Microbiol.*, **156**, 865–873.

108 Marchessault, R.H., Bluhm, T.L., Deslandes, Y., Hamer, G.K., Orts, W.J., Sundarajan, P.R., Taylor, M.G., Bloemberg, S., and Holden, D.A. (1998) Poly(β-hydroxyalkanoates): biorefinery polymers in search of applications. *Makromol. Chem. Macromol. Symp.*, **19**, 235–254.

109 Lafferty, R.M., Korsatko, B., and Korsatko, W. (1988) *Microbial Production of Poly-B-Hydroxtbutyric Acid*, vol. 6b (eds H.-J. Rehm and G. Rees), VCH Verlagsgesellschaft, Weinheim, pp. 135–176.

110 Holmes, P.A. (1985) Applications of PHB – a microbially produced biodegradable thermoplastic. *Phys. Technol.*, **16**, 32–36.

111 Ouyang, S.P., Liu, Q., Fang, L., and Chen, G.Q. (2007) Construction of pha-operon-defined knockout mutants of *Pseudomonas putida* KT2442 and their applications in poly(hydroxyalkanoate) production. *Macromol. Biosci.*, **7**, 227–233.

112 Liu, W. and Chen, G.Q. (2007) Production and characterization of medium-chain-length polyhydroxyalkanoate with high 3-hydroxytetradecanoate monomer content by fadB and fadA knockout mutant of *Pseudomonas putida* KT2442 *Appl. Microbiol. Biotechnol.*, **76**, 1153–1159.

113 Chen, S., Qian Liu, Q., Wang, H., Zhu, B., Yu, F., Chen, G., and Inoue, Y. (2009) Polymorphic crystallization of

fractionated microbial medium-chain-length polyhydroxyalkanoates. *Polymer*, **50**, 4378–4388.

114 Keshavarz, T. and Roy, I. (2010) Polyhydroxyalkanoates: bioplastics with a green agenda. *Curr. Opin. Microbiol.*, **13**, 321–326.

115 Francis, L., Meng, D., Knowles, J.C., Roy, I., and Boccaccini, A.R. (2010) Multi-functional P(3HB) microsphere/45S5 Bioglass®−based composite scaffolds for bone tissue engineering. *Acta Biomater.*, **6**, 2773–2786.

116 Hu, Y., Wei, X., Zhao, W., Liu, Y., and Chen, G. (2009) Biocompatibility of poly(3-hydroxybutyrate-*co*-3-hydroxyvalerate-*co*-3-hydroxyhexanoate) with bone marrow mesenchymal stem cells. *Acta Biomater.*, **5**, 1115–1125.

117 Xu, X., Li, X., Peng, S., Xiao, J., Liu, C., Fang, G., Chen, K., and Chen, G. (2010) The behaviour of neural stem cells on polyhydroxyalkanoate nanofiber scaffolds. *Biomaterials*, **31**, 3967–3975.

3
Polyanhydrides

Avi Domb, Jay Prakash Jain, and Neeraj Kumar

3.1
Introduction

Polyanhydrides (PAs) were unearthed by Bucher and Slade in as early as 1909, but were not regarded as a "good polymer" until first utilized by Langer in early 1980s for controlled drug delivery [1]. Today, polyanhydride can be regarded as "designer polymers" because they can be synthesized in such a way to produce polymers with various degrees of crystallinity, degradation behavior, branching, crosslinking, etc., and have been used in various forms to deliver diverse active agents and other biomedical applications. For a surface-eroding device, the polymer must be hydrophobic but contain water labile linkages.

Some polyanhydrides are nontoxic and degrade finally into diacids which are either excreted as such in feces/urine or undergo extensive metabolism to form carbon dioxide and water in the body. The safety of polyanhydrides is evident from clinically used Glidel (wafers for BCNU delivery in glimoa) and Septacin (beads for delivery of gentamicin in osteomyelitis) products. Success in these products led to the development of an array of devices for various applications ranging from delivery of bioactive molecules to tissue engineering.

There has been a need to develop more rational approaches for creating improved biomaterials for drug delivery, especially biodegradable polymers. For such polymers, to maximize control over release, it is often desirable for a system to degrade only from its surface. These surface-eroding polymers are expected to release the drug at a constant release rate, thus the rate is directly proportional to the polymer erosion rate. For a surface-eroding device, the polymer must be hydrophobic but contain water labile linkages. Polyanhydrides are believed to predominantly undergo surface erosion due to (i) the high water lability of the anhydride bonds on the surface and (ii) hydrophobicity, which restricts water penetration into the bulk. A decrease in the device thickness throughout the erosion process, maintenance of the structural integrity, and the nearly zero-order degradation kinetics suggest the dominancy of heterogeneous surface erosion [2–4].

Handbook of Biodegradable Polymers: Synthesis, Characterization and Applications, First Edition. Edited by Andreas Lendlein, Adam Sisson.
© 2011 Wiley-VCH Verlag GmbH & Co. KGaA. Published 2011 by Wiley-VCH Verlag GmbH & Co. KGaA.

High hydrolytic reactivity of the anhydride linkage provides an intrinsic advantage in versatility and control of degradation rates. By varying the type of monomer and their ratios, surface-eroding polymers with degradation times of 1 week to several years can be designed and synthesized. The hydrolytic degradation rates can be obtained varying several thousand folds by simple changes in the polymer backbone and by altering the hydrophobic and hydrophilic balance of the polymer [5–7]. Aliphatic polyanhydrides degrade in a few days, while some aromatic polyanhydrides degrade over a few years. Degradation rates of copolymers of aliphatic and aromatic polyanhydrides vary between these extremes, and this feature of polyanhydrides gives an opportunity for making a drug delivery system which can provide the release of drugs for a desired time length of treatment.

3.2 Types of Polyanhydride

Bucher and Slade synthesized aromatic polyanhydrides [8]; however, these were first explored by Conix after almost 50 years to form fibers for textile applications [9]. Hill and Carothers [10, 11] had worked in the 1930s on aliphatic PAs of adipic and sebacic acid (SA); because of hydrolytic instability, no further development was carried on these polymers until they were explored by Langer in the 1980s for drug delivery [1, 12]. Heterocyclic PAs were also developed in the meantime by Yoda *et al.* with good film and fiber-forming properties [13]. Once the degradable and biocompatible nature of PAs was uncovered, various types of copolymers were prepared thereon and utilized in drug delivery. One of the simplest classifications for PAs can be homo- and hetero-PAs; however, in the development of erodible materials, the use of copolymers (heteropolymers) is important for their different erosion rates, enabling the achievement of different target times for release, and this is possible by using different monomers and their ratio. In most PA copolymers, the aliphatic chain used is composed of polysebacic acid (PSA) and thus these are classified on the basis of the other part of the copolymer, which in turn governs the polymer properties. All the polyanhydrides with their representative chemical structure are shown in Table 3.1.

3.2.1 Aromatic Polyanhydrides

Aromatic homopolyanhydrides are insoluble in common organic solvents and melt at temperatures above 200 °C [6, 15]. These properties limit the use of purely aromatic polyanhydrides, since they cannot be fabricated into films or microspheres using solvent or melt techniques. Fully aromatic polymers that are soluble in chlorinated hydrocarbons and melted at temperatures below 100 °C were obtained by copolymerization of aromatic diacids such as isophthalic acid (IPA), terephthalic acid (TA), 1,3-bis(carboxyphenoxy)propane (CPP), or 1,3-bis (carboxyphenoxy)hexane (CPH).

Table 3.1 Types of the polyanhydrides and their representative chemical structures.

Polymer	Structure	Reference
Alipahtic polyanhydrides		
PSA	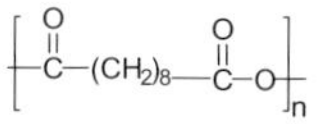	[7]
Aromatic polyanhydrides		
P(CPV)	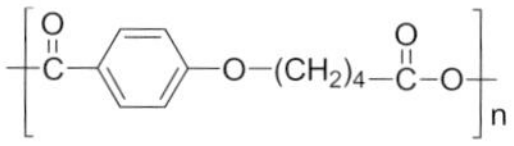	[14]
P(TA-IPA)	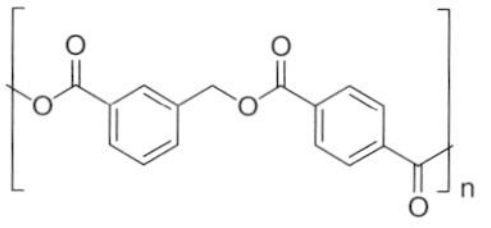	[15]
P(CPP-IPA)	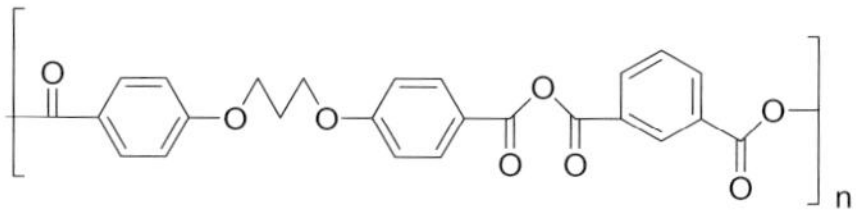	[15]
Aromatic polyanhydrides		
P(CPP-SA)	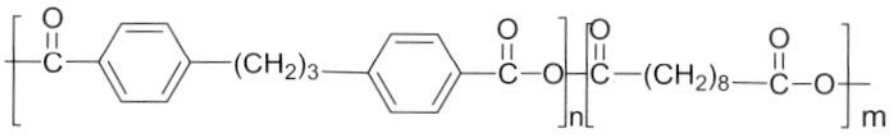	[14]
P(CPM-SA)	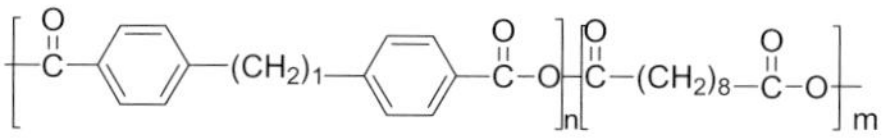	[14]
P(CPH-SA)	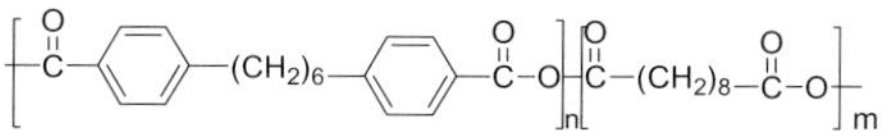	[14]
Fatty acid polyanhydrides		
P(RA-SA)	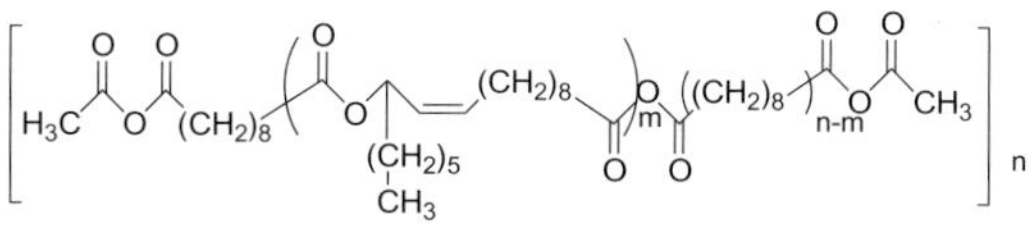	[16]
P(EAD-SA)	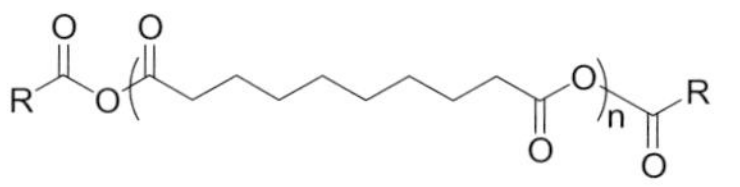 n=25-50	[17]
Stearic acid terminated with PSA	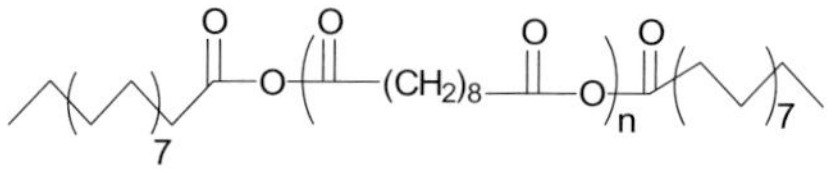	[18]

(*Continued*)

Table 3.1 (*Continued*)

Polymer	Structure	Reference
Pegylated polyanhydride		
P(CPOEG-5)	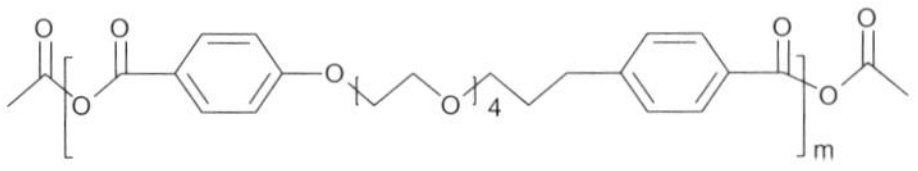	[19]
Recinoleic acid-based copolyesters	(a) Ring-opening polymerization using RA lactones (P[RA:LA]) 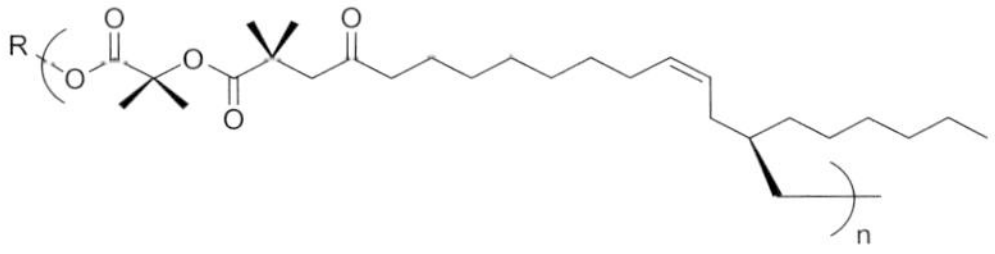 (b) Transesterification (P[RA:LA]) 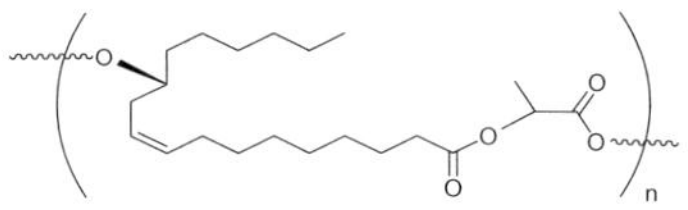 (c) Melt condensation (P[RA:LA]) 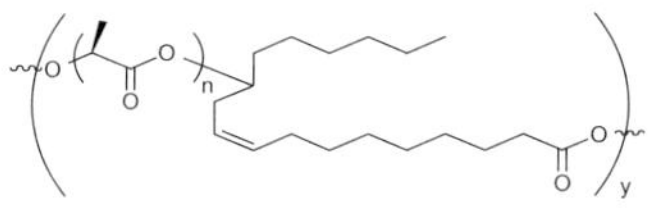	[20, 21]
Succinic acid functional polymers	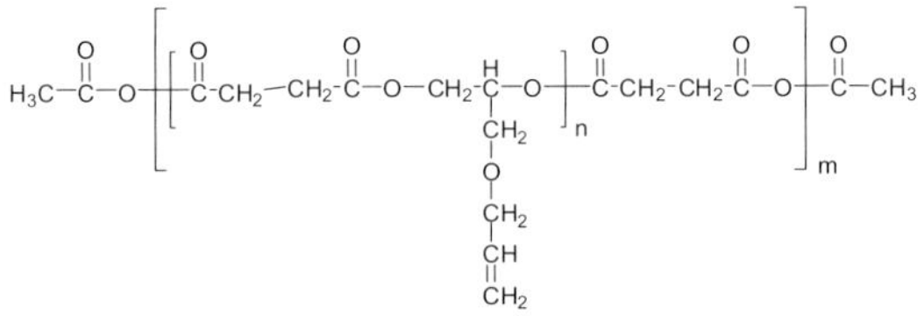	[22, 23]
Amino acid-based crosslinked polyanhydrides (with anhydride linkages only)	Crosslinked polymer of *N*-trimellitylimido-amino acid and SA using 1,3,5-benzenetricarboxylic acid as crosslinker	[24]
Salicylate based	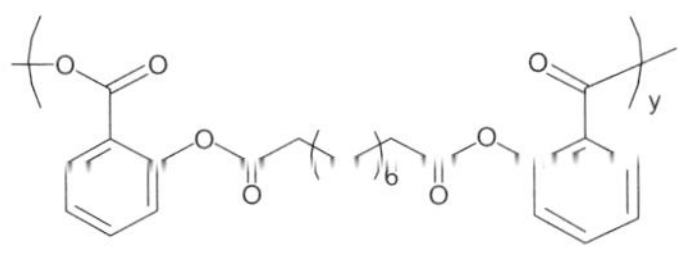	[25]

3.2.2
Aliphatic–Aromatic Polyanhydrides

Polyanhydrides of diacid monomers containing aliphatic and aromatic moieties, such as poly([*p*-carboxyphenoxy]alkanoic anhydride), were synthesized by either melt or solution polymerization, with molecular weights reaching up to 44,600 Da [6]. The polymers of carboxyphenoxy alkanoic acid having methylene groups ($n = 3$, 5, and 7) were soluble in chlorinated hydrocarbon solvents and melted at temperatures below 100 °C. These polymers displayed a zero-order hydrolytic degradation profile for 2–10 weeks. The length of the alkanoic chain positively correlates to the degradation time [6].

3.2.3
Poly(Ester-Anhydrides) and Poly(Ether-Anhydrides)

4,4-Alkane and oxa-alkanedioxydibenzoic acids were used for the synthesis of polyanhydrides. These polyanhydrides melted at a temperature range of 98–176 °C and had M_w up to 12,900 kDa. Di- and triblock copolymers of poly(caprolactone) (PCL), polylactic acid (PLA), and polyhydroxybutyrate (PHB) have been prepared from carboxylic acid-terminated low M_w prepolymers copolymerized with SA prepolymers by melt condensation. Similarly, di-, tri-, and brush copolymers of polyethylene glycol (PEG) with poly(sebacic anhydride) (PSA) have been prepared by melt copolymerization of carboxylic acid-terminated PEG [26, 27].

3.2.4
Fatty Acid-Based Polyanhydrides

These polyanhydrides were synthesized from dimer and trimer unsaturated fatty acids [17, 28–30]. The dimers of oleic acid and eurecic acid are liquid oils containing two carboxylic acids, which are available for polymerization; correspondingly, the homopolymers are viscous liquids. Copolymerization with increasing amounts of SA forms solid polymers with increased melting points. The polymers are soluble in tetrahydrofuran, 2-butanone, and acetone. Polyanhydrides synthesized from nonlinear hydrophobic fatty acid esters based on ricinoleic acid (RA), maleic acid, and SA possessed desired physicochemical properties such as low melting point and hydrophobicity and good flexibility [31] in addition to biocompatibility [32] and biodegradability [33].

3.2.5
RA-Based Polyanhydrides

Incorporation of the fatty acid in the biodegradable polymer backbone is advantageous but it is restricted by monofunctionality of most naturally occurring fatty acids. The unsaturated monofunctional fatty acids first need to be converted to dimers for further polymerization. The dimer contains a branched C–C linkage which cannot be metabolized by the body and the dimer may remain in the body for 6 months [34]. RA (*cis*-12-hydroxyoctadeca-9-eonoic acid) was found to be the most appropriate alternative for the synthesis of the fatty acid-based polyanhydrides. It is one of the few commercially available fatty acids which have the additional 12-hydroxy group. The advantage of RA is that it is a bifunctional fatty acid containing a hydroxyl group along the acid group and, therefore, can be incorporated into the polyanhydride backbone by the formation of an ester bond.

RA-based polymers are the newest addition to the polyanhydride series which were first investigated in the late 1990s [16]. However, polymers produced were for solid implant that need surgical intervention for application to the body system. Recent work is more focused on converting this solid form to liquid injectable form which can form solid or semisolid implants after administration by injection [16]. For this, the first series of efforts were made with SA as the other monomer and this also included two subtypes; one is insertion of RA in preformed SA chains [35] and second is usual melt condensation carried out at lower temperature in one-pot synthesis, where dicarboxylic acid derivative of RA and SA are condensed together to from random copolymer rather than block copolymer [36]. Both of these efforts lead to the formation of polymers in the liquid injectable state. Although the common physicochemical properties such as low melting point, hydrophobicity, flexibility, biocompatibility, and biodegradability desired for a drug carrier possessed by all RA-based polyanhydrides, the liquid state was achieved only with the polymer having more than 70% of RA content.

Low molecular weight polymers synthesized by one pot-low temperature condensation method afforded the release of anticancer drug, methotrexate for around 10 days [36]. Although a change in the ratio of RA maleate (RAM) to SA was having a role, the faster release from the higher RAM containing polymer was elucidated on the basis of polymer crystallinity, which hinders the release by inhibiting water penetration in the device which decreases with increase in RAM content [16]. Similar kinds of results were found in polymers obtained by insertion of RA in SA chains [35, 37], where these polymers were loaded with cisplatin (5% w/w) and paclitaxel (5–20% w/w) and drug release was faster with the pasty polymers. *In vivo* evaluation of bupivacaine-loaded P(SA:RA)(2:8) injectable polymer was made in terms of efficacy and toxicity for producing motor and sensory block when injected near the sciatic nerve [38]. Single injection of 10% bupivacaine in the polymer caused motor and sensory block that lasted 30 h without causing any adverse effects.

Ricinoleic lactones were utilized for the synthesis of copolyester by ring-opening polymerization (ROP) [20]. RA lactones were synthesized by using dicyclohexylcarbodimide and (dimethylamino)pyridine as catalysts. Various macrolactones were obtained, mono- to haxalactone depending on the number of RA moieties which participate in the lactone ring formation. Polymerization of the RA lactones with catalysts commonly used for ring-opening polymerization of lactones, under specific reaction conditions, resulted in oligomers. Polymerization of chromatography-purified dilactone with $Sn(Oct)_2$ resulted in the formation of longer oligomers (weight average $M_w = 5700$). However, copolymerization with lactide resulted in copolymers of low molecular weight. Polymers with molecular weights in the range 5000–16,000 were obtained with melting temperatures of 100–130 °C for copolymers containing 10–50% (w/w) RA residues. The polymers were off-white in color that became yellow with an increase of the RA content. The molecular weights of the polymers decreased with an increase in the content of the RA lactone. It was hypothesized that more reactive lactide activated first by catalyst polymerizes and only in the end some RA lactones react. The reaction was terminated because of the RA lactones' low reactivity. This low reactivity can be attributed to the low ring strain and to the steric hindrance of the ester bond by the fatty acid side chain. *In vitro* degradation of RA–LA copolymers showed that copolymerization with RA had some effect on the degradation rate and the polymer physical properties, which is related to the low incorporation of RA in the polymer. Addition of RA to PLA is expected to improve the hydrophobicity of the polymer and thus drug release profile.

In continuation of the above study, synthesis methods other than ROP like transesterification and melt condensation were also utilized [21]. The liquid state of the polymer, which makes it a potential candidate for directly injectable drug delivery carrier, was achieved when RA content increased more than 15% and 50% in case of melt condensation and transesterfication, respectively.

Polymers synthesized by all three methods were compared for the release of hydrophilic and hydrophobic drugs viz. 5-FU and triamcinolone, respectively. 5-FU release was faster in all cases with the total release lasting for 17 days from polymers prepared by transesterification and melt condensation. Slower 5-FU release was obtained from polymer prepared by ROP (40% in 17 days). The same pattern was observed for triamcinolone, where release was obtained only 5% in 17 days from ROP polymer in contrast to the 30% from polymer synthesized by transesterification. The difference was attributed to the diblock nature of ROP polymer, its high crystallinity, and melting point, all of which inhibit water penetration and thus degradation, which finally shows up in release profiles [21].

3.2.6
Amino Acid-Based Polyanhydrides

Amino acid-based PAs were first reported in 1990s by Domb [39]. However, recent progress in this class has been made in terms of producing crosslinked PAs which are suitable for *in vivo* use [24, 40]. Earlier, alanine-containing crosslinked PAs in which linkages were produced by irradiation of methacrylated end groups which when hydrolyzed gave rise to nonbiodegradable products having limited biocompatibility. To overcome these limitations, crosslinked amino acid PAs were produced having exclusively anhydride bonds which are hydrolabile in nature. Crosslinked amino acid-containing polyanhydrides based on *N*-trimellitylimido-β-alanine (TMA-ala) or *N*-trimellitylimido-glycine (TMA-gly) and SA were synthesized by copolycondensation using 1,3,5-benzenetricarboxylic acid prepolymer as a crosslinking agent. Crosslinking was confirmed by single melting peak of the polymer in differential scanning calorimeter (DSC) studies [40]. Monomeric SA prepolymer was prepared to prevent phase separation and produce homogeneous polymeric matrix. *p*-Nitroaniline was incorporated in the polymer matrix by compression-molding in the form of a disk. They were then placed in buffer (0.1 M, 7.4 pH) and release of *p*-nitroaniline as well as TMA-gly was measured and found to be similar to its linear counterpart of the polymer (TMA-gly:SA 30:70) [41, 42], indicating that crosslinking has little effect on the degradation behavior of this particular polymer, possibly due to its high hydrophilicity and low degree of crosslinking. Thus, this system gives opportunity to further evaluate the degree of crosslinking and control over the same to produce material useful for varied applications.

In another study by Zhang *et al.* [43], the effect of type of amide bonds present in the PA backbone and its blending with polyesters like PLA on degradation has also been studied. Polymers of *N,N′*-bis(L-alanine)-sebacoylamide (BSAM) and P(1,6-bis[*p*-carboxyphenoxy] hexane [CPH]-BSAM) were synthesized and blended with PLA. Hydrolytic degradation of polyanhydrides and their blends with PLA were evaluated in 0.1 M phosphate buffer pH 7.4 at 37 °C. The results indicate that the existence of amide bonds in the main chain of polymers slow down the degradation rate, and this tendency increases with the increasing amount of these. The copolymers and their blends with PLA possess excellent physical and mechanical properties, thus making them more widely used in drug delivery and nerve regeneration.

3.2.7
Photopolymerizable Polyanhydrides

Photocrosslinking is preferred over chemical crosslinking which utilizes chemicals that can cause adverse effects. Fiber-optic cables are used to provide photons immediately after introduction of the polymer system to the desired site via injection. The other main advantages of photoinitiated polymerizations over other crosslinking techniques are spatial and temporal control of the polymerization which allows the precise control of polymer formation by directing and shuttering the light source. The reactions are rapid enough to overcome oxygen inhibition and moisture effects and can be controlled to occur over a time frame of seconds to minutes. Ease of fashioning and flexibility during implantation in terms of physical and mechanical properties of materials without major modifications to the backbone chemistry, which can alter biocompatibility, is added advantage. However, the principal limitation to more extensive use of photopolymerizations in biotechnology and medicine is the lack of biocompatible monomers and/or oligomers that photopolymerize to form degradable polymer networks [44, 45]

Anhydride monomers with reactive methacrylate functionalities have been developed and used for the preparation of PAs which shows *in-situ* crosslinking on exposure to light. These systems were demonstrated to be biocompatible and were used for bone augmentation applications [46].

Shastri *et al.* have prepared a new family of photochemically cured PAs which can produce semi-interpenetrating degradable networks and evaluate them for biocompatibility in subcutaneous tissue in rats. These systems appear to undergo degradation primarily by surface erosion. They observed that the inflammatory response to these implants was minimal at both short (3 and 6 weeks) and long (28 weeks) time points. Further, the fibrotic response was largely absent throughout the duration of this study. For reference, linear PA controls were tested and showed a foreign body response culminating in the formation of relatively nonvascular fibrous capsule several cell layers thick, which became thicker over time, a response similar to what is typically observed in Food and Drug Administration (FDA)-approved implantable polymeric device systems [47].

In another study, Poshusta *et al.* examined cell–polymer interactions in subcutaneous and bony tissue after implantation of *in situ* forming and surface-eroding photopolymerized disks of several polyanhydride compositions in rats. Varied histological responses were observed depending on the degrading polymer composition. It was shown that 50/50 poly(MSA)/poly(CPP:CPH) showed a cellular response that was similar to PLA controls. A model defect created in the proximal tibia was used to assess the effects of the photopolymerization reaction on local bony tissue. At 7 days, new bone spicules in the fibrous callous were found to be present which indicated healing of the polymer-treated defect with no adverse effects from the photopolymerization reaction [45].

Weiner *et al.* have recently evaluated the potential of photocrosslinked PAs networks as an injectable delivery system for sustained release of bioactive molecules. Crosslinked networks composed of sebacic acid dimethacrylate (MSA), 1,6-bis-carboxyphenoxyhexane dimethacrylate, and PEG diacrylate, supplemented with calcium carbonate were examined for *in vitro* release of two model proteins (horseradish peroxidase and bovine serum albumin labeled with fluorescein isothiocyanate). Release of protein ranging from 1 week to 4 months was achieved. In general, a more hydrophobic network resulted in slower rates of protein release. These results suggest that this system may be useful as an injectable delivery system for long-term delivery of macromolecules [48].

3.2.8 Salicylate-Based Polyanhydrides

Erdmann *et al.* [49] have reported salicylic acid-based polymers in year 2000 and these have then been investigated extensively in the last few years. These salicylate-based polyanhydride-esters were collectively referred to as PolyAspirin because they hydrolytically degrade into salicylic acid, a nonsteroidal anti-inflammatory drug. Salicylic acid-based polymers are unique example of polymer therapeutics wherein the drug (salicylate derivative) is an integrated part of the polymer backbone. Aminosalicylic acid is a useful bioactive agent for inflammatory bowl disease and the drug needs to be specifically delivered at the site of action, that is, colon. Synthesis of this category of the polymers can be carried out by usual preparation of prepolymers and then melt condensation to produce high molecular weight polymer [25]. Salicylic acid-based poly(ester-anhydride)s have also been tested for healing of long bone defects in rats with 5-mm mid-diaphyseal defects in femurs.

Microspheres of the polymer were packed into the defect and compared with collagen sponge for reduction in bone loss. Though initially there was no significant reduction in bone loss, after 8 weeks significant reduction in the bone, weight loss was observed in the polymer group [50]. In another study, polymers prepared from salicylic acid derivative were evaluated for cytotoxicity using L929 fibroblast cells in serum-containing medium on parameters like cell viability, proliferation, and morphology and these were found normal for most of the polymers evaluated [51].

All salicylate-based polymers degrade to produce salicylic acid and all of these were found to follow primarily surface erosion patterns [52]. Furthermore, effect of media on degradation rate was studied and found to increase marginally better (14%) in media containing actively growing bacterial culture than sterile media. A significant reduction in formation of *Pseudomonas aeruginosa* biofilm in a long-term (3 day) study with the salicylic acid containing polymers was demonstrated and a pathway was postulated using *P. aeruginosa* pMHLAS, containing a fluorescent reporter gene which involved inhibition of the las quorum sensing system [53]. In another study, a clear difference was seen between bacterial strains that form biofilms at the air–liquid interface (top-forming) and those that form at the surface–liquid interface (bottom-forming). The results lead to the conclusion that the polymers may not interfere with attachment; rather, the polymers likely affect another mechanisms essential for biofilm formation in Salmonella [54].

3.2.9 Succinic Acid-Based Polyanhydrides

Succinic acid is one of the naturally occurring substances of living tissues, and polymers prepared using this acid can be inherently biodegradable and biocompatible. Inclusion of succinic acid in the polymer chain has been made for various functions. Initially, it has been used to convert monocarboxylic monomers to dicarboxylic, as in case of RA, to help polymerization reaction [55–57]. In earlier work, succinic acid was directly used as one of the monomer units, for example, Ben-Sabat *et al.* having synthesized copolyanhydrides from trimers of fumaric acid, succinic acid, and propylene glycol [22]. These polymers were found to degrade and release the entrapped drug substance in a week's time, and *in vivo* testing in rat proved the polymer safe for further investigations as drug delivery carrier. Succinic acid derivatives were utilized more widely to synthesize unsaturated and functional polymers. Copolymers of 2-hexadecylsuccinic acid and SA were prepared using usual melt-condensation method and demonstrated to be potential drug carriers for localized drug delivery [58]. In another study, hydroxyl-group functional polylactones were prepared and converted to acid-terminated polyesters in a reaction with a series of alkenylsuccinic anhydrides containing 8, 12, or 18 carbons in their alkenyl chains [55]. These polyester units were then condensed in high molecular weight polymer. Polymer hydrolysis was found to decrease by the presence of alkenyl chain in case of low molecular weight precursors but converse was the case with polymers with high molecular weight prepolymers. There was no pronounced effect of differences in length of the alkenyl group in degradation rate. Recently, succinic acid-based functional polymers have been synthesized with allyl pendent group which can be utilized for further copolymerization or attachment of other moieties to perform specialized function [23].

Synthesis of these functional polymers was carried out in three steps; initially carboxyl-terminated functional oligoesters with molecular weight 300–1000 Da were obtained by melt condensation of allyl glycidyl ether with an excess of succinic acid then the macromer with carboxyl end obtained were converted to mixed anhydride groups by refluxing in acetic anhydride, and finally, melt polycondesation of ester-anhydride prepolymers was carried out to form the polymer. Influence of molecular weight of initial oligoesters as well as of parameters of the process on selected properties of poly(ester-anhydride)s was examined. The hydrolytic degradation was monitored by the determination of mass loss and the ester-to-anhydride groups ratio. These poly(ester-anhydride)s display a two-phase degradation profile with a rapid initial degradation of anhydride bonds followed by relatively slower degradation of oligoester.

3.2.10
Blends

Blending, or mixing, appropriate polymers can alter the physical and mechanical properties of polyanhydrides. Blends of poly(trimethylene carbonate) with poly(adipic anhydride) were found biocompatible in both *in vitro* and *in vivo* experiments [59–61]. Blends were prepared by dissolving each polymer in methylene chloride followed by separately mixing in varying proportions using solvent-mixing technique [62]. The results indicate that the blend may be a promising candidate for controlled drug delivery [59–61] and varying the proportion of poly(trimethylene carbonate) and poly(adipic anhydride) can control the erosion rate of the polymer blend. Low molecular weight polyesters such as PLA, PHB, and PCL are miscible with polyanhydrides, whereas high molecular weight polyesters ($M_w > 10{,}000$ Da) are not compatible with polyanhydrides. Uniform blends of PCL with 10–90% by weight of poly(dodecanedioic anhydride) were prepared by melt mixing at 120 °C and exhibited good mechanical strength. Hydrolysis studies indicated that the anhydride component degraded and was released from the blend composition without affecting the PCL degradation [62]. Combinatorial methods have also been developed to study the phase behavior of biodegradable polyanhydrides for drug delivery applications [63].

3.3
Synthesis

Polyanhydrides have been synthesized by various techniques, viz. melt condensation, ROP, interfacial condensation, dehydrochlorination, and dehydrative coupling agents (Scheme 3.1) [64, 65]. Linear and crosslinked polyanhydrides have also been made using photoinitiated thiol-ene chemistry [66]. Solution

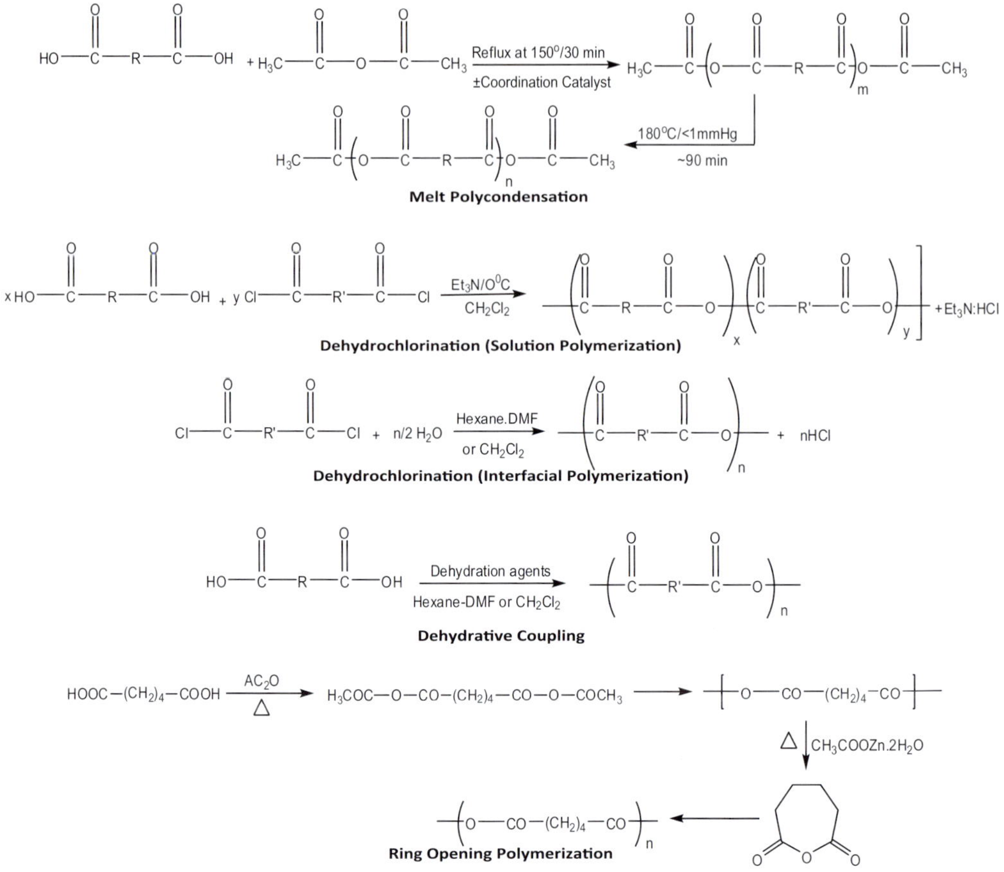

Scheme 3.1 General synthesis schemes for polyanhydrides.

polymerization in general yielded low molecular weight polymers. A variety of catalysts have been used in the synthesis of a range of polyanhydrides by melt condensation. Particularly, coordination catalysts facilitate anhydride interchange in the polymerization and enhance the nucleophilicity of the carbonyl carbon. Significantly higher molecular weights in shorter reaction time were achieved by utilizing cadmium acetate, earth metal oxides, and $ZnEt_2 \cdot H_2O$. Except for calcium carbonate, which is a safe natural material, the use of these catalysts for the production of medical grade polymers is limited due to their potential toxicity [7].

Since melt condensation occurs at high temperatures, it is not suitable for heat-sensitive monomers, which require milder reaction conditions. A variety of solution polymerizations at ambient temperature have been reported [65, 67]. Polyanhydride formation can be effected at room temperature by dehydrochlorination between a diacid chloride and a dicarboxylic acid. In an attempt to prepare copolyanhydrides of regular structures, polycondensation was conducted in organic solvent pairs such as pyridine–benzene and pyridine–ether. The reaction between a diacid chloride and a diacid ethyl ester in the presence of zinc chloride was also studied. The formation of polyanhydrides in these reactions was confirmed only by IR spectroscopy. The polycondensation was achieved between acyl chloride and carboxylic acid in a single solvent in the presence of an acid acceptor such as triethylamine. Polymerization took place on contact of the monomers and was essentially complete within an hour as monitored by gel permeation chromatography. The degree of polymerization was in the range of 20–30 as determined by vapor phase osmometry. Comparable results in terms of molecular weights and yields were obtained for polymerization conducted in solvents such as dichloromethane, chloroform, benzene, and ethyl ether. The degree of polymerization was influenced by the mode of addition. Adding the diacid solution drop wise to the diacid chloride solution consistently produced higher molecular weights and yield as compared to the reverse order of addition. This could be understood as the terephthaloyl chloride complex which with triethylamine forms an ionic salt. Since a slight excess of acid acceptors was often used to solubilize the acid monomers, some of the terephthaloyl chloride would be lost due to complexation. The unbalanced stoichiometry therefore accounted for the inefficient polymerization. Adding the acyl chloride in a single portion yields satisfactory results, which suggests that the rate of dehydrochlorination is comparable to the rate of acid chloride–amine complexation. The inconvenience with this homogeneous Schotten–Baumann condensation in solution is the need to obtain the highly purified diacid chloride monomer. Stringent stoichiometric conditions must also be met.

ROP was used for the synthesis of poly(adipic anhydride) from cyclic adipic anhydride (oxepane-2,7-dione) using cationic (e.g., $AlCl_3$ and $BF_3 \cdot [C_2H_5]O$), anionic (e.g., CH_3COO^- K^+ and NaH), and coordination-type inhibitors such as stannous-2-ethylhexanoate and dibutyltinoxide [68]. ROP takes place in two steps in which the first step is the preparation of the cyclic monomer and the second step is the polymerization of the cyclic monomers as shown in Scheme 3.1. A one-step polymerization using diacyl chloride, phosgene, or diphosgene as coupling agents and various acid acceptors was reported [67]. Although phosgene and diphosgene are equally efficient, diphosgene as a liquid is preferred because of its ease of handling and lower vapor pressure. The polymers have similar molecular weights with regard to the type of amine bases used. The heterogeneous acid acceptor, poly(4-vinylpyridine) (PVP), produced satisfactory results, whereas the nonamine heterogeneous base K_2CO_3 yielded a lower molecular weight. When using soluble amines as acid acceptors, they form a soluble intermediate complex of acid–amine which improves the interaction with the coupling agent under homogeneous conditions. Although the PVP is insoluble in the reaction medium, it swells and forms a similar acid–PVP complex. K_2CO_3, however, forms a heterogeneous mixture with the acid and thus presumably reacts more slowly with the coupling agents to form the polymer.

3.4 Properties

Thermal. Because crystallinity is an important factor in controlling polymer erosion, the effect of polymer composition on crystallinity was studied [69, 70]. Almost all polyanhydrides show some degree of crystallinity as manifested by their crystalline melting points. Polymers based on SA, CPP, CPH, and FA were particularly investigated. Some of the important physicochemical properties of P(CPP:SA) and P(FAD:SA) are given in Table 3.2. Homopolyanhydrides of aromatic and aliphatic diacids, for example, poly(CPP) and poly(FA), were crystalline (>50% crystallinity), whereas the copolymers possess a less degree of crystallinity, which increases by enhancing the mole ratio of either aliphatic or aromatic diacid monomers [15]. The heat of fusion (ΔH) values for poly(CPP-SA) demonstrated a sharp decrease from 36.6 to 2.0 cal/g as CPP is gradually added up to 40%, while an increase in ΔH value was observed up to 26.5 cal/g on further addition of CPP [71]. The trend of decreasing crystallinity, as one monomer is added, was noted using X-ray diffraction or DSC methods. The decrease in crystallinity is a direct result of the random presence of other units in the polymer chain. A detailed analysis of the copolymers of SA with the aromatic and unsaturated monomers, CPP, CPH, FA, and trimellitic-amino acid derivative, showed that copolymers with high ratios of SA and CPP, TMA-gly, or CPH were crystalline while copolymers with equal ratios of SA and CPP or CPH were amorphous [69]. In contrasts, the poly(FA-SA) series displayed high crystallinity regardless of comonomer ratio [72]. Aliphatic polyanhydrides generally melt at lower temperatures than do aromatic polyanhydrides. The melting point of aromatic–aliphatic copolymers is proportional to the aromatic content in the copolymer. Introduction of fatty acids in copolymers also lowers the melting point of the bulk polymer. Thermal properties along with the molecular weight of representative fatty acid-based polyanhydrides are given in Table 3.3.

Inclusion of an aromatic amide linkage in the backbone is found to increase the transition temperatures. The formation of intermolecular hydrogen bonds is believed to cause this high crystallinity. Polyanhydrides-*co*-amide also have high thermal stability [39].

Mechanical. Polyanhydrides show poor mechanical properties in comparison to other polymers such as polyesters. Mechanical properties of various polyanhydrides and their copolyanhydrides were tested as transparent and flexible films made by melt compression and solvent casting. It was observed that increasing the CPP content in copolymer composition increases the tensile strength as well as elongation of various polyanhydrides tested [7]. Despite the low molecular weight ($M_n = 6400$) of poly(CPP-SA) (60:40), it has a higher tensile strength of 981 MPa (100 kgf/cm^2) than it has in the 20:80 composition ($M_n = 18,900$), 441 MPa (45 kgf/cm^2). Decreasing the M_n of films of the same CPP content (60%) from 12,100 to 6400 results in lower tensile strength. The elongation at break of these films ranges from 17% to 23%. Table 3.2 shows the mechanical

Table 3.2 Physicochemical properties of P(CPP:SA) and P(EAD:SA).

Analysis	Units	Conditions	Typical data/information								Reference
			Poly(EAD-SA)				Poly(CPP-SA)				
Thermal properties	Mol%	DSC – 10 °C/min	*0:100*	*8:92*	*22:78*	*100:0*	*0:100*	*22:78*	*46:54*	*100:0*	[15, 73]
	K	T_m	358.0	348.0	337.0	293.0	359.0	339.0	458.0	513.0	
	K	T_g	333.1	283.0	283.0	273.0	333.1	320.0	274.8	369.0	
	kJ/kg	ΔH	150.7	250.2	13.0	4.0	150.7	64.0	13.0	110.9	
Crystallinity	%	W_c	*0:100*	*8:92*	*22:78*	*100:0*	*0:10*	*22:7*	*46:5*	*100:*	
			66	54	35	<5	66.0	35.0	14.2	61.4	
Mechanical			*22:78*				*40:60*				[17, 74]
Tensile strength	MPa	Film by melt	42				14.9				
Tensile modulus	MPa	22:78, $M_w = 155$ kDa	4.5								
Elongation yield	%		85								
Elongation at break	%		14								
Solubility	% w/v	Equilibrium	*50:50*				*20:80*				
			Dichloromethane > 30				>30				
			Chloroform > 30				>30				
			Tetrahydrofuran > 10				2.1				
			2-Butanone 5.0				0.1				
			4-Methyl-2-pentanone 3.1				0.02				
			Acetone 3.0				0.00				
			Ethyl acetate 2.4				0.02				

(*Continued*)

Table 3.2 (*Continued*)

Analysis	Units	Conditions	Typical data/information		Reference
			Poly(EAD-SA)	Poly(CPP-SA)	
Spectral					[17, 75, 76]
^{1}H NMR	ppm	1% (w/v) in $CDCl_3$, 22 °C	2.33 (t, 4H), 1.32 (m, 8H), 1.23 (s, 60H), 0.8–0.9 (T, 8H)		
FTIR	cm^{-1}	Film on NaCl plate	1740; 1810	1740; 1770; 1810	
Raman	cm^{-1}			1723; 1765; 1804	
UV-wavelength	nm		253	265	
Molecular weight	10^4 g/mol	Gel permeation chromatography-polystyrene standard	$M_w = 3–30$; $M_n = 1–3$	$M_w = 3–20$, $M_n = 0.5–3$	[73, 77]
Viscosity	dL/g	25 °C, in CH_2Cl_2	$\eta_{sp} = 0.2–1.4$	$\eta_{sp} = 0.2–0.9$	
Mark–Houwink constants	mL/g	23 °C, in CH_2Cl_2	$k = 3.46$; $a = 0.634$	$k = 3.88$; $a = 0.658$	
Surface and bulk					[78, 79]
XPS		Film by spin casting	Quantitative elemental and functional group information, molecular specificity, and surface morphology of polymer	Quantitative elemental and functional group information, molecular specificity, and surface morphology of polymer	
TOF-SIMS		on aluminum sheet,			
AFM		0.1% (w/v) in $CHCl_3$			

Table 3.3 Properties of some representative fatty acid-based polyanhydrides.

Polymer	M_w	M_n	T_m (°C)	ΔH (J/g)c	Crystallinity	Reference
Fatty acid-terminated polymers						[18, 28, 80]
P(OCTA:SA) 30:70	7800	5600	70.2	72.5	–	
P(LAUA:SA) 30:70	5300	3900	71.3	78.5	–	
P(MYR:SA) 30:70	4800	3600	78.3	82.7	–	
P(OA:SA) 30:70	6300	4500	73.1	59.9	–	
P(StA:SA) 30:70	7400	5400	77.8	103.7	–	
P(OA:SA) 30:70	6350	3206	71.3	59.9	–	
P(LA:SA) 30:70	5807	3367	69.5	57.4	–	
P(LitA:SA) 30:70	6575	3804	68.35	95.2	–	
P(RA:SA) 30:70	7961	4952	71.29	60.1	–	
P(RAStA:SA) 30:70	7300	5000	79.0	64.6	–	
Dimer acid (DA)-based polymer						[81, 82]
P(DA-DDDA) 50:50	24,220	21,625	70.1	–	16	
P(DA-TA) 50:50	38,561	35,904	80.2	–	48	
P(DA-SA) 50:50	26,000	11,000	67.1	–	–	
C_{12}-,C_{13}-, C_{14}-, C_{15}-based polymer				–	–	[83]
P(DDDA-TA) 50:50	29,600	24,200	75.1	–	28	
P(BA-PA) 50:50	25,700	22,800	72.4	–	29	
RA-based polymer						[16, 36, 84]
P(RAM:SA) 50:50	31,200	12,800	59.3	–	–	
P(RAS:SA) 50:50	48,700	21,700	61.1	–	–	
P(HSAS:SA) 50:50	41,000	19,700	70.4	–	–	
P(RAM:SA) 50:50 (one pot low M_w)	3768	1983	41.06	–	–	
P(RA-PSA) 5:5 (RA insertion in SA chain)	19,000	8000	55.7	–	–	
Ricinoleic acid (RA)-based copolyesters						[85]
P(L-LA-RA) 50:50 (ROP)	9800	7300	105	–	–	
P(L-LA-RA) 50:50 (melt condensation)	4500	3500	Liquid at RT	–	–	
P(L-LA-RA) 50:50 (transesterification)	8200	5600	Liquid at RT	–	–	

properties of fatty acid-based polyanhydrides. Films of fatty acid polyanhydrides were transparent and flexible with a tensile strength of 4–19 MPa and elongation at break in the range of 77–115%. The terpolymer of (fatty acid trimer [FAT]-CPP-SA] in a 1:1:1 weight ratio formed the strongest film. The polymer had a tensile strength of 2.5–3.2 MPa and yield stress at break of around 20% in comparison to poly(EAD-SA) (1:1) and PSA, which had tensile strengths of 5.7 and 7.2 MPa and yield stress at break of 10% and 1.5%, respectively. Thus, introduction of nonlinear fatty acid structures in polyanhydrides provides hydrophobicity and flexibility to the polymers.

Stability. The stability of polyanhydrides in solid state and dry chloroform solution was studied [86]. Aromatic polymers such as poly(CPP) and poly(1,1-bis[*p*-carboxyphenoxy] methane) maintained their original molecular weight for at least 1 year in the solid state. In contrast, aliphatic polyanhydrides, such as PSA, showed decreased molecular weight over time. The decrease in molecular weight shows a first-order kinetics, with activation energies of 7.5 kcal/(mol K). The decrease in molecular weight was explained by an internal anhydride interchange mechanism, as revealed from elemental and spectral analyses. This mechanism was supported by the fact that the decrease in molecular weight was reversible and heating the depolymerized polymer at 180 °C for 20 min yielded the original high molecular weight polymers. However, under similar conditions, the hydrolyzed polymer did not increase in molecular weight [86]. In many cases, it was observed that the stability of polymers in the solid state or in organic solution did not correlate with its hydrolytic stability [86]. A similar decrease in molecular weight as function of time was also observed among the aliphatic–aromatic copolyanhydrides and imide-containing polyanhydrides [6, 87]. Gamma-irradiation technique is typically used to sterilize polyanhydrides [88]. Aliphatic and aromatic homo- and copolymers were irradiated at 2.5 Mrad dose and the change in properties was monitored before and after irradiation. Properties such as molecular weight, melting temperature, and heat of fusion remained the same, and ^{1}HNMRand FTIR spectra of the polymer were also similar before and after irradiation [34, 89]. Using the same concept, these studies were extended for saturated and unsaturated polyanhydrides [90]. RA-based copolymers with SA and poly(CPP:SA) were irradiated under dry ice and at room temperature, while poly(FA:SA) was irradiated only at room temperature. Saturated polyanhydrides are stable enough during irradiation; however, the presence of double bonds conjugated to an anhydride bond creates an unstable structure and leads to the formation of free radicals [90]. These free radical polyanhydrides degrade into less conjugated polyanhydrides. The outcome of this process is self-depolymerization via inter- and/or intramolecular anhydride interchange to form polymers with lowered molecular weight. In general, polymers with high melting points and crystallinity give the highest yield of room temperature observable radicals. These endogenous free radicals were used to study processes of water penetration and polymer degradation *in vivo* [88]. The detection of gamma-sterilization-induced free radicals *in vivo* using EPR could be of significance because changes in the mobility of the radicals can be used to study drug release kinetics in a noninvasive and continuous fashion, without introducing paramagnetic species [34].

3.5
In Vitro Degradation and Erosion of Polyanhydrides

Polyanhydrides are made of sparingly water-soluble diacid monomers connected to each other by anhydride bonds, which are hydrolytically very labile and split readily into two carboxylic acids in the presence of water molecules. Hydrolysis of the anhydride bond is base catalyzed, and thus, pH of the surrounding media can significantly affect the rate of degradation of the polymer. The diffusion of oligomers and monomers formed by polymer degradation depends on pH of the surrounding medium and solubilities of these compounds in the medium. Since polyanhydrides degrade into carboxylic acids, solubilities of these degradation products are more at higher pH and hence erosion is higher at higher pH [91]. At low pH, these degradation products are in their unionized form, difficult to solubilize in surrounding biological media at implantation site and thus polyanhydrides in general degrade more rapidly in basic media than in acidic media [92]. Degradation of the polymer designates the process of polymer chain cleavage [93], while erosion is the sum of all processes that lead to the loss of mass from a polyanhydride matrix [92]. Erosion of the polymer matrices depends on processes such as rate of degradation, swelling, porosity, and ease of diffusion of oligomers and monomers from the matrices. Such erosions maintain constant surface area and hence lead to zero-order drug release [93]. Although polyanhydride degrades by surface erosion, there are many factors that influence the mechanism and rate of degradation, for example, type of monomers and their composition is one of the most important attributes. Aliphatic homopolymer like PSA are usually highly crystalline (about 66%) with unfavorable mechanical properties [94]. The *in situ* AFM images have provided the evidence that amorphous polymer areas erode faster than crystalline ones [77]. All aliphatic polyanhydrides are rigid, crystalline materials, and their melting point increases with their monomer chain length. They usually erode fast and therefore are not much used alone for pharmaceutical applications except some aliphatic polyanhydride such as P(FA:SA) having bioadhesive properties. Aromatic polyanhydrides are high melting polymers and degrade slowly. P(CPP) has a melting point of approximately 240 °C and its degradation rate is extremely slow [14, 15]. Combined properties of aliphatic and aromatic polyanhydride have been used to get the copolymer with improved mechanical characteristics and adjustable erosion times. The most successful polyanhydride is a copolymer of P(CPP:SA) and has been reported to erode at a constant rate [2, 95]. Erosion velocity of P(CPP:SA) decreases with increasing CPP content. Erosion zones in P(CPP:SA) are highly porous and separated from noneroded polymer by erosion fronts which move at constant velocity from the surface of a matrix into its center [92, 96]. P(FAD:SA) has also showed erosion zone but due to low solubility of FAD, the erosion zone mainly consisted of a semisolid mixture of FAD and FAD salts, instead of porous erosion zone. The semisolid layer forms a permeation barrier and SA acid was found to precipitate inside the erosion zone; this ultimately leads to slow release of SA as well as drug. Later polyanhydrides based on RA were reported to undergo sharp decreases in molecular weight during first 24 h of erosion *in vitro* and lost 40% of their anhydride bonds in 48 h [16]. Polyanhydride chains terminated with linear fatty acid like lauric, oleic, or stearic acid also show exponential loss of molecular weight and erosion behavior similar to RA-based polymer [18]. The increase in amount of fatty acid and the chain length induced the bulk erosion properties of polyanhydrides [92]. The photo-crosslinked polyanhydride obtained from MSA, MCPP, and 1,6-bis-carboxyphenoxyhexane dimethacrylate showed linear erosion profiles, when eroded *in vitro* [33, 97, 98]. Increase in the hydrophilicity of polyanhydride by increasing PEG content in the polymer enhances the degradation rate even though it maintains the surface-eroding property of polyanhydride [27]. Another important factor which affects the polyanhydride degradation and erosion is geometry of the matrix. It is very interesting to understand the macroscopic and microscopic degradation properties of the polyanhydrides at the molecular level. It is reported that erosion of matrices is strongly related to their geometry and rate of degradation for bigger matrices was lower than smaller ones due to smaller surface area [27, 99–101]. For example, during *in vitro* erosion of microspheres made of p(FAD-SA) 8:92, p(FAD-SA) 25:75, and p(FAD-SA) 44:56 with average diameters below 100 μm, SA was released completely in 100 h, while the release time was in weeks from matrix form of the polymer [102]. Some theoretical models have been proposed which allowed description and prediction of the erosion behavior of polyanhydride matrices [103]. Empirical models are based on the assumption of linear moving erosion front [104–106]. Monte Carlo based models offered the advantages of degradation modeling of the polymer as a random event that obeyed first-order kinetics rather than describing the degradation of individual bonds [100, 107–109].

3.6 *In Vivo* Degradation and Elimination of Polyanhydrides

Polyanhydrides were initially developed in matrix form as implantable drug carrier systems. Thus, it is critical to understand the processes involved in degradation and erosion in an *in vivo* environment and the differences between *in vitro* and *in vivo* degradation of polyanhydrides. Surface erosion of polyanhydrides depends on the penetration of water into the matrix system to hydrolyze the anhydride bonds. After hydrolysis, matrices degrade into degradation products of polyanhydrides and solubilize in the biological environment of the implantation site and are eliminated. Polyanhydrides are composed of sparingly water-soluble diacid monomers and thus elimination *via* solubilization in biological environment is a slow process [110]. Aliphatic monomers such as SA will most likely participate in the β-oxidation pathway yielding acetyl-coA which could be used in a typical biosynthetic pathway, while aromatic monomers are eliminated without further metabolic transformation [111]. Dang *et al.* studied surface erosion of Gliadel wafers during *in vivo* degradation in rat brain as well as during *in vitro* degradation in phosphate-buffered saline [2]. Morphological changes of the wafer during erosion were studied and SEM was used to present a visual proof of the erosion process. The wafer cross section before and after implantation in the brains of rats for various time periods has been studied. Before implantation, the surface of a BCNU-loaded polyanhydride wafer appeared very uniform with spray-dried microspheres densely packed together on the outer surface. Two hours after the wafer implantation, the porous structure extended approximately 20–30 μm from the surface into the interior of the wafer with outer thin layer of the wafer being eroded in the beginning and rest remained intact. Cross section of the degrading wafer followed dynamic process of water penetration from the surface to interior. One day following implantation, the wafer surface became highly porous and porosity decreased toward the region closer to the interior of the wafer. Higher magnification of the erosion zone revealed that the eroded microspheres had a dense structure at the external surface, while the materials from the inner core had already eroded and disappeared. As the advancing waterfront erodes deeper layers of the wafer, the porosity of the wafer increases resulting in increased numbers of channels and pores for water to access the interior of the wafer. Five days after wafer implantation, the entire cross section of the wafer displayed a uniformly high porosity without any individual microspheres being present. It indicates that water had penetrated through the whole wafer and degraded the interior as well as the exterior of the wafer. These results indicated that SEM analysis and weight loss studies were in a good correlation of *in vitro–in vivo* degradation behavior. Domb *et al.* studied the metabolic disposition and elimination process of (P[CPP-SA] 20:80) by implantation in adult Sprague-Dawley rat brain using radiolabeled polymers [112]. The results clearly showed that P(CPP-SA) 20:80 copolymer is extensively hydrolyzed 7 days postimplantation and revealed that the anhydride bonds in the copolymer are gradually degraded to give water-soluble SA monomer which are extensively metabolized in the body and excreted mostly as carbon dioxide. The elimination of the CPP component was slow due to its minimal solubility. The main route of elimination of insoluble CPP is by macrophages and inflammatory cells after its disintegration into small fragments.

3.7
Toxicological Aspects of Polyanhydrides

The toxicological aspect of polyanhydrides deals with the host response in terms of cytotoxicity, allergic responses, irritation, inflammation, and systemic and chronic toxicity. Cytotoxicity tests are the first in a sequential program of tests for assessing the biocompatibility of a polymer for which, tissue culture methods are used [111]. In a study, bovine aortic endothelial cells and bovine smooth muscle cells were used to evaluate the *in vitro* biocompatibility of three polyanhydrides

P(CPP-SA) 45:55, P(TA-SA) 50:50, and P(TA). These cultured mammalian cells are sensitive to the changes in growth medium and substrate [113]. The study showed the absence of acute toxicity of these polymers or their degradation products to sensitive mammalian cells. Chemical carcinogenesis usually proceeds by a mutagenic route; therefore, mutagenicity testing has been used as a rapid screening test for neoplastic transformation. The *in vitro* results for mutagenicity and the corresponding cytotoxicity of the degradation products of polyanhydrides particularly P(CPP-SA) showed that they are noncytotoxic, nonmutagenic, and have a very low teratogenic potential [113].

Polyanhydrides for intramuscular or dermal applications are tested for local tissue irritation and inflammation by muscle and skin tests. Leong *et al.* studied the local tissue response of polyanhydrides (P[CPP] and PTA-SA 50:50) by implantation of polymer samples into the cornea of rabbits [113]. No observable inflammatory characteristics were reported for the entire 6 weeks implantation period of polymers in rabbit corneas. The clarity of the corneas was maintained throughout, and proliferation of new blood vessels was absent. Histological examination confirmed the absence of inflammatory cells throughout the corneas. Laurencin *et al.* administered high doses of P(CPP-SA) 20:80 subcutaneously in rats to study the acute systemic toxicity of the polyanhydrides [32]. Polymer implants in the form of disks were administered subcutaneously for a period of 8 weeks at two different doses in two groups. One group was implanted with one matrix each and the other group implanted with three matrices each, whereas the control group received no polymer matrices. The systemic toxicological effects and effects on individual organs were evaluated based on blood clinical chemistry, hematological parameters, and histological evaluation of the organ sites and implant sites. Pre-necropsy examination of all rats in the study showed no changes in physical appearance or activity due to implantation of polyanhydride matrices. Gross examination of the body cavities and tissues at the time of necropsy did not show any evidence of changes due to polymer implantation, and histological examination of all organ tissues revealed no histomorphological evidence of induced systemic toxicity of the polyanhydride copolymer implantation.

There are no reports available regarding the long-term carcinogenicity studies on polyanhydrides or their degradation products. However, Leong *et al.* showed from histological examination that subcutaneous implantation of P(CPP) in rats over a 6-month period showed no evidence of tumor formation [113]. The brain biocompatibility of P(CPP-SA) 20:80 was established in rat brain by Tamorgo *et al.* [114]. They experimentally proved that none of the animals showed any behavioral changes or neurological deficits suggestive of either systemic or localized toxicity from biodegradable polyanhydrides P(CPP-SA) 20:80 after implantation in rat brain. Brem *et al.* have also evaluated the brain biocompatibility of polyanhydride P(CPP-SA) 50:50 by implantation in rabbit brain [115]. The animals were evaluated daily after the surgery for behavioral changes such as decreased alertness, passivity, impaired grooming, restlessness, irritability, fearfulness, and focal motor neurological deficits. None of the animals showed any behavioral changes or neurological deficits (suggestive of toxicity) and all the animals survived till they were sacrificed. It was concluded that P(CPP:SA) 50:50, a polyanhydride matrix that can be used for the interstitial delivery of drugs in the brain is biocompatible in the rat brain. Thus, the various types of *in vitro* and *in vivo* toxicity studies on polyanhydrides show that these polymers are well tolerated by the body and can be considered biocompatible.

Various fatty acid-based polyanhydrides have also been found biodegradable, biocompatible, and nontoxic in various *in vivo* studies. *In vivo* biodegradation and biocompatibility studies in rats of the 30% stearic acid terminated P(SA) revealed that these polymers are biocompatible and gradually degrade and eliminate within 10 weeks [18]. Fatty acid dimer-based polymers have been thoroughly investigated for their biocompatibility. Brem *et al.* also studied the *in vivo* biocompatibility of P(FAD-SA) 50:50 in rat brain [30]. All animals survived to the scheduled date of sacrifice with no evidence of behavioral changes or neurological deficits suggestive of toxicity. The biocompatibility of P(DA-SA) (50:50) was preliminary evaluated in rabbits brain and it was found that all the experimental rabbits survived healthily and actively to the date of their sacrifice, and histopathological examination indicated that the copolymer is well tolerated by the brain tissue of rabbit [116].

Toxicity of ricinolic acid-based polyanhydride was studied in rats by implanting the polymer strips in subcutaneous, muscle, and brain tissues. It has been found that all animals were healthy throughout the experiment, and the implantation site or any other organ tested did not show any abnormal gross histopathological changes. Blood chemistry and blood count levels were similar for the treated, untreated, and control rats [16, 117–119]. Injectable P(SA:RA) 2:8 loaded with bupivacaine was evaluated for the efficacy and toxicity in producing motor and sensory block when injected near the sciatic nerve[38]. Histological evaluation of the sciatic nerves surrounding tissues (fat and muscle) and the major organs at day 3 and 7 did not showed evidence of active inflammatory reaction or tissue irritation. All the examined organs (lung, liver, heart, brain, and spleen) were normal throughout the period. In all these polymers, fatty acid components undergo extensive metabolism in the body and are mainly excreted in the form of carbon dioxide and minimally through urine and feces. The *in vitro* [120] and *in vivo* toxicity data point to the fact that these polymers are well tolerated by the tissues and can be generally considered to be a biocompatible class of polymers.

3.8 Fabrication of Delivery Systems

Two basic techniques can be utilized for incorporating drug into the polymer matrix, viz. melt mixing and using common organic solvent. Polyanhydrides have low melting point and good solubility in common organic solvents, for example, methylene chloride and chloroform allow for the easy dispersion of a drug into their matrix [121]. Polymer slabs loaded with drug can also be prepared by compression molding a powder containing the drug [122]. Similarly, one can injection mold the drug–polymer formulation into beads or rods [123]. Polymer films can be prepared by solvent casting the polymer solution containing the drug onto Teflon-coated dish [124]. Microsphere-based delivery systems can be formulated by the common techniques including solvent removal, hot-melt encapsulation, and spray drying [125–132]. Some recent studies report nanoparticles formulation with polyanhydride and thus increasing the spectrum of application for polyanhydrides [133–137]. However, it is essential that all processes be performed under anhydrous conditions to minimize hydrolysis of the polymer.

3.9 Production and World Market

Polyanhydrides are not commercially available. One polyanhydride composition, poly(CPP-SA) (20:80), is manufactured at Guilford Pharmaceuticals in Baltimore on a kilogram scale, as part of the Gliadel implant for the treatment of brain tumors. Poly(dimer eurecic acid-*co*-sebacic acid) was developed for large-scale production by Abbott Lab (Chicago) for the fabrication of the Septacin implant for prevention of bone infections. The Septacin product was manufactured by injection molding of the polymer–drug composition. The development of this product was stopped for marketing reasons. Samples of polyanhydrides may be obtained by a request from the corresponding author.

3.10 Biomedical Applications

Polyanhydrides themselves and there hetero-copolymers with -amide, -ester, etc., have been used for diverse biomedical applications. Polyanhydrides find major application in controlled drug delivery. Delivery of chemotherapeutic drugs in cancer is the major area of research for localized delivery using polyanhydrides. There are about 60% of cancer patients with localized disease and it has been estimated that around 32% of localized cancer patients face recurrence, following initial treatment. Most of the anticancer drugs which are in clinical use do not have specific effects on invasiveness or the tendency to metastasize but they are only antiproliferative [138], and therefore, these drugs affect all the rapidly dividing cells including normal tissues and show dose limiting toxic effects.

First-order targeting is increased delivery of drug to the body compartment, while second-order targeting is increased drug delivery to tumor cells; and intracellular delivery is third-order targeting [138]. First- and second-order targeting is achieved by local delivery using polyanhydrides through systems like implant, surgical paste, microspheres, etc. Drug delivery in brain tumor (Glioblastoma multiforme) is important aspect as many of the anticancer drugs are large, ionically charged or hydrophilic, and not able to cross the BBB; intolerably high systemic drug levels are required to achieve the therapeutic doses within CNS [139, 140]. Localized delivery resolves the problem associated with permeability of chemotherapeutic agents through BBB [141]. Various polyanhydrides and drug combinations have been used to obtain the optimum release profile and treat the brain tumor or glioma. Gliadel wafer is one of the most successful delivery systems using polyanhydride and is commercially available. Gliadel has been approved in 1996 by the US-FDA for the use as an adjunct to surgery to prolong survival in patients with recurrent Glioblastoma multiforme for whom surgical resection is indicated [142]. An additional approval from US-FDA in February 2003 has been granted for the use of Gliadel in patients with newly diagnosed high-grade malignant gliomas, as an adjunct to surgery and radiation. A study has been performed by Frazier *et al.* [143] to find the efficacy of local delivery of minocycline and systemic BCNU on intracranial glioma. Minocycline, an antiangiogenic agent, was incorporated in P(CPP:SA) at a ratio of 50:50 by weight and found that the combination of intracranial minocycline and systemic BCNU extended median survival by 82% compared to BCNU alone ($p < 0.0001$) and 200% compared to no treatment ($p < 0.004$). Polyanhydride matrix has been used to deliver heparin and cortisone acetate as antineoplastic agent. They have reported the inhibition of growth of 9L glioma and found out that in the presence of heparin and cortisone, and of cortisone alone, there was a 4.5- and 2.3-fold reduction, respectively, in the growth of 9L glioma [144]. A potential paclitaxel (taxol) formulation in polymeric disk of P(CPP:SA, 20:80) with 20–40% of taxol loading, and maintained concentration of 75–125 ng taxol/mg brain tissue, within 1–3 mm radius of the disk [145]. Another polyanhydride system for delivery of taxol has been formulated using P(FAD:SA, 50:50) but, due to the hydrophobic nature of the FAD, the release rate was very slow and therapeutic concentration could not be achieved [146]. 4-HC, a hydrophilic derivative of cyclophosphamide, with and without *t*-buthionine sulphoxine (inhibiting glutathione synthesis, which catalyzes inactivators of alkylating agents), was incorporated in P(FAD:SA) and found to be effective in rat intracranial 9L gliosarcoma and F98 glioma model [147–149]. Fluorodeoxyuridine, an antimetabolite, has also been optimally released from P(FAD-SA) polymer *in vitro* and *in vivo* [150]. Adriamycin incorporated in P(CPP:SA) has shown improved median survival in rat intracranial 9L glioma model [151]. Fifty percent 5-iodo-2′-deoxyuridine containing P(CPP:SA, 20:80) have been used successfully for radiosensitization of experimental human malignant gliomas [152] Methotrexate–dextran conjugate (to improve stability and inhibit degradation of MTX) when incorporated in P(FAD:SA) offered significant improvement over controls in rat intracranial 9L glioma [151]. Recently, antineoplastic RANse encapsulated in P(RA-SA) implants showed promising efficacy against 9L glioma, while evading neurotoxicity in the cerebellum. The controlled release of Amphibinases forms the potential for a new therapy against brain tumors [153]. Carboplatin and camptothesin are the other anticancer molecules which have shown promising results when incorporated in the P(CPP:SA) [154, 155]. Further, various chemotherapeutic drugs such as ciplatin, methotreaxte, etc., have been delivered using fatty acid-based polyanhydride in squamous cell carcinoma of head and neck. [16, 35, 156, 157].

Osteomyelitis is another disease condition where polyanhydride implant (Septacin) has been found successful in efficacious delivery of gentamicin clinically

[158–160]. Lately, another polyanhydride, poly(OAD/LOAD:SA) indicated its usefulness in delivering gentamicin and to treat chronic osteomyelitis [161]. Blends of PSA with PLA have also found potential as carriers for delivery of ofloxacin in osteomyelitis [162]. Because of controlled release behavior of polyanhydride implants, they find good application in regional anesthesia for clinical areas involving acute or chronic pain including postsurgical pain. Implants of P(CPP:SA) and P(FAD-SA) have been used for delivery of local anesthetics like bupivacaine [92], dibucaine, etc. [122].

Flexibility to control degradation rate and period with polyanhydrides gives opportunity to utilize them in various disease conditions. One such case is restenosis where drug release is required at local site for around 6 months. P(FAD:SA) have been used perivascularly to release heparin to microvascular anastomoses [124, 163]. It was found that the vessel potency rates were significantly greater in vessels treated with polyanhydride–heparin compared to controls, after surgery [163, 164]. PLA have been used to coat the P(FAD:SA) sheets to improve the release profile and strength of the films. [164]. Glaucoma is another disease where delivery of antifibrotic agent can prolong filtration surgery. Disks of P(CPP:SA) containing different drugs such as 5-fluorocil, 5-fluorouridine, taxol, and etoposide have been evaluated and some of the devices were very successful in maintaining intraocular pressure to the postsurgery level [165–168]. Polyanhydride microspheres have also been tried to deliver the drug in controlled manner in vitreoretinal disorders, to avoid the repeated intravite injection to achieve intraocular drug levels within the therapeutic range [169].

Besides the conditions described above, polyanhydrides have been used for many other applications. Delivery of macromolecules especially DNA, proteins, and peptides *via* polymer is an important issue. Delivery of DNA molecules for gene therapy is a challenge and nonviral carriers are always under search. Photocrosslinked polyanhydrides could allow repeated transfection, with an appropriate amount of DNA for the rate of local cell division and the cells capacity for DNA uptake [170, 171]. Bioadhesive nanoparticles of polyanhydrides were also found to have potential for oral delivery of DNA [172]. Polyanhydride matrices can be used for controlled delivery of proteins or polymer–drug conjugates. [173]. Polyanhydride-*co*-imides have been used for controlled release of bovine serum albumin as a model compound which suggest that polyanhydrides may be appropriate for delivery of many therapeutic proteins, including vaccine antigens [174–177]. Moreover, stability and activity of proteins and peptides can also be maintained by using polyanhydride as a carrier [128, 135, 178, 179]. Lucas *et al.* have done early trials on localized protein delivery, where they have incorporated water-soluble protein-possessing chondrogenic stimulating activity in polyanhydride polymeric vehicle. The delivery system was capable of inducing cartilage and bone up to 50% of the time. It was concluded that polyanhydride could be used as a controlled release delivery vehicle for soluble bioactive factors that interacts with local cell population [180]. PSA-*b*-PEG and (P[TMA-Glycine-*co*-SA]-b-PEG) were used as isolating layers for their good processing properties at room temperature. These polymers were advantageous for pulsatile protein delivery due to their pH sensitivity and appropriate erosion duration [181]. Kubek *et al.* incorporated an endogenous neuropeptide thyrotropin-releasing hormone (protirelin), and were first to provide evidence in support of *in situ* pharmacotherapy for potential delivery in intractable epilepsy and possibly other neurological disorders. [182] Cai *et al.* have synthesized a novel polyanhydride, P([CBF]-ASA), with 5-aminosalicylic acid (5-ASA) incorporated in the backbone. They hypothesized the potential of colon-specific delivery of 5-ASA moieties considering high drug loading (50.2% of 5-ASA moieties in the backbone) and degradation characteristics [183]. Localized intracerebral delivery of neurotransmitters using SA copolymer has also been tried and was concluded by authors that intracerebral polymeric drug delivery successfully reversed lesion-induced memory deficit and has potential as a neurological treatment for Alzheimer's disease and other neurological disorders [184].

References

1 Rosen, H.B., Chang, J., Wnek, G.E., Linhardt, R.J., and Langer, R. (1983) *Biomaterials*, **4**, 131.
2 Dang, W., Daviau, T., and Brem, H. (1996) *Pharm. Res.*, **13**, 683.
3 Jain, J.P., Modi, S., Domb, A.J., and Kumar, N. (2005) *J. Control. Release*, **103**, 541.
4 Kumar, N., Langer, R.S., and Domb, A.J. (2002) *Adv. Drug Deliv. Rev.*, **54**, 889.
5 Brem, H., Tamargo, R.J., Olivi, A., Pinn, M., Weingart, J.D., Wharam, M., and Epstein, J.I. (1994) *J. Neurosurg.*, **80**, 283.
6 Domb, A.J., Gallardo, C.F., and Langer, R. (1989) *Macromolecules*, **22**, 3200.
7 Domb, A.J. and Langer, R. (1987) *J. Polym. Sci. A Polym. Chem.*, **25**, 3373.
8 Bucher, J.E. and Slade, W.C. (1909) *J. Am. Chem. Soc.*, **31**, 1319.
9 Conix, A. (1958) *J. Polym. Sci. A*, **29**, 343.
10 Hill, J.W. (1930) *J. Am. Chem. Soc.*, **52**, 4110.
11 Hill, J.W. and Carothers, H.W. (1932) *J. Am. Chem. Soc.*, **54**, 5169.
12 Leong, K.W., Kost, J., Mathiowitz, E., and Langer, R. (1986) *Biomaterials*, **7**, 364.
13 Yoda, N. (1962) *Makromol. Chem.*, **55**, 174.
14 Domb, A.J. and Langer, R. (1989) *Macromolecules*, **22**, 3200.
15 Domb, A.J. (1992) *Macromolecules*, **25**, 12.
16 Teomim, D., Nyska, A., and Domb, A.J. (1999) *J. Biomed. Mater. Res.*, **45**, 258.
17 Domb, A.J. and Maniar, M. (1993) *J. Polym. Sci. A Polym. Chem.*, **31**, 1275.
18 Teomim, D. and Domb, A.J. (2001) *Biomacromolecules*, **2**, 37.
19 Vogel, B.M. and Mallapragada, S.K. (2005) *Biomaterials*, **26**, 721.
20 Slivniak, R. and Domb, A.J. (2005) *Biomacromolecules*, **6**, 1679.
21 Slivniak, R., Ezra, A., and Domb, A.J. (2006) *Pharm. Res.*, **23**, 1306.
22 Ben-Shabat, S., Elmalak, O., Nyska, A., and Domb, A.J. (2005) *Isr. J. Chem.*, **45**, 411.
23 Katarzyna, J. (2007) *Macromol. Symp.*, **254**, 109.
24 Cheng, G., Aponte, M.A., and Ramirez, C.A. (2003) *PMSE Preprints*, **89**, 618.
25 Anastasiou, T.J. and Uhrich, K.E. (2003) *J. Polym. Sci. A Polym. Chem.*, **41**, 3667.
26 Gref, R., Domb, A., Quellec, P., Blunk, T., Mueller, R.H., Verbavatz, J.M., and Langer, R. (1995) *Adv. Drug Deliv. Rev.*, **16**, 215.
27 Hou, S., McCauley, L.K., and Ma, P.X. (2007) *Macromol. Biosci.*, **7**, 620.
28 Teomim, D. and Domb, A.J. (1999) *J. Polym. Sci. A Polym. Chem.*, **37**, 3337.
29 Buahin, K.G., Judy, K.D., and Hartke, C. (1993) *Polym. Adv. Technol.*, **3**, 311.
30 Brem, H., Domb, A., Lenartz, D., Dureza, C., Olivi, A., and Epstein, J.I. (1992) *J. Control. Release*, **19**, 325.

31 Zhang, T., Gu, M., and Yu, X. (2001) *J. Biomater. Sci. Polym. Ed.*, **12**, 491.

32 Laurencin, C., Domb, A., Morris, C., Brown, V., Chasin, M., McConnell, R., Lange, N., and Langer, R. (1990) *J. Biomed. Mater. Res.*, **24**, 1463.

33 Domb, A.J. and Nudelman, R. (1995) *J. Polym. Sci. A Polym. Chem.*, **33**, 717.

34 Mader, K., Cremmilleux, Y., Domb, A.J., Dunn, J.F., and Swartz, H.M. (1997) *Pharm. Res.*, **14**, 820.

35 Krasko, M.Y., Shikanov, A., Ezra, A., and Domb, A.J. (2003) *J. Polym. Sci. A Polym. Chem.*, **41**, 1059.

36 Jain, J.P., Modi, S., and Kumar, N. (2008) *J. Biomed. Mater. Res. A*, **84**, 740.

37 Shikanov, A., Ezra, A., and Domb, A.J. (2005) *J. Control. Release*, **105**, 52.

38 Shikanov, A., Domb, A.J., and Weiniger, C.F. (2007) *J. Control. Release*, **117**, 97.

39 Domb, A.J. (1990) *Biomaterials*, **11**, 686.

40 Cheng, G., Aponte, M.A., and Ramírez, C.A. (2004) *Polymer*, **45**, 3157.

41 Uhrich, K.E., Larrier, D.R., Laurencin, C.T., and Langer, R. (1996) *J. Polym. Sci. A Polym. Chem.*, **34**, 1261.

42 Uhrich, K.E., Thomas, T.T., Laurencin, C.T., and Langer, R. (1997) *J. Appl. Polym. Sci.*, **63**, 1401.

43 Zhang, Z.-Q., Su, X.-M., He, H.-P., and Qu, F.-Q. (2004) *J. Polym. Sci. A Polym. Chem.*, **42**, 4311.

44 Anseth, K.S., Shastri, V.R., and Langer, R. (1999) *Nat. Biotechnol.*, **17**, 156.

45 Poshusta, A.K., Burdick, J.A., Mortisen, D.J., Padera, R.F., Ruehlman, D., Yaszemski, M.J., and Anseth, K.S. (2003) *J. Biomed. Mater. Res.*, **64A**, 62.

46 Young, J.S., Gonzales, K.D., and Anseth, K.S. (2000) *Biomaterials*, **21**, 1181.

47 Shastri, V.P., Padera, R.F., Tarcha, P., and Langer, R. (2004) *Biomaterials*, **25**, 715.

48 Weiner, A.A., Bock, E.A., Gipson, M.E., and Shastri, V.P. (2008) *Biomaterials*, **29**, 2400.

49 Erdmann, L. and Uhrich, K.E. (2000) *Biomaterials*, **21**, 1941.

50 Harten, R.D., Svach, D.J., Schmeltzer, R., and Uhrich, K.E. (2005) *J. Biomed. Mater. Res. A*, **72**, 354.

51 Schmeltzer, R.C., Schmalenberg, K.E., and Uhrich, K.E. (2005) *Biomacromolecules*, **6**, 359.

52 Whitaker-Brothers, K. and Uhrich, K. (2006) *J. Biomed. Mater. Res. A*, **76**, 470.

53 Bryers, J.D., Jarvis, R.A., Lebo, J., Prudencio, A., Kyriakides, T.R., and Uhrich, K. (2006) *Biomaterials*, **27**, 5039.

54 Rosenberg, L.E., Carbone, A.L., Romling, U., Uhrich, K.E., and Chikindas, M.L. (2008) *Lett. Appl. Microbiol.*, **46**, 593.

55 Korhonen, H., Hakala, R.A., Helminen, A.O., and Seppala, J.V. (2006) *Macromol. Biosci.*, **6**, 496.

56 Korhonen, H., Helminen, A.O., and Seppälä, J.V. (2004) *Macromol. Chem. Phys.*, **205**, 937.

57 Wiggins, J.S. and Storey, R.F. (2005) *Polymer Prepr.*, **46**, 333.

58 Hamdan, Y., Jiang, X., Huang, K., and Yu, K. (2007) *Am. J. Appl. Sci.*, **4**, 128.

59 Dinarvand, R., Alimorad, M.M., Amanlou, M., and Akbari, H. (2005) *J. Biomed. Mater. Res. A*, **75**, 185.

60 Edlund, U., Albertsson, A.C., Singh, S.K., Fogelberg, I., and Lundgren, B.O. (2000) *Biomaterials*, **21**, 945.

61 Edlund, U. and Albertsson, A.-C. (1999) *J. Appl. Polym. Sci.*, **72**, 227.

62 Domb, A.J. (1993) *J. Polym. Sci. A Polym. Chem.*, **31**, 1973.

63 Thorstenson, J.B., Petersen, L.K., and Narasimhan, B. (2009) *J. Comb. Chem.*, **11**, 820.

64 Domb, A.J., Amselem, S., Shah, J., and Maniar, M. (1993) *Adv. Polym. Sci.*, **107**, 93.

65 Leong, K.W., Simonte, V., and Langer, R. (1987) *Macromolecules*, **20**, 705.

66 Shipp, D.A., McQuinn, C.W., Rutherglen, B.G., and McBath, R.A. (2009) *Chem. Commun.*, **42**, 6415.

67 Domb, A.J., Ron, E., and Langer, R. (1988) *Macromolecules*, **21**, 1925.

68 Albertsson, A.-C. and Lundmark, S. (1988) *J. Macromol. Sci. A Chem.*, **25**, 247.

69 Staubli, A., Mathiowitz, E., Lucarelli, M., and Langer, R. (1991) *Macromolecules*, **24**, 2283.

70 Uhrich, K.E., Gupta, A., Thomas, T.T., Laurencin, C.T., and Langer, R. (1995) *Macromolecules*, **28**, 2184.

71 Ron, E., Mathiowitz, E., Mathiowitz, G., Domb, A.J., and Langer, R. (1991) *Macromolecules*, **24**, 2278.

72 Domb, A.J., Mathiowitz, E., Ron, E., Giannos, S., and Langer, R. (1991) *J. Polym. Sci. A Polym. Chem.*, **29**, 571.

73 D'Emanuele, A., Hill, J., Tamada, J.A., Domb, A.J., and Langer, R. (1992) *Pharm. Res.*, **9**, 1279.

74 Davies, M.C., Khan, M.A., Domb, A., Langer, R., Watts, J.F., and Paul, A.J. (1991) *J. Appl. Polym. Sci.*, **42**, 1597.

75 McCann, D.L., Heatley, F., and D'Emanuele, A. (1999) *Polymer*, **40**, 2151.

76 Tudor, A.M., Melia, C.D., Davies, M.C., Hendra, P.J., Church, S., Domb, A.J., and Langer, R. (1991) *Spectrochim. Acta A Mol. Biomol. Spectrosc.*, **47A**, 1335.

77 Shakesheff, K.M., Davis, M.C., Domb, J., Jeckson, D.E., Roberts, C.J., Tendler, S.J.B., and Williams, P.M. (1995) *Macromolecules*, **28**, 1108.

78 Shakesheff, K.M., Davies, M.C., Roberts, C.J., Tendler, S.J.B., Shard, A.G., and Domb, A. (1994) *Langmuir*, **10**, 4417.

79 Shard, A.G., Shakesheff, K.M., Roberts, C.J., Tendler, S.J.B., and Davies, M.C. (1997) *Handbook of Biodegradable Polymers*, vol. 7 (eds J.K.A.J. Domb and D.M. Wiseman), Harwood Academic Publishers, Amsterdam, p. 417.

80 Krasko, M.Y., Shikanov, A., Kumar, N., and Domb, A.J. (2002) *Polym. Adv. Technol.*, **13**, 960.

81 Hui-Bi, X., Zhi-Bin, Z., and Kai-Xun, H. (2001) *Polym. Bull.*, **46**, 435.

82 Wen-Xun, G. and Kai-Xun, H. (2004) *Polym. Degrad. Stab.*, **84**, 375.

83 Guo, W.-X., Huang, K.-X., Tang, R., and Chi, Q. (2004) *Polymer*, **45**, 5743.

84 Shikanov, A. and Domb, A.J. (2006) *Biomacromolecules*, **7**, 288–296.

85 Slivniak, R., Ezra, A., and Domb, A.J. (2006) *Pharm. Res.*, **23**, 1306–1312.

86 Domb, A.J. and Langer, R. (1989) *Macromolecules*, **22**, 2117.

87 Staubli, A., Ron, E., and Langer, R. (1990) *J. Am. Chem. Soc.*, **112**, 4419.

88 Mader, K., Domb, A., and Swartz, H.M. (1996) *Appl. Radiat. Isot.*, **47**, 1669.

89 Mader, K., Bacic, G., Domb, A., Elmalak, O., Langer, R., and Swartz, H.M. (1997) *J. Pharm. Sci.*, **86**, 126.

90 Teomim, D., Mader, K., Bentolila, A., Magora, A., and Domb, A.J. (2001) *Biomacromolecules*, **2**, 1015.

91 Shieh, L., Tamada, J., Chen, I., Pang, J., Domb, A., and Langer, R. (1994) *J. Biomed. Mater. Res.*, **28**, 1465.

92 Gopferich, A. and Tessmar, J. (2002) *Adv. Drug Deliv. Rev.*, **54**, 911.

93 Tamada, J.A. and Langer, R. (1993) *Proc. Natl. Acad. Sci. USA*, **90**, 552.

94 Mathiowitz, E., Ron, E., Mathiowitz, G., Amato, C., and Langer, R. (1990) *Macromolecules*, **23**, 3212.

95 Leach, K.J. and Mathiowitz, E. (1998) *Biomaterials*, **19**, 1973.

96 Albertsson, A.C. and Liu, Y. (1997) *J. Macromol. Sci. Pure Appl. Chem.*, **A34**, 1457.

97 Burkoth, A.K., Burdick, J., and Anseth, K.S. (2000) *J. Biomed. Mater. Res.*, **51**, 352.

98 San Roman, J., Lopez Madruga, E., and Pargada, L. (1987) *Polym. Degrad. Stab.*, **19**, 161.

99 Akbari, H., D'Emanuele, A., and Attwood, D. (1998) *Int. J. Pharm.*, **160**, 83.

100 Gopferich, A. and Langer, R. (1995) *AIChE J.*, **41**, 2292.

101 Mathiowitz, E., Bernstein, H., Giannos, S., Dor, P., Turek, T., and Langer, R. (1992) *J. Appl. Polym. Sci.*, **45**, 125.

102 Tabata, Y. and Langer, R. (1993) *Pharm. Res.*, **10**, 391.

103 Siepmann, J. and Gopferich, A. (2001) *Adv. Drug Deliv. Rev.*, **48**, 229.

104 Hopfenberg, H.B. (1976) *Controlled Release Polymeric Formulations*, vol. 33 (eds D.R. Paul and F.W. Harris), American Chemical Society, Washington, DC, p. 26.

105 Cooney, D.O. (1972) *AIChE J.*, **18**, 446.

106 Gopferich, A., Karydas, D., and Langer, R. (1995) *Eur. J. Pharm. Biopharm.*, **41**, 81.

107 Zygourakis, K. (1990) *Chem. Eng. Sci.*, **45**, 2359.

108 Zygourakis, K. and Markenscoff, P.A. (1996) *Biomaterials*, **17**, 125.

109 Gopferich, A. and Langer, R. (1993) *Macromolecules*, **26**, 4105.

110 Domb, A.J. and Nudelman, R. (1995) *Biomaterials*, **16**, 319.

111 Katti, D.S., Lakshmi, S., Langer, R., and Laurencin, C.T. (2002) *Adv. Drug Deliv. Rev.*, **54**, 933.

112 Domb, A.J., Rock, M., Schwartz, J., Perkin, C., Yipchuk, G., Broxup, B., and Villemure, J.G. (1994) *Biomaterials*, **15**, 681.

113 Leong, K.W., D`Amore, P., Marletta, M., and Langer, R. (1986) *J. Biomed. Mater. Res.*, **20**, 51.

114 Tamargo, R.J., Epstein, J.I., Reinhard, C.S., Chasin, M., and Brem, H. (1989) *J. Biomed. Mater. Res.*, **23**, 253.

115 Brem, H., Kader, A., Epstein, J.I., Tamargo, R.J., Domb, A., Langer, R., and Leong, K.W. (1989) *Sel. Cancer Ther.*, **5**, 55.

116 Xu, H.B., Zhou, Z.B., Huang, K.X., Lei, T., Zhang, T., and Liu, Z.L. (2001) *Polym. Bull.*, **46**, 435.

117 Jain, J.P., Modi, S., and Kumar, N. (2007) *J. Biomed. Mater. Res. A*, **84A**(3), 740–752.

118 Shikanov, A., Vaisman, B., Krasko, M.Y., Nyska, A., and Domb, A.J. (2004) *J. Biomed. Mater. Res.*, **69A**, 47.

119 Vaisman, B., Motiei, M., Nyska, A., and Domb, A.J. (2010) *J. Biomed. Mater. Res. A*, **92**, 419.

120 Petersen, L.K., Xue, L., Wannemuehler, M.J., Rajan, K., and Narasimhan, B. (2009) *Biomaterials*, **30**, 5131.

121 Domb, A.J. (1994) *Polymeric Site-Specific Pharmacotherapy* (ed. A.J. Domb), John Wiley & Sons, Inc., New York, p. 1.

122 Masters, D.B., Berde, C.B., Dutta, S., Turek, T., and Langer, R. (1993) *Pharm. Res.*, **10**, 1527.

123 Deng, J.S., Li, L., Tian, Y., Meisters, M., Chang, H.C., Stephens, D., Chen, S., and Robinson, D. (2001) *Pharm. Dev. Technol.*, **6**, 541.

124 Teomim, D., Fishbien, I., Golomb, G., Orloff, L., Mayberg, M., and Domb, A.J. (1999) *J. Control. Release*, **60**, 129.

125 Sun, L., Zhou, S., Wang, W., Su, Q., Li, X., and Weng, J. (2009) *J. Mater. Sci. Mater. Med.*, **20**, 2035.

126 Lopac, S.K., Torres, M.P., Wilson-Welder, J.H., Wannemuehler, M.J., and Narasimhan, B. (2009) *J. Biomed. Mater. Res. B Appl. Biomater.*, **91**, 938.

127 Berkland, C., Pollauf, E., Varde, N., Pack, D.W., and Kim, K.K. (2007) *Pharm. Res.*, **24**, 1007.

128 Determan, A.S., Wilson, J.H., Kipper, M.J., Wannemuehler, M.J., and Narasimhan, B. (2006) *Biomaterials*, **27**, 3312.

129 Pfeifer, B.A., Burdick, J.A., and Langer, R. (2005) *Biomaterials*, **26**, 117.

130 Berkland, C., Pollauf, E., Pack, D.W., and Kim, K.K. (2004) *J. Control. Release*, **96**, 101.

131 Mathiowitz, E., Saltzman, W.M., Domb, A., Dor, P., and Langer, R. (1988) *J. Appl. Polym. Sci.*, **35**, 755.

132 Mathiowitz, E. and Langer, R. (1987) *J. Control. Release*, **5**, 13.

133 Lee, W.C. and Chu, I.M. (2008) *J. Biomed. Mater. Res. B Appl. Biomater.*, **84**, 138.

134 Petersen, L.K., Sackett, C.K., and Narasimhan, B. (2009) *J. Comb. Chem.*, **12**, 51.

135 Petersen, L.K., Sackett, C.K., and Narasimhan, B. (2010) *Acta Biomater.*, **6**, 3873.

136 Ulery, B.D., Phanse, Y., Sinha, A., Wannemuehler, M.J., Narasimhan, B., and Bellaire, B.H. (2009) *Pharm. Res.*, **26**, 683.

137 Yoncheva, K., Guembe, L., Campanero, M.A., and Irache, J.M. (2007) *Int. J. Pharm.*, **334**, 156.

138 Dhanikula, A.B. and Panchagnula, R. (1999) *Int. J. Pharm.*, **183**, 85.

139 Greig, N.H. (1987) *Cancer Treat. Rev.*, **14**, 1.

140 Abbott, N.J. and Romero, I.A. (1996) *Mol. Med. Today*, **2**, 106.

141 Lesniak, M.S. and Brem, H. (2004) *Nat. Rev. Drug Discov.*, **3**, 499.

142 Aoki, T., Hashimoto, N., and Matsutani, M. (2007) *Exp. Opin. Pharmacother.*, **8**, 3133.

143 Frazier, J.L., Wang, P.P., Case, D., Tyler, B.M., Pradilla, G., Weingart, J.D., and Brem, H. (2003) *J. Neurooncol.*, **64**, 203.

144 Tamargo, R.J., Leong, K.W., and Brem, H. (1990) *J. Neurooncol.*, **9**, 131.

145 Walter, K.A., Cahan, M.A., Gur, A., Tyler, B., Hilton, J., Colvin, O.M., Burger, P.C., Domb, A., and Brem, H. (1994) *Cancer Res.*, **54**, 2207.

146 Park, E.S., Maniar, M., and Shah, J.C. (1998) *J. Control. Release*, **52**, 179.

147 Judy, K.D., Olivi, A., Buahin, K.G., Domb, A., Epstein, J.I., Colvin, O.M., and Brem, H. (1995) *J. Neurosurg.*, **82**, 481.

148 Sipos, E.P., Witham, T.F., Ratan, R., Burger, P.C., Baraban, J., Li, K.W.,

Piantadosi, S., and Brem, H. (2001) *Neurosurgery*, **48**, 392.
149 Colvin, O.M., Friedman, H.S., Gamcsik, M.P., Fenselau, C., and Hilton, J. (1993) *Adv. Enzyme Regul.*, **33**, 19.
150 Choti, M.A., Saenz, J., Yang, X., and Brem, H. (1995) *Proc. Am. Assoc. Cancer Res.*, **36**, 309.
151 Wang, P.P., Frazier, J., and Brem, H. (2002) *Adv. Drug Deliv. Rev.*, **54**, 987.
152 Williams, J.A., Dillehay, L.E., Tabassi, K., Sipos, E., Fahlman, C., and Brem, H. (1997) *J. Neurooncol.*, **32**, 181.
153 Slager, J., Tyler, B., Shikanov, A., Domb, A.J., Shogen, K., Sidransky, D., and Brem, H. (2009) *Pharm. Res.*, **26**, 1838.
154 Olivi, A., Ewend, M.G., Utsuki, T., Tyler, B., Domb, A.J., Brat, D.J., and Brem, H. (1996) *Cancer Chemother. Pharmacol.*, **39**, 90.
155 Storm, P.B., Moriarity, J.L., Tyler, B., Burger, P.C., Brem, H., and Weingart, J. (2002) *J. Neurooncol.*, **56**, 209.
156 Shikani, A.H., Eisele, D.W., and Domb, A.J. (1994) *Arch. Otolaryngol. Head Neck Surg.*, **120**, 1242.
157 Shikani, A.H. and Domb, A.J. (2000) *Laryngoscope*, **110**, 907.
158 Tian, Y., Li, L., Gao, X., Deng, J., Stephens, D., Robinson, D., and Chang, H. (2002) *Drug Dev. Ind. Pharm.*, **28**, 897.
159 Stephens, D., Li, L., Robinson, D., Chen, S., Chang, H., Liu, R.M., Tian, Y., Ginsburg, E.J., Gao, X., and Stultz, T. (2000) *J. Control. Release*, **63**, 305.
160 Li, L.C., Deng, J., and Stephens, D. (2002) *Adv. Drug Deliv. Rev.*, **54**, 963.
161 Yang, X.F., Zeng, F.D., Zhou, Z.B., Huang, K.X., and Xu, H.B. (2003) *Acta Pharmacol. Sin.*, **24**, 306.
162 Chen, L., Wang, H., Wang, J., Chen, M., and Shang, L. (2007) *J. Biomed. Mater. Res. B Appl. Biomater.*, **83**, 589.
163 Orloff, L.A., Glenn, M.G., Domb, A.J., and Esclamado, R.A. (1995) *Surgery*, **117**, 554.
164 Orloff, L.A., Domb, A.J., Teomim, D., Fishbein, I., and Golomb, G. (1997) *Adv. Drug Deliv. Rev.*, **24**, 3.
165 Lee, D.A., Leong, K.W., Panek, W.C., Eng, C.T., and Glasgow, B.J. (1988) *Invest. Ophthalmol. Vis. Sci.*, **29**, 1692.
166 Jampel, H.D., Leong, K.W., Dunkelburger, G.R., and Quigley, H.A. (1990) *Arch. Ophthalmol.*, **108**, 430.
167 Jampel, H.D., Thibault, D., Leong, K.W., Uppal, P., and Quigley, H.A. (1993) *Invest. Ophthalmol. Vis. Sci.*, **34**, 3076.
168 Uppal, P., Jampel, H.D., Quigley, H.A., and Leong, K.W. (1994) *J. Ocul. Pharmacol.*, **10**, 471.
169 Herrero-Vanrell, R., and Refojo, M.F. (2001) *Adv. Drug Deliv. Rev.*, **52**, 5.
170 Pfeifer, B.A., Burdick, J.A., Little, S.R., and Langer, R. (2005) *Int. J. Pharm.*, **304**, 210.
171 Quick, D.J., Macdonald, K.K., and Anseth, K.S. (2004) *J. Control. Release*, **97**, 333.
172 Yoncheva, K., Centelles, M.N., and Irache, J.M. (2008) *J. Microencapsul.*, **25**, 82.
173 Dang, W. and Saltzman, W.M. (1994) *J. Biomater. Sci. Polym. Ed.*, **6**, 297.
174 Carrillo-Conde, B., Schiltz, E., Yu, J., Chris Minion, F., Phillips, G.J., Wannemuehler, M.J., and Narasimhan, B. (2010) *Acta Biomater.*, **6**, 3110.
175 Martins, D.C.R., Gamazo, C., and Irache, J.M. (2009) *Eur. J. Pharm. Sci.*, **37**, 563.
176 Kipper, M.J., Wilson, J.H., Wannemuehler, M.J., and Narasimhan, B. (2006) *J. Biomed. Mater. Res. A*, **76**, 798.
177 Hanes, J., Chiba, M., and Langer, R. (1998) *Biomaterials*, **19**, 163.
178 Manoharan, C. and Singh, J. (2009) *J. Pharm. Sci.*, **98**, 4237.
179 Tabata, Y., Gutta, S., and Langer, R. (1993) *Pharm. Res.*, **10**, 487.
180 Lucas, P.A., Laurencin, C., Syftestad, G.T., Domb, A., Goldberg, V.M., Caplan, A.I., and Langer, R. (1990) *J. Biomed. Mater. Res.*, **24**, 901.
181 Jiang, H.L. and Zhu, K.J. (2000) *Int. J. Pharm.*, **194**, 51.
182 Kubek, M.J., Liang, D., Byrd, K.E., and Domb, A.J. (1998) *Brain Res.*, **809**, 189.
183 Cai, Q.X., Zhu, K.J., Chen, D., and Gao, L.P. (2003) *Eur. J. Pharm. Biopharm.*, **55**, 203.
184 Howard, M.A., 3rd, Gross, A., Grady, M.S., Langer, R.S., Mathiowitz, E., Winn, H.R., and Mayberg, M.R. (1989) *J. Neurosurg.*, **71**, 105.

4
Poly(Ortho Esters)

Jorge Heller[1)]

4.1
Introduction

Poly(ortho esters) (POE), developed following the use of poly(glycolic acid) and poly(glycolic acid-*co*-lactic acid) copolymers, were first described in 1970 and have been under development since then. They were the first new biodegradable polymers synthesized specifically for drug delivery applications. Four different families have been developed as shown in Scheme 4.1.

POE I was developed at the ALZA Corporation in the 1970s [1–4] and its synthesis is shown in Scheme 4.2. POE I undergoes a hydrolysis as shown in Scheme 4.3. Since a primary hydrolysis product is butyrolactone that rapidly hydrolyzes to butyric acid, and since poly(ortho esters) are acid-labile, the polymer undergoes an uncontrolled autocatalyzed hydrolysis resulting in rapid disintegration. To prevent that, a base such as sodium carbonate must be added. The need to use a base to stabilize the polymer, a difficult synthesis and unsatisfactory mechanical properties, has prevented the commercialization and this polymer is no longer under development.

POE III is synthesized as also shown in Scheme 4.2 [5]. It has been extensively investigated in ocular applications [6]. Although the polymer has been found to be highly biocompatible and excellent drug release has been achieved, difficulties in achieving a reproducible synthesis and an inability to scale up the reaction have prevented its commercialization and this polymer system is also no longer under development.

However, POE II and POE IV are highly successful polymers that are currently undergoing commercialization and as of this writing, POE IV has completed a

1) Deceased.

Handbook of Biodegradable Polymers: Synthesis, Characterization and Applications, First Edition. Edited by Andreas Lendlein, Adam Sisson.
© 2011 Wiley-VCH Verlag GmbH & Co. KGaA. Published 2011 by Wiley-VCH Verlag GmbH & Co. KGaA.

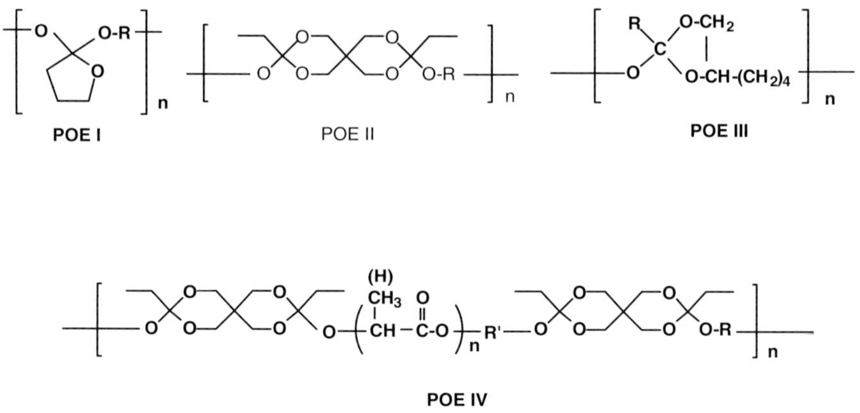

Scheme 4.1 The four families of poly(ortho esters).

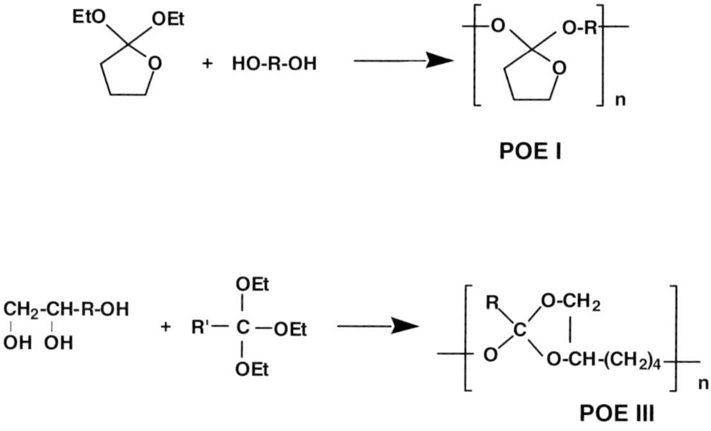

Scheme 4.2 Synthesis of POE I and POE III.

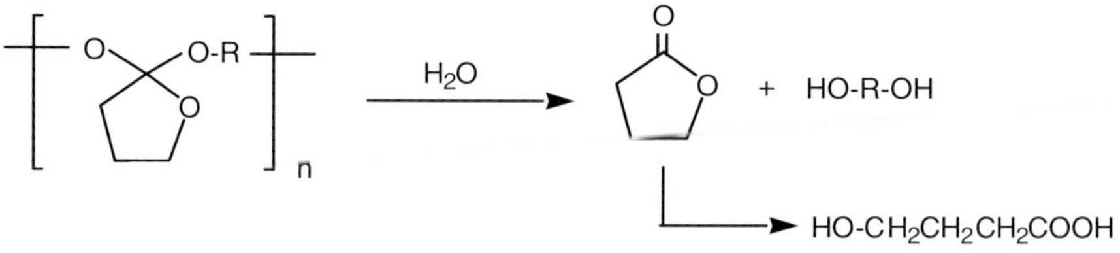

Scheme 4.3 Hydrolysis of POE I.

Phase III clinical trial for the prevention of chemotherapy-induced immediate and delayed nausea and vomiting using the delivery of granisetron. It has also completed a Phase II clinical trial to treat postoperative pain using the delivery of mepivacaine. The developments of various proprietary products using POE II are also ongoing.

4.2
POE II

Even though ortho ester linkages are very hydrolytically labile, and indeed a water-soluble poly(ortho ester) will completely hydrolyze in a matter of hours, a hydrophobic polymer has a very long lifetime in water, as shown in Figure 4.1 [7].

Erosion rates of POE II can be adjusted by incorporating into the polymer acidic excipients such a suberic acid, but this method was never very successful. However, because the polymer is so labile in an aqueous environment, erosion rates can also be manipulated by controlling the hydrophilicity of the polymer by using hydrophilic diols such as triethylene glycol (TEG).

4.2.1
Polymer Synthesis

POE II is prepared by the reaction between the diketene acetal 3,9 diethylidene-2,4,8,10-tetraoxaspiro [5.5] undecane (DETOSU) as shown in Scheme 4.4 [8, 9].

DETOSU is not commercially available and is prepared by the rearrangement of diallyl pentaerytritol as shown in Scheme 4.5 by using either *n*-BuLi in ethylene

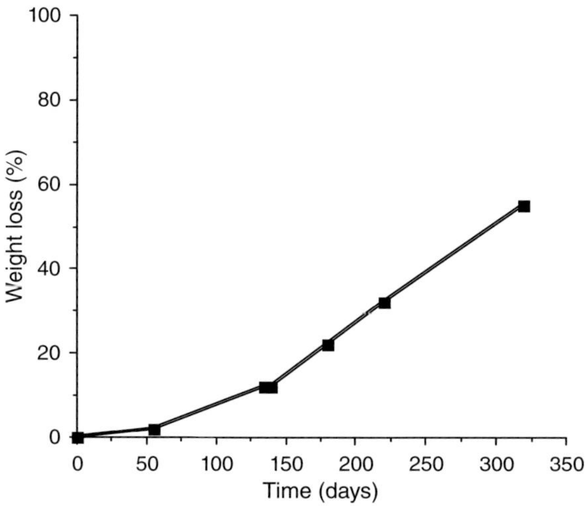

Figure 4.1 Weight loss as a function of time for a polymer prepared from 3,9-dimethylene-2,4,8,10-tetraoxaspiro [5.5] undecane and HD; 0.05 M phosphate buffer, pH 7.4, 37 °C.

Scheme 4.4 Synthesis of POE II.

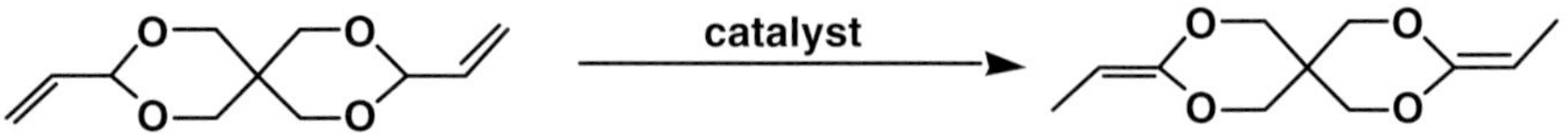

Scheme 4.5 Synthesis of 3,9-diethylidene-2,4,8,10-tetraoxaspiro[5.5]undecane (DETOSU).

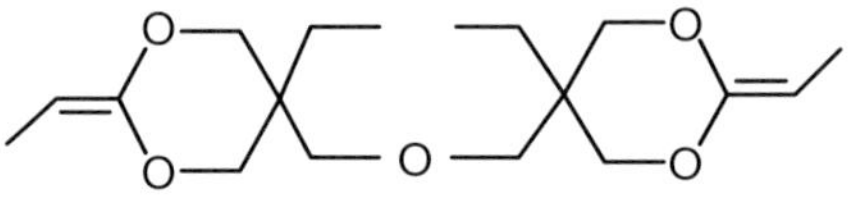

Scheme 4.6 Structure of di(5-methyl-2-ethylidene[1.3]dioxan-5yl)methyl ether.

diamine [9], KOtBu in ethylene diamine [10], a photochemical rearrangement [11], or an $Ru(PPh_3)_3Cl_2$ catalyzed rearrangement [12].

4.2.1.1 Rearrangement Procedure Using an $Ru(PPh_3)_3Cl_2$ Na_2CO_3 Catalyst

A round-bottom flask was charged with 224 g of 3,9-divinyl-2,4,8,10-tetraoxaspiro[5.5] undecane, 0.8 g dichlorotris(triphenylphosphine)ruthenium ($Ru(PPh_3)_3Cl_2$) and 0.8 mg Na_2CO_3. The mixture was heated at 120 °C under nitrogen for a minimum of 16 h. The progress of the reaction was followed by ^1H NMR in D_2O. After cooling to room temperature, the product was distilled under reduced pressure and purified by recrystallization from *n*-pentane containing a few drops of triethylamine. To obtain a polymerization-grade product, two more recrystallizations were required.

4.2.1.2 Alternate Diketene Acetals

Even though the great majority of the work was carried out using DETOSU, another diketene acetal, di(5-methyl-2-ethylidene[1.3]dioxan-5yl)methyl ether, was briefly investigated. The structure of this diketene acetal is shown in Scheme 4.6.

Polymers prepared using this diketene acetal will be discussed under Section 7.1.

4.2.1.3 Typical Polymer Synthesis Procedure

In a dry-box, 2.163 g (15 mmol) of *trans*-cyclohexanedimethanol (CDM), 4.727 g (40 mmol) of 1,6-hexanediol (HD), and 6.760 g (45 mmol) of TEG were dissolved in 40 g of tetrahydrofuran (THF). Then, 21.437 g (101 mmol) of 3,9-diethylidene-2,4,8,10-tetraoxaspiro [5.5] undecane were weighed into a round-bottom flask and added to the diols solution with the aid of 20 g THF, in several portions. The flask was removed from the dry-box, rapidly connected to a condenser and nitrogen inlet, and a few drops of *p*-toluenesulfonic acid solution (10 mg/mL) added. After the exotherm subsided, the solution was slowly poured into 3 L of methanol, containing about 1000 ppm of triethylamine. After isolation by filtration and drying in a vacuum oven at 40 °C for 24 h, the yield was 32.1 g (89.7%).

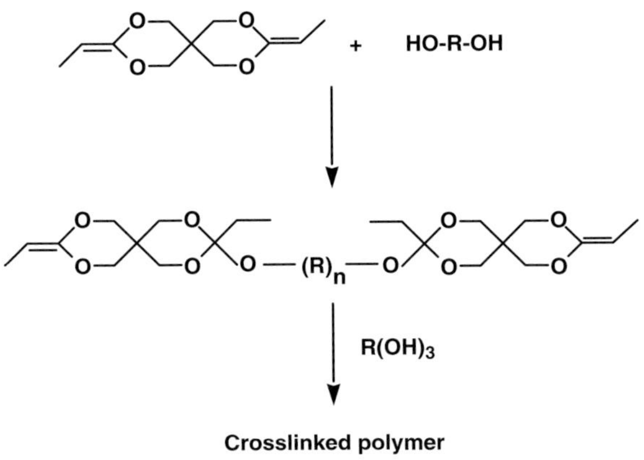

Scheme 4.7 Synthesis of crosslinked POE II.

4.2.2 Drug Delivery

4.2.2.1 Development of Ivermectin Containing Strands to Prevent Heartworm Infestation in Dogs

The most extensive investigation of POE II for drug delivery was carried out at the former Interx Laboratories of Merck (Kansas City, MO). In this application, a crosslinked polymer was used.

A crosslinked POE II can be prepared as shown in Scheme 4.7 [13]. Briefly, a prepolymer of DETOSU and a diol is prepared so that the prepolymer has ketene acetal end-groups. This prepolymer is then reacted with a triol, or polyols having a functionality greater than 2 to form a crosslinked network.

In this particular instance, the objective was to develop an ivermectin device capable of preventing heartworm infestation in dogs for at least 6 months [14]. Since ivermectin is not stable at 140–155 °C, extrusion of strands was not a viable method so that a device based on a crosslinked POE II was developed. Ivermectin has three hydroxyl groups and can thus compete with the crosslinker, 1,2,6-hexanetriol, for the ketene acetal end-groups. Consequently, in the final device, ivermectin is chemically bound to the matrix.

4.2.2.2 Experimental Procedure

The poly(ortho ester) matrix was prepared by a two-step procedure involving the preparation of a low molecular weight prepolymer followed by a crosslinking reaction. HD (3.72 g, 31.5 mmol) was dissolved in 20 mL of freshly distilled (from sodium) THF. DETOSU (10.03 g, 47.3 mmol) in a 50 mol% excess over HD was added via an oven-dried syringe. The mixture was refluxed 1 h under nitrogen to form a ketene acetal end-capped prepolymer. The THF was removed at room temperature under reduced pressure (ca. 4 mmHg). An aliquot (3.122 g) of the prepolymer was triturated with 0.151 g of magnesium hydroxide (hydrolytic stabilizer) and 0.329 g of ivermectin. The crosslinking agent, *n*-hexane-1,2,6-triol (0.289 g), was mixed with the composite and quickly extruded into 1/32″ ID FEP tubing (Cole-Parmer) and cured at 70 °C for 16 h. The tubing was cut and removed to yield highly flexible elastomeric matrices which were cut to length. Drug-free matrices were prepared in similar fashion.

4.2.2.3 Results

The behavior of the strands was investigated in dogs and the rate of ivermectin release was estimated from an implant retrieval study since plasma levels were below assay detection limits. The *in vivo* release rate was approximately 38 µg/month/cm of device. A correlation of the amount of drug remaining in the device with the amount of residual polymer suggested that erosion was a major determinant in the release of ivermectin, as would be expected for a system where ivermectin is chemically bound to the polymer.

On the basis of the data obtained, it was concluded that the crosslinked strands are capable of providing canine heartworm profilaxis for more than 6 months [14]. Unfortunately, it was not possible to develop devices that would erode in a predictable and reproducible fashion so that this system was never commercialized.

4.3 POE IV

POE IV was developed to overcome difficulties in controlling the rate of erosion of POE II and to make it more generally useful.

4.3.1 Polymer Synthesis

POE IV is prepared by the reaction between the diketene acetal DETOSU, a diol, or mixture of diols and a latent acid diol, as shown in Scheme 4.8 [15]. An alternate diketene acetal described under Section 4.2.1 and a diol can also be used.

4.3.1.1 Typical Polymer Synthesis Procedure

In a dry-box, 2.163 g (15 mmol) of *trans*-CDM, 4.727 g (40 mmol) of HD, 6.007 g (40 mmol) of TEG, and 1.041 g (5 mmol) of the triethylene glycol glycolide (TEG-GL) were dissolved in 40 g of THF. Then, 21.437 g (101 mmol) of 3,9-diethylidene-2,4,8,10-tetraoxaspiro [5.5] undecane were weighed into a round-bottom flask and added to the diols solution with the aid of 20 g THF, in several portions. The flask was removed from the dry-box, rapidly connected to a condenser and nitrogen inlet, and a few drops of *p*-toluenesulfonic acid solution (10 mg/mL) added. After the exotherm subsided, the solution was slowly poured into 3 L of methanol, containing about 1000 ppm of triethylamine. After isolation by filtration and drying in a vacuum oven at 40 °C for 24 h, the yield was 32.1 g (89.7%).

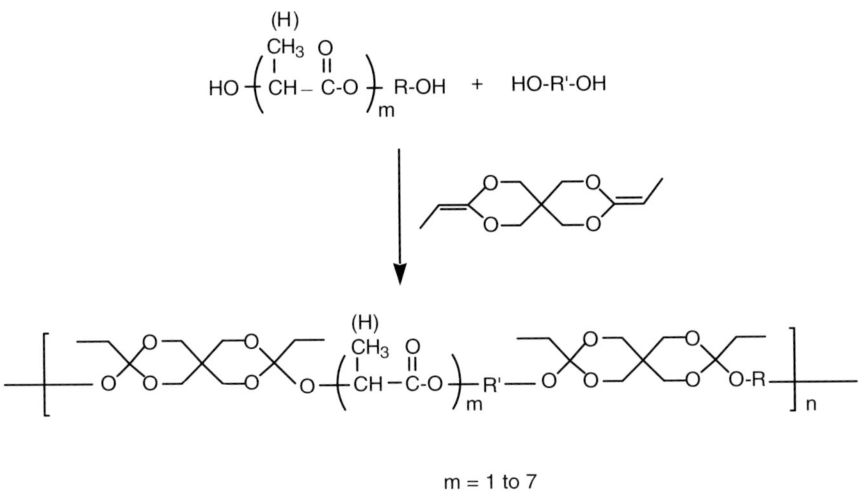

Scheme 4.8 Synthesis of POE IV.

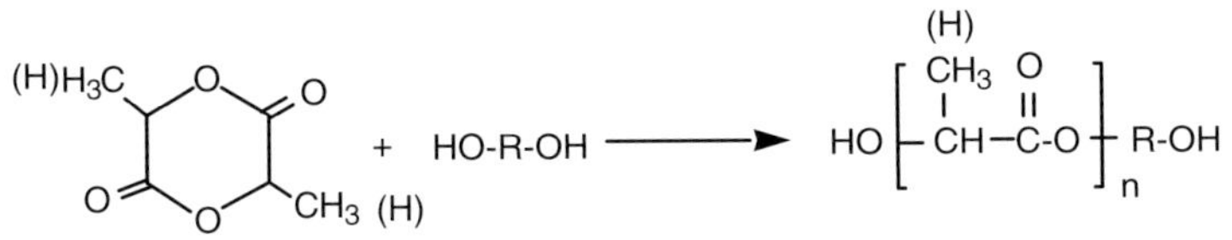

Scheme 4.9 Synthesis of latent acid based on lactide and a diol.

4.3.1.2 **Latent Acid**

The latent acid is prepared by an uncatalyzed high-temperature reaction between a diol and ether lactide, or glycolide as shown in Scheme 4.9 [16]. Mainly due to transesterification reactions, a mixture of products is obtained, as shown in Figure 4.2 [17]. The exact structure of the latent acid is not important and it is the total concentration of the α-hydroxy acid segments in the polymer that controls erosion rate.

4.3.1.3 **Experimental Procedure**

Into a round-bottom flask sealed with a rubber septum, 7.25 g (50 mmol) of DL-lactide and 8.713 g (50 mmol) of 1,10-decanediol were introduced under an argon atmosphere. The mixture was vigorously stirred for 3 days at 160 °C. The viscous diol–lactide was used without further purification.

4.3.2
Mechanical Properties

The ability to use diols having different structures allows the preparation of polymers having an extraordinarily broad range of physical properties, and materials

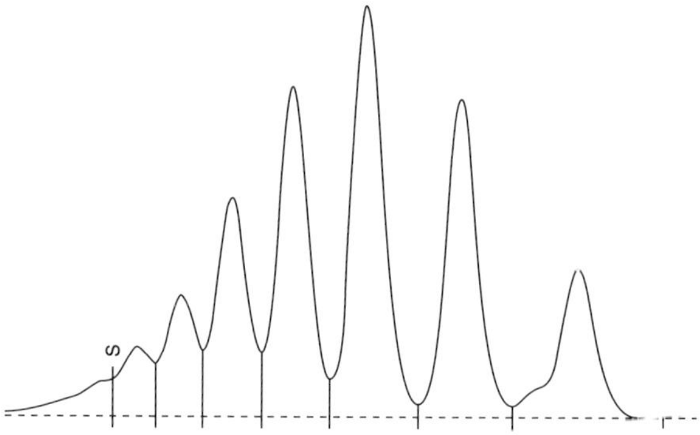

Figure 4.2 Gel permeation chromatograph of reaction products between lactide and TEG. Reprinted from [17], p. 1022, with permission from Elsevier.

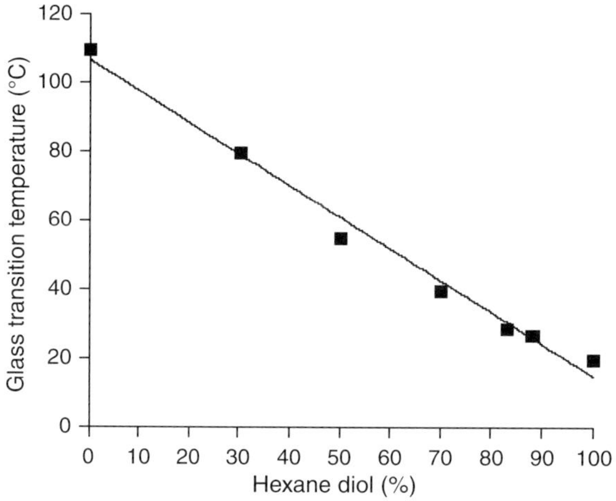

Figure 4.3 Glass transition temperature of 3,9-diethylidene-2,4,8,10-tetraoxaspiro [5.5] undecane, *trans*-CDM, HD polymer as a function of mol% HD.

ranging from hard, solid materials to viscous ointment-like materials can be prepared.

One of the more useful methods of achieving control over mechanical properties is to use a mixture of a rigid diol, for example, *trans*-CDM, and a flexible diol, for example, HD. When the glass transition temperature is determined for mixtures ranging from pure rigid diol to pure flexible diol, the plot shown in Figure 4.3 is obtained [18]. When linear, aliphatic diols having varying number of methylene groups are used, the plot shown in Figure 4.4 is obtained [19].

Figures 4.3 and 4.4 have been generated with POE II that has no latent acid in the polymer backbone. When POE IV is used, the latent acid in the polymer backbone does have a significant effect on the glass transition temperature, as shown

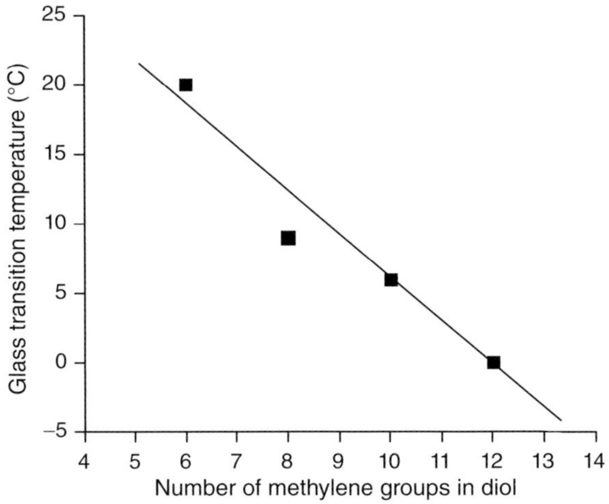

Figure 4.4 Effect of diol chain length on the glass transition temperature of polymers prepared from 3,9-diethylidene-2,4,8,10-tetraoxaspiro [5.5] undecane and α,ω-diols.Reprinted from [19], p. 47, with permission from Harwood Academic Publishers.

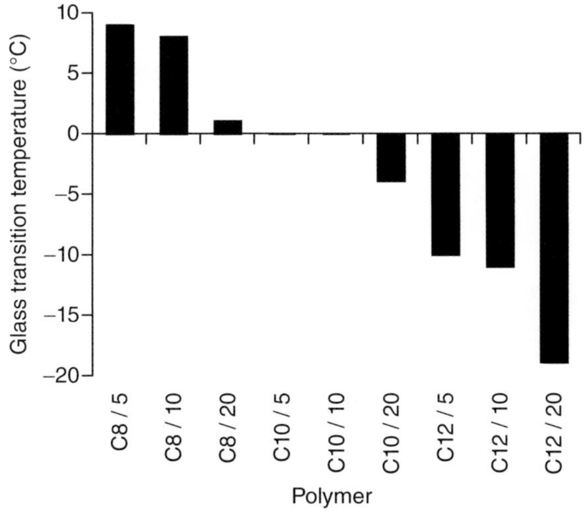

Figure 4.5 Glass transition temperatures for poly(ortho ester) prepared from 3,9-diethylidene-2,4,8,10-tetraoxaspiro [5.5] undecane and *n*-octanediol, *n*-decanediol, and *n*-dodecanediol, each with 5, 10, and 20 mol% of the corresponding lactide. Reprinted from [17], p. 1025, with permission from Elsevier.

in Figure 4.5 [17]. Thus, both the diol structures, the latent acid diol structure and their ratios must be considered when designing polymers having desired thermal and mechanical characteristics.

The ability to vary mechanical properties by proper choice of diols allows the synthesis of a wide range of materials, but the two most useful ones are solid polymers and gel-like materials.

4.4 Solid Polymers

4.4.1 Fabrication

A successful POE drug delivery system requires the development of suitable fabrication methods that can produce devices able to achieve the desired drug release profiles. Desired release profiles that are free from drug burst and are reasonably linear can best be achieved with devices that are fabricated to minimize internal porosity, and where the drug is uniformly dispersed in the matrix with minimal particle-to-particle contact.

There are many different types of solid devices used in controlled drug delivery. The two most often used are microspheres and strands prepared by an extrusion process. Of these, strands prepared by extrusion have a number of significant advantages. Dominant among these is the ability to fabricate devices without the use of solvents, and the ability to prepare dense devices with drugs that are uniformly dispersed along the length of the strand.

Extrusion requires the use of moderately elevated temperatures and a typical small-scale extrusion requires about 20–30 min. For this reason, it was of interest to investigate potential changes in molecular weight as a function of time by sectioning the entire extruded strand into 10 mm pieces and determining the molecular weight of selected pieces. Results of one such study are shown in Figure 4.6 [20].

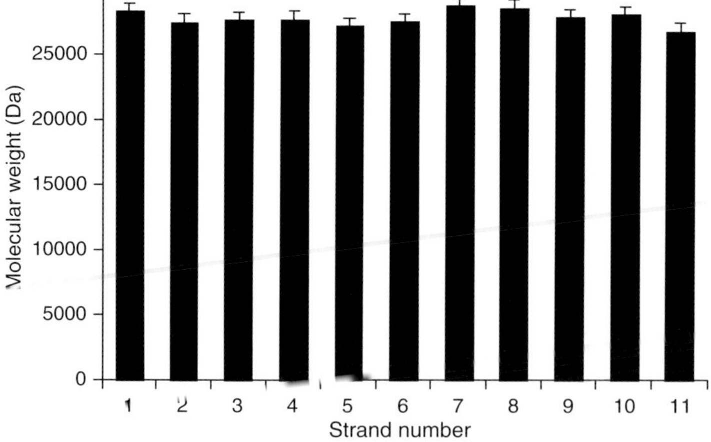

Figure 4.6 Molecular weight of each segment of an entire extruded strand cut into 10×1 cm sections along the entire length of the strand. Polymer prepared from 3,9 diethylidene-2,4,8,10-tetraoxaspiro [5.5] undecane, *cis/trans*-CDM, TEG, 1,10-decanediol, and TEG-GL (100/40/10/49.9/0.1). Reprinted from [20], p. 98, with permission from CRC Press.

In this particular case, there was no significant change in molecular weight along the length of the extruded strand despite a long exposure to 95 °C, the extrusion temperature. Based on these studies, POE IV is found to be suitable for the preparation of extruded strands, provided that the temperature does not exceed about 100 °C.

Because POE II and IV are readily soluble in solvents such as methylene chloride, ethyl acetate, or THF, microspheres can be easily prepared using conventional procedures.

4.4.2 Polymer Storage Stability

As shown in Figure 4.7, poly(ortho esters) have excellent stability and are stable at room temperature, when stored under anhydrous conditions [17]. The particular polymer used in this stability study was a hydrophilic polymer containing 40 mol% latent acid that had been ground to produce microparticles thus greatly increasing surface area. This is a very rapidly eroding polymer that will completely erode in a matter of a few days if placed in an aqueous buffer. Despite this high reactivity, when stored under anhydrous conditions, it is stable for a number of months.

4.4.3 Polymer Sterilization

The polymer is also relatively stable when sterilized by irradiation [17]. As shown in Figure 4.8, there is a decrease in molecular weight after irradiation, but the

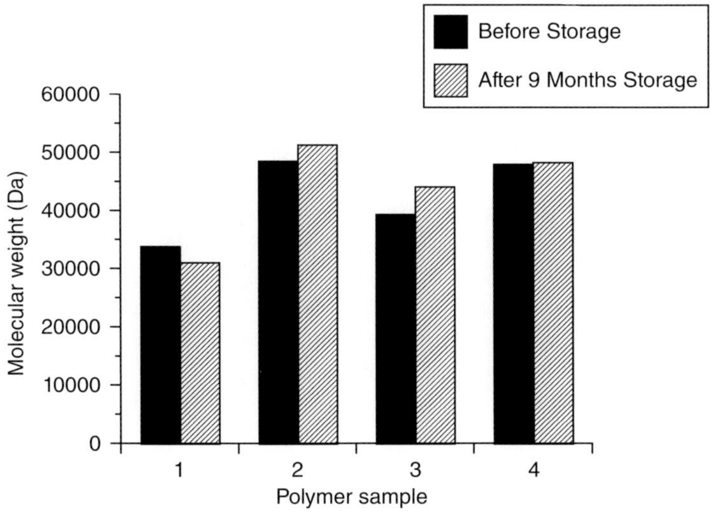

Figure 4.7 Stability of a polymer prepared from 3,9 diethylidene-2,4,8,10-tetraoxaspiro [5.5] undecane, *cis/trans*-CDM, TEG, and TEG-GL (100/35/25/40) stored at room temperature and under anhydrous conditions. Reprinted from [17], p. 1026, with permission from Elsevier.

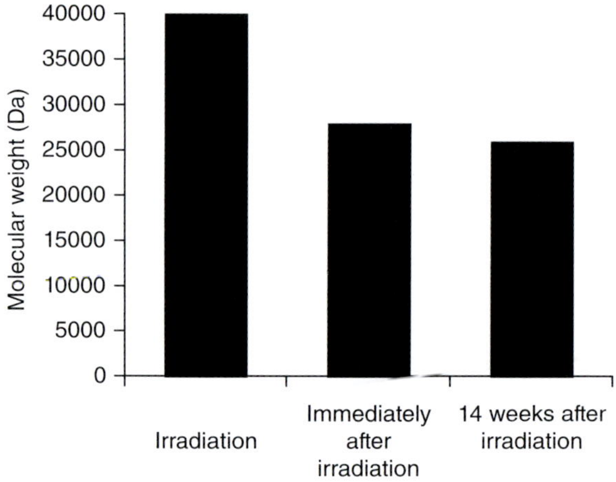

Figure 4.8 Effect of β-irradiation at 24 kGy on a poly(ortho ester) prepared from 3,9 diethylidene-2,4,8,10-tetraoxaspiro [5.5] undecane, *trans*-CDM, HD, TEG, and TEG-GL (15/40/40/5). Polymer stored postirradiation at 5 °C in a dessicator. Reprinted from [17], p. 1026, with permission from Elsevier.

decrease is of the same order of magnitude as that observed with other bioerodible polymers. Because irradiation generates free radicals that in a solid matrix can be long-lived, stability studies were extended to 3 months to determine if postirradiattion chain cleavage takes place. Thus, the polymer is stable after the initial drop in molecular weight and electron paramagnetic resonance (EPR) studies have shown that free radicals dissipate in less than 24 h.

4.4.4 Polymer Hydrolysis

The hydrolysis proceeds in three consecutive steps [21]. In the first step, the low short latent acid segment, either glycolic acid or lactic acid, in the polymer backbone hydrolyzes to generate a polymer fragment containing a carboxylic acid end-group that will catalyze ortho ester hydrolysis. A second cleavage produces free glycolic, or lactic acid that also catalyzes hydrolysis of the ortho ester links. Further hydrolysis of the polymer then proceeds to first generate the diol, or mixture of diols used in the synthesis and pentaerythritol dipropionate, followed by ester hydrolysis to produce pentaerythritol and propionic acid.

Scheme 4.10 shows details of polymer hydrolysis [21]. In this particular case, and for simplicity sake, we have depicted the latent acid as a dimer of lactic acid.

The most significant finding of the hydrolysis study is the linearity of weight loss and the concomitant release of lactic and propionic acid as shown in Figure 4.9. While linear rate of weight loss alone does not necessarily indicate surface erosion [22], the concomitant linear weight loss and release of lactic acid argues

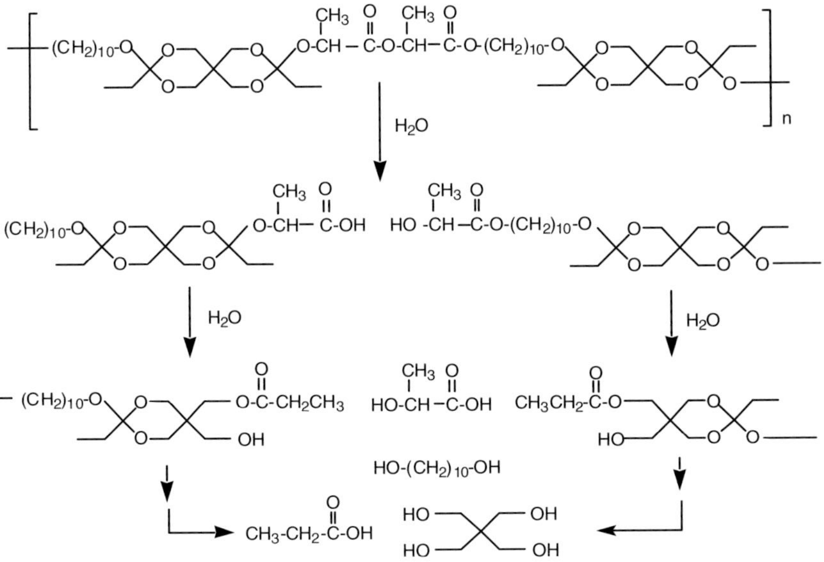

Scheme 4.10 Details of polymer hydrolysis. For simplicity, the latent acid has been depicted as a dimer of lactic acid.

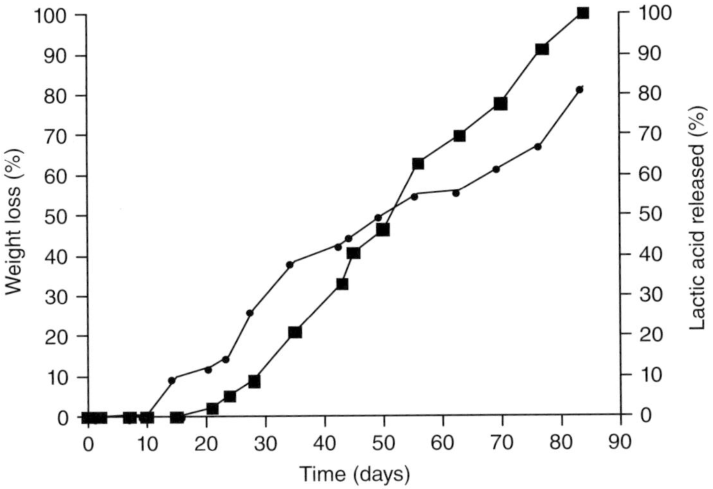

Figure 4.9 The relationship between lactic acid release (J) and weight loss (B) for a poly(ortho ester) prepared from 3,9 diethylidene-2,4,8,10-tetraoxaspiro [5.5] undecane, and a 100/70/30 mixture of 1,10-decanediol and 1,10-decanediol lactide; 0.13 M, pH 7.4 sodium phosphate buffer at 37 °C. Reprinted from [21], p. 304, with permission from American Chemical Society.

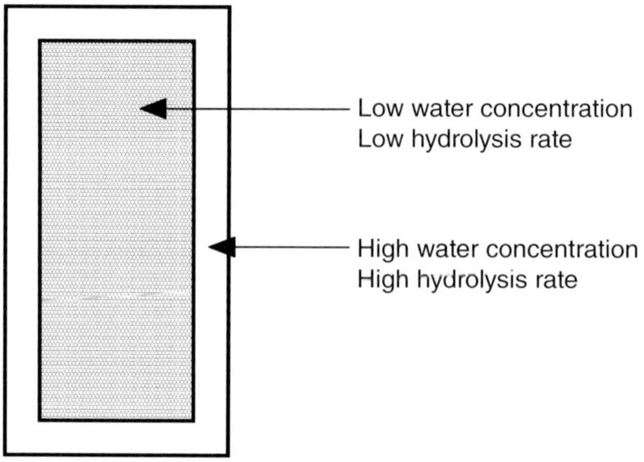

Figure 4.10 Schematic of proposed erosion mechanism.

convincingly for a process confined predominantly to the surface layers of the polymer matrix.

The erosion process is shown schematically in Figure 4.10 [20]. Surface erosion demands a much higher rate of hydrolysis in the surface layers of a solid device as compared to the interior of the device, and pure surface erosion can only take place if no hydrolysis occurs in the interior of the device. This can only take place if no water penetrates the polymer and since no polymer is so hydrophobic that no water can penetrate the matrix, some hydrolysis will always take place in the interior of the matrix. However, there is a significant difference in hydrolysis rates between surface and interior due to differences in water concentration. In the surface layers, the concentration of water is fairly high, and the rate of hydrolysis is also high. But there is a progressively lower concentration of water with deeper layers, so the rate of hydrolysis will also progressively decrease with end-result being that erosion is confined predominantly to the surface layers.

Implicit in the use of latent acid is an expectation that an increased amount of latent acid in the polymer backbone would translate into increased rate of polymer erosion. That this is actually the case can be seen in Figure 4.11 where the latent acid content was varied from 5 to 0.1 mol% [23]. Clearly, there is a correlation between latent acid content and polymer erosion rate.

An erosion process that is confined predominantly to the surface layers has a number of important benefits. First, if the drug is well immobilized in the matrix, its release is controlled by polymer erosion so that an ability to control polymer erosion translates into an ability to control rate of drug release. Second, because drug release is controlled by polymer erosion, drug release and polymer erosion take place concomitantly and when drug release has been completed, no polymer remains. Third, because most of the hydrolysis occurs in the outer layers of the device, acidic hydrolysis products can diffuse away from the device and do not accumulate in the bulk material. Thus, the interior of the matrix does not become

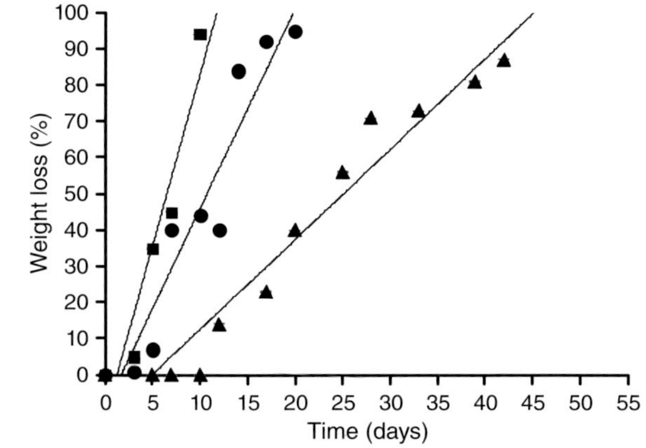

Figure 4.11 Effect of latent acid content on erosion rates for a polymer prepared from 3,9-diethylidene-2,4,8,10-tetraoxaspiro [5.5] undecane, CDM, decanediol, TEG, and TEG-GL. (B) 40/45/10/5; (J) 40/49/10/1; (H) 40/49.9/10/0.1.Reprinted from [23], p. 1629, with permission from American Chemical Society.

highly acidic, as is the case of poly(lactide-*co*-glycolide) copolymers, or poly(lactic acid) [24], and acid-sensitive drugs can be released without loss of activity.

4.4.5
Drug Delivery

A great many studies have been carried out and only the more important ones will be described here.

4.4.5.1 Release of Bovine Serum Albumin from Extruded Strands

As discussed in Section 4.1., POEs can be readily extruded and since the glass transition temperature can be adjusted to any desired value, extrusion temperatures can be tailored for the protein of interest. Of particular interest is a procedure by which finely ground polymer and a micronized protein are intimately mixed and then extruded into thin strands at temperatures low enough so that protein activity is not compromised.

Figure 4.12 shows release of FITC-BSA from extruded strands and the weight loss of the strands as a function of time [25]. Three features are notable. First, there is only a minimal burst despite the fact that 15 wt% of a water-soluble material has been incorporated. Second, there is a significant lag before release of FITC-BSA begins. And third, release is linear and concomitant with weight loss.

While a long induction period may be desirable in some applications, for example, in vaccine delivery, for general protein delivery it is not desirable and an investigation to eliminate the induction period was carried out. One means of accomplishing this is to use an AB block copolymer of poly(ortho ester) and polyethylene glycol. When such a block was used, BSA release kinetics shown in Figure 4.13 was obtained [25].

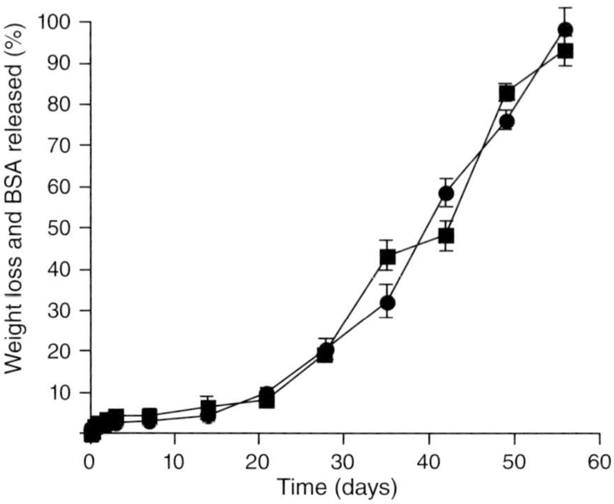

Figure 4.12 Release of FITC-BSA (H) and weight loss (B) from a poly(ortho ester) prepared from 3,9-diethylidene-2,4,8,10-tetraoxaspiro [5.5] undecane, 1,4-pentanediol, and HD glycolide (100/95/5). Strands, 1 × 10 mm, extruded at 70 °C. 0.01 M PBS, pH 7.4, 37 °C. FITC-BSA loading 15 wt%. Reprinted from [25], p. 34, with permission from Elsevier.

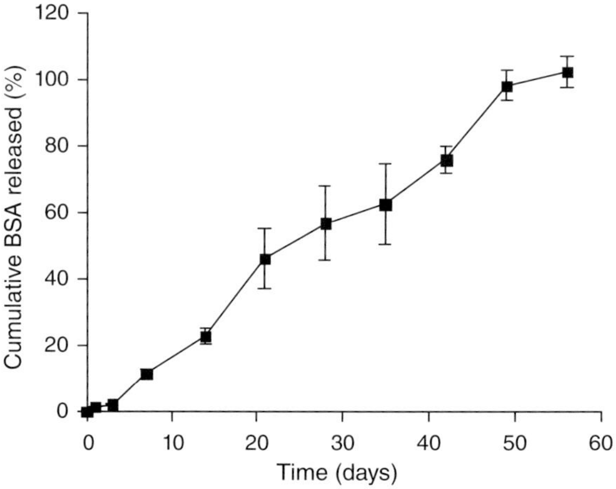

Figure 4.13 Release of FITC-BSA from an AB block copolymer containing 6 wt% 2 kDa polyethylene glycol. Poly(ortho ester) prepared from 3,9-diethylidene-2,4,8,10-tetraoxaspiro [5.5] undecane, 1,3-propanediol, and TEG-GL (100/85/15). Strands, 1 × 10 mm, extruded at 70 °C. 0.01 M PBS, pH 7.4, 37 °C. FITC-BSA loading 15 wt%. Reprinted from [25], p. 36, with permission from Elsevier.

Clearly, this represents a significant improvement and demonstrates the potential of AB, or ABA block copolymers of POE and polyethylene glycol as matrices for the controlled release of proteins. However, this potential has not yet been fully exploited. The AB block copolymer was prepared as shown in Scheme 4.11 [25].

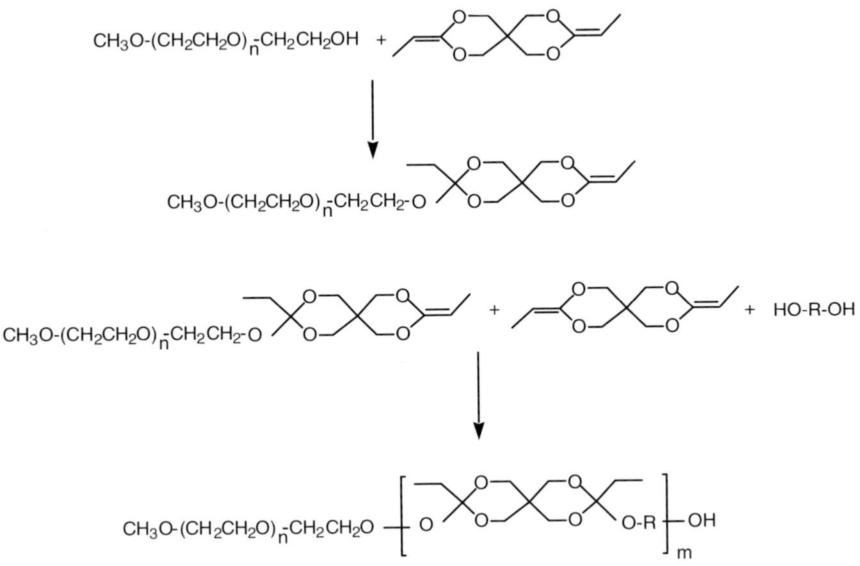

Scheme 4.11 Synthesis of AB block copolymer.

4.4.5.2 Experimental Procedure

Under anhydrous condition, a mixture of 3,9-diethylidene-2,4,8,10-tetraoxaspiro [5.5] undecane (21.225 g, 100 mmol) and the monomethyl ether of a 2-kDa polyethylene glycol (2 g, 1 mmol) was dissolved in 50 mL of THF. To the solution was added 0.05 mL of a *p*-toluenesulfonic acid solution (20 mg/mL) in THF. The solution was stirred using a magnetic stirrer and warmed to about 50 °C. After 15 min, another 0.05 mL of a *p*-toluenesulfonic acid solution was added, and the reaction was allowed to proceed at 50 °C for an additional 30 min. Next, 1,3-propanediol (6.436 g, 84.5875 mmol) and 3,9-diethylidene-2,4,8,10-tetraoxaspiro [5.5] undecane (3.108 g, 11.684 mmol) were added to the solution with the aid of 20 mL THF. After the addition of another portion of 0.05 mL *p*-toluenesulfonic acid solution, the reaction was stirred for an additional 1 h. The reaction mixture was then added to 600 mL hexane and the precipitated polymer was collected and dried overnight in a vacuum oven at about 40 °C.

4.4.6 Delivery of DNA Plasmid

The delivery of DNA from poly(ortho esters) is of particular interest because an erosion-controlled release as well as an essentially neutral pH in the interior of the matrix have been demonstrated. Thus, the incorporation of DNA into the polymer and its subsequent release should not adversely affect DNA integrity.

Microspheres of 5 μm were prepared by a double emulsion method [26]. When the microspheres were placed in a pH 7.2 buffer and release of DNA

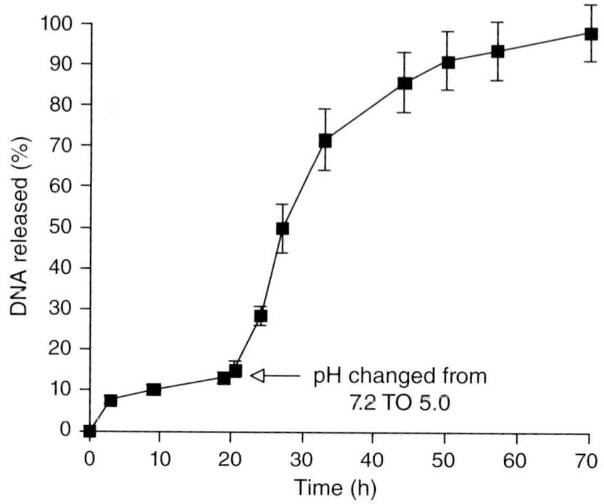

Figure 4.14 Release of DNA from a poly(ortho ester) prepared from 3,9-diethylidene-2,4,8,10-tetraoxaspiro [5.5] undecane, TEG, 1,2-propane diol, and TEG-GL (100/35/15/45/5). Microspheres, 5 μm, phosphate buffer (pH 7.4) or sodium acetate buffer (pH 5.0) at 37 °C.

followed using the pico green method, release was slow, corresponding to slow polymer erosion, as shown in Figure 4.14 [27]. However, when the pH of the buffer was changed to 5.0, the pH of endosomes, an immediate and significant acceleration of DNA release was noted. Since a lowered pH results in an increased erosion rate, this plot provides unequivocal evidence of an erosion-controlled DNA release.

4.4.6.1 DNA Plasmid Stability

As analyzed by gel electophoresis, DNA plasmid retained its active conformation (supercoiled and relaxed) when released from the POE microspheres placed into a pH 7.4 buffer and also that remained in the microspheres. However, when the microspheres were placed into a pH 5.0 buffer, significant damage to the DNA plasmid was noted. This is consistent with the known acid sensitivity of DNA. The DNA plasmid remaining in the microspheres dispersed in the pH 5.0 buffer retained its active conformation.

This is a significant finding indicating that the internal pH within the poly(ortho ester) matrix must be above pH 5.0 and that the microspheres are able to protect the DNA plasmid from a low pH external environment.

4.4.6.2 Microencapsulation Procedure

Microspheres were prepared by a modified water-in-oil-water double emulsion, solvent evaporation procedure. The two phases consisting of 250 μL of DNA solution (250 mg of DNA) and 7 mL of methylene chloride containing 200 μg POE were emulsified by sonication for 10 s at room temperature. The primary emulsion temperature was then lowered below the freezing point of the aqueous inner phase by liquid nitrogen immersion, and 50 mL of a 5% PVA solution (4–7 °C) was added and homogenized at 5000–9000 rpm for 14 s. After homogenizing, the resulting emulsion was diluted in 100 mL of 1% PVA and the system stirred magnetically for 3 h to allow for evaporation of the organic solvent. Microspheres were finally collected by centrifugation and washed three times with water to remove excess PVA. All PVA solutions were adjusted to the osmotic pressure of the inner aqueous phase using agents such as saccharides. The microspheres were resuspended in approximately 1 mL of water, frozen in liquid nitrogen and lyophilized at room temperature for 24 h.

4.4.7 Delivery of 5-Fluorouracil

Unlike BSA or DNA that are high molecular weight, water-soluble molecules, 5-fluorouracil (5FU) is a small water-soluble molecule. Therefore, there is the potential for significant diffusion from the polymer. However, as shown in Figure 4.15, when 5FU release from thin wafers and weight loss of the wafers was determined, within experimental error, both processes occurred concomitantly suggesting that the dominant drug release mechanism was polymer erosion [28].

This is an encouraging result and indicates that a wide range of therapeutic agents can be delivered from poly(ortho esters), as has been validated in numerous studies.

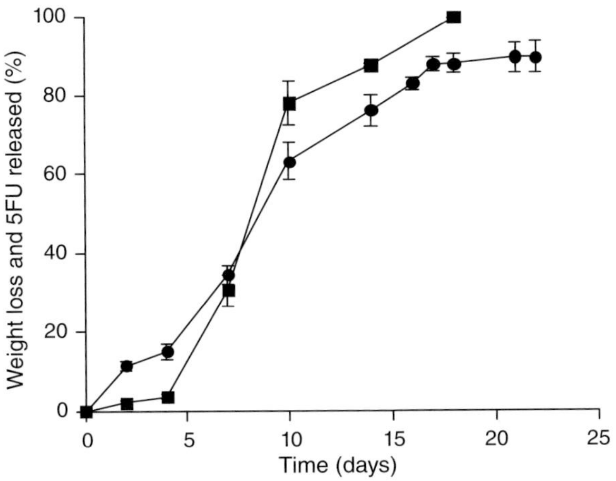

Figure 4.15 Polymer weight loss (H) and 5-fluorouracil (5-FU) release (B) from a polymer prepared from 3,9-diethylidene-2,4,8,10-tetraoxaspiro [5.5] undecane, 1,3-propanediol, and TEG-GL (90/10). Drug loading 20 wt%. 0.05 M phosphate buffer, pH 7.4, 37 °C. Reprinted from [28], p. 126, with permission from Elsevier.

4.5 Gel-Like Materials

To prepare gel-like materials, it is necessary to use highly flexible diols and their viscosity must be limited by having molecular weights no higher than about 6 kDa. To limit toxicology studies required for regulatory approval, only two diols, TEG and 1,10-decanediol, were used. Polymers based on TEG produce hydrophilic materials, while polymers based on 1,10-decanediol produce hydrophobic materials [29].

The most significant advantage of gel-like materials is the ability to incorporate therapeutic agents at ambient temperature and without the use of solvents by a simple mixing procedure. Mixing can be accomplished on a small scale by using a mortar and pestle, but on a somewhat larger scale it is better carried out using a three roll mill [19].

4.5.1 Polymer Molecular Weight Control

Polymer molecular weight control can be achieved by using an excess of diol relative to the diketene acetal, or by using a chain-stopper. When a chain-stopper is used, a calculated amount of a monofunctional alcohol is used [30]. As shown in Scheme 4.12, when *n*-decanol is used as a chain-stopper in combination with 1,10-decanediol, both polymer ends have *n*-decanol residues so that a chain-stopped material is somewhat more hydrophobic relative to a stoichiometry-controlled material that has terminal hydroxyl groups.

The use of chain-stoppers allows excellent and reproducible molecular weight control by varying the ratios of 1,10-decanediol to *n*-decanol as shown in Figure 4.16 [30]. The existence of terminal methyl groups has been established by ^{1}H NMR studies [30].

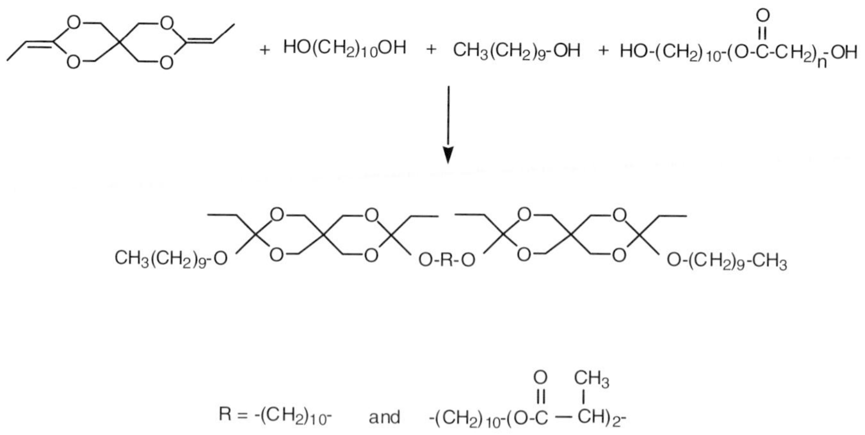

Scheme 4.12 Use of a monofunctional alcohol as chain stopper to control molecular weight.

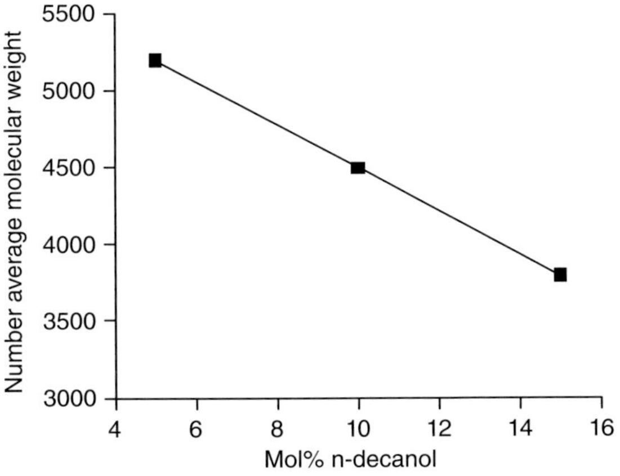

Figure 4.16 Effect of *n*-decanol on the molecular weight of a poly(ortho ester) prepared from 3,9-diethylidene-2,4,8,10-tetraoxaspiro [5.5] undecane, 1,10-decanediol and 1,10-decanediol lactide (100/70/30).

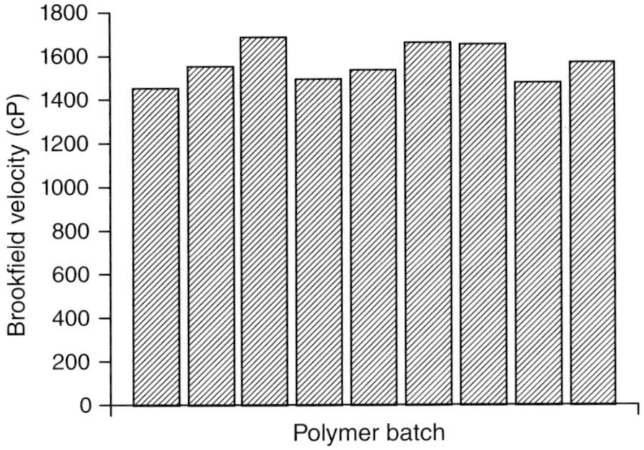

Figure 4.17 Variation in Brookfield viscosity for nine typical preparations at 25 °C. Injectable formulation prepared from 3,9-diethylidene-2,4,8,10-tetraoxaspiro[5.5] undecane, TEG, and TEG-GL (60/50/50). Formulation contains 20 wt% monomethoxy polyethylene glycol, molecular weight 550. Reprinted from [31], p. 4399, with permission from Elsevier.

Because the principal means of administration of gel-like materials is by injection, preparation of materials having reproducible viscosities is important. Synthesis reproducibility as measured by Brookfield viscosity for a number of different preparations is shown in Figure 4.17 [31]. Clearly, the synthesis is sufficiently reproducible to assure that the same viscosity materials can be repeatedly prepared.

4.5.2
Polymer Stability

Figure 4.18 shows changes in molecular weight of a gel-like polymer after storage at room temperature under anhydrous conditions for 9 months [31]. As with the solid polymers, within experimental error, there is no change.

Figure 4.19 shows the effect of irradiating the polymer at a dose of 22.9 to 25.6 kGy [31]. Within experimental error, no changes in molecular weight could be detected. This is consistent with previous finding for POE III and indicates that low molecular weight polymers, unlike their high molecular weight analogs, do not significantly change molecular weight on irradiation.

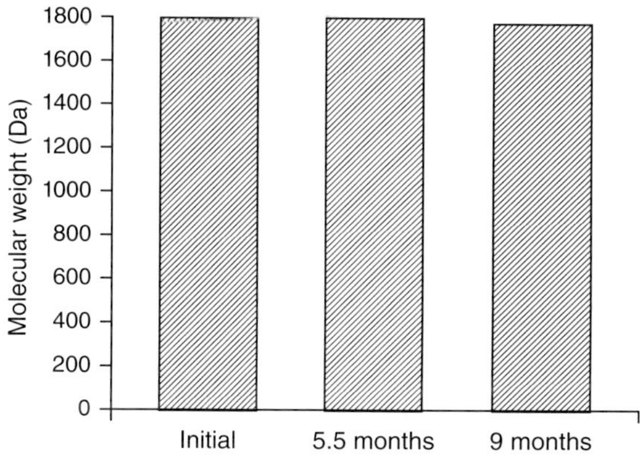

Figure 4.18 Room temperature stability for a polymer prepared from 3,9-diethylidene-2,4,8,10-tetraoxaspiro[5.5]undecane, TEG, and TEG-GL (60/50/50). Material stored under anhydrous conditions. Reprinted from [31], p. 4399, with permission from Elsevier.

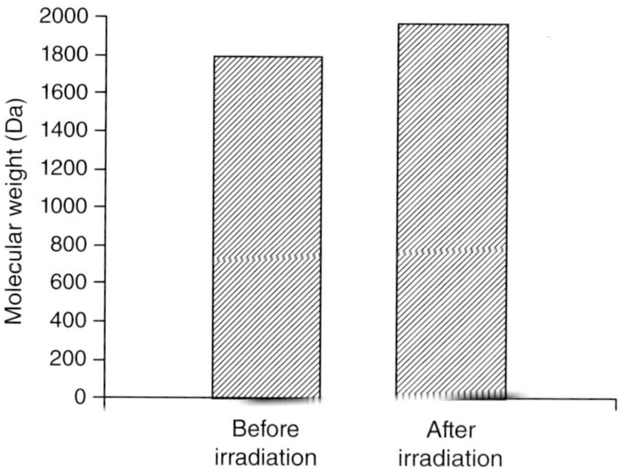

Figure 4.19 Stability of a polymer prepared from 3,9-diethylidene-2,4,8,10-tetraoxaspiro[5.5] undecane, TEG, and TEG-GL (60/50/50) irradiated at 24 kGy. Reprinted from [31], p. 4400, with permission from Elsevier.

4.5.3 Drug Delivery

Gel-like materials based on POE IV constitute AP Pharma's "Biochronomer" delivery technology, and a Phase III clinical trial for the delivery of granisetron, an established 5-HT_3 receptor antagonist, to prevent chemotherapy-induced nausea and vomiting (CINV) has just been completed. In addition, a Phase II clinical trial to treat postoperative pain using the analgesic agent mepivacaine has also been completed.

4.5.3.1 Development of APF 112 Mepivacaine Delivery System

Following surgery, currently used local anesthetics using a simple injection are only effective for a few hours. An important advance would be the development of a system that would result in the sustained delivery of a local anesthetic for a few days thus reducing the need for opiate use with their well-know side-effects. Further, if the delivery system is placed within the surgical incision, it should be possible to maintain a high local concentration without a concomitant high systemic concentration. This is important in view of the toxicity that mepivacaine, the local anesthetic used, shares with other amide local anesthetics [32].

4.5.3.2 Formulation Used

The structure of the gel-like material used is shown in Scheme 4.13. In order to improve injectability and ease of handling, the molecular weight of the polymer was limited to about 6 kDa and methoxy polyethylene glycol having a molecular weight of 550 Da was used as an excipient.

The actual composition of the clinical formulation designated as APF 112 was 77.6 wt% polymer, 19.4 wt% methoxy polyethylene glycol, and 3 wt% mepivacaine.

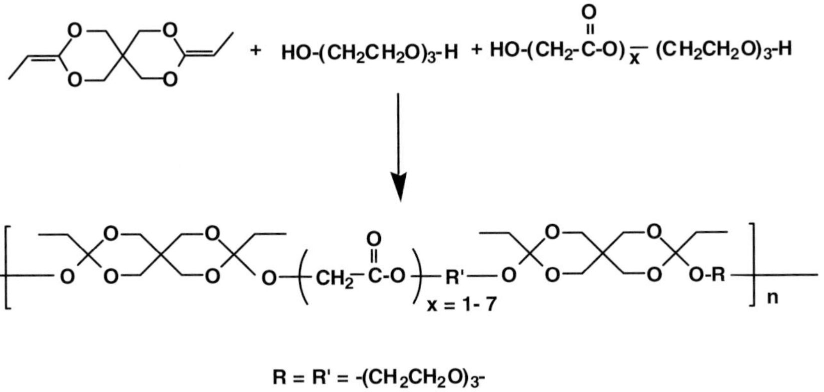

Scheme 4.13 Structure of AP 530 used in clinical trials.

4.5.4
Preclinical Toxicology

Two types of studies were carried out. In one study, the polymer was hydrolyzed and the hydrolysate tested and the other study utilized the actual formulation [33].

4.5.4.1 Polymer Hydrolysate

Hydrolyzing the polymer into its hydrolysis products simulates the instantaneous erosion of an implant and thus represents a worst case scenario.

The hydrolysate was prepared by hydrolyzing the polymer in phosphate-buffered saline (PBS) at 80 °C for 24 h, adjusting the pH to 7.4 with NaOH, adding the methoxy polyethylene glycol, mixing thoroughly, adding deionized water to adjust osmolarity and finally filtering through a 0.45-μm filter. The solution was then injected subcutaneously into male and female Sprague-Dawley rats and into male and female beagle dogs. In the rat study, the doses used were 0, 1, 3, and 10 mL/kg and in beagle dogs, the dose was 0, 0.05, 0.1, and 0.2 mL/kg. Both animal species were observed for 14 days, and no adverse effects by clinical observation and gross necropsies were found. In addition, no histological evidence of systemic toxicity was observed in all organs evaluated.

4.5.4.2 Wound Instillation

The following incisional wound instillation study was carried out in rats. A 1-cm full-thickness incision was made, a subcutaneous pocket thus created by blunt dissection, the APF 112 formulation administered into the subcutaneous pocket, the skin closed with 4–0 nylon sutures, which were removed after 7 days.

The study was carried out using Sprague-Dawley male and female rats using 500 and 1000 μL in a single dose and the rats sacrificed at day 8. Both doses were well tolerated, but the 1000 μL dose resulted in some leakage and wound distension.

4.5.5
Phase II Clinical Trial

The objectives of this trial was to evaluate the safety and tolerability of APF 112 when administered into the surgical incision during inguinal hernia repair, a moderately to severely painful procedure. Results indicated excellent safety and tolerability, and pharmacokinetics showed sustained release of mepivacaine over 72 h [34].

However, due to an unexpectedly low level of pain displayed by the control group in the study, it was not possible to demonstrate that APF 112 is effective in controlling postsurgical pain.

4.5.6
Development of APF 530 Granisetron Delivery System

4.5.6.1 Preclinical Toxicology

Since APF 530 uses the same polymer as that used in APF 112, no polymer hydrolysate studies were needed.

4.5.6.2 Rat Study

Male and female Sprague-Dawley rats ($N = 20$/sex/group) were administered APF 530 as a single total subcutaneous dose of 0.25 or 1.0 mL/animal. The 1-mL dose was administered at four sites at 0.25 mL/site. For rats, a 0.25-mL dose/site was the maximum feasible dose for the polymer formulation based on leakage from the injection site. The total mass of granisetron administered in the APF 530 formulation was approximately 5 and 20 mg/animal. The 5-mg dose was approximately 14–19 and 21–28 mg/kg of granisetron in males and females, respectively. The 20-mg dose was approximately 57–77 and 85–113 mg/kg of granisetron in male and females, respectively.

Additional animals ($N = 20$/sex/group) were administered 1 mL/animal of saline control divided equally into four sites, or aqueous granisetron at an intravenous dose of 9 mL/kg, or an subcutaneous dose of 1 mL/animal (0.25 mL/site). Saline control and test formulations were administered through a 16-gage needle. Five rats/sex/group were sacrificed on days 4, 8, 15, and 29.

Administration of APF 530 was well tolerated both locally and systemically. Histopathological evaluation of the APF 530 injection sites revealed several reversible changes consistent with the injection of a biodegradable polymer. By day 29, the response to the polymer had resolved without any residual or untoward effects.

4.5.6.3 Dog Study

A study in beagle dogs was also conducted to further characterize the systemic and local toxicity profile of APF 530. Male and female beagle dogs ($N = 6$/sex/group) were administered APF 530 at a single total subcutaneous dose volume of 1.0 or 4 mL/animal. For beagle dogs, a 1-mL/site is the maximum feasible dose for the polymer formulation based on leakage from the injection site. For the 1-mL dose, two sites received 0.25 mL and one site received 0.5 mL. For the 4-mL dose volume, 1-mL was administered at four separate sites. The total mass dose of granisetron administered in the APF 530 formulation was approximately 20 and 80 mg/animal, or approximately 1.5–2.5 and 6–10 mg/kg of granisetron, respectively. Additional animals ($N = 6$/sex/group) were administered aqueous granisetron at an intravenous dose of 3 mL/kg or a subcutaneous dose of 4 mg/animal (1 mL/site), or 2.75 mL/animal of saline control divided into four sites (0.25 and 0.5 mL in one site. 1 mL in two sites). The total mass dose of aqueous granisetron administered subcutaneously translated to approximately 0.3–0.5 mg/kg. Saline control and test formulations were administered through a 16-gage needle.

Administration of APF 530 was well tolerated both locally and systemically. Histopathological evaluation of the APF 530 injection sites revealed several reversible changes consistent with the injection of a biodegradable polymer. All effects appeared to be resolving by day 15.

4.5.6.4 Phase II and Phase III Clinical Trials

Phase II and Phase III clinical trials have been completed.

In a Phase II clinical trial, the safety, tolerability, and pharmacokinetics in cancer patients were evaluated. In addition, efficacy end-points were evaluated relating to emetic events and the use of additional medication.

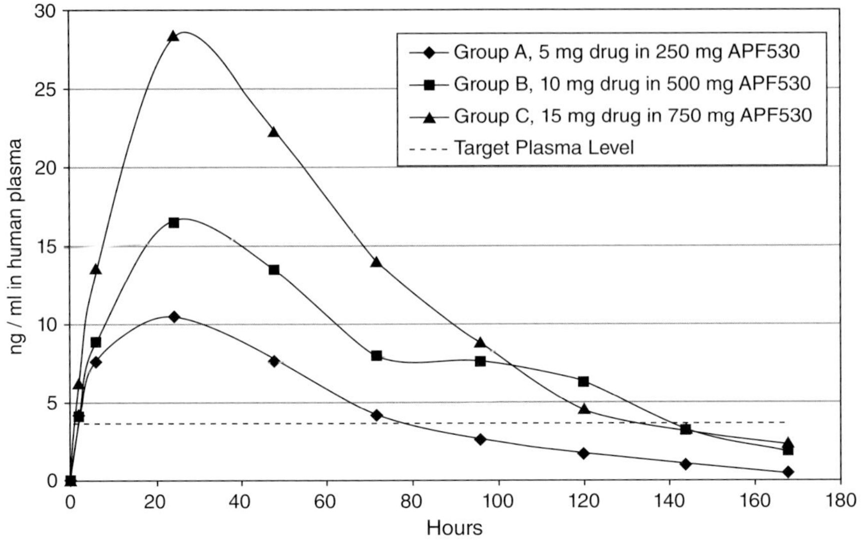

Figure 4.20 Granisetron plasma levels in patients from Phase II clinical trial of APF 530. From AP Pharma 2006 Annual Report.

A pharmacokinetic evaluation of three dose groups, 250, 500, and 750 mg injection doses corresponding to 5, 10, and 15 mg of granisetron, respectively, has demonstrated plasma levels of granisetron shown in Figure 4.20. On the basis of this study, a 10-mg dose was selected for a Phase III clinical trial [34].

A Phase III clinical trial compared APF 530 to Aloxi which contains the 5-HT_3 antagonist palonesetron, and is administered either as an intravenous single dose 30 min prior to chemotherapy, or as an oral dose 1 h prior to chemotherapy. The Phase IIII clinical trial involved 1395 patients in 103 centers. The trial assessed acute and delayed onset of CINV for highly, or moderately emetogenic chemotherapy. As of this writing, APF 530 demonstrated equivalence, but not superiority to Aloxi.

4.6
Polymers Based on an Alternate Diketene Acetal

Preparation of gel-like materials requires very flexible polymers. Because DETOSU is a very rigid molecule, an attempt was made to replace DETOSU with another, less rigid, diketene acetal. The structure of this diketene acetal was shown under Section 4.2.1.

Only very preliminary data are available. Preparation of this diketene acetal is shown in Scheme 4.14.

Scheme 4.15 shows glass transition temperatures for polymers prepared using the rigid diol, *trans*-CDM with both DETOSU and the more flexible alternate

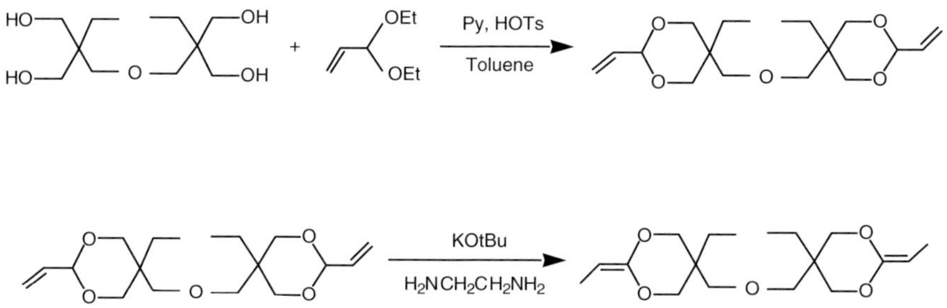

Scheme 4.14 Synthesis of di(5-methyl-2-ethylidene[1.3]dioxan-5yl)methyl ether.

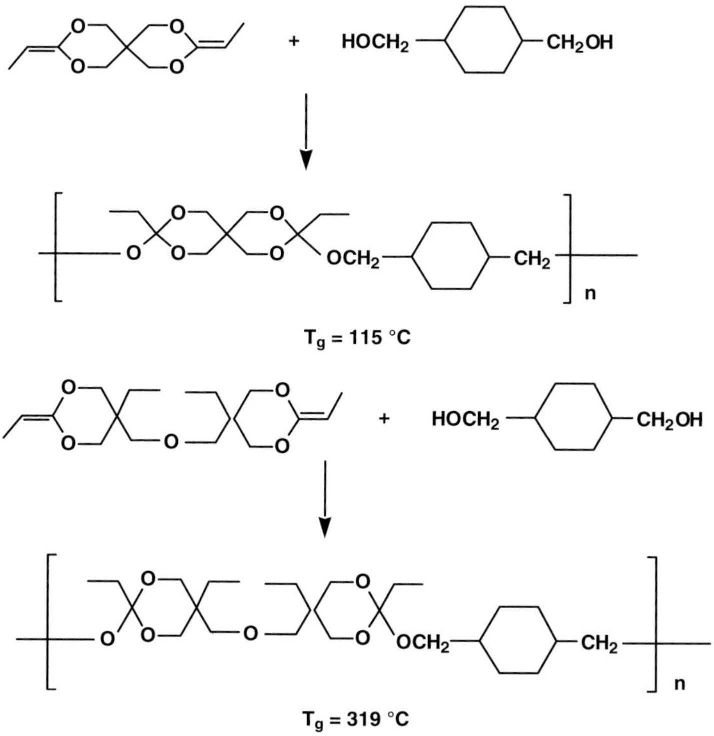

Scheme 4.15 Glass transition temperatures for polymers prepared from DETOSU and di(5-methyl-2-ethylidene[1.3]dioxan-5yl)methyl ether, each with *trans*-CDM.

diketene acetal. Since glass transition temperatures are a direct indication of chain flexibility, it is clear that polymers prepared using the alternate diketene acetal are significantly more flexible and that use of the alternate diketene acetal should produce useful materials at higher molecular weight than those based on DETOSU.

4.7 Conclusions

Poly(ortho ester) have been under development since 1970, and while it is a very well-understood system, its commercialization has been slow in coming. This was primarily due to the fact that much of the early poly(ortho ester) work was carried in an academic setting at the former Stanford Research Institute, now SRI International.

Beginning in 1985, serious attempts by the former Interx Laboratories of Merck to develop a 6-month ivermectin delivery implant based on POE II to prevent heart-worm infestation in dogs was initiated. However, even though desired ivermectin blood levels have been achieved for many months and in a clinical trial the formulation was 100% effective in preventing heart-worm infestations in dogs, it was not possible to prepare devices that had reproducible erosion times. This irreproducibility problem eventually doomed commercialization.

Beginning in 1994, the fourth family of poly(ortho ester), POE IV, was developed at Advanced Polymer Systems, now AP Pharma. This polymer system is currently under active development at AP Pharma for a number of applications, and a Phase II clinical trial using mepivacaine for postoperative pain control was completed. This trial demonstrated for the first time that a specific family of poly(ortho ester) has a benign toxicology and that it is a suitable system for use in humans.

Based on the benign toxicology of this particular family of poly(ortho esters), the development of a granisetron delivery system to control chemotherapy-induced nausea and vomiting was initiated. This system has recently completed a Phase III clinical trial and work preparatory to an NDA filing is underway.

In addition, a number of proprietary systems based on POE II are also under development.

Thus, after a very long induction period, a number of delivery systems based on poly(ortho esters) are on their way of becoming a commercial reality.

References

1 Choi, N.S. and Heller, J. (1978) US Patent 4, 093,709.

2 Choi, N.S. and Heller, J. (1978) US Patent 4, 131,648.

3 Choi, N.S. and Heller, J. (1979) US Patent 4, 138,344.

4 Choi, N.S. and Heller, J. (1979) US Patent 4, 180,646.

5 Heller, J., Ng, S.Y., Fritzinger, B.K., and Roskos, K.V. (1990) *Biomaterials*, **11**, 235–237.

6 Heller, J. (2005) *Adv. Drug Deliv. Revs.*, **57**, 2053–2062.

7 Heller, J., Penhale, D.W.H., Helwing, R.F., and Fritzinger, B.K. (1981) *Polym. Eng. Sci.*, **21**, 727–731.

8 Heller, J., Penhale, D.W.H., and Helwing, R.F. (1980) *J. Polym. Sci. Polym. Lett. Edn.*, **18**, 82–83.

9 Ng, S.Y., Penhale, D.W.H., and Heller, J. (1992) *Macromol. Synth.*, **11**, 23–26.

10 Helwing, R.F. (1983) US Patent 4, 523, 335.

11 Newsome, P.W., Frisbee, A.R., Itov, Z., Morgans, A.J., and Noe, R.A. (2002) US Patent 6, 863,782 B2.

12 Crivello, J.V., Malik, R., and Lai, Y.-L. (1996) *J. Polym. Sci. A Polym. Chem.*, **34**, 3091–3102.

13 Heller, J., Fritzinger, B.K., Ng, S.Y., and Penhale, D.W.H. (1985) *J. Control. Release*, **1**, 233–238.

14 Shih, C., Fix, J., and Seward, R.L. (1993) *J. Control. Release*, **25**, 155–162.

15 Ng, S.Y., Vandamme, T., Taylor, M.S., and Heller, J. (1997) *Macromolecules*, **30**, 770–772.

16 Schwach-Abdellaoui, K., Heller, J., and Gurny, R. (1999) *J. Biomater. Sci. Polym. Edn.*, **10**, 375–389.

17 Heller, J., Barr, J., Ng, S.Y., Schwach-Abdellaoui, K., and Gurny, R. (2002) *Adv. Drug Deliv. Revs.*, **54**, 1015–1039.

18 Heller, J., Penhale, D.W.H., Fritzinger, B.K., Rose, J.E., and Helwing, R.F. (1983) *Contracept. Deliv. Syst.*, **4**, 43–53.

19 Heller, J., Rime, A.-F., Rao, S.S., Fritzinger, B.K., and Heller, N. (1995) *Trends and Future Perspectives in Peptides and Protein Drug Delivery*, Harwood Academic Publishers, Switzerland.

20 Heller, J., Barr, J., Shah, D.T., Ng, S.Y., Shen, H.-R., and Baxter, B.C. (2006) *Scaffolding in Tissue Engineering*, CRC Press, Boca Raton, FL.

21 Schwach-Abdellaoui, K., Heller, J., and Gurny, R. (1999) *Macromolecules*, **32**, 301–307.

22 Shah, S.S., Cha, Y., and Pitt, C.G. (1992) *J. Control. Release*, **18**, 261–270.

23 Heller, J. and Barr, J. (2004) *Biomacromolecules*, **5**, 1625–1632.

24 Fu, K., Pack, D.W., Klibanov, A.M., and Langer, R. (2000) *Pharm. Res.*, **17**, 100–106.

25 Rothen-Weinhold, A., Schwach-Abdellaoui, K., Barr, J., Ng, S.Y., Shen, H.-R., Gurny, R., and Heller, J. (2001) *J. Control. Release*, **71**, 31–37.

26 Ando, S., Putnam, D., Pack, D.W., and Langer, R. (1999) *J. Pharm. Sci.*, **88**, 126–130.

27 Wang, C., Ge, Q., Ting, D., Shen, H.-R., Chen, J., Eisen, H.N., Heller, J., Langer, R., and Putnam, D. (2004) *Nat. Mater.*, **3**, 190–196.

28 Heller, J., Barr, J., Ng, S.Y., Shen, H.-R., Schwach-Abdellaoui, K., Einmahl, S., Rothen-Weinhold, A., and Gurny, R. (2000) *Eur. J. Pharm. Biopharm.*, **50**, 121–128.

29 Heller, J., Barr, J., Ng, S.Y., and Shen, H.-R. (2002) *Drug Deliv.*, **2**, 38–43.

30 Schwach-Abdellaoui, K., Heller, J., Barr, J., and Gurny, R. (2002) *Int. J. Polym. Anal. Charact.*, **7**, 145–161.

31 Heller, J., Barr, J., Ng, S.Y., Shen, H.-R., Schwach-Abdellaoui, K., Gurny, R., Vivien-Castioni, N., Loup, P.J., Baehni, P., and Mombelli, A. (2002) *Biomaterials*, **23**, 4397–4404.

32 Tucker, G.T. (1979) *Clin. Pharmacokinet.*, **4**, 241–248.

33 Heller, J. and Barr, J. (2005) *Exp. Opin. Drug Deliv.*, **2**, 169–183.

34 Pharma, A.P. (2006) Annual Report.

5
Biodegradable Polymers Composed of Naturally Occurring α-Amino Acids

Ramaz Katsarava and Zaza Gomurashvili

5.1
Introduction

The synthesis and study of biodegradable polymers is at the forefront of modern polymer chemistry because of the technological challenge and commercial potential. For many medical, agricultural, and environmental purposes, it is important to have biodegradable polymers that degrade under the action of physiological environment or in soil. Biodegradable polymers have become increasingly important for the development of surgical and pharmaceutical devices like wound closure devices, vascular grafts, nerve guidance tubes, absorbable bone plates, orthopedic pins and screws, body-wall/hernia repair, sustained/controlled drug delivery systems, to name a few. Different materials with tailored properties are required for each of these applications. Therefore, biodegradable polymers with a variety of hydrophilicity/hydrophobicity, permeability, morphology, degradation rates, chemical, and mechanical properties are needed.

The limitation for many synthetic biodegradable polymers as biomedical materials is the potential toxicity of the degradation products. Therefore, research was focused toward the materials entirely composed of naturally occurring and nontoxic ("physiological") building blocks. Such polymers release metabolic components upon biodegradation, which are digested by cells and reveal certain nutritious values, in parallel with high biocompatibility.

In the light of this, heterochain polymers composed of α-hydroxy acids (α-HAs) and α-amino acids (α-AAs) are considered as promising representatives of synthetic resorbable biomaterials, especially the latter because after biodegradation the release products are essential α-AAs and their derivatives.

Well-characterized aliphatic polyesters (PEs), for example, PGA, PLA, PLGA, PDLLA [1], are far from perfect: the synthesis of PEs requires dry conditions, which is rather complex and costly. The shelf-life of the PEs is rather short. Also, aliphatic PEs reveal useful material properties only at high molecular weights (100,000 Da and higher) due to weak intermolecular forces. They show low hydrophilicity and hence do not actively interact with the surrounding tissues in a desirable manner after implantation that diminishes the biocompatibility [2].

Handbook of Biodegradable Polymers: Synthesis, Characterization and Applications, First Edition. Edited by Andreas Lendlein, Adam Sisson.
© 2011 Wiley-VCH Verlag GmbH & Co. KGaA. Published 2011 by Wiley-VCH Verlag GmbH & Co. KGaA.

On the other hand, α-AA-based polymers have strong hydrogen bonds due to amide linkages that increase both intermolecular forces (that means desirable material properties at much lower molecular weights) and hydrophilicity, and hence biocompatibility [3].

The earliest representatives of α-AAs based synthetic polymers were poly(α-amino acids) (PAAs). The most common method for the synthesis of high-molecular-weight PAAs is ring-opening polymerization of N-carboxyanhydrides. In spite of expectations, PAAs that belong to the class of polyamides (Nylons-2) and contain only amide bonds in the backbones turned out to be less suitable as biodegradable materials for biomedical engineering use for many reasons, such as difficult and costly manufacturing processes because of unstable N-carboxyanhydrides, insolubility in common organic solvents, thermal degradation on melting, and poor processability. The rates of degradation under physiological conditions are often too slow to be useful as biodegradable biomaterials. These limitations of PAAs could be somewhat reduced by the synthesis of copolymers containing two or more α-AAs. However, this originates immunogenicity, and the biodegradation rate was still low due to the polyamide (PA) nature of the polymers [3].

Therefore, the research efforts were redirected to the synthesis of α-AAs-based polymers that contain easily cleavable (degradable) chemical bonds in the backbones with molecular architecture that diminishes (or at all excludes) immunogenicity.

How could macrochains using α-AAs as building blocks be constructed? Let us consider the structure of α-AAs as a vector directed from N-terminus to C-terminus (Figure 5.1).

Linear macrochains on the basis of α-AAs can be constructed using both α-functional groups (H_2N and COOH), or one α-(H_2N or COOH) and one lateral functional group F (which could be NH_2, COOH, or OH). Hence, the orientation of α-AAs in macrochains can be diverse (Figure 5.2).

This multifunctionality along with a high number of naturally occurring α-AAs opens unlimited synthetic possibilities for constructing various macrochains.

Among the various possible orientations of α-AAs in the polymeric backbones, the directional, "head-to-tail" orientation is conventional observed in biopolymers, proteins, and polypeptides. This orientation determines their primary and secondary structures that, in turn, determine their biochemical properties including immunogenicity. The same is true for synthetic poly-α-AAs [3]. All the said polymers belong to the class of polypeptides, in fact AB type polyamides.

More promising for biomedical applications are synthetic polymers in which the α-AAs have nonconventional orientations–adirectional ("head-to-head" and

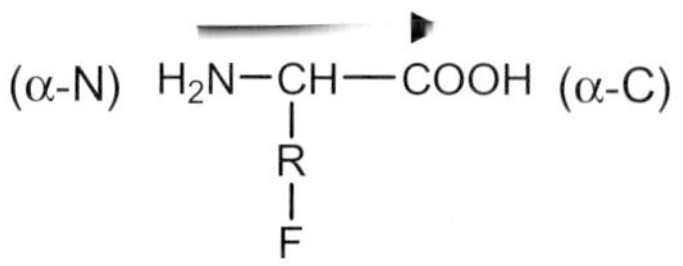

Figure 5.1 The general structure of α-AAs.

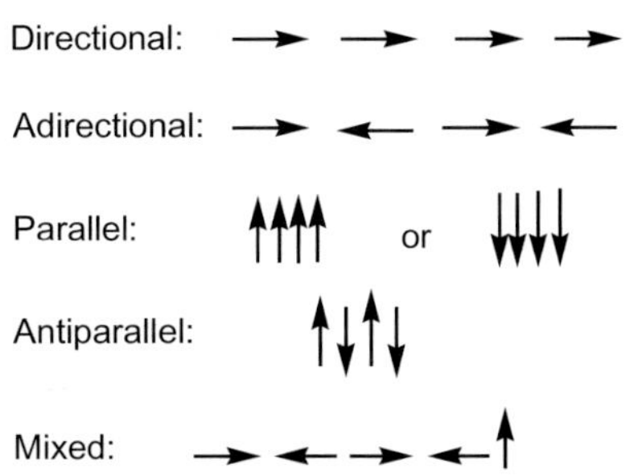

Figure 5.2 The possible orientations of α-AAs in the polymeric backbones.

"tail-to-tail"), parallel, antiparallel, or mixed (Figure 5.2). These could be polymers of other classes – polyurethanes and polyureas along with the said polyamides. To render the polymers easily cleavable (in most cases hydrolysable), the labile chemical bonds have to be incorporated into the polymeric backbones to provide desirable rates of biodegradation. Preference should be given to ester bonds taking into account both biodegradation rates and the stability (shelf life). The new polymers comprising different types of heterolinks such as ester, urethane, urea, along with peptide (amide) bonds, with the nonconventional orientation of α-AAs are expected to diminish the immunogenicity of the polymers by "confusing nature" due to "unrecognizable" structures of macromolecules.

5.2 Amino Acid-Based Biodegradable Polymers (AABBPs)

5.2.1 Monomers for Synthesizing AABBPs

In this chapter, three classes of AABBPs containing ester bonds as biodegradable sites are considered. These are AA-BB polycondensation polymers with nonconventional "head-to-head" and "tail-to-tail" orientation of α-AAs in the polymeric backbones – poly(ester amide)s (PEAs), poly(ester urethane)s (PEURs), and poly(ester urea)s (PEUs). The PEAs are composed of three building blocks: (i) α-AAs, (ii) fatty diols, and (iii) dicarboxylic acids. They allow manipulation of polymer properties in a wide range. PEURs and PEUs are also composed of three types of building blocks – two blocks are (i) α-AAs and (ii) diols; however, the third block is (iii) carbonic acid.

5.2.1.1 Key Bis-Nucleophilic Monomers

Key monomers for synthesizing all three classes of AABBPs are bis-nucleophiles that represent dimerized α-AAs-bis-(α-amino acyl)-alkylene diester (tosic acid salt of amino acid/alkylene diester, TAAD). These compounds are stable in the salt form, commonly as di-*p*-toluenesulfonic acid (TosOH) salts. They are generally

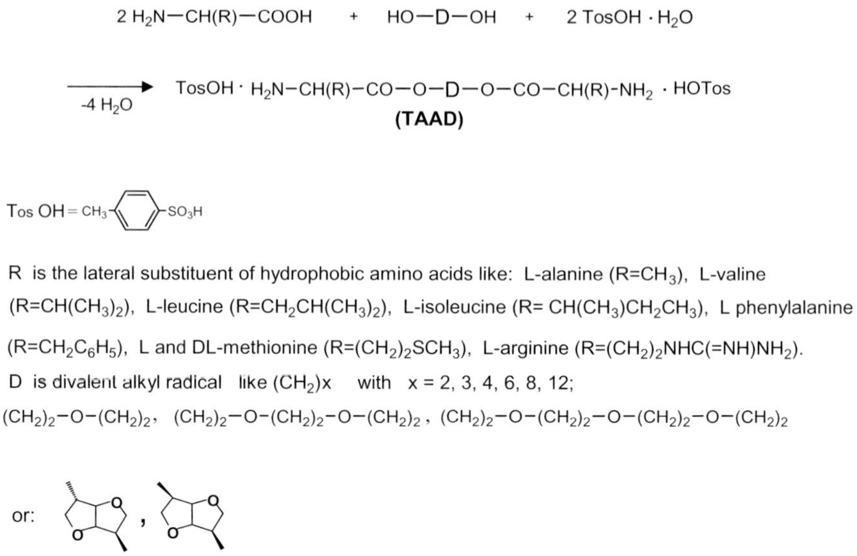

Scheme 5.1 Synthesis of TAADs.

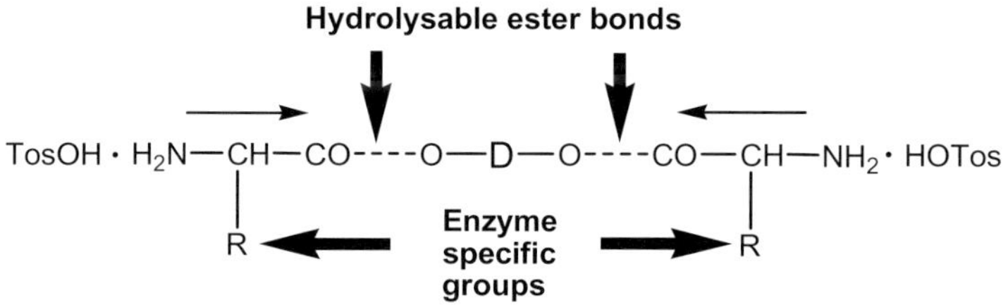

Figure 5.3 Structural peculiarities of TAADs.

prepared by direct condensation of α-AAs (2 mol) with fatty diols (1 mol) in refluxed benzene or toluene in the presence of TosOH monohydrate (2 mol), Scheme 5.1.

The presence of TosOH•H_2O (2 mol) serves as both the reaction catalyst and amino group protector, preventing undesirable side reactions including amine interaction with inherent ester groups of TAAD.

This strategy allows us to generate diamine monomer with two inherent biodegradable (hydrolysable) ester bonds, along with enzyme specific groups, Figure 5.3, and the nonconventional "head-to-head" orientation of α-AAs put at a monomer stage

The first synthesis of TAAD according to this very simple procedure was reported by Huang and coworkers [4], on the basis of L-phenylalnine and 1,2-ethanediol. Later, TAADs were obtained from other hydrophibic α-amino acids: glycine [5–9], alanine [10–13], valine [14], leucine [6, 14–21], isoleucine, norleucine, methionine [14], phenylalanine [6, 7, 14–31], and arginine [32–35]. Accordingly, arginine-based TAADs are tetra-(TosOH) salts.

Various aliphatic α,ω-alkylene diols [8–17, 21, 24, 27–35], dianhydrohexitols [25, 26], and di-, tri, and tetraethylene glycols [35, 36] were used by different authors for synthesizing TAADs.

The obtained di- or tetra-TAADs are stable compounds. The most of these monomers were purified by recrystallizing from water or organic solvents. The yields of pure, polycondensation grade products ranged within 60–90%.

5.2.1.2 Bis-Electrophiles

For successful synthesis of AABBPs with tailored architecture, the selection of suitable bis-electrophilic monomer(s) is also important–counterpartners of TAADs. The syntheses of various bis-electrophiles are discussed below as detailed as possible within the bounds of this chapter.

Dicarboxylic acids HO–CO-A-CO–OH can be incorporated into the PEA backbones by means of either dichlorides Cl–CO-A-CO–Cl (dicarboxylic acid dichloride, DDC) or active diesters R_1-CO-A-CO-R_1 (dicarboxylic acid active diester, DAD) as bis-electrophilic monomers (for A and R_1, see Scheme 5.2).

Many DDCs are commercial products. DADs are obtained using three synthetic methods: (i) by interaction of DDCs with various hydroxyl compounds HOR_1 (activating agents), Scheme 5.2 [14, 15, 20, 24, 25], or by direct interaction of dicarboxylic acids (ii) with HOR_1 in the presence of various condensing (coupling) agents, Figure 5.4 [15, 21, 37], or (iii) with various *trans*-esterifying agents that are derivatives of HOR_1, Scheme 5.3 [37].

All three methods give DADs in a good yield ranged from 60% to 90%.

Monomers for synthesizing PEURs. The third building block of PEURs–carbonic acid–can be incorporated into the polymeric backbones by means of either bis-chloroformates Cl–CO–O-D_1-O–CO–Cl (diol bis-chloroformate, BCF) or active bis-carbonates R_1-CO–O-D_1-O–CO-R_1 (DBCs) as bis-electrophilic monomers (D_1 can be the same as D in Scheme 5.1).

Diol bis-carbonates (DBCs) can be obtained using two synthetic methods: (i) by interaction of BCFs with hydroxyl compounds HOR_1, Scheme 5.4 [38], or (ii) by interaction of diols with mono-chloroformates of hydroxyl compounds Cl–CO–O-R_1, Scheme 5.5 [39].

The building block for PEUs, carbonic acid, can be incorporated into the polymeric backbones by means of polycondensation using either phosgene (derivatives), or active carbonates (AC) obtained according to Scheme 5.6 or related compounds [40].

5.2.2 AABBPs' Synthesis Methods

PEAs. The synthesis of PEAs on the basis of TAADs can be carried out at a low temperature via interfacial polycondensation (IP) and solution polycondensation (SP). The IP and SP reactions proceed according to Figure 5.5 in the presence of acid acceptor (HCl and/or TosOH).

The selection of the polycondensation method depends on the nature of bis-electrophilic monomer. The IP is suitable method when DDCs are used.

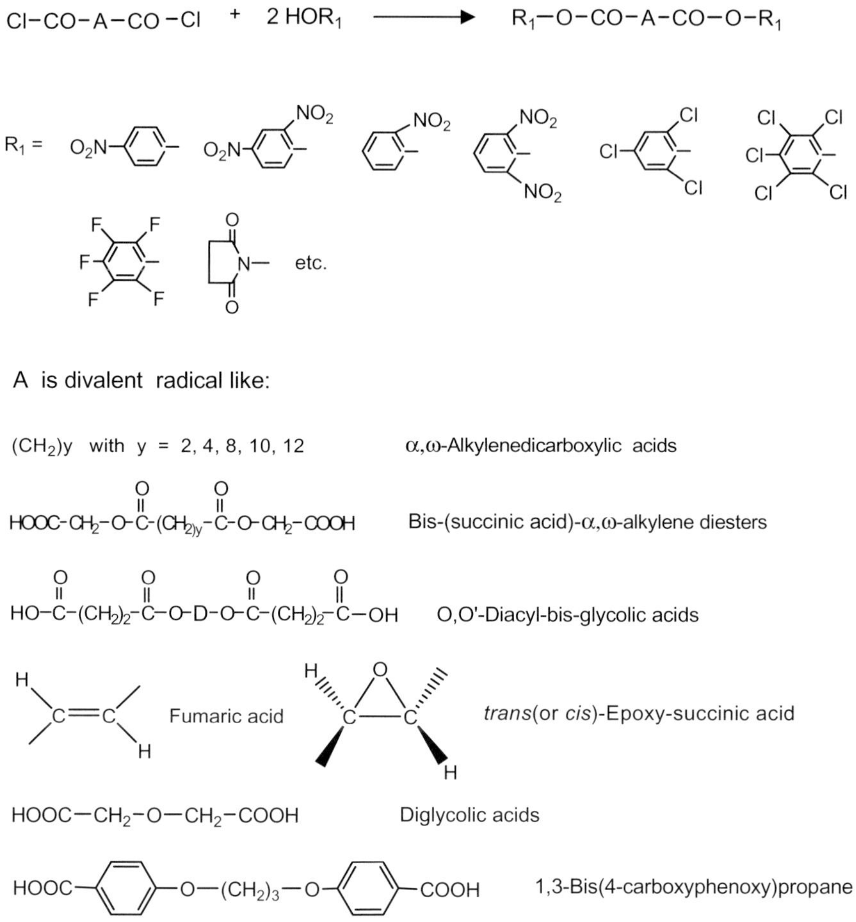

Scheme 5.2 Synthesis of DADs by interacting DDC with HOR1, method (i).

$$HOOC{-}A{-}COOH + 2\,HOR_1 \xrightarrow{\text{Condensing agent, } B_N} R_1{-}O{-}OC{-}A{-}CO{-}O{-}R_1$$

Condensing agent = $SOCl_2$, $(CF_3CO)_2O$, $C_6H_{11}{-}N{=}C{=}N{-}C_6H_{11}$

B_N= Tertiary amine (Pyridin, NEt_3, etc.)

Figure 5.4 Synthesis of DADs from free dicarboxylic acids using condensing agents, method (ii).

However, this method results into high-molecular-weight PEAs only with the hydrophobic diacids like sebacic acid with $y = 8$, or higher (Scheme 5.2) or aromatic DDCs, such as terephthaloyl chloride [8–13]. It has to be noted that DDCs are less suitable monomers for SP with aliphatic diamines since these electrophiles enter into numerous undesirable side reactions with tertiary amines [41] that are

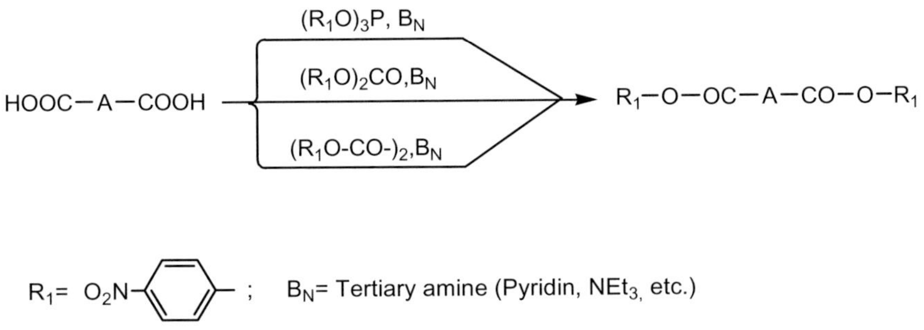

Scheme 5.3 Synthesis of DADs from free dicarboxylic acids using *trans*-esterifying agents, method (iii).

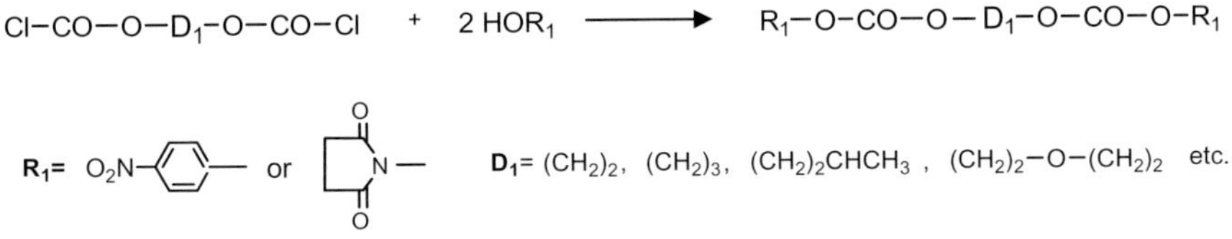

Scheme 5.4 Synthesis of DBCs by interacting BCFs with activating agents HOR1, method (i).

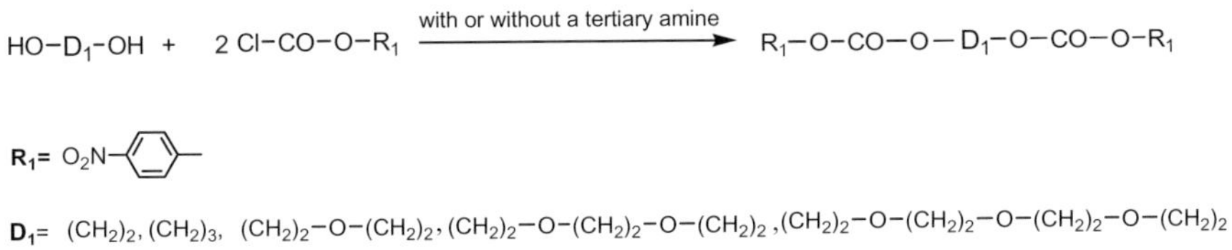

Scheme 5.5 Synthesis of DBCs by interacting diols with *p*-nitrophenyl-chloroformate, method (ii).

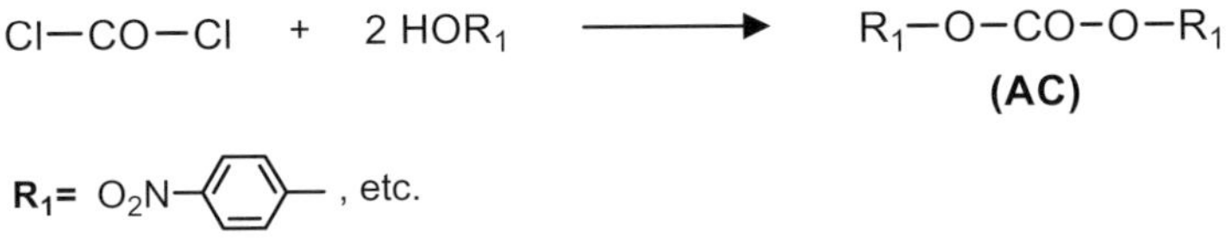

Scheme 5.6 Synthesis of active carbonates (AC).

used as acceptors of liberated hydrogen chloride; for TAADs, tertiary amines are used to remove TosOH from amino groups as well. These side reactions cause the limitation of the chain growth resulting in the formation of low-molecular-weight polymers with poor material properties.

For hydrolytically less stable DDCs, or DDCs that are unavailable at polycondensation purity (like short-chain succinic ($y = 2$), adipic ($y = 4$), fumaric, and

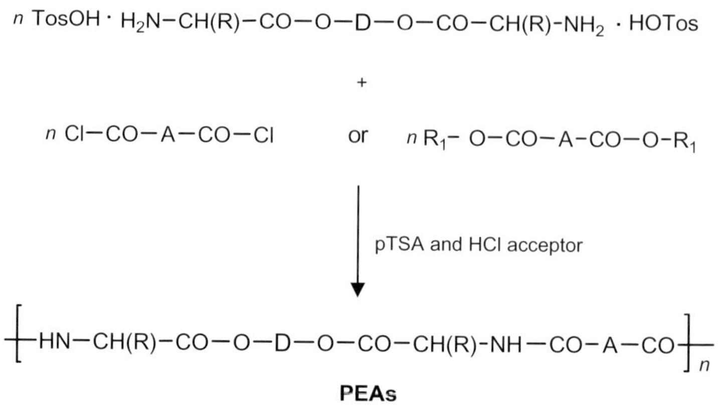

Figure 5.5 Synthesis of PEAs by IP and SP (AP).

epoxy-succinic acids, as well as bis-(succinic acid)-α,ω-alkylene diesters and *O,O′*-diacyl-bis-glycolic acids, see Scheme 5.2), the preference should be given to SP using DADs as bis-electrophilic monomers. The SP via active diesters of various classes – DADs, DBCs, and ACs – is called "active polycondensation" (AP) [42] to distinguish it from traditional polycondensation methods. Hereafter we use the term AP for polycondensation with participating active diester of diacid. The AP with DADs is normally carried out in polar aprotic solvents DMA, DMSO, etc., or in common organic solvents like chloroform, THF, etc., at 20–80 °C using mostly triethylamine (TEA) as TosOH acceptor [14–16, 20–22, 24, 25, 27, 28, 32, 33, 42–47]. It was shown that DADs are stable against both amide-type solvents and tertiary amines [48] under the conditions of AP that minimizes undesirable side reactions and results in the formation of high-molecular-weight polymers.

It has to be noted that PEAs composed of the same three building blocks – α-AA (glycine), fatty diols, and dicarboxylic acids – were synthesized recently [5] using the third method – thermal polycondensation (TP) in melt, in the presence of titanium butoxyde as a catalyst at 160–220 °C.

The advantage of TP is the possibility to process polymers from melt directly after the polycondensation, that is, without the separation and purification of the resulting polymers. However, the method is less suitable for thermally sensitive and unstable monomers including optically active ones since high reaction temperature can cause racemization and destruction. The use of metalorganic catalyst is one of the drawbacks as well.

The AABBPs type PEURs can be synthesized on the basis of TAADs under the conditions of either IP or AP similar to Figure 5.5 using as bis-electrophilic monomers BCFs instead of DDCs, and DBCs instead of DADs.

Like for the PEA, the PEUR synthesis by IP is less suitable with short-chain DBC due to their hydrolytic instability that results in low-molecular-weight polymers. Kohn *et al.* [49–51] suggest that more appropriate monomers for polyurethane synthesis via IP are DBCs that are hydrolytically more stable. The results are high-molecular-weight lysine based poly(ether urethane)s even on the basis of water-soluble monomers–bis-succinimidyl carbonates of PEGs (PEG-based DBCs). The same approach seems promising for the synthesis of PEURs on the basis of TAADs.

DBCs were very effective as bis-electrophiles in AP as well. They resulted in the high-molecular-weight PEURs [16, 52] having excellent film-forming properties. The conditions of AP with DBCs are the same as for DADs above.

The PEUs on the basis of TAADs can also be synthesized via IP or AP similar to Figure 5.5 using phosgenes (mono, di, or tri) as bis-electrophilic monomers instead of DDCs [53], and ACs instead of DADs [52]. In contrast to PEAs and PEURs above, IP unambiguously led to high-molecular-weight PEUs.

5.2.3
AABBPs: Synthesis, Structure, and Transformations

5.2.3.1 Poly(ester amide)s

Regular PEAs. We consider as "nonfunctional" those PEAs that have no functional groups except two terminal reactive groups–normally one nucleohpile and one electrophile, Figure 5.6.

According to the polycondensation theory of Kricheldorf [54, 55], a substantial portion of macromolecules obtained via AP have no terminal functional groups, since they form macrocycles.

The first "nonfunctional" regular PEAs representing AABBPs [6, 7, 14, 17, 18, 24–26] was synthesized via AP of TAADs with active diesters of α,ω-alkylenedicarboxylic acids [A = $(CH_2)y$], according to Figure 5.5 above.

Polysuccinates. Recently [44] a new class of nonfunctional AABBPs–PEAs based on succinic acid (rather alkylene disuccinates) with higher density of cleavable ester bonds–were synthesized by AP of TAADs with active di-*p*-nitrophenyl esters of bis-(succinic acid)-α,ω-alkylene diesters (Scheme 5.2). Their general structure is given in Figure 5.7.

Polysuccinates have two additional ester bonds (in total four ester bonds) as compared with regular PEAs above, having in total two ester bonds per repeating

Figure 5.6 Regular PEAs composed of α-AAs, diols and α,ω-alkylenedicarboxylic acids.

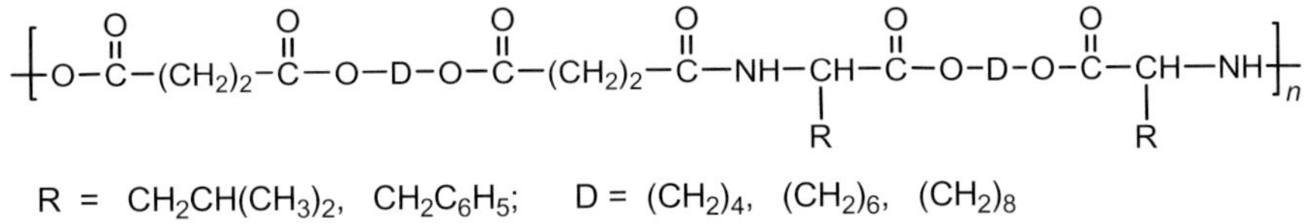

R = $CH_2CH(CH_3)_2$, $CH_2C_6H_5$; D = $(CH_2)_4$, $(CH_2)_6$, $(CH_2)_8$

Figure 5.7 PEAs on the basis of bis-(succinic acid)-α, ω-alkylene diesters.

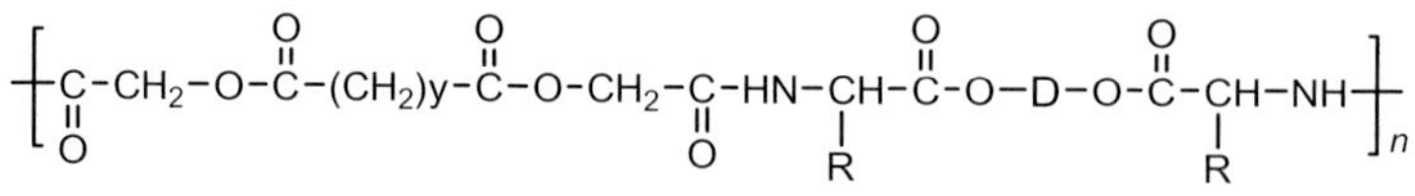

R = $CH_2CH(CH_3)_2$, $CH_2C_6H_5$; D = $(CH_2)_8$, $(CH_2)_{12}$; y = 4, 8.

Figure 5.8 AA-BB PDPs composed of glycolic acid, α-AAs, and dicarboxylic acids.

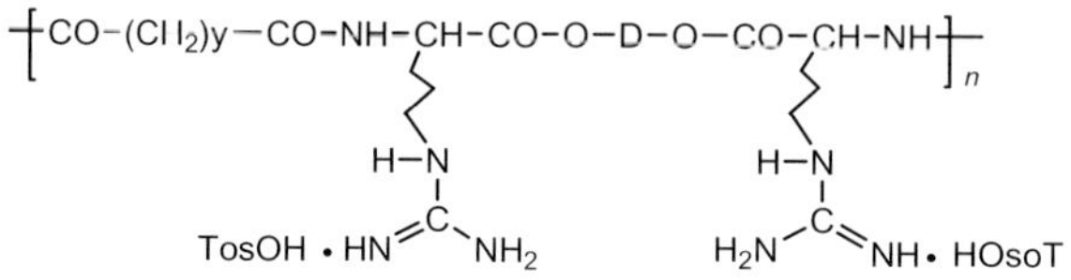

y = 2, 4, 8; D = $(CH_2)_3$, $(CH_2)_4$, $(CH_2)_6$

Figure 5.9 Arginine-based cationic ABBPs-PEAs.

unit, and showed increased biodegradation rates. Additionally, the enhanced hydrolysis of polysuccinates is linked with intramolecular catalysis (see Ref. [56] and references cited therein).

Poly(depsipeptide)s (PDPs). Very recently [21, 43] a new class of nonfunctional AABBPs–AA-BB-type PDPs–were obtained by AP of TAADs with active di-*p*-nitrophenyl esters of *O,O′*-diacyl-bis-glycolic acids (Scheme 5.2) and have the general structure given in Figure 5.8.

PDPs also have two additional and highly polarized (close by nature to the ester bonds in poly(glycolic acid)) ester bonds (in total four ester bonds) as compared with regular PEAs above, containing two ester bonds per elemental links, and hence showed increased biodegradation rates.

Functional PEAs. Polyacids. Katsarava and Chu [15, 16] synthesized functional *co*-PEAs containing a variable amount of lateral carboxyl groups, applying di-TosOH salt of L-lysine benzyl ester as a comonomer. The goal *co*-PEAs were obtained by selective catalytic hydrogenolysis (debenzylation) of benzyl ester pre-polymer using Pd catalyst. Free lateral COOH groups can be used for numerous chemical transformations and *co*-PEAs are suitable drug carriers that will be discussed below. It has to be noted that lysine has parallel orientation (Figure 5.2), whereas other amino acids' orientation is adirectional, that is, in whole α-AAs' orientation in this types of polymers is mixed.

Polycations. Arginine-based TAADs are tetra-*TosOH* salts that act as bifunctional nucleophilic monomer (via two α-amino groups). This allows to synthesize the linear and soluble polycationic PEAs (Figure 5.9) by AP of L-arginine-based TAADs with di-*p*-nitrophenyl esters of α,ω-alkylenedicarboxylic acids [33–35].

The arginine-based PEA composed of succinic acid and 1,3-propanediol (the less hydrophobic one among the PEAs obtained) was water soluble at room temperature. Very recently Memanishvili *et al.* [35] obtained arginine-based poly(ether ester amide)s, PEEAs, and poly(ether ester urethane)s, PEEURs, and poly(ether ester

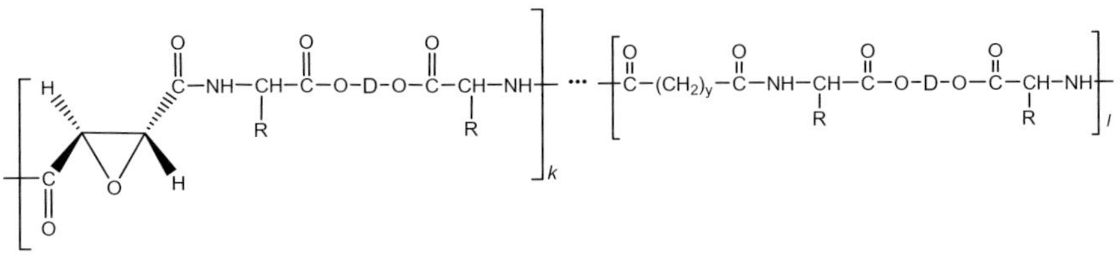

k = 1.0, 0.8, 0.6; l = 1 - k

R = $CH_2CH(CH_3)_2$, $CH_2C_6H_5$;   D = $(CH_2)_4$, $(CH_2)_6$, $(CH_2)_8$;   y = 4, 8.

Figure 5.10 Epoxy-PEAs on the basis of *trans*-epoxy-succinic acid.

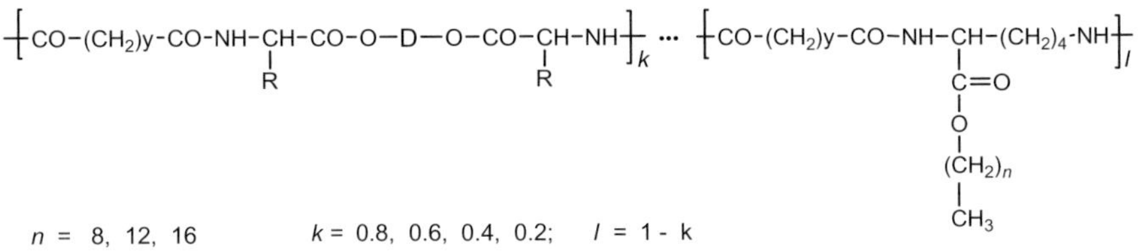

n = 8, 12, 16 k = 0.8, 0.6, 0.4, 0.2; l = 1 - k

R = $CH_2CH(CH_3)_2$, $CH_2C_6H_5$;   D = $(CH_2)_6$;   y = 4, 8.

Figure 5.11 Brush-like PEAs containing long-chain *n*-alkyl substituents.

urea)s, PEEUs, having polyethylene glycol like polymeric backbones and showing enhanced water solubility as compared with the said arginine-based PEAs.

Biodegradable cationic PEAs were also obtained [57] by covalent conjugation to PEA-polyacids with arginine methyl ester and agmatine.

Unsaturated PEAs. One of the most convenient and universal ways to render biodegradable polymers functional is the incorporation of unsaturated double bonds in the polymeric backbones [58]. Unsaturated PEAs (UPEAs) containing a variable amount of double bonds in the backbones were obtained by Katsarava, Chu, and coworkers [19, 27–30, 45] using TAADs on the basis of 1,4-butendiol, or DAD based on fumaric acid as monomers/comonomers in combination with saturated TAADs and DADs.

The unsaturated double bonds can be subjected to various chemical and photochemical transformations.

Epoxy-PEAs. Very recently Katsarava, Tugushi, and coworkers [20, 46] have synthesized a new class of functional biodegradable polymers – epoxy-PEAs – using active di-*p*-nitrophenyl ester of *trans*-epoxy-succinic acid as a monomer, or comonomer in combination with DADs containing α,ω-alkylenedicarboxylic acids in AP with TAADs. They have the structure given in Figure 5.10.

The epoxy groups of PEAs can be subjected to various chemical transformations under mild conditions, as well as thermal or chemical curing.

Brush-like PEAs. PEAs with a brush-like architecture containing long-chain alkyl substituents were obtained by Katsarava and coworkers [59] using L-lysine *n*-alkyl esters as comonomer in a mixture with TAADs in AP with DADs. They have the structure shown in Figure 5.11.

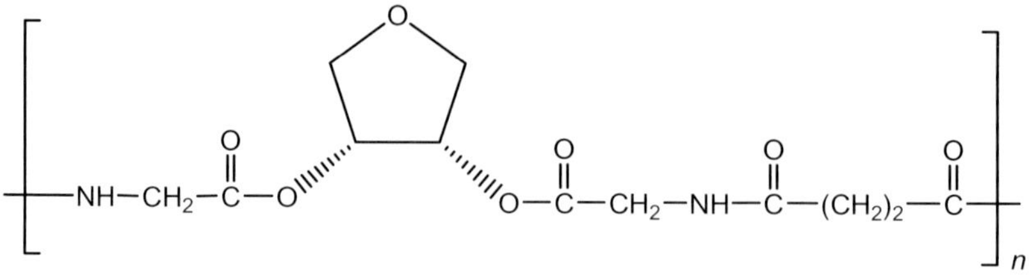

Scheme 5.7 Water-soluble PEAs on the basis of 1,4-anhydroerythritol, glycine, and succinic acid.

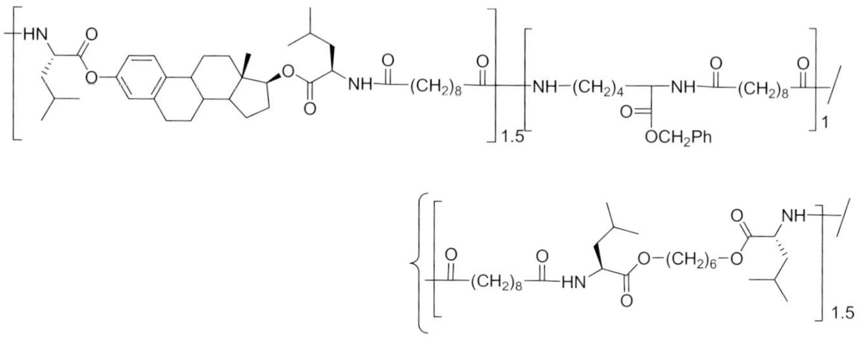

Scheme 5.8 Biodegradable polymeric drug composed of 17β-estradiol, 1,6-hexanediol, L-leucine, L-lysine benzyl ester, and sebacic acid.

These PEAs are suitable for constructing devices with sustained/controlled release in which a drug is attached to the macromolecules via hydrophobic forces.

Hydroxyl-containing and water-soluble PEAs. PEAs containing free OH groups were obtained by Gomurashvili *et al.* [60] by AP of TAADs composed of unsubstituted α-AA glycine and glycerol with di-*p*-nitrophenyl esters of succinic, glutaric, adipic, and diglycolic acids. Depending on the synthetic strategy used, three types of hydroxyl-containing polymers were synthesized: PEAs with pending primary hydroxyls, with pending secondary hydroxyls, or a copolymer containing both primary and secondary glycerol hydroxyls (not shown here). PEAs composed of short aliphatic diacids such as succinic, glutaric, and diglycolic acids are water soluble.

Water-soluble PEAs, having the structure given in Scheme 5.7, were also obtained by AP of TAAD composed of 1,4-anhydroerythritol and glycine with di-*p*-nitrophenyl succinate [60].

Polymeric drugs. The strategy of the synthesis of AABBPs allows constructing biodegradable polymeric drugs. For example, therapeutic copolymers composed of sebacic acid, L-leucine, 1,6 hexanediol, 17β-estradiol, and L-lysine benzyl ester (M_w up to 82,000 Da) was obtained by Gomurashvili *et al.* [61] via AP of di-*p*-nitrophenyl sebacate with three comonomers – two TAADs composed of L-leucine/1,6-hexanediol and L-leucine/17β-estradiol, and di-*p*-toluensulfonic acid salt of L-lysine benzyl ester, Scheme 5.8.

5.2.3.2 Poly(ester urethane)s

Regular PEURs. This class of AABBPs was synthesized for the first time by Katsarava and coworkers [52] by AP of TAADs with DBCs as discussed above.

These polymers, like the regular PEAs above, have only two terminal functional groups and are considered nonfunctional.

Functional PEURs. PEURs containing a variable amount of lateral carboxyl (COOH) groups were obtained by Katsarava and Chu [16] similar to PEAs discussed above. The only difference consists in the use of DBCs instead of DADs in AP with α-AAs-based comonomers for synthesizing benzyl ester prepolymer.

5.2.3.3 Poly(ester urea)s

Historically PEUs were the first examples of AABBPs synthesized by Huang and coworkers [4] by interaction of bis-(L-phenylalanine)-1,2-ethylene diester as free base (separated from corresponding di-*TosOH* salt) with aromatic diisocyanates. As a result, low-molecular-weight powdery PEUs were obtained. The main cause of low-molecular-weight polymers presumably is a high tendency of alkyl esters of α-AAs to enter into various undesirable self-condensation reactions [62] with the formation of diketopiperazines and other cyclic and linear unidentified products. This leads to imbalance of stoichiometry and contributes to the limitation of chain growth. In spite of this, Huang's study initiated a rational synthesis of a large variety of key monomers–TAADs–and showed the suitability of the incorporation of enzyme-specific α-AAs and ester bonds into macro-chains for constructing biodegradable biomaterials.

The synthesis of PEUs by AP of TAADs with ACs in DMA solution was carried out by Katsarava and coworkers [52]. However, recently Katsarava *et al.* [53] have found that high-molecular-weight PEUs having excellent material properties could be synthesized via IP of TAAD with phosgene or triphosgene using a two-phase system chloroform/water+Na_2CO_3 similar to Figure 5.5. These results are quite contrary to the synthesis of PEAs and PEURs above where the synthesis of high-molecular-weight polymers on the basis of short-chain DDCs or BCFs is problematic. This is because in the synthesis of PEAs and PEURs, the hydrolysis of bis-electrophiles–DDCs and BCFs–generates mono-functional impurities that cause the termination of the chain growth, whereas in case of phosgene no mono-functional compound is formed since it hydrolyses with the liberation of CO_2 and HCl.

5.2.3.4 Transformation of AABBPs

All the AABBPs containing lateral functional groups can be subjected to various chemical transformations that modify their properties, for chemical attachment of drugs, bioactive substances, etc.

Free COOH groups in polyacids containing L-lysine residues can be used for chemical modification with condensing agents. For example, 4-amino-2,2,6,6-tetramethyl-piperidinyloxy free radical (4-amino-TEMPO) was covalently attached to functional *co*-PEA (Scheme 5.9) using carbonyldiimidazole (Im_2CO) as a condensing agent [15–17, 61].

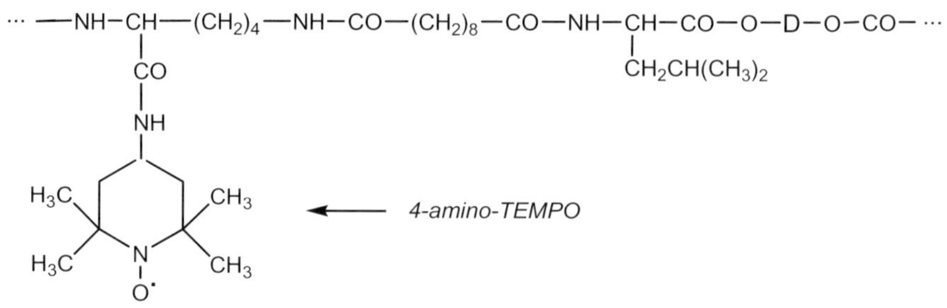

Scheme 5.9 Functional *co*-PEA containing covalently attached 4-amino-TEMPO.

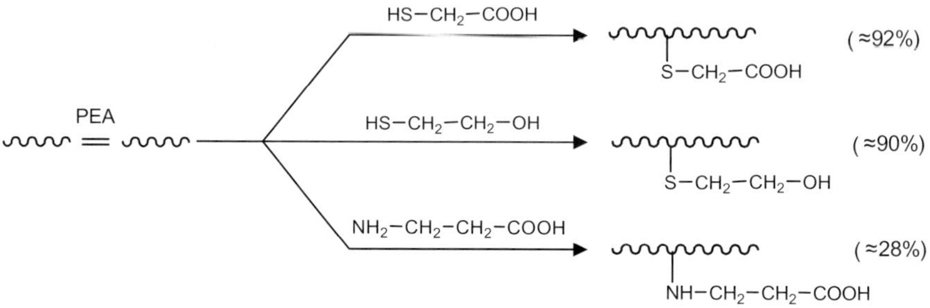

Figure 5.12 Chemical transformations of fumaric acid based UPEA.

The covalent attachment of mono-ethanolamine to these polymers increases their hydrophilicity and water solubility (depending on mole portion of lysine residue in the polymers backbones). The obtained polyols can be used for further transformations, for example, to obtain chemically and photochemically active polymers by attaching unsaturated acids like acrylic, methacrylic, etc. [63, 64].

UPEAs, containing active double bonds of fumaric acid's residue can be functionalized by their interaction under mild conditions with thiol- and amino-compounds [19] as is shown in Figure 5.12.

As one can conclude, the thio-compounds are far more active in these transformations.

The epoxy-PEAs given in Figure 5.10 contain activated (by two adjacent electron-withdrawing carbonyl groups) oxirane cycles and interact under mild conditions (DMA, 20–60 °C) with various compounds of both nucleophilic and electrophilic nature [20].

Due to the high activity of epoxy-PEAs, they can be considered as "ready for use" carriers – in contrast to polyacids above, because they interact with 4-amino-TEMPO in DMA solution at 60 °C without using condensing agent, as shown in Scheme 5.10.

Both UPEAs and epoxy-PEAs are also subjected to chemical, thermal and photochemical curing that allows one to regulate their properties, for example, to increase mechanical characteristics and decrease biodegradation rate.

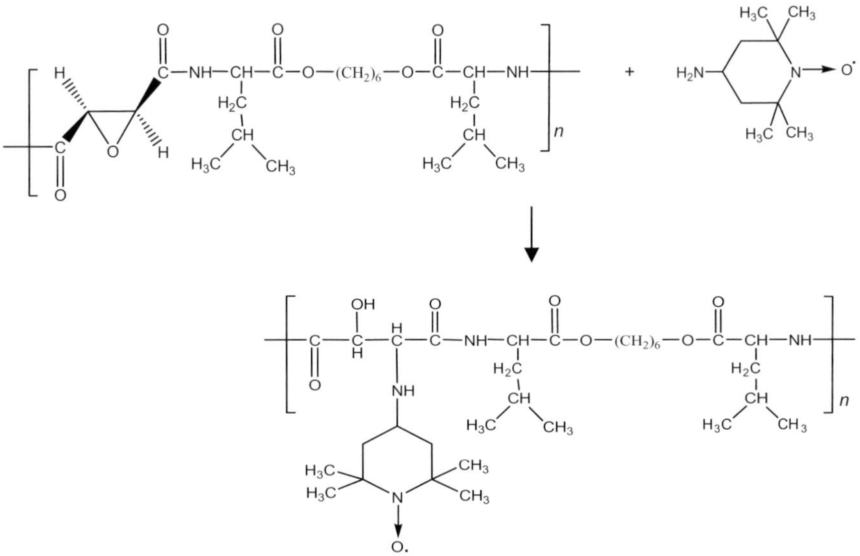

Scheme 5.10 Covalent attachment of 4-amino-TEMPO to epoxy-PEA.

5.2.4 Properties of AABBPs

5.2.4.1 MWs, Thermal, Mechanical Properties, and Solubility

All the AABBPs obtained via AP [14] have high molecular weights (M_w = 24,000–180,000 Da, GPC) and narrow polydispersity (1.20–1.81).

DSC study of AABBPs showed that these polymers have a wide range of glass transition temperature (T_g from 5 °C to 102 °C), some of them (PEAs and PEUs) are semicrystalline with T_m = 103–153 °C [14, 53]. It was shown that T_g of the polymers can be increased by incorporating rigid fragments into macrochains such are dianhydrohexitols [25, 26] or aromatic diacid-1,3-bis(4-carboxyphenoxy)propane [61].

The chemical structure affects the mechanical properties of AABBPs, which varies in a wide range: tensile strength from 15–20 (PEURs and some PEAs) to 80–100 MPa (PEUs and some PEAs), elongation at break from 8–100 (PEUs and some PEAs) to 800–1000% (PEURs and some PEAs), Young's modulus up to 2–6 GPa (PEUs and some PEAs).

The AABBPs are soluble in common organic solvents such as DMF, THF, methylene chloride, chloroform, some of them in dioxane, acetone, and ethanol. The low melting temperatures and solubility of AABBPs in common solvents substantially facilitate their processing into different shapes.

5.2.4.2 Biodegradation of AABBPs

Katsarava, Chu *et al.* studied *in vitro* biodegradation of AABBPs under the conditions close to physiological (t = 37 °C and pH 7.4) using both potentiometric titration (PT) [14, 24, 52, 65] and weight loss (WL) [66]. Among the synthesized AABBPs, the PEAs are the most studied polymers for both *in vitro* and *in vivo* biodegradation.

Titration is a facile and fast method to assess the tendency of polymers to hydrolytic degradation, especially for the polymers with labile ester linkages that provide rather high rates of chain scission. During ester hydrolysis generated COOH groups will be neutralized automatically by an alkaline solution, which consumption profile represents the kinetic curve of biodegradation.

This method has an advantage over WL because gravimetric measurements at early stage of biodegradation (the first 1–3 h, the time normally used for short-term assessment) are complicated due to the water absorption, particularly for those polymers having high water affinity [66].

A systematic *in vitro* biodegradation study of regular PEAs using PT method was carried out in the presence of hydrolases: trypsin, α-chymotrypsin, and lipase [65]. The spontaneous immobilization (absorption) of key enzymes from buffer solutions onto the PEA film surfaces was observed. The surface-immobilized enzymes extend the erosion of polymer and also catalyze the hydrolysis of both low- (ATEE) and high-molecular-weight (protein) external substrates. It is found that the enzyme surface absorption is reversible by nature. A kinetic method for a quantitative determination of the enzyme desorbed from the film surface was developed. The enzymes could also be impregnated into the PEAs to make them "self-destructive" at a target rate. A comparison of the PEAs and polylactide (PDLLA) *in vitro* biodegradation data showed that PEAs exhibited a higher tendency toward enzyme-catalyzed biodegradation than PDLLA.

For complete understanding of the PEA biodegradation phenomena, it would require the data from WL method, particularly in the late stage for slowly degrading biomaterials.

A systematic *in vitro* WL study of PEAs was carried out in the presence of hydrolases such as α-chymotrypsin, lipase, and a complex of proteases of Papaya [65]. The last enzyme was used for modeling the catalytic action of nonspecific proteases. It was found that the PEAs, in the presence of enzyme solutions, were biodegraded by surface erosion mechanism (close to the zero order kinetics) without compromising bulk properties: no change of polymer molecular weight and polydispersity was observed. The WL method also confirmed the catalytic action of the spontaneously immobilized enzymes and the effectiveness of impregnated enzymes, making polymers "self-destructive" at a target rate. The erosion rates of the PEAs were studied in the enzyme-catalyzed three cases: enzyme in solution, surface immobilized enzyme, and impregnated enzyme. The results ranged from 10^{-3} to 10^{-1} mg/cm^2/h, and are comparable with the erosion rates of polyanhydrides [67], the fastest biodegradable polymers. The WL also demonstrated that PEAs exhibited a higher tendency toward enzyme-catalyzed biodegradation than PDLLA.

The analysis of biodegradation products showed that the hydrolases mediated biodegradation of PEAs takes place preferentially by cleaving the ester bonds in the polymeric backbones. *N,N′*-adipoyl-bis-L-phenylalanine was separated as one of the main products of biodegradation of PEAs composed of adipic acid, phenylalanine, and 1,4-butanediol (PEA **4F4**) [66].

Based on the ^{1}H NMR study, Puiggali *et al.* [10] also concluded that the hydrolytic degradation of PEA composed of sebacic acid, L-alanine, and 1,2-dodecanediol (PEA **8A12** obtained by IP) takes place in the ester bonds and amide groups remain unchanged.

Nagata [13] also studied the enzymatic degradation of PEAs stereopolymers derived from L- and D-alanine, using proteolytic enzymes (proteinase-K, papain, and α-chymotrypsin), and lipase, and also confirmed that the degradation of PEAs with this group of hydrolases proceeds via the hydrolysis of the ester linkages and amide groups remain unchanged. The *in vitro* biodegradation mechanism of PEAs predominately via ester bonds hydrolysis was also suggested by Saotome *et al.* [6].

It has to be noted that the biodegradation rate as well as mechanical and physical–chemical properties of AABBPs can be manipulated in the widest range not only by changing their stereochemical composition (i.e., using L- and D-isomers of one α-AA [10, 13]) but also by preparing copolymers with two or more α-AAs [7, 19, 66], two or more dicarboxylic acids [19, 28]. An alternative way to tailor the properties of AABBPs lies in blending the polymers [18]. The blending of AABBPs of various classes looks possible as well because high affinity of macromolecules can guarantee their compatibility.

A preliminary *in vivo* biodegradation study of selected PEA (**4F4**) films in rats, with and without impregnated lipase [66], showed that PEAs impregnated with lipase were completely absorbed within 1–2 months, or within 3–6 months for the lipase-free samples. These findings prompt to suggest that new PEAs may have a great potential for designing drug-sustained/controlled release devices as well as implantable surgical devices.

5.2.4.3 Biocompatibility of AABBPs

Among AABBPs, only few PEAs were studied for biocompatibility. For example, the PEA composed of adipic acid, L-phenylalanine, and 1,4-butanediol (PEA **4F4**) supported the growth of human osteosarcoma and fibroblasts cells and showed the material to be biocompatible (Y. Shved and R. Katsarava, unpublished results). Aqueous solutions of model biodegradation products – *N,N′*-adipoyl-bis-L-phenylalanine and 1,4-butanediol – at 1:1 mol ratio were subcutaneously injected to rats. No acute or chronic toxicity was observed [68]. LD_{50} could not be determined since it was higher than 1500-fold excess (6 g/kg) of an average therapeutic dose, confirming high biocompatibility of the PEA and its biodegradation products.

Elastomeric functional *co*-PEA on the basis of sebacic acid (1.00 mol), TAAD composed of L-leucine and 1,6-hexanediol (0.75 mol), and L-lysine (0.25 mol) [*co*-PEA $\mathbf{8(L6)_{0.75}K_{0.25}}$] [15, 16] showed excellent blood and tissue compatibility in both *in vitro* [69] and *in vivo* (pigs) [70] tests. The same *co*-PEA selectively supported the *in vitro* growth of epithelial cells [69]. The *in vivo* biocompatibility was tested in porcine coronary arteries, comparing the polymer-coated stents with bare metal stents in 10 pigs [70]. All animals survived till sacrificed 28 days later. Prior to sacrifice, angiography revealed identical diameter stenosis in both groups. Histology confirmed similar injury scores, inflammatory reaction, and area stenosis.

These results support the notion that polymer has a high potential for cardiovascular applications [71].

Recently Yamanouchi *et al.* [34] reported that the arginine-based PEAs showed good cell compatibility over a wide range of dosages and had minimal adverse effects on the cell morphology, viability, and apoptosis. Very recently, Memanishvili *et al.* [35] showed that arginine-based PEEAs, PEEURs, and PEEUs, having PEG-type polymeric backbones, possess higher cell compatibility than the said arginine-based PEAs.

The above-mentioned biological studies of several biodegradable AABBPs indicate that this family of biodegradable polymers is biocompatible. However, these studies are rather sporadic and comprise mostly the assessment of biocompatibility. So far there is no systematic study of biocompatibility and/or tissue regeneration potential. In particular, there is no relevant data about tissue regeneration mechanisms and the influence of factors like chemical composition and biodegradation rate that determines discharging of degradation products into surrounding environment that can activate macrophages to produce cell growth factors, mediators, and so forth, for accelerated wound healing [72, 73]. Therefore, for wide practical applications of this very promising family of biodegradable polymers, it is indispensable to carry out a comprehensive study of the interaction of AABBPs of various chemical compositions with living organism to assess their biocompatibility (including immune response), and tissue regeneration capability.

5.2.5
Some Applications of AABBPs

Selected representatives of PEAs were used for constructing biodegradable hydrogels, nanoformulations, drug-eluting devises and coatings, and so forth.

Chu and Guo used the UPEAs for obtaining hybrid hydrogels through photochemical conjugation with either PEG diacrylate [30] or polysaccharides containing unsaturated double bonds (e.g., methacryloyl dextrane) [29]. The biodegradable hybrid hydrogels are promising for many biomedical and pharmaceutical applications, such as drug delivery systems and tissue engineering and so forth.

Legashvili *et al.* [31] used brush-like *co*-PEAs (Figure 5.11) to obtain molecular complexes with PEGs that are promising as nanocarriers of drugs.

Yamanouchi *et al.* [34] evaluated complexation of a novel family of synthetic biodegradable L-arginine-based PEAs (Figure 5.9) with DNA, for their capability to transfect rat vascular smooth muscle cells, a major cell type participating in vascular diseases. Arg-PEAs showed high binding capacity toward plasmid DNA. The binding activity was inversely correlated to the number of methylene groups in the diol segment of Arg-PEAs. All Arg-PEAs transfected smooth muscle cells with an efficiency that was comparable to the commercial transfection reagent Superfect. However, unlike Superfect, Arg-PEAs, after a wide range of dosages, had minimal adverse effects on cell morphology, viability, and apoptosis. The authors demonstrated that Arg-PEAs were able to deliver DNA into nearly 100% of cells under optimal polymer-to-DNA weight ratios, and the high level of delivery was achieved through an active endocytosis mechanism. A large portion of DNA delivered, however, was trapped in acidic endocytotic compartments, and subsequently was not expressed. These results suggest that with further modification to enhance their endosome escape, Arg-PEAs can be attractive candidates for nonviral gene carriers owning to their high cellular uptake nature and reliable cellular biocompatibility.

Katsarava *et al.* [18] used PEA and their blends for constructing various medical biocomposites. One of them, registered as "PhagoBioDerm" in Republic of Georgia and is produced as elastic films, represents novel wound-dressing device (artificial skin). Product consists of lytic bacteriophages, antibiotics, pain killer, and proteolytic enzymes. PhagoBioDerm showed an excellent therapeutic effect in the management of infected wounds and ulcers (of both trophic and diabetic origin) [74] and in the complex treatment of infected local radiation injuries caused by the exposure to ^{90}Sr [75].

Recently, Katsarava *et al.* [76] have developed bactericidal wound dressing that represents an alcohol solution of biodegradable *co*-PEA containing silver sulfadiazine and other antimicrobials. The preparation sprayed onto the wound forms a thin, elastic, and transparent film that accelerates healing of superficial wounds, ulcers, and burns.

The functional biodegradable *co*-PEAs $\mathbf{8(L6)_{0,75}K_{0,25}}$ with covalently attached 4-aminoTEMPO (Scheme 5.9) revealed high elastic properties and excellent adhesion to stainless steel, and is being used as vascular stent coating. Currently the polymer-coated stents are under clinical trials[1]. The results of this study suggest that the polymer is biocompatible and should not elicit an inflammatory reaction. Therefore, MediVas LLC (San Diego, CA) uses the biodegradable *co*-PEAs as bioactive wound dressings [77], wound care polymer compositions [78], vaccine delivery compositions [79], polymer particle delivery compositions [80], delivery of ophthalmologic agents to the exterior or interior of the eye [81], therapeutic polymers [82], and so forth.

5.2.6 AABBPs versus Biodegradable Polyesters

Here in brief are listed some advantageous properties of AABBPs over aliphatic PEs like polyglycolic and poly(lactic acids), their copolymers, poly(caprolactone), and so forth:

- polycondensation synthesis without using any toxic catalyst;
- higher hydrophilicity and, hence, better compatibility with tissues;
- longer shelf-life;
- a wide range of desirable mechanical properties at lower molecular weights;

1) Medivas' polymer technology was licensed to DSM Biomedical, http://www.dsm.com/en_US/html/dbm/homepage.htm.

- a variable hydrophobicity/hydrophilicity balance suitable for constructing drug-sustained/controlled release devices;
- an erosive mechanism and *in vitro* biodegradation rates ranged from 10^{-3} to 10^{-1} mg/(cm^2 h) that can be regulated by impregnating enzymes;
- fusibility (<150 °C) and solubility in common organic solvents (ethanol, THF, chloroform, methylene chloride, etc.) and ease of processing into different forms and shapes; and
- excellent adhesion to plastic, metallic, and glass surfaces that is important for their use as coatings.

5.3 Conclusion and Perspectives

The concise review highlights how dimerized and polyfunctional forms of naturally occurring α-AAs are suitable building blocks for constructing new bioanalogous macromolecular systems with non-natural orientation of α-AAs' residues in the polymeric backbones. This nonconventional architecture of polymer chain is expected to provide low immunogenicity of α-AAs-based polymeric biomaterials by "confusing nature." For practical biomedical applications as biodegradable biomaterials, the most promising are polymers containing hydrolysable ester bonds in the backbones called as amino acid based biodegradable polymers (AABBPs). There are various classes of high-molecular-weight AABBPs: PEAs, PEURs, and PEUs obtained with α-AAs and other nontoxic building blocks such as fatty diols, dicarboxylic acids, α-Has, and carbonic acid. The selection of appropriate building blocks under optimal polycondensation methods allows the synthesis of AABBPs with tailored material properties. Listed AABBPs contain H-bond-forming chemical units that enhance their mechanical characteristics, hydrophilicity, and biocompatibility. The AABBPs exhibit some obvious advantages over existing and commercially successful aliphatic PEs: polyglycolic and poly(lactic acids), their copolymers, poly(caprolactone), and so forth.

The most extensively studied AABBPs to date are PEAs because of ample availability, low cost of starting monomers, and desirable material properties that can be tailored in a wide range. PEAs have been successfully tested, for example, as medicated wound dressing, also, in animals and humans for cardiovascular applications. *Ex vivo* cell-based assays have strongly supported recent human trial data indicating that PEAs are blood and tissue compatible, with advantageous properties for implantation into tissue. Tremendous value that PEAs have in the health science and industry is the mechanism by which they degrade, and drug release profile. PEAs' biodegradation and mechanism of drug release is believed to proceed by surface erosion and primarily follows zero-order kinetics. These unique material properties have shown PEAs' multiuse potential as a new family of biodegradable biomaterials as drug delivery platforms or as components of resorbable surgical implants.

Other AABBPs – PEURs and PEUs – also have potential for numerous biomedical applications. However, for wider applications of these polymers, as well as for expanding scopes of PEAs' applications, it is important to carry out a comprehensive study on the whole family of AABBPs, to determine how they interact with living organism, to assess their biocompatibility (including immune response), and tissue regeneration capability. This study has to answer the question "are high biocompatibility, toxicological safety, and accelerated tissue regeneration inherent characteristics of AABBPs?"

References

1 Shalaby, S.W. and Johnson, R.A. (1994) Synthetic absorbable polyesters, in *Biomedical Polymers: Designed-to Degrade Systems* (ed. S.W. Shalaby), Hanser, New York, pp. 1–34.

2 Jacoby, M. (2001) Custom-made biomaterials. *Chem. Eng. News*, **5** (2), 30–32.

3 Nathan, A. and Kohn, J. (1994) Amino acid derived polymers, in *Biomedical Polymers: Design-to-Degrade Systems* (ed. S.W. Shalaby), Hanser, New York, pp. 117–151.

4 Huang, S.J., Bansleben, D.A., and Knox, J.R. (1979) Biodegradable polymers: chymotrypsin degradation of a low molecular weight poly(ester-urea) containing phenylalanine. *J. Appl. Polym. Sci.*, **23**, 429–437.

5 Asin, L., Armelin, E., Montane, J., Rodriguez-Galan, A., and Puiggali, J. (2001) Sequential poly(ester amides) based on glycine, diols, and dicarboxylic acids: thermal polyesterification versus interfacial polyimidation. *J. Polym. Sci. A: Polym. Chem.*, **39**, 4283–4293.

6 Saotome, Y., Miyazawa, T., and Endo, T. (1991) Novel enzymatically degradable polymers comprising α-amino acid, 1,2-ethanediol and adipic acid. *Chem. Lett.*, 21–24.

7 Saotome, Y., Tashiro, M., Miyazawa, T., and Endo, T. (1991) Enzymatic degrading solubilization of a polymer comprising glycine, phenylalanine, 1,2-ethanediol, and adipic acid. *Chem. Lett.*, 153–154.

8 Parades, N., Rodriguez-Galan, A., and Puiggali, J. (1998) Synthesis and characterization of a family of biodegradable poly(ester amides) derived from glycine. *J. Polym. Sci. Part A: Polym. Chem.*, **36**, 1271–1282.

9 Parades, N., Casas, M.T., Puiggali, J., and Lotz, B. (1999) Structural data on the packing of poly(ester amide)s derived from glycine, hexanediol, and odd-numbered dicarboxylic acids. *J. Polym. Sci. Part A: Polym. Chem.*, **37**, 2521–2533.

10 Paredes, N., Rodrıguez-Galan, A., Puiggalı, J., and Peraire, J. (1998) Studies on the biodegradation and biocompatibility of a new poly(ester amide) derived from L-alanine. *J. Appl. Polym. Sci.*, **69**, 1537–1549.

11 Rodrıguez-Galan, M., Pelfort, J.E., Aceituno, J., and Puiggalı (1999) Comparative studies on the degradability of poly(ester amides) derived from L- and L,D-alanine. *J. Appl. Polym. Sci.*, **74**, 2312–2320.

12 Rodrıguez-Galan, A., Fuentes, L., and Puiggalı, J. (2000) Studies on the degradability of a poly(ester amide) derived from L-alanine, 1,12-dodecanediol and 1,12-dodecanedioic acid. *Polymer*, **41**, 5967–5970.

13 Nagata, M. (1999) Synthesis and enzymatic degradation of poly(ester-amide) stereocopolymers derived from alanine. *Macromol. Chem. Phys.*, **200**, 2059–2064.

14 Katsarava, R., Arabuli, N., Beridze, V., Kharadze, D., Chu, C.C., and Won, C.Y. (1999) Amino acid-based bioanalogous polymers. Synthesis, and study of regular poly(ester amides) based on bis(α-amino acid) α,ω-alkylene diesters, and aliphatic dicarboxylic acids. *J. Polym Sci. Part A: Polym. Chem.*, **37**, 391–407.

15 Jokhadze, G., Machaidze, M., Panosyan, H., Chu, C.C., and Katsarava, R. (2007) Synthesis and characterization of functional elastomeric poly(ester amide) co-polymers. *J. Biomater. Sci. Polym. Ed.*, **18**, 411–438.

16 Chu, C.C. and Katsarava, R. (2003, 2007, 2008) Elastomeric functional biodegradable copolyester amides and copolyester urethanes. U.S. Patents: 6,503,538; 7,304,122; 7,408,018.

17 Gomurashvili, Z., Zhang, H., Da, J., Jenkins, T.D., Katsarava, R., and Turnell, W.G. (2006) From drug-eluting stents to bio-pharmaceuticals: poly(ester amide) a versatile new bioabsorbable polymer. ACS 232nd National Meeting, 10–14 September, San Francisco, CA.

18 Katsarava, R. and Alavidze, Z. (2004) Polymer Blends as Biodegradable Matrices for Preparing Biocomposites. US 6,703,040 B2.

19 Chkhaidze, E. (2008) Biodegradable unsaturated poly(ester amide)s composed of α-amino acids: synthesis and transformations, PhD thesis, Technical University of Georgia, Tbilisi, Georgia

20 Katsarava, R., Tugushi, D., Zavradashvili, N., Gomurashvili, Z., Jokhadze, G., Gverdtsiteli, M., and Samkharadze, M. (2007) Amino acid based epoxy-poly(ester amide)s – a new class of biodegradable functional polymers: synthesis and characterization. Polymers in Medicine and Biology, Santa Rosa, CA.

21 Ochkhikidze, N., Razmadze, E., Tugushi, D., Kupatadze, N., Gomurashvili, Z., and Katsarava, R. (2008) AABB-poly(depsipeptide)s – a new class of amino acid based biodegradable polymers. International Symposium "Polycondensation-2008", 8–11 September, Tokyo, Japan.

22 Katsarava, R., Kharadze, D., Omiadze, T., Tsitlanadze, G., Goguadze, T., Arabuli, N., and Gomurashvili, Z. (1994) New biodegradable polymers on the basis of *N,N*-diacyl-bis-phenylalanine. *Vysokomol. Soedin. Ser. A*, **36** (5), 1162–1167.

23 Kharadze, D., Kirmelashvili, L., Medzmariashvili, N., Tsitlanadze, G., Tugushi, D., Chu, C.C., and Katsarava, R. (1999) Synthesis and α-chymotrypsinolysis of regular poly(ester amides) based on phenylalanine, diols, and terephthalic acid. *Polymer Sci. (Russia) Ser. A*, **41** (2), 883–890.

24 Arabuli, A., Tsitlanadze, G., Edilashvili, L., Kharadze, D., Goguadze, T., Beridze, V., Gomurashvili, Z., and Katsarava, R. (1994) Heterochain polymers based on natural amino acids. Synthesis and enzymatic hydrolysis of regular poly(ester amide)s based on bis(L-phenylalanine)-α,ω-alkylene diesters and adipic acid. *Macromol. Chem. Phys.*, **195**, 2279–2289.

25 Gomurashvili, Z., Katsarava, R., and Kricheldorf, H.R. (2000) Amino acid based bioanalogous polymers. Synthesis and study of new poly(ester amide)s composed of hydrophobic α-amino acids and dianhydrohexitoles. *J. Macromol. Sci. Pure Appl. Chem.*, **37**, 215–227.

26 Okada, M., Yamada, M., Yokoe, M., and Aoi, K. (2001) Biodegradable polymers based on renewable resources. V. Synthesis and biodegradation behavior of poly(ester amides) composed of 1,4:3,6-dianhydro-D-glucitol, α-amino acid, and aliphatic dicarboxylic acid units. *J. Appl. Polym. Sci.*, **81**, 2721–2726.

27 Guo, K., Chu, C.C., Chkhaidze, E., and Katsarava, R. (2005) Synthesis and characterization of novel biodegradable unsaturated poly(ester amide)s. *J. Polym. Sci. Part A: Polym. Chem.*, **43**, 1463–1477.

28 Guo, K. and Chu, C.C. (2007) Synthesis, characterization, and biodegradation of copolymers of unsaturated and saturated poly(ester amide)s. *J. Polym. Sci. Part A: Polym. Chem.*, **45**, 1595–1606.

29 Guo, K. (2005) Biodegradable unsaturated poly(ester amide)s and their hydrogels: synthesis, characterization, biodegradation and application as drug carriers. PhD thesis, Cornell University, Ithaca, NY.

30 Guo, K. and Chu, C.C. (2005) Synthesis and characterization of novel biodegradable unsaturated poly(ester amide)/poly(ethylene glycol) diacrylate hydrogels. *J. Polym. Sci.: Part A: Polym. Chem.*, **43**, 3932–3944.

31 Legashvili, I., Nepharidze, N., Katsarava, R., Sannigrahi, B., and Khan, I.M. (2007) Non-covalent nano-adduct of co-poly(ester amide) and poly(ethylene glycol): preparation, characterization and model

drug-release studies. *J. Biomater. Sci. Polym. Ed.*, **18**, 673–685.

32 Song, H. (2007) L-arginine based biodegradable poly(ester amide)s, their synthesis, characterization, fabrications, and applications as drug and gene carriers. PhD thesis, Cornell University, Ithaca, NY.

33 Chu, C.C., Mutschler-Chu, M.A., Song, H., Liu, B., and Gomurashvili, Z. (2009) Biodegradable cationic polymer gene transfer compositions and methods of use. PCT Application WO 2009/015143.

34 Yamanouchi, D., Wu, J., Lazar, A.N., Craig Kent, K., Chu, C.C., and Liu, B. (2008) Biodegradable arginine-based poly(ester-amide)s as non-viral gene delivery reagents. *Biomaterials*, **29**, 3269–3277.

35 Memanishvili, T., Kupatadze, N., Tugushi, D., Torchilin, V.P., and Katsarava, R. (2010) Biodegradable arginine-based polymers with PEG-like backbones as potential non-viral gene delivery system. 1st Russian-Hellenic Symposium with International Participation *"Biomaterials and Bionanomaterials: Recent Advances and Safety-Toxicology Issues"*, May 3–9, Iraklion, Crete, Greece.

36 Guo, K. and Chu, C.C. (2007) Synthesis, characterization and biodegradation of novel poly(ether ester amide)s based on L-phenylalanine and oligoethylene glycol. *Biomacromolecules*, **8**, 2851–2861.

37 Katsarava, R., Kharadze, D., and Avalishvili, L. (1982) Synthesis of active diesters directly from free dicarboxylic acids. *Izv. Akad. Nauk Gruz. SSR Ser. Khim.*, **8** (1), 102–109.

38 Katsarava, R., Kartvelishvili, T., Kharadze, D., and Patsuria, M. (1987) Synthesis of polyurethanes by polycondesation of active bis-carbonates of diols with diamines under mild conditions. *Vysokomol. Soed. Ser. A*, **29**, 2069–2078.

39 Katsarava, R., Kartvelishvili, T., Khosruashvili, T., and Beridze, V. (1995) Synthesis of polyurethanes by polycondensation of active biscarbonates of diols with hexamethylenediamine and its derivatives. *Macromol. Chem. Phys.*, **196**, 3061–3074.

40 Katsarava, R.D., Kartvelishvili, T.M., Japaridze, N.N., Goguadze, T.A., and Khosruashvili, T. (1993) Synthesis of polyureas by polycondensation of diamines with active derivatives of carbonic acid. *Macromol. Chem.*, **194**, 3209–3228.

41 Morgan, P.W. (1965) *Condensation Polymers: By Interfacial and Solution Methods*, Interscience, New York.

42 Katsarava, R. (1991) Progress and problems in active polycondensation. *Russian Chem. Rev. (The British Library)*, **60** (7), 722–737.

43 Katsarava, R., Ochkhikidze, N., Tugushi, D., and Gomurashvili, Z. (2008) AABB-poly(depsipeptide) biodegradable polymers and methods of use (to MediVas LLC), U.S. Provisional Application # 61/088,678.

44 Katsarava, R., Mazanashvili, N., Mchedlishvili, N., and Gomurashvili, Z. (2007) Alkylene-dicarboxylate-containing biodegradable poly(ester-amides) and methods of use (to MediVas LLC), U.S. Application # 11/728,100.

45 Chu, C.C., Katsarava, R., and Guo, K. (2006) Unsaturated poly(ester-amide) biomaterials (Cornell University), *U.S. Application* # 11/587,530.

46 Katsarava, R., Tugushi, D., Zavradashvili, N., and Gomurashvili, Z. (2007) Epoxy-containing poly(ester-amides) and methods of use (to MediVas LLC), U.S. Application # 11/893,719.

47 Katsarava, R. (2003) Active polycondensation: from peptide chemistry to amino acid based biodegradable polymers. *Macromol. Symp.*, **199**, 419–429.

48 Katsarava, R. and Kharadze, D. (1991) The study of stability of active phenyl esters of carboxylic acids in aprotic polar solvents. *Zh. Obshch. Khim.*, **61** (11), 2413–2418.

49 Nathan, A., Bolikal, D., Vyavahare, N., Zalipsky, S., and Kohn, J. (1992) Hydrogels based on water-soluble poly(ether urethanes) derived from L-lysine and poly(ethylene glycol). *Macromolecules*, **25**, 4476–4484.

50 Nathan, A., Zalipsky, S., Ertel, S.I., Agathos, S.N., Yarmush, M.L., and Kohn, J. (1993) Copolymers of lysine and polyethylene glycol: a new family of

functionalized drug carriers. *Bioconjug. Chem.*, **4**, 54–62.

51 Vyavahare, N. and Kohn, J. (1994) Photocrosslinked hydrogels based on copolymers of poly(ethylene glycol) and lysine. *J. Polym. Sci. Part A: Polym. Chem.*, **32**, 1271–1281.

52 Kartvelishvili, T., Tsitlanadze, G., Edilashvili, L., Japaridze, N., and Katsarava, R. (1997) Amino acid based bioanalogous polymers. Novel regular poly(ester urethane)s and poly(ester urea) s based on bis-(L-phenylalanine) α,ω-alkylene diesters. *Macromol. Chem. Phys.*, **198**, 1921–1932.

53 Katsarava, R., Tugushi, D., and Gomurashvili, Z. (2006) Poly(ester urea) polymers and methods of use U.S. Patent Application 20070128250 A1.

54 Kricheldorf, H.R. and Shwarz, G. (2003) Cyclic polymers by kinetically controlled step-growth polymerization. *Macromol. Rapid Commun.*, **24**, 359–381.

55 Kricheldorf, H.R. (2008) The role of self-dilution in step-growth polymerizations. *Macromol. Rapid Commun.*, **29**, 1695–1704.

56 Katsarava, R., Kharadze, D., and Avalishvili, L. (1986) Synthesis of high-molecular-weight polysuccinamides by polycondensation of active succinates with diamines. *Macromol. Chem.*, **187**, 2053–2062.

57 Turnell, W., Cruz-Aranda, G.A., Wu, M.M., Chantung, R.L., Gomurashvili, Z., and DeFife, K.M. (2009) Cationic alpha-amino acid-containing biodegradable polymer gene transfer compositions. US Patent Application # 2009/0068743.

58 Lou, X., Detrembleur, C., Leconte, P., and Jerome, R. (2002) Novel unsaturated ε-caprolactone polymerizable by ring-opening and ring-opening metathesis mechanisms. *e-Polymers*, **(34)**, 1–12.

59 Nepharidze, N., Machaidze, M., Zavradashvili, N., Mazanashvili, N., Tabidze, V., Katsarava, R., and Tugushi, D. (2006) Biodegradable copoly(ester amide)s having hydrophobic lateral substituents. *Polim. Med.*, **2**, 27–33.

60 Gomurashshvili, Z., Turnell, W.G., Vassilev, V., and Chowdary, N.S. (2007) Biodegradable water soluble polymers. US Patent Application # 2007/0282011 A1.

61 Gomurashvili, Z., Zhang, H., Da, J., Jenkins, T.D., Hughes, J., Wu, M., Lambert, L., Grako, K.A., DeFife, K.M., Macpherson, K., Vassilev, V., Katsarava, R., and Turnell, W.G. (2008) *From Drug-Eluting Stents to Biopharmaceuticals: Poly(Ester Amide) A Versatile New Bioabsorbable Biopolymer. ACS Symposium Series 977: Polymers for Biomedical Applications* (eds A. Mahapatro and A.S. Kulshrestha), Oxford University Press, Oxford, pp. 10–26.

62 Greenstein, J.P. and Winitz, M. (1961) *Chemistry of the Amino Acids*, John Wiley & Sons, Inc., New York, London.

63 Zavradashvili, N. (2008) New functional biodegradable poly(ester amide)s composed of α-amino acids and their potential application in biomedicine. PhD thesis, Technical University of Georgia, Tbilisi, Georgia.

64 Zavradashvili, N., Jokhadze, G., Kviria, T., and Katsarava, R. (2008) Thermally- and photo-chemically curable biodegradable poly(ester amide)s with double bond moieties in lateral chains, in *Chemistry of Advanced Compounds and Materials* (eds N. Lekishvili and G.E. Zaikov), NOVA Science Publishers, Inc., New York, pp. 173–179.

65 Tsitlanadze, G., Kviria, T., Chu, C.C., and Katsarava, R. (2004) *In vitro* enzymatic biodgradation of amino acid based poly(ester amide)s biomaterials. *J. Mater. Sci. Mater. Med.*, **15**, 185–190.

66 Tsitlanadze, G., Machaidze, M., Kviria, T., Djavakhishvili, N., Chu, C.C., and Katsarava, R. (2004) Biodegradation of amino acid based poly(ester amides): *in vitro* weight loss and preliminary *in vivo* studies. *J. Biomater. Sci. Polym. Ed.*, **15**, 1–24.

67 Leong, K.W., Brott, B.C., and Langer, R. (1985) Bioerodible polyanhydrides as drug-carrier matrices. I: characterization, degradation, and release characteristics. *J. Biomed. Mater. Res.*, **19**, 941–948.

68 Kshutashvili, T., Tuskia, L., and Shubladze, I. (1998) Preclinical Trials of "PhagoBioDerm", Institute of

Experimental and Clinical Medicine, Medical University of Tbilisi.
69 DeFife, K.M., Grako, K., Cruz-Aranda, G., Price, S., Chantung, R., Macpherson, K., Khoshabeh, R., Gopalan, S., and Turnell, W.G. (2009) Poly(ester amide) copolymers promote blood and tissue compatibility. *J. Biomater. Sci.*, **20**, 1495–1511.
70 Lee, S.H., Szinai, I., Carpenter, K., Katsarava, R., Jokhadze, G., Chu, C.C., Huang, Y., Verbeken, E., Bramwell, O., De Scheerder, I., and Hong, M.K. (2002) In-vivo biocompatibility evaluation of stents coated with a new biodegradable elastomeric and functional polymer. *Coron. Artery Dis.*, **13**, 237–241.
71 Carpenter, K.W., Zhang, H., McCarthy, B.J., Szinai, I., Turnell, W.G., Gopalan, S.M., and Katsarava, R. (2004) Bioactive stents and methods for use thereof, US Patent Application 20040170685 A1.
72 Pratt, L.M. and Chu, C.C. (1994) Dimethyltitanocene-induced surface chemical degradation of synthetic biodegradable polyesters. *J. Polym. Sci. Part A: Polym. Chem.*, **32**, 949–960.
73 Greisler, H.P., Henderson, S.C., and Lam, T.M. (1993) Basic fibroblast growth factor production *in vitro* by macrophages exposed to dacron and polyglactin 910. *J. Biomater. Sci. Polym. Ed.*, **4**, 415–426.
74 Markosishvili, K., Tsitlanadze, G., Katsarava, R., Morris, J.G., and Sulakvelidze, A. (2002) A novel sustained-release matrix based on biodegradable poly(ester amide)s and impregnated with bacteriophages and an antibiotic shows promise in management of infected venous stasis ulcers and other poorly healing wounds. *Int. J. Dermatol.*, **41**, 453–458.
75 Jikia, D., Chkhaidze, N., Imedashvili, E., Mgaloblishvili, I., Tsitlanadze, G., Katsarava, R., Morris, J.G., Jr., and Sulakvelidze, A. (2005) The use of a novel biodegradable preparation capable of the sustained release of bacteriophages and ciprofloxacin, in the complex treatment of multidrug-resistant *Staphylococcus aureus*-infected local radiation injuries caused by exposure to Sr90. *Clin. Exp. Dermatol.*, **30**, 23–26.
76 Katsavara R., PI (2007) Medical Bactericidal Glue GF-6. CRDF/STEP Project # BPG - 01/08 2007.
77 Carpenter, K.W., Turnell, W.G., Defife, K.M., Grako, K.A., and Katsarava, R. (2008) Bioactive wound dressings and implantable devices and methods of use. U.S. Patent Application 20080020015 A1.
78 Carpenter, K.W., Zhang, H., Mccarthy, B.J., Szinai, I., Turnell, W.G., Gopalan, S.M., and Katsarava, R. (2006) Wound care polymer compositions and methods for use thereof. U.S. Patent Application 20060188486 A1.
79 Turnell, W.G., Vassilev, V.P., Defife, K.M., Li, H., Gomurashvili, Z., and Katsarava, R. (2006) Vaccine delivery compositions. U.S. Patent Application 20060188469 A1.
80 Turnell, W.G., Li, H., Gomurashvili, Z., and Katsarava, R. (2006) Polymer particle delivery compositions and methods of use, U.S. Patent Application US20060177416.
81 Landis, G.C., Turnell, W.G., and Yumin, Y. (2007) Delivery of ophthalmologic agents to the exterior or interior of the eye. U.S. Patent Application 20070292476.
82 Turnell, W.G., Gomurashvili, Z., and Katsarava, R. (2006) Therapeutic polymers and methods. U.S. Patent Application 20060286064 A1.

6
Biodegradable Polyurethanes and Poly(ester amide)s

Alfonso Rodríguez-Galán, Lourdes Franco, and Jordi Puiggalí

Abbreviations

BDI	1,4-buthylenediisocyanate
BDO	1,4-butanediol
DSC	differential scanning calorimetry
DMSO	dimethyl sulfoxide
DMPA	dimethylol propionic acid
DUD	diurethanediol
DMTA	dynamic mechanical thermal analysis
EM	electron microscopy
ED	ethylene diamine
H12MDI	dicyclohexylmethane diisocyanate
HS	hard segment
HDI	1,6-hexamethylene diisocyanate
IR	infrared spectroscopy
IPDI	isophorone diisocyanate
LDI	lysin methyl ester diisocyanate
MDI	diphenylmethane diisocyanate
NMR	nuclear magnetic resonance
PCL	polycaprolactone
PCUs	polycarbonate-based polyurethanes
PDA	propanediamine
PDMO	poly(decamethylene glycol)
PEAs	poly(ester amide)s
PEEA	poly(ether ester amide)
PEO	poly(ethylene glycol)
PEUs	polyester-based polyurethanes
PHMO	poly(hexamethylene glycol)
POMO	poly(octamethylene glycol)
PTMO	polytetramethylene oxide glycol
PURs	polyurethanes
SAXS	small-angle X-ray scattering

Handbook of Biodegradable Polymers: Synthesis, Characterization and Applications, First Edition. Edited by Andreas Lendlein, Adam Sisson.
© 2011 Wiley-VCH Verlag GmbH & Co. KGaA. Published 2011 by Wiley-VCH Verlag GmbH & Co. KGaA.

TMDI	trimethylhexamethylene diisocyanate
WAXD	wide-angle X-ray diffraction

6.1
Chemistry and Properties of Biodegradable Polyurethanes

Polyurethanes (PURs) were first used for industrial applications in the 1940s, but the development of biocompatible polymers did not start until the 1960s. PURs have since then remained one of the most popular groups of biomaterials employed in medical devices. Toughness, durability, biocompatibility, and biostability are some of the characteristics that make PURs interesting for a wide variety of long-term implantable devices. However, the number of applications requiring biodegradability instead of biostability is on the rise, and consequently also the demand for new PURs with a controlled degradation rate.

Biodegradable PURs employed as thermoplastics are basically synthesized using a diisocyanate, a diol, and a chain-extension agent as main raw components [1, 2] (Tables 6.1–6.3, Figure 6.1). Although both aromatic and aliphatic diisocyanates have an applied interest, it should be pointed out that the putative carcinogenic nature of aromatic compounds [3, 4] is leading to an increasing use of HDI, BDI, and LDI, whose ultimate degradation products are more likely to be nontoxic (e.g., lysine).

The diol component commonly chosen is a low-molecular-weight polymer with hydroxyl end groups and a backbone that, in the case of biodegradable PURs, may correspond to a polyether, polyester, or polycarbonate [5]. The first gave rise to the

Table 6.1 Diisocyanate raw materials.

OCN–$CH_2CH_2CH_2CH_2CH_2CH_2$–NCO *1,6-Hexamethylene diisocyanate (HDI)*	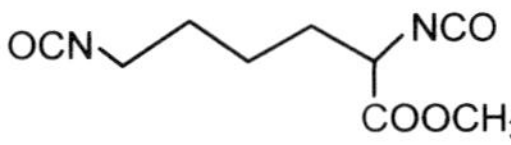 *Lysine methyl ester diisocyanate (LDI)*
OCN–$CH_2CH_2CH_2CH_2$–NCO *1,4-Butylenediisocyanate (BDI)*	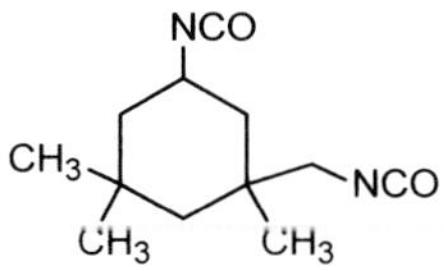 *Isophorone diisocyanate (IPDI)*
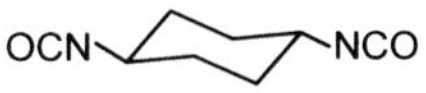 *trans-1,4-Cyclohexylene diisocyanate*	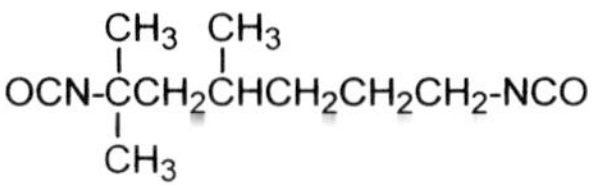 *2,2,4-Trimethylhexamethylene diisocyanate (TMDI)*
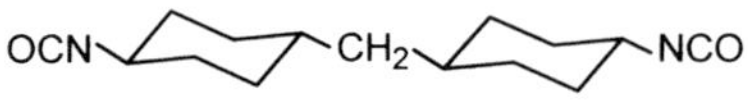 *Dicyclohexylmethane diisocyanate (H12MDI)*	

Table 6.2 Macrodiol raw materials.

Polyether-based PURs		Polyesther-based PURs	
Poly(ethylene glycol), PEO	$HO[(CH_2)_2O]_nH$	*Polyglycolide*	$HO[CH_2COO]_n–R–[OOCCH_2]_mOH$
Poly(tetramethylene glycol), PTMO	$HO[(CH_2)_4O]_nH$	*Poly(D,L-lactide)*	$HO[CH_2COO]_n-R-[OOCCH_2]_mOH$ $HO[CH(CH_3)COO]_n-R-[OOCCH(CH_3)]_mOH$
Poly(hexamethylene glycol), PHMO	$HO[(CH_2)_6O]_nH$	*Poly(ε-caprolactone)*	$HO[(CH_2)_5COO]_n–R–[OOC(CH_2)_5]_mOH$
Poly(octamethylene glycol), POMO	$HO[(CH_2)_8O]_nH$		
Poly(decamethylene glycol), PDMO	$HO[(CH_2)_{10}O]_nH$	***Poly(ethylene adipate)*** m = 2 ***Poly(propylene adipate)*** m = 3 ***Poly(butylene adipate)*** m = 4	$HO[(CH_2)_mOOC(CH_2)_4COO]_nH$

Table 6.3 Chain extender raw materials.

Diols		Diamines	
Ethylene glycol	$HOCH_2CH_2OH$	*Ethylene diamine (ED)*	$H_2NCH_2CH_2NH_2$
1,4-Butanediol (BDO)	$HOCH_2CH_2CH_2CH_2OH$	*1,3-Propanediamine (1,3-PDA)*	$H_2NCH_2CH_2CH_2NH_2$
1,3-Butanediol	$HOCH_2CH_2CH(OH)CH_3$	*1,2-Propanediamine (1,2-PDA)*	$H_2NCH_2CH(NH_2)CH_3$
2,2-Dimethyl-propanediol	$HOCH_2C(CH_3)_2CH_2OH$	*1,4-Butanediamine*	$H_2NCH_2CH_2CH_2CH_2NH_2$
 *1,4-Cyclohexanedimethanol*		 *1,4-Cyclohexanedimethanol-L-phenylalanine diester*	

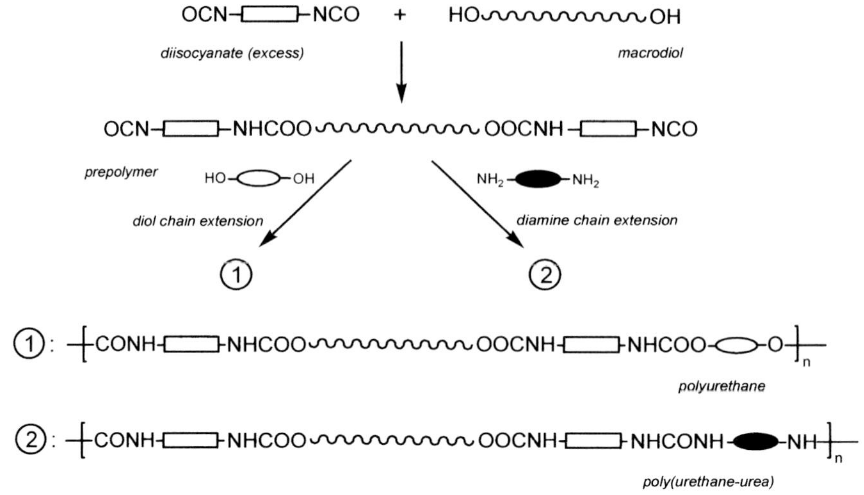

Figure 6.1 Schematic representation showing the two steps involved in the synthesis of segmented polyurethanes.

so-called polyether-based urethanes, which have been the most common so far. Nevertheless, in recent years polyester-based PURs have begun to be developed due to their increased biodegradability. Selected macrodiols are all viscous liquids with a number average molecular weight ranging between 400 and 5000 g/mol. Polyester diols can be prepared by ring-opening polymerization of a cyclic lactone [6] or condensation between a dicarboxylic acid and an excess of a diol. In some cases, the polyester diol, which is characterized by a hydrophobic character, is mixed with the more hydrophilic polyethylene glycol (PEG) before performing the reaction with the corresponding diisocyanate. This way, PURs with an increased biodegradation rate and enhanced cell attachment can be obtained. Note that these characteristics can be easily tailored by a simple change in the composition of the mixture [7].

The reaction between the diol and the diisocyanate is carried out with an excess of the latter (keeping the isocyanate/hydroxyl molar ratio usually close to 2:1) in order to obtain a reactive prepolymer with isocyanate end groups. Catalysts (typically tertiary amines, stannous octoate, or dibutyltin dilaurate) and high temperatures (60–90 °C) are required to increase the reaction rate. A thermoplastic PUR material characterized by a segmented architecture is finally obtained by reaction of the terminal isocyanate groups with a chain extender (Figure 6.1) which may be either a diol or a diamine with low molecular weight [8]. In the first case, urethane bonds are formed and the final polymer is usually thermally processable, whereas in the second case new urea bonds are formed and the resulting poly(urethane/urea) is usually only suitable for solvent casting.

Some secondary reactions, which generally result in branched or cross-linked polymers, can also occur under certain conditions [9]. The most usual are

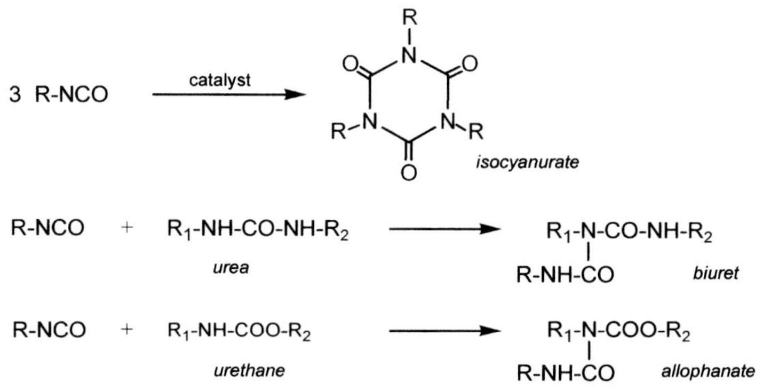

Scheme 6.1 Characteristic secondary reactions observed in the synthesis of polyurethanes.

(Scheme 6.1) (i) trimerization of isocyanate groups leading to isocyanurates, (ii) formation of biuret linkages from urea groups, and (iii) formation of allophanate units by reaction between an isocyanate group and the NH of a urethane group. This last reaction may sometimes be of interest since mechanical properties can be improved by a small number of crosslinking bonds. A great advantage is that the allophanate formation reaction is thermally reversible, and so it is feasible to obtain thermally processable materials.

From an industrial point of view, PUR synthesis can be performed in a single step by mixing all reagents or following the above two-step methodology [8–10]. In the first case, bulk polymerization can be carried out by a single batch procedure or by a semicontinuous process using reactive extruders or injection-molding machines. The two-step procedure has two main advantages: (i) the polymer architecture can be well controlled, and (ii) polymers with a heterogeneous composition, which are obtained when nonpolar macrodiols are involved, can be avoided. This synthesis can be accomplished in bulk or in solvents (typically *N,N*-dimethylacetamide and *N,N*-dimethylformamide) [11] although the latter option is commercially less attractive.

The mechanical properties of segmented PURs are highly interesting due to the microphase separation (Figure 6.2a) of their two constitutive segments [12]: nonpolar soft segments and more polar hard segments derived from the diisocyanate and the chain extender. The soft microdomain is amorphous and often has a glass transition temperature lower than 0 °C, resulting in rubber characteristics like extensibility and softness. In contrast, hard segments can crystallize as a consequence of the strong hydrogen-bond intermolecular interactions that can be established between their urethane or urea groups. These ordered domains act as physical crosslinks providing cohesive strength to the polymer matrix and allowing the material to resist flow when stress is applied. Segmented PURs can be considered thermoplastic elastomers since physical crosslinks can be easily disrupted by heating the polymer above the melting temperature of hard segment domains or by dissolving the material in aprotic solvents like dimethylformamide.

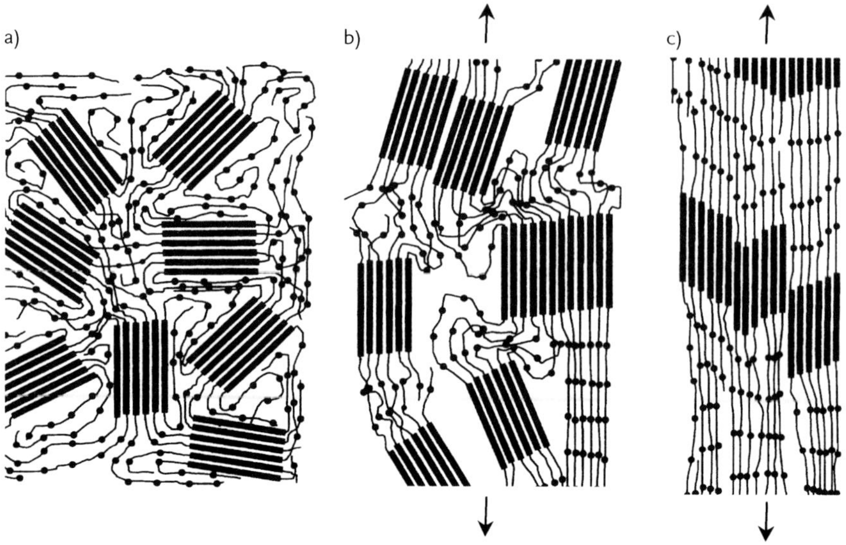

Figure 6.2 Representation of the characteristic microphase separation in a segmented polyurethane (a) and the influence of stretching into orientation and crystallization of microdomains (b, c). Moderate (b) and high extension (c) are represented. The thick strokes represent hard segments and the thin strokes soft segments.

Thermoset PURs can be prepared by inducing chemical crosslinks, either in the hard segment or the soft segment, or both. The resulting material has greater strength and durability, and worse phase separation. Crosslinking is achieved by using intermediates with a functionality higher than two (e.g., trimethylolpropane, glycerol, and 1,2,6-hexanetriol) (Scheme 6.2). These networks can have rigid or flexible characteristics, mainly depending on the density of chemical crosslinks, and may give rise to biodegradable foams useful for many applications such as scaffolds [13–15]. In fact, the reaction of water with an isocyanate group leads to the formation of carbon dioxide gas, which can be used as a blowing agent in the creation of pores.

Several factors must be considered when designing PUR materials with targeted properties [16]: (i) harder and stiffer polymers with higher tear strength and lower elongation at break can be prepared by increasing the chain extender to diol ratio and/or decreasing the molecular weight of the macrodiol unit; (ii) diamine chain extenders lead to hard segments with higher melting temperature and harder mechanical properties; (iii) aromatic diisocyanates increase chain stiffness and facilitate aggregation of the hard phase by π-electron association; and (iv) variation in the number of substitutions and spacing between and within branch chains affects the flexibility of molecular chains. The knowledge of the hard segment content is thus an easy way to predict mechanical properties of PURs: soft material (HS < 15 wt%), rubbery elastomer (15 wt% < HS< 40 wt%), tough elastomer (40% < HS < 65 wt%), and strong engineering polymer (HS > 65 wt%) [17].

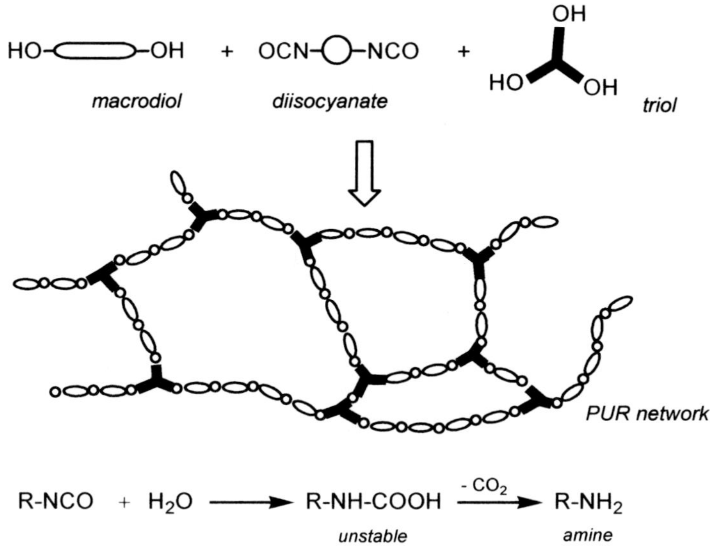

Scheme 6.2 Synthesis of PURs' networks and reaction conducing to CO_2 as blowing agent.

Indeed, understanding the morphology is crucial for the design of materials with specific properties. Molecular organization of PURs has been investigated by several techniques, including differential scanning calorimetry (DSC), wide-angle X-ray diffraction (WAXD), small-angle X-ray scattering (SAXS), infrared spectroscopy (IR), electron microscopy (EM), dynamic mechanical thermal analysis (DMTA), and nuclear magnetic resonance (NMR) [18].

DSC experiments show that PURs have several thermal transitions, the interpretation of which is rather complex [19]. Glass transitions of both hard and soft amorphous microphases can be detected. The T_g value of the soft domain, which appears at the lowest temperature, may be used to evaluate the number of hard segments in this domain since T_g should increase when the degree of mixing is raised. However, a quantitative analysis is problematic due to the influence of factors like restrictions on the motion of soft segments caused by the presence of microcrystals. In addition, DSC traces can show multiple endothermic peaks which may be ascribed to morphological effects and be broadly divided into loss of long- and short-range order.

Early explanations about these multiple endotherms were based on the disruption of different kinds of hydrogen-bonding interactions [20, 21]. However, infrared thermal analysis led to discarding a clear relationship between endothermic peaks and these interactions [22]. Hydrogen bonding plays a significant role in the design of biostable or biodegradable materials as it is a determinant factor of their hydrolytic stability. Susceptibility to hydrolytic degradation is clearly enhanced when the carbonyl groups in the hydrolyzable group do not act as hydrogen-bond acceptors. The knowledge of hydrogen-bond distribution in PUR materials is thus essential to obtain materials with a specific degradation rate.

Molecular ordering crystallization may be favored by subjecting a PUR chain to stress [23]. Thus, at a moderate extension (e.g., 250%) macrodiols of the soft segment become partially aligned and crystallized. When the extension is increased, further crystallization occurs and hard segments turn into the direction of elongation and form paracrystalline layer lattice crystals (Figure 6.2).

6.2
Biodegradation Mechanisms of Polyurethanes

Susceptibility of PURs to biodegradation is an inherent feature of their chemistry [24, 25]. It was detected by the industrial manufacturing community before systematic biodegradation studies were conducted in the 1980s. In fact, degradation of PURs may initiate during fabrication due to high temperatures, the presence of liquids, and the difficulty to completely remove moisture from the reaction mixture [26].

Microorganisms can be easily grown in appropriate cellular media following well-established technologies that allow using enzymes segregated outside cells, even in industrial applications. Biodegradation is governed by organism type, polymer characteristics, and the pretreatment performed on the sample. During degradation, the polymer is first converted into its monomers, which should then be mineralized. It is clear that polymers are too large to pass through cell membranes, so they must first be depolymerized into smaller compounds which may then be absorbed and biodegraded within microbial cells [27] (Figure 6.3). Complete mineralization can thus be achieved, the end products being biomass, CO_2, and water when aerobic microorganisms are involved, plus CH_4 when anaerobic

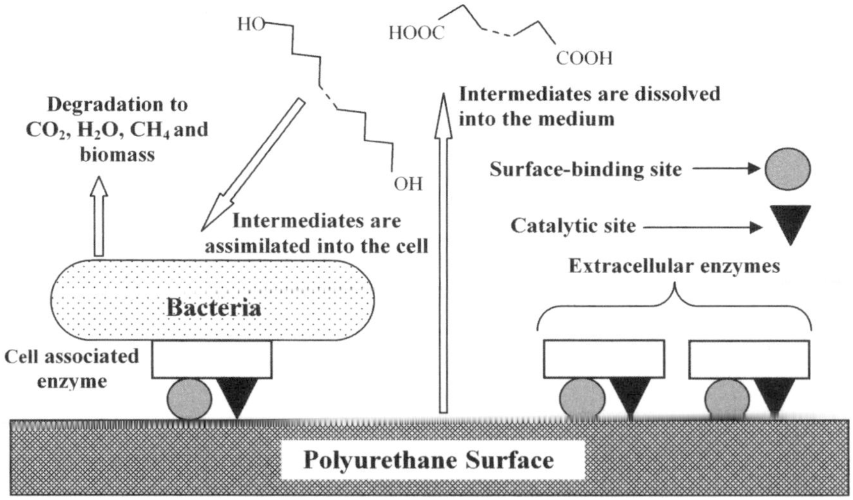

Figure 6.3 Proposed model for the degradation of PURs by the action of a cell-associated enzyme and extracellular enzymes.

conditions are used [28]. Degradation processes can be roughly classified into those involving urethane bonds and those involving the macrodiol units of both polyester and polyether types [24].

It is well known that low-molecular-weight urethanes may be easily degraded by some microorganisms, hydrolysis being catalyzed by enzymes with an estearase activity [29]. Although cleavage of urethane bonds has also been reported for polymers [30], it is not clear whether these bonds were hydrolyzed directly or after a first degradation step, resulting in lower molecular weight compounds.

Degradation of polyester-based PURs by microorganisms mainly occurs by hydrolysis of their ester bonds. It has been stated that aliphatic polyesters used in the synthesis of PURs (e.g., polyethylene adipate or poly(caprolactone)) are easily degraded by microorganisms or estereolytic enzymes like lipase [31]. It has also been reported that PURs prepared from high-molecular-weight polyesters degrade faster than those prepared from low-molecular weight polyesters [32].

Experiments show that a large variety of fungi can be highly effective in degrading PURs [32, 33]. Systematic studies on the effects of fungi are relatively scarce but point to a remarkable influence of the specific diisocyanate used in the synthesis, as well as an improvement of resistance to degradation by the presence of side chains in the polyester segment. In general, degradation by fungi requires the addition of several nutrients such as gelatin. A degradation mechanism of polyester PURs, based on extracellular estearases, has been proposed: a synergic effect is obtained by random action throughout the polymer chain of endoenzymes and successive monomer scission from the chain ends by exoenzymes [34].

Both Gram-positive and Gram-negative bacteria have been reported as PUR degraders, although few detailed works have been performed until now. Kay *et al.* [35] investigated the ability of 16 kinds of bacteria to degrade polyester PURs following their burial in soil for 28 days. In all cases, IR led to determining that the ester segments were the main site of attack because of the hydrolytic cleavage of the ester bonds. The bacterial attack usually proceeded by the binding of cells to the polymer surface with subsequent floc formation and degradation of the substrate to metabolites. Estearase and/or protease activities were identified and two kinds of enzymes were observed: (i) a cell-associated membrane bound polyurethanase and (ii) an extracellular polyurethanase [36] (Figure 6.3). The former provides cell-mediated access to the hydrophobic polymer surface, and must consequently be characterized by both a surface-binding domain and a catalytic domain. Note that enzyme molecules can easily attack water-soluble substrates, resulting in a high degradation rate. However, when the substrate is insoluble, it seems necessary to improve the contact between the enzyme and the substrate by means of a binding domain. Adherence of the bacteria enzyme to the polymer substrate must be followed by hydrolysis to soluble compounds, which will then be metabolized by the cell. This mechanism would decrease competition between degrading bacteria and other cells, as well as allowing adequate access to metabolites. The soluble extracellular enzymes should stick on the polymer surface and also hydrolyze the polymer into smaller units, facilitating the metabolization of soluble products and providing easy access of enzymes to the partially degraded polymer.

Studies on the dependence of the degrading activity upon enzyme concentration indicate that activity increased to a saturation value that remained constant when an excess of the enzyme was present [29]. This observation contrasts with the decrease in activity reported for depolymerases with a similar two-domain structure (e.g., polyhydroxyalkanoate depolymerase) [37]. It has been suggested that both domains of polyurethanases are either located in three-dimensionally close positions or separated by a flexible linker. In the former case, the catalytic domain can access the polymer substrate even if the surface is saturated with the maximum number of enzymes molecules per unit surface. It might be possible to obtain new solid polyester degrading enzymes by adding new binding domains to estearases, which are ineffective in solid substrate degradation.

Unlike polyester derivatives, polyether-based PURs are quite resistant to degradation by microorganisms [32]. *Sthaphylococcus epidermidis* was reported to degrade some kinds of polyether derivatives although the degradation rate was very slow. This feature was interpreted according to a degradation mechanism involving an exo-type depolymerization that differed from the endo-type depolymerization typical of polyester-based PURs [38]. Despite this, polyether urethane materials are known to be susceptible to a degradative phenomenon involving crack formation and propagation, which is considered environmental stress cracking [39]. This seems to be the result of a residual polymer surface stress introduced during fabrication and not sufficiently reduced by subsequent annealing.

6.3 Applications of Biodegradable Polyurethanes

Nowadays PURs play a dominant role in the design of medical devices with excellent performance in life-saving areas. PURs are highly interesting for internal (*in vivo*) uses, particularly for short-term applications like catheters or long-term applications like implants. External (*in vitro*) uses like controlled drug delivery systems must also be considered. Biodegradable properties are only required for some of their biomedical applications.

6.3.1 Scaffolds

Degradation characteristics are of special interest for design of scaffolds for *in vivo* tissue engineering. The advantages of these devices lie in that they do not have to be removed surgically once they are no longer needed, and that problems such as stress shielding may be avoided by adapting the degradation rate to the specific application. Scaffolds can be prepared by a wide range of well-established techniques such as salt leaching/freeze drying, thermally induced phase separation, and even electrospinning. Features like suitable mechanical properties, overall porosity, pore size, and interconnectivity are basic to develop materials for scaffold applications. Thus, literature data indicate that a correct cell in-growth requires a pore size of about 150–350 μm, and that minimum interconnecting openings should be larger than 10–12 μm in order to facilitate the transport of nutrients and cellular waste products, as well cell diffusion in the scaffold [40]. In general, the design of degradable devices for reconstruction must meet several biological and mechanical criteria, such as (i) high initial strength to prevent mechanical failure of the implant prior to tissue in-growth; (ii) a moderate degradation rate to induce in-growth of organized tissue since rapid degradation may cause failure of host tissue, whereas stress yielding may occur if the degradation rate is too slow, and (iii) good blood compatibility. It is worth noting that in segmented PURs, surface composition varies due to the mobility of soft segments and the trend to minimize interfacial free energy. Thus, the interface should be enriched of polar hard segments when the environment is polar (e.g., blood or water) and of nonpolar soft segments when the environment is nonpolar (e.g., air or vacuum). This is important because the host response is strongly influenced by the surface composition of the material [41].

6.3.1.1 **Cardiovascular Applications**

Soft-tissue engineering requires elastic scaffolds as these can be adapted to mechanical conditions during tissue development. Thus, scaffolds for cardiovascular tissue engineering should have high elongation at break and high tensile strength. These properties can be achieved using, for example, segmented PURs derived from macrodiols such as PCL and PCL-*b*-PEO-*b*-PCL, a diisocyanate like BDI or LDI and a chain extender like 1,4-butanediamine (putrescine) [42]. The last compound, which forms during polymer degradation, is essential for cell growth and differentiation. *In vivo* studies have revealed the promising applications of PUR scaffolds [43].

6.3.1.2 **Musculoskeletal Applications**

Common uses of PURs in musculoskeletal tissue regeneration include (i) anterior cruciate ligament reconstruction. PURs prepared from MDI, 1,3-PDA, and a poly(caprolactone) diol of a M_n molecular weight of 530 have been found suitable since they exhibit high tensile strength, a high modulus, and retention of mechanical properties when degraded adequately to the time required for the application [16, 44]. (ii) Meniscal and fibrocartilage reconstruction. In this case, high shear stresses to which prostheses are exposed, and consequently problems associated with stress hysteresis, must be born in mind. Some examples of PUR materials are those prepared from 1,4-*trans*-cyclohexane diisocyanate, a poly(caprolactone) macrodiol and a mixture of ciclohexanedimethanol and glycerol, which act as chain extenders [45]. (iii) Bone-tissue engineering. Scaffolds are prepared with an aliphatic diisocyanate (BDI or LDI), a polyethylene oxide macrodiol, and a diurea diol chain extender synthesized by coupling two equivalents of tyrosine or tyramine with one equivalent of BDI. The aromatic rings of tyrosine or tyramine units increase the rigidity of the hard segment; furthermore, these units favor *in vitro* attachment and proliferation of viable human osteoblast-like cells [46].

6.3.1.3 Neurological Applications

Once the nervous system is impaired, recovery is difficult and malfunctions in other parts of the body may occur. In order to increase the prospects of axonal regeneration and functional recovery, research has focused on the design of "nerve guidance channels" or "nerve conduits," which can be made using a biodegradable and porous channel wall. A biodegradable PUR based on hexamethylene diisocyanate, poly(ε-caprolactone), and dianhydro-D-sorbitol has been used to prepare tubular scaffolds by extrusion of the polymer solution in dimethyl formamide into a water-coagulation bath [47]. The implants had an uniaxially oriented pore structure with a pore size ranging between 2 (the pore wall) and 75 × 700 μm (elongated pores in the implant lumen), whereas the skin of tubular implants remained nonporous. *In vivo* results suggest that these scaffolds support peripheral nerve regeneration.

6.3.2 Drug Delivery Systems

Smooth delivery of a consistent dosage over time can be achieved by combining an active ingredient with a carefully developed and selected polymer. Oral and transdermal patches are the preferred routes of delivery for both the pharmaceutical industry and patients. However, the possibilities of nanotechnology should also be considered as smaller doses can be used and higher levels of bioavailability are reached. Furthermore, nanoparticles have proved capable of crossing blood barriers easily.

PURs are receiving attention as new polymer matrices for drug delivery since they allow direct link between drugs and polymers and encapsulation into micro- or nanoparticles, as shown in the following representative examples:

- Sivak *et al.* [48] developed a biodegradable and biocompatible PUR drug delivery system based on lysine diisocyanate and glycerol for controlled release of 7-*tert*-butyldimethylsilyl-10-hydroxy-camptothecin. This special type of anticancer agent was covalently bonded to the polymer matrix to improve its efficiency. Ghosh *et al.* reported the synthesis and release characteristics of a novel PEG-based PUR bearing covalently bonded ibuprofen, an anti-inflammatory drug. Ibuprofen was first reacted with butane diol diglycidyl ether, and the PUR was then obtained by reaction with 2,4-toluidene diisocyanate [49]. Antibacterial agents have also been incorporated into PURs in an effort to address bacterial infection associated with the use of medical devices. Thus, Woot *et al.* [50] synthesized a novel biodegradable polymer using 1,6-hexane diisocyanate, poly(caprolactone) diol and a fluoroquinolone antibiotic, that is, ciprofloxacin.
- Campos *et al.* reported the development of PUR-based microparticles prepared by emulsion polymerization using a poly(caprolactone) macrodiol and poly(propylene glycol)-tolylene 2,4-diisocyanate terminated or poly(propylene oxide)-based tri-isocyanated prepolymers [51]. The microparticles had a spherical shape and smooth surface.

- The encapsulation of organic liquids in PUR nanocapsules prepared by interfacial miniemulsion polycondensation of isophorone diisocyanate and propanetriol has also been reported [52]. Encapsulation efficiency was found to be dependent on the water solubility, interfacial tension against water, and compatibility with PUR of the liquids.

6.3.3 Other Biomedical Applications

Bone cements and hydrogels are other examples of biomedical applications where biodegradable PURs have a promising future. Guelcher *et al.* [53] studied biodegradable PURs as an alternative to acrylic bone cements. New PUR networks have been prepared by two-component reactive liquid molding of low-viscosity prepolymers derived from lysine polyisocyanates (lysine methyl ester diisocyanate and lysine triisocyanate) and poly(3-caprolactone-*co*-DL-lactide-*co*-glycolide) triols.

Loh *et al.* developed thermoresponsive multiblock poly(ester urethane)s synthesized from poly(ε-caprolactone), PEG, and poly(propylene glycol) using 1,6-hexamethlyene diisocyanate (HDI) as a coupling agent [54]. Bulk hydrophilicity of the obtained copolymers could be controlled either by adjusting the composition of the copolymer or by changing the temperature of the environment. These materials give rise to a hydrogel-like material without using toxic crosslinking agents. Furthermore, films of these samples are thermoresponsive since they form highly swollen hydrogel-like materials when soaked in cold water and shrink when soaked in warm water, these changes being reversible.

6.4 New Polymerization Trends to Obtain Degradable Polyurethanes

New synthetic routes are currently being studied to obtain green and sustainable PURs [55]. New polymers should avoid diisocyanate reactants, improve chemical recyclability, and enhance biodegradability while keeping high-performance properties of typical PURs.

6.4.1 Polyurethanes Obtained without Using Diisocynates

Several approaches can be considered although none has been used at an industrial level. The first representative example is the reaction between a cyclic carbonate and an amine rendering the urethane bond. In particular, the polyaddition reaction between L-lysine and a bifunctional five-membered cyclic carbonate in the presence of a strong base was described by Kihara *et al.* [56]. An alternative route (Scheme 6.3a) was reported by Rokicki *et al.* [57], who obtained PURs by reaction of ethylene carbonate with 1,6-hexanediamine or 1,4-butanediamine at room temperature and without any catalyst, the authors obtained a diurethanediol

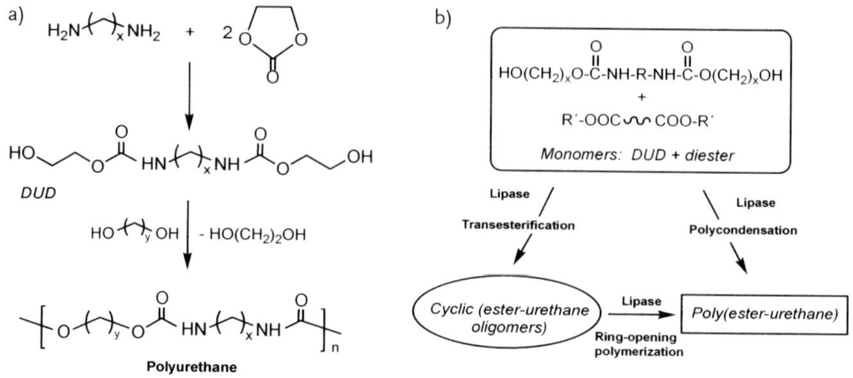

Scheme 6.3 (a) Synthesis of polyurethanes from cyclic carbonates. (b) Enzymatic synthesis of polyester-based polyurethanes by direct polymerization and cyclization with subsequent ring-opening polymerization.

(DUD) which was polycondensated with α,ω-diols containing six or more carbon atoms and rendered samples with M_n molecular weights between 1600 and 3500. The process required the presence of tin catalysts, such as $Bu_2Sn(OCH_3)_2$ or Bu_2SnO, to prevent thermal decomposition of the urethane group.

Other authors have prepared PURs with a regular sequence by ring-opening polymerization of cyclic trimethylene urethane and tetramethylene urethane at 100 °C [58]. A more recent route was developed by Schmitz *et al.* [59], who studied the copolymerization of equimolar amounts of 2,2-dimethyltrimethylene carbonate and tetramethylene urea. The obtained copolymers had microstructures that were strongly dependent on reaction conditions and the catalyst used.

6.4.2
Enzymatic Synthesis of Polyurethanes

Enzymes have remarkable characteristics compared with conventional chemist catalysts, like high catalyst activity, stereoselectivity, and lack of side reactions. Furthermore, enzymes are interesting as a renewable and naturally occurring catalyst. Because enzymatic polymerization can be considered the reverse reaction of degradation, polyester-based PEUs obtained using a biological catalyst are expected to be biodegraded by the same enzyme.

Many authors have reported the enzymatic synthesis of PEUs by enzymatic polyesterification [55]. Thus, lipase has been employed to promote the esterification reaction between a low-molecular-weight DUD and various diacids/diesters. Two strategies can be applied (Scheme 6.3b): (i) direct polycondensation of the DUD and the appropriate diester, and (ii) ring-opening polymerization of a cyclic ester–urethane oligomer which has been previously prepared by a lipase-induced transesterification reaction between a biodegradable DUD and a diester. The latter method gives higher molecular weights.

Soeda *et al.* [60] prepared polyester PURs by both enzymatic polycondensation of the DUD with the diethylester of a dicarboxylic acid and enzymatic cyclization and subsequent ring-opening polymerization of the cyclic ester–urethane monomer. Molecular weights and polymer yields increased with increasing temperature and the number of methylene groups of the dicarboxilate moiety (malonate, succinate, glutarate, and adipate). This PEU showed chemical recyclability.

McCabe and Taylor [61] proposed an enzymatic synthesis using also a DUD and a mixture of diol and dicarboxylic acid monomers. Thus, bis(hydroxyethyl)carbamate was dissolved in 1,4-butanediol (BDO) and reacted with adipic acid using lipase as catalyst under reduced pressure to render a polymer with a M_w molecular weight of 9350.

Although the carbonate bond is more resistant to hydrolysis against alkaline media than the ester linkage, polycarbonate-based PURs (PCUs) can also be degraded by enzymes. An enzymatically biodegradable PCU has been synthesized by reaction of a low-molecular-weight DUD with diethyl carbonate in the presence of *Candida antarctica* lipase [55].

6.4.3 Polyurethanes from Vegetable Oils

PURs can also be prepared with polyols derived from vegetable oils such as castor oil, soybean oil, palm oil, and rapeseed oil, which are most commonly employed for large-scale products [62]. The obtained PURs are ideal for foam applications when the hydroxyl functionality of the polyol is higher than two. It is also desirable that the molecular weight of the polyol ranges between 3000 and 6000. PURs from vegetable oils have some advantages derived from their good oxidative thermal stability. For example, soybean oil-based polyisocyanurate foams have been reported to exhibit better thermal stability, lower flammability, greater rigidity (modulus), and higher compression strength than those based on propylene oxide polyols of the same molecular weight and functionality. Furthermore, these PURs are assumed to be biodegradable since this is a characteristic of vegetable oils. Shogren *et al.* [63] studied the degradation of polymers prepared from soybean oil, triolein, and linsed oil by respirometry. Rapid degradation of a small fraction of PURs and a trend toward less degradation as the polyol functionality increased (linsed < soy < triolein) were observed, although differences were small. However, it is also clear that the hydrophobic character of fatty acid chains may reduce susceptibility to hydrolysis.

6.4.4 Polyurethanes from Sugars

Polysaccharides (e.g., starch, cellulose, chitin, or chitosan) are naturally occurring polymers obtained from renewable sources which tend to degrade in biologically active environments like soil, sewage, and marine locations where bacteria are active. Much effort has been made to develop PURs based on the above biopolymers as an increase in biodegradability is expected to be achieved.

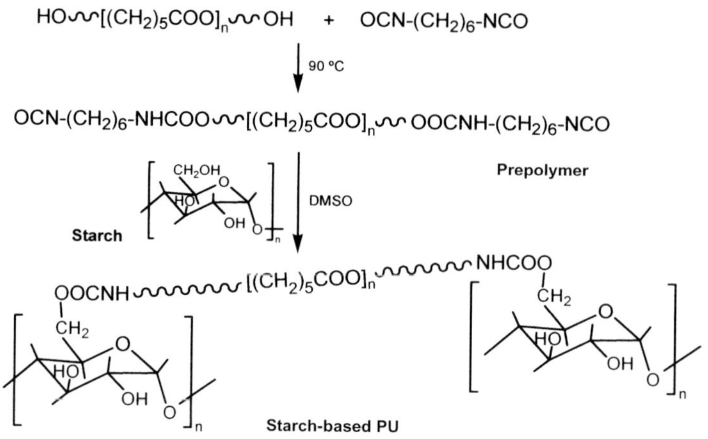

Scheme 6.4 Synthesis of starch-based polyurethanes.

Starch, which is the second largest biomass on earth, and synthetic plastics do not mix easily. This problem can be overcome by chemically linking the synthetic and the natural polymer. Barikani and Mohammadia [64] used the hydroxyl functionality of the biopolymer and grafted a prepolymer derived from HDI and a PCL macrodiol onto starch (Scheme 6.4). SEM micrographs confirmed that the starch granules were completely coated by PUR.

It was found that larger amounts of prepolymer led to an increase in hydrophobicity and a decrease in glass transition temperature.

Current research is aimed at combining natural products containing cellulose with different materials like plastics to obtain new materials that can be tailored according to final use requirements. Since 1998, new biocomposite wood-replacement panels made from wheat straw and a PUR resin have been commercialized by Dow Polyurethanes [65]. The resulting biocomposite material can be used for kitchen counters, shelving, ready-to-assemble furniture, cabinets, door cores, and floor underlays.

Chitin, which is widely distributed in nature mainly as the skeletal material of crustaceans, is structurally similar to cellulose as it has an acetamide group in place of a hydroxyl group. Chitin-based PUR elastomers with potential as biomedical implants and tunable mechanical properties have been synthesized by step-growth polymerization techniques using PCL and MDI [66]. The prepolymer was extended with different mass ratios of chitin and BDO. The mechanical properties of these polymers were improved by increasing chitin content, which furthermore lowered the cytotoxicity of samples.

Chitosan, commercially produced by deacetylation of chitin, has several applications in the biomedical field (e.g., in bandages and other hemostatic agents). A number of graft and block PUR derivatives have been investigated: (i) PUR prepolymers prepared from PEG and isophorone diisocyanate have been successfully grafted onto chitosan [67]; (ii) Xu *et al.* described a novel blood-compatible waterborne PUR using chitosan as the chain extender. The prepolymer was obtained from poly(tetramethylene oxide glycol) (PTMO), isophorone diisocyanate (IPDI), and 2,2'-dimethylol propionic acid (DMPA) [68]. The authors concluded that the addition of chitosan lends a remarkable anticoagulative character to the final polymer.

6.5 Aliphatic Poly(ester amide)s: A Family of Biodegradable Thermoplastics with Interest as New Biomaterials

Poly(ester amide)s (PEAs) have been regarded as a new promising family of biodegradable polymers since the 1990s, although the first syntheses were reported in 1979 when a polyamide (i.e., nylon 6) and a polyester (i.e., poly(caprolactone)) were subjected to amide–ester interchange reactions induced by temperature [69]. PEAs can combine a degradable character due to the existence of hydrolyzable ester groups (–COO–) with relatively good thermal and mechanical properties afforded by strong intermolecular hydrogen-bonding interactions established between their amide groups (–NHCO–). Initially, the main interest was the technical potential of PEAs due to their good fiber-forming properties and ease of processability. Indeed, the degradable characteristics and low cost of raw materials made the commercialization of a random polymer derived from 1,4-butanediol, adipic acid, and ε-caprolactam as a commodity material feasible (BAK 1095) [70]. PEAs are currently being intensively investigated as a new class of promising materials for biomedical applications such as those mentioned above for PURs.

PEAs can be synthesized by different chemical methods which allow polymers with segmented, random, and regular distributions to be obtained.

Segmented PEAs are basically similar to those described for PURs, that is, they may have a microphase separated structure with amide-rich hard domains acting as physical crosslinkers and ester-rich soft domains, which confer flexibility and extensibility upon the polymer. Bera *et al.* [71] reported the synthesis of segmented PEAs by reaction of an alternating ester–amide oligomer with an oligoester prepared from 1,2-ethanediol and dimethyl adipate. The ester–amide compound was obtained by reaction of adipic acid with a bisamide-diol derived from 1,6-diaminohexane and γ-butyrolactone. By decreasing the oligoester soft block content from 80 to 20 mol%, the modulus increased from 70 to 600 MPa and the yield stress increased from 70 to 600 MPa. *In vivo* assays have revealed both nontoxicity and biodegradability characteristics. Direct polycondensation of bisamide-diols with aliphatic diols and dimethyl adipate has also been extensively studied (Figure 6.4a). Stapert *et al.* [72] prepared PEAs by reaction of dimethyl adipate, 1,4-butanediol, and a bisamide-diol derived from 1,4-diaminobutane and ε-caprolactone. The mechanical properties of final samples were easily tuned by changing the molar ratio of the two diols. Thus, the elastic modulus varied from 70 to 524 MPa when the hard segment content was increased from 10% to 85%. *In vitro* and *in vivo* assays performed with such polymers indicated good

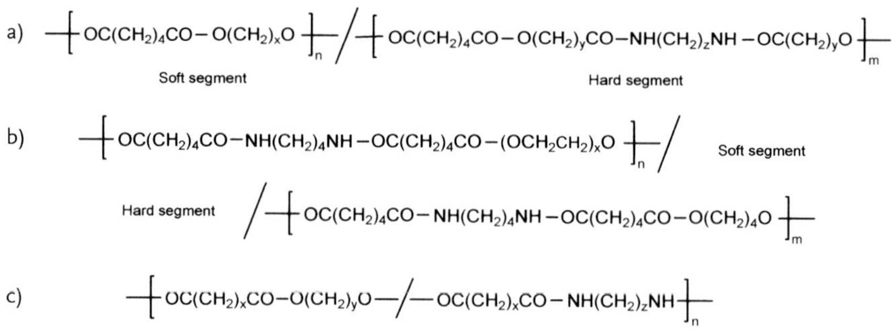

Figure 6.4 Representative segmented and random poly(ester amide)s.

biocompatibility and a relatively slow degradation rate. Additionally, it has been demonstrated that closed-cell foams of these materials can be obtained using CO_2 as a blowing agent [73]. Poly(ether ester amide) (PEEA) copolymers based on PEG, 1,4-butanediol, and dimethyl-7,12-diaza-6,13-dione-1,18-octadecanedioate [74] (Figure 6.4b) have also been evaluated as scaffold materials for tissue engineering and have proved to sustain adhesion and growth of endothelial cells.

Random PEAs can be easily obtained by a two-step procedure firstly described by Castaldo *et al.* [75] (Figure 6.4c). In this work, an oligoester mixture was prepared by reaction of a diol (1,6-hexanediol, 1,10-decanediol, or 1,12-dodecanediol) with an excess of sebacoyl dichloride. Oligoesters and the excess of dichloride were then reacted with a stoichiometric amount of the appropriate diamine (1,6-hexanediamine, 1,10-decanediamine, 1,12-dodecanediamine). Thermal properties turned out to be strongly dependent on the final amide ratio, as also found for polymers based on adipic acid, 1,6-hexanediamine, and 1,4-butanediol. PEAs related to nylons 6 or 66 and poly(caprolactone) have also been studied and their enzymatic degradation has been demonstrated [76, 77].

PEAs with a regular microstructure can be obtained by using different kinds of monomers. Thus, polymers differing in the arrangement of amide and ester groups within their molecular chain can be derived. Obviously, the amide/methylene ratio may also be easily varied depending on the methylene content of the involved monomers. These materials are generally crystalline and render oriented fiber X-ray diffraction patterns as well as single-crystal electron diffraction patterns. Analysis of their structure has shown a complex unit cell arrangement which reflects the packing preferences of amide and ester groups [78]. Biodegradable PEAs with a regular structure can be classified on the basis of their main representative chemical units:

- Derivatives of α-amino acids and α-hydroxy acids (polydepsipeptides). They can combine useful properties of poly(α-hydroxy acid)s and poly(α-amino acids)s and are considered as particularly attractive reabsorbable materials. Polymers can be prepared by ring-opening polymerization of morpholine-2,5-diones. However, the synthesis of these monomers is complex and polymerization usually requires severe reaction conditions resulting in unexpected by-products from side reactions. This, together with the usual low molecular

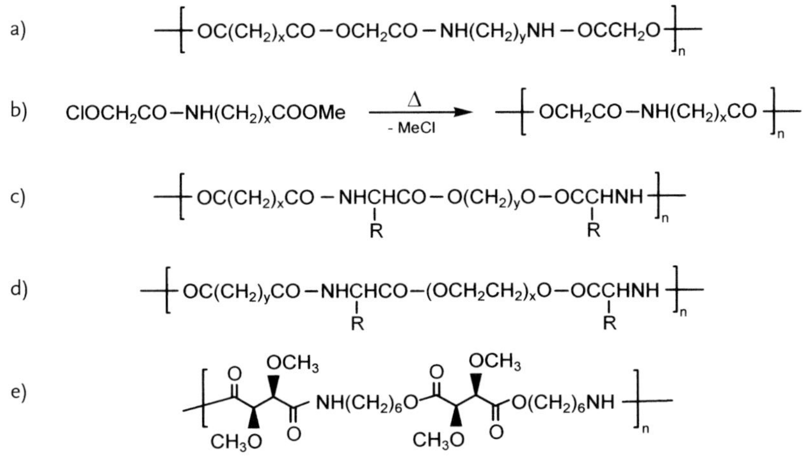

Scheme 6.5 Representative regular poly(ester amide)s.

weight of samples, hinders the use of such polymers in the biomedical field, where material purity becomes essential.

- Derivatives of α-hydroxy acids. Alternating PEAs with potential interest as bioabsorbable sutures have been obtained by solution polymerization of an acid dichloride such as succinyl dichloride with a bisamide-diol prepared from glycolic acid and diaminoalkanes containing 2–12 methylene groups (Scheme 6.5a). Their mechanical properties, degradability, and biocompatibility have been evaluated and results are highly promising [79, 80]. A different synthetic route based on the formation of metal halide salts as the driving force of a thermal polycondensation reaction has recently been proposed [81]. This method has also been successfully applied to prepare alternating copolymers of glycolic acid and ω-amino acids (Scheme 6.5b).
- Derivatives of α-amino acids. Polymers derived from naturally occurring units should be preferred for biomedical applications since degradation products are nontoxic and easily metabolized by the organism. For this reason, poly(α-amino acids) were polyamides extensively studied as an interesting alternative but were finally discarded due to inherent problems like production costs, insolubility in common organic solvents, thermal instability, and processing difficulties. In contrast, PEAs incorporating α-amino acid units have been developed and extensively studied [82, 83] (Scheme 6.5c). They can be obtained by interfacial polymerization of an acid dichloride and the *p*-toluensulfonic salt of a bis-(α-amino acid) α,ω-alkylene diester [84, 85]. Alternatively, thermal polyesterification between a diol and a bisamide diester derived from acid dichloride and α-amino acid methyl ester units has been proposed [86]. This kind of polymers has proved to be biodegradable, biocompatible, and with the mechanical, thermal, and degradation properties that may be tuned by changing the methylene/amide ratio. Moreover, the enzymatic degradation rate has been found to be easily modified by the stereochemical composition of the

polymer (i.e., the ratio between L- and D-amino acid units) [87]. Some of these polymers have been studied as matrices in the form of microspheres for drug delivery systems [88]. Similar PEAs have also been prepared by using an unsaturated dicarboxylic acid or an oligomer derived from an α-amino acid and oligoethylen glycol [89] (Scheme 6.5d).

- Carbohydrate derivatives. Carbohydrates like arabinose, xylose, and tartaric acid [90] have been reacted with amino alcohols to render new PEAs whose degradation properties were investigated [90] (Scheme 6.5e). The same procedure has been extended to succinic and glutaric acid derivatives [91, 92]. Polymers can be directional or adirectional depending on the synthesis procedure. The hydrolytic degradation rate has proved to be strongly influenced by chain microstructure since formation of amide rings can accelerate the process.

Acknowledgments

The authors want to indicate the financial support given by CICYT and FEDER (grant MAT 2006-02406).

References

1 Oertel, G. (1994) *Polyurethane Handbook*, Hanser Gardner Publications, Berlin.

2 Szycher, M. (1999) *Szicher's Handbook of Polyurethanes*, CRC Press, Boca Raton, FL.

3 Cardy, R.H. (1979) *J. Natl. Cancer Inst.*, **62**, 1107–1116.

4 Schoental, R. (1968) *Nature*, **219**, 1162–1163.

5 Urbanski, J., Czerwinski, W., Janicka, K., Majewska, F., and Zowall, H. (1977) *Handbook of Analysis of Synthetic Polymers and Plastics*, Ellis Horwood Limited, Chichester.

6 Sawnhey, A.S. and Hubbell, J.A. (1990) *J. Biomed. Mater. Res.*, **24**, 1397–1411.

7 Gorna, K. and Gogolewski, S. (2002) *J. Biomed. Mater. Res.*, **60**, 592–606.

8 Herpburn, C. (1991) *Polyurethane Elastomers*, 2nd edn (ed. C. Hepburn), Elsevier, London, p. 29.

9 Wirpsza, Z. (1993) *Polyurethanes: Chemistry, Technology and Applications* (ed. Z. Wirpsza), Ellis Horwood, New York, p. 139.

10 Dieterich, D., Grigat, E., Hahn, W., Hespe, H., and Schmelzer, H.G. (1985) *Polyurethane Handbook*, 2nd edn (ed. G. Oertel), Hanser, Munich, p. 11.

11 Lyman, D.J. (1960) *J. Polym. Sci.*, **45**, 49–59.

12 Spaans, C.J., de Groot, J.H., Dekens, F.G., and Pennings, A.J. (1998) *Polym. Bull.*, **41**, 131–138.

13 Gorna, K. and Gogolewski, S. (2003) *J. Biomed. Mater. Res.*, **67A**, 813–827.

14 Wiggins, J.S. and Storey, R.F. (1992) *Polym Prepr.*, **33**, 516–517.

15 Guelcher, S.A. (2008) *Tissue Eng.*, **14**, 3–17.

16 Gisselfalt, K., Edberg, B., and Flodin, P.S. (2002) *Biomacromolecules*, **3**, 951–958.

17 Zdrahala, R.J. and Zdrahala, I.J. (1999) *J. Biomater. Appl.*, **14**, 67–90.

18 Gunatillake, P.A., Martin, D.J., Meijs, G.F., McCarthy, S.J., and Adhikari, R. (2003) *Aust. J. Chem.*, **56**, 545–557.

19 Martin, D.J., Maijs, G.F., Gunatillake, P.A., McCarthy, S.J., and Renwick, G.M. (1997) *J. Appl. Polym. Sci.*, **64**, 803–817.

20 Seymour, R.W. and Cooper, S.L. (1971) *Polym. Lett.*, **9**, 689–694.

21 Clough, S.B., Schneider, N.S., and King, A.O. (1968) *J. Macromol. Sci. Phys.*, **B2**, 641–648.

22 Seymour, R.W. and Cooper, S.L. (1973) *Macromolecules*, **6**, 48–53.

23 Bonart, R. (1968) *J. Macromol. Sci. Phys.*, **B2**, 115–138.

24 Nakajima-Kambe, T., Shigeno-Akutsu, Y., Nomura, N., Onuma, F., and Ankara, T. (1999) *Appl. Microbiol. Biotechnol.*, **51**, 134–140.

25 Stokes, K. and McVennes, R. (1995) *J. Biomater. Appl.*, **9**, 321–335.

26 Aithal, U.S., Aminabhavi, T.M., and Balundgi, R.H. (1990) *J. Macromol. Sci.*, **C30**, 43–105.

27 Shah, A.A., Hasan, F., Hameed, A., and Ahmed, S. (2008) *Biotechnol. Adv.*, **26**, 246–265.

28 Barlaz, M.A., Ham, R.K., and Schaefer, D.M. (1989) *J. Environ. Eng.*, **115**, 1088–1102.

29 Oshiro, T., Shinji, M., Morita, Y., Takayama, Y., and Izumi, Y. (1997) *Appl. Microbiol. Biotechnol.*, **48**, 546–548.

30 Jansen, B., Schumacher-Perdreau, F., Peters, G., and Pulverer, G. (1991) *Zentralbl. Bakteriol.*, **276**, 36–45.

31 Tokiwa, Y. and Suzuki, T. (1977) *Nature*, **270**, 76–78.

32 Darby, R.T. and Kaplan, A.M. (1968) *Appl. Microbiol.*, **16**, 900–905.

33 Kaplan, A.M., Darby, R.T., Greenberger, M., and Rodgers, M.R. (1968) *Dev. Ind. Microbiol.*, **82**, 362–371.

34 Wales, D.S. and Sagan, B.R. (1988) *Biodeterioration*, 7th edn (eds D.R. Houghton, R.N. Smith, and H.O.W. Eggins), Elsevier Applied Science, London, pp. 351–358.

35 Kay, M.J., Morton, L.H.G., and Prince, E.L. (1991) *Int. Biodeterior. Bull.*, **27**, 205–222.

36 Howard, G.T. (2002) *Int. Biodeterior. Biodegradation*, **49**, 245–252.

37 Mukai, K., Yamada, D., and Hawai (1993) *Int. J. Biol. Macromol.*, **15**, 361–366.

38 Y., F., Okamoto, T., Suzuki, and T. (1985) *J. Ferment. Technol.*, **63**, 239–244.

39 Phillips, R.E., Smith, M.C., and Thomas, R.J. (1995) *J. Biomater. Appl.*, **9**, 321–335.

40 Heijkants, R.G.J.C., van Calk, R.V., van Tienen, T.G., de Groot, J.H., Pennings, A.J., Buma, P., Veth, R.P.H., and Schouten, A.J. (2008) *J. Biomed. Mater. Res. A*, **87A**, 921–932.

41 Santerre, J.P., Woodhouse, K., Laroche, G., and Labow, R.S. (2005) *Biomaterials*, **26**, 7457–7470.

42 Guan, J., Sacks, M.S., Beckman, E.J., and Wagner, W.R. (2004) *Biomaterials*, **25**, 85–96.

43 Fujimoto, K.L., Guan, J., Oshima, H., Sakai, T., and Wagner, W.R. (2007) *Ann. Thorac. Surg.*, **83**, 648–654.

44 Gisselfaet, K. and Helgee, B. (2003) *Macromol. Mater. Eng.*, **288**, 265–271.

45 de Groot, J.H., de Vrijer, R., Pennings, A.J., Klompmaker, J., Veth, R.P.H., and Jansen, H.W.B. (1996) *Biomaterials*, **17**, 163–173.

46 Kavlock, K.D., Pechar, T.W., Hollinger, J.O., Guelcher, S.A., and Goldstein, A.S. (2007) *Acta Biomater.*, **3**, 475–484.

47 Hausner, T., Schmidhammer, R., Zandieh, S., Hopf, R., Schultz, A., Gogolewski, S., Hertz, H., and Redl, H. (2007) *Acta Neurochir. Suppl.*, **100**, 69–72.

48 Sivak, W.N., Pollack, I.F., Petoud, S., Zamboni, W.C., Zhang, J., and Beckman, E.J. (2008) *Acta Biomater.*, **4**, 852–862.

49 Ghosh, S. and Mandal, S.M. (2008) *J. Macromol. Sci. Pure Appl. Chem.*, **45**, 445–448.

50 Woo, G.L.Y., Mittelman, M.W., and Santerre, J.P. (2000) *Biomaterials*, **21**, 1235–1246.

51 Campos, E., Cordeiro, R., Alves, P., Rasteiro, M.G., and Gil, M.H. (2008) *J. Microencap.*, **25**, 154–169.

52 Johnsen, H. and Schmid, R.B. (2007) *J. Microencapsul.*, **24**, 731–742.

53 Guelcher, S.A., Srinivasan, A., Dumas, J.E., Didier, J.E., McBride, S., and Hollinger, J.O. (2008) *Biomaterials*, **29**, 1762–1775.

54 Loh, X.J., Colin Sng, K.B., and Li, J. (2008) *Biomaterials*, **29**, 3185–3194.

55 Matsumura, S., Soeda, Y., and Toshima, K. (2006) *Appl. Microbiol. Biotechnol.*, **70**, 12–20.

56 Kihara, N., Kushioda, Y., and Endo, T. (1996) *J. Polym. Sci. Part A: Polym. Chem.*, **34**, 2173–2179.

57 Rokicki, G. and Piotrowska, A. (2002) *Polymer*, **43**, 2927–2935.
58 Kusan, J., Keul, H., and Höcker, H. (2001) *Macromolecules*, **34**, 389–395.
59 Schmitz, F., Keul, H., and Höcker, H. (1998) *Polymer*, **39**, 3179–3186.
60 Soeda, Y. and Matsumura, S. (2005) *Macromol. Biosci.*, **5**, 277–288.
61 McCabe, R.W. and Taylor, A. (2004) *Green Chem.*, **6**, 151–155.
62 Petrovic, Z.S. (2008) *Polymer Rev.*, **48**, 109–155.
63 Shogren, R.L., Petrovic, Z., Liu, Z., and Erhan, S.Z.J. (2004) *Polym. Environ.*, **12**, 173–178.
64 Barikani, M. and Mohammadi, M. (2007) *Carbohydr. Polym.*, **68**, 773–780.
65 Flieger, M., Kantorová, M., Prell, A., Rezanka, T., and Votruba, J. (2003) *Folia Microbiol.*, **48**, 27–44.
66 Zia, K.M., Zuber, M., Bhatti, I.A., Barikani, M., and Sheikh, M.A. (2009) *Int. J. Biol. Macromol.*, **44**, 18–22.
67 Silva, S.S., Menezes, S.M.C., and Garcia, R.B. (2003) *Eur. Polym. J.*, **39**, 1515–1519.
68 Xu, D., Meng, Z., Han, M., Xi, K., Jia, X., Yu, X., and Chen, Q. (2008) *J. Appl. Polym. Sci.*, **109**, 240–246.
69 Tokiwa, Y., Suzuki, T., and Ando, T. (1979) *J. Appl. Polym. Sci.*, **24**, 1701–1711.
70 Grigat, E., Koch, R., and Timmermann, R. (1998) *Polym. Degrad. Stab.*, **59**, 223–226.
71 Bera, S. and Jedlinski, Z. (1993) *J. Polym. Sci., Part A. Polym. Chem.*, **31**, 731–739.
72 Stapert, H.R., Bouwens, A.W., Dijkstra, P.J., and Feijen, J. (1999) *Macromol. Chem. Phys.*, **200**, 1921–1929.
73 Lips, P.A.M., Velthoen, I.W., Dijkstra, P.J., Wessling, M., and Feijen, J. (2005) *Polymer*, **46**, 9396–9403.
74 Deschamps, A.A., van Apeldoorn, A.A., de Bruin, J.D., Grijpma, D.W., and Feijen, J. (2003) *Biomaterials*, **24**, 2643–2652.
75 Castaldo, L., de Candia, F., Maglio, G., Palumbo, R., and Strazza, G. (1982) *J. Appl. Polym. Sci.*, **27**, 1809–1822.
76 Gonsalves, K.E., Chen, X., and Cameron, J.A. (1992) *Macromolecules*, **25**, 3309–3312.
77 Arvanitoyannis, I., Nakayama, A., Kawasaki, N., and Yamamoto, N. (1995) *Polymer*, **36**, 857–866.
78 Paredes, N., Casas, M.T., Puiggalí, J., and Lotz, B. (1999) *J. Polym. Sci. Part B: Polym. Phys.*, **37**, 2521–2533.
79 Barrows, T.H. (1986) *Polymers in Medicine II* (ed. E. Chiellini), Plenum, New York, pp. 85–90.
80 Horton, V.L., Blegen, P.E., and Barrows, T.H. (1988) *Progress in Biomedical Polymers* (eds C.G. Gebelijn and R.L. Dunn), Plenum, New York, pp. 263–282.
81 Vera, M., Rodríguez-Galán, A., and Puiggalí, J. (2004) *Macromol. Rapid Commun.*, **25**, 812–817.
82 Katsavara, R., Beridze, V., Arbuli, N., Kharadze, D., Chu, C.C., and Won, C.Y. (1999) *J. Polym Sci. Part A: Polym. Chem.*, **37**, 391–407.
83 Guo, K. and Chu, C.C. (2007) *J. Polym Sci. Part A: Polym. Chem.*, **45**, 1595–1606.
84 Saotome, Y., Tashiro, M., Miyazawa, T., and Endo, T. (1991) *Chem. Lett.*, 21–25.
85 Paredes, N., Rodríguez-Galán, A., and Puiggalí, J. (1998) *J. Polym. Sci. Part A: Polym. Chem.*, **36**, 1271–1282.
86 Asín, L., Armelin, E., Montané, J., Rodríguez-Galán, A., and Puiggalí, J. (2001) *J. Polym. Sci. Part A: Polym. Chem.*, **39**, 4283–4293.
87 Rodríguez-Galán, A., Pelfort, M., Aceituno, J.E., and Puiggalí, J. (1999) *J. Appl. Polym. Sci.*, **74**, 2312–2320.
88 Vera, M., Puiggalí, J., and Coudane, J. (2006) *J. Microencapsul.*, **23**, 686–697.
89 Guo, K. and Chu, C.C. (2007) *Biomacromolecules*, **8**, 2851–2861.
90 Bueno, M., Molina, I., Zamora, F., and Galbis, J.A. (1997) *Macromolecules*, **30**, 3197–3203.
91 Villuendas, I., Molina, I., Regaño, C., Bueno, M., Martínez de Ilarduya, A., Galbís, J., and Muñoz-Guerra, S. (1999) *Macromolecules*, **32**, 8033–8040.
92 Vera, M., Almontassir, A., Rodríguez-Galán, A., and Puiggalí, J. (2003) *Macromolecules*, **36**, 9784–9796.

7
Carbohydrates

Gerald Dräger, Andreas Krause, Lena Möller, and Severian Dumitriu

7.1
Introduction

Polysaccharides are an integral part of the living matter. Due to this huge presence in organisms, they are highly biocompatible and biodegradable and therefore idealy match the basic characteristics for polymers used as biomaterials.

All polysaccharides used derive from natural sources. Biodegradation is defined as an event which takes place in the natural environment and living organisms. Since polysaccharides are ubiquitous in nature and present a valuable carbon and energy source in the life cycle of organisms, their biodegradation is a highly evolved process using effective and usually specific enzymes. This makes polysaccharides a promiscuous basis for the development of biodegradable polymers.

Polysaccharide-based biomaterials are of great interest in several biomedical fields such as drug delivery, tissue engineering, or wound healing. Important properties of the polysaccharides include controllable biological activity, biodegradability, and their ability to form hydrogels. Polysaccharides are also used as additives in the food industry and in many technical applications. Here the main focus lies on the superb rheological properties of many polysaccharides together with their biodegradability and their positive environmental and toxicological effects.

Several important and up-to-date reviews have to be mentioned and should be considered in order to gain insight in this complex topic. In 2008, Rinaudo summarized the main properties and current applications of some polysaccharides as biomaterials [1]. The application of biodegradable systems in tissue engineering and regenerative medicine with a strong focus on carbohydrates is summarized by Reis and coworkers [2]. Polysaccharides-based nanoparticles as drug-delivery systems are reviewed by Liu *et al.* [3], whereas Coviello *et al.* focused on polysaccharide hydrogels for modified release formulations [4]. In this chapter, we summarize the basic properties, modifications, and applications of biodegradable polysaccharides. We deliberately omit starch and pectin since there are numerous reviews and books dealing solely with these materials.

Handbook of Biodegradable Polymers: Synthesis, Characterization and Applications, First Edition. Edited by Andreas Lendlein, Adam Sisson.
© 2011 Wiley-VCH Verlag GmbH & Co. KGaA. Published 2011 by Wiley-VCH Verlag GmbH & Co. KGaA.

7.2
Alginate

Alginate belongs to the family of linear (unbranched), nonrepeating copolymers. It consists of variable amounts of β-D-mannuronic acid (M) and its C5-epimer α-L-guluronic acid (G) which are linked via β-(1,4)-glycosidic bonds. The glycosidic bonds of mannuronic acid are connected to the following unit by a diequatorial $^4C_1$ linkage, while guluronic acids are diaxial $^1C_4$ linked. Alginate can be regarded as a true block copolymer composed of homopolymeric M and G regions, called M- and G-blocks, respectively, interspersed with regions of alternating structure [5] (Scheme 7.1).

The physicochemical properties of alginate have been found to be highly affected by the M/G ratio as well as by the structure of the alternating zones. In terms of specific medical applications, alginate materials with a high guluronic acid ratio exhibit a much better compatibility [6]. The first protocol to hydrolyze the glycosidic bonds of alginate has been published by Haug *et al.* in the 1960s and is based on a pH-dependent acid-catalyzed hydrolysis, which leads to a fragmentation of the polymeric chain. Breaking the glycosidic bonds of both building blocks selectively could be achieved due to different pK_a values of mannuronic acid (pK_a: 3.38) and guluronic acid (pK_a: 3.65) [7]. Therefore, polyguluronic acid can be separated by precipitation in aqueous conditions after protonating the carboxyl groups.

Alginate can be extracted from marine brown algae or it can be produced by bacteria. Both species produce alginate as an exopolymeric polysaccharide during their growth phase. Isolated alginates from marine brown algae like *Laminaria hyperborea* or *lessonia*, gained by harvesting brown seaweeds from coastal regions, tend to vary in their constitution due to seasonal and environmental changes. Like chitin in shellfish, alginates in algae have structure-forming functions. This is due to the intracellular formed gel matrix, which is responsible for mechanical strength, flexibility, and form. Alginates in bacteria are synthesized only by two genera, *Pseudomonas* and *Azotobacter*, and have been extensively studied over the last 40 years. While primarily synthesized in the form of polymannuronic acid, the biosynthesis undergoes chemical modifications comprising acetylation and epimerization, which occurs during periplasmic transfer and before final export through the outer membrane. Extracted alginate from *Pseudomonas* contains only M blocks

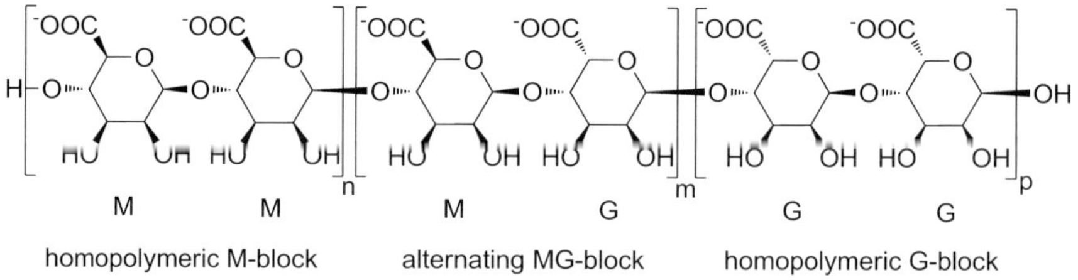

Scheme 7.1 Chemical structure of alginate with mannuronic acid (M), alternating, and guluronic acid (G) blocks.

and may be O-2 and/or O-3 acetylated. The G units are introduced by mannuronan C-5 epimerases. The genetic modification of alginate-producing microorganisms could enable biotechnological production of new alginates with unique, tailor-made properties, suitable for medical and industrial applications [5].

Depolymerization of alginate is catalyzed by different lyases. The depolymerization occurs by cutting the polymeric chain via β-elimination, generating a molecule containing 4-desoxy-L-erythro-hex-4-enepyranosyluronate at the nonreducing end. Such type of lyases have been found in organisms using alginate as a carbon source, in bacteriophages specific for alginate-producing organisms, and in alginate-producing bacteria [8]. In recent times, recombinant alginat lyases with different preferences for the glycosidic cleavage were published [9].

Alginate is a well-known polysaccharide widely used due to its gelling properties in aqueous solutions. The gelling is related to the interactions between the carboxylic acid moieties and bivalent counterions, such as calcium, lead, and copper. It is also possible to obtain an alginic acid gel by lowering the environmental pH value. Like DNA, alginate is a negatively charged polymer, imparting material properties ranging from viscous solutions to gel-like structures in the presence of divalent cations. Divalent ions at concentration of >0.1% (w/w) are sufficient for gel formation. The gelling process takes place by complexation of divalent cations between two alginate chains; primarily G building blocks interact with present cations. Since calcium ions interact with carboxyl functions of four G-units, the formed structure induces helical chains. This coordination geometry is generally known as "egg box" model [10, 11] (Scheme 7.2).

The fact that G-units are responsible for gelation leads to the attribute that alginate with a higher G content show higher moduli. Enriched high-G gels have more regular, stiff structures with short elastic segments. They obtain a more ridged, static network compared to the more dynamic and entangled structure of low-G gels with their long elastic segments [12]. Interactions with univalent cations in solution have been investigated by Seale *et al.* by circular dichroism and

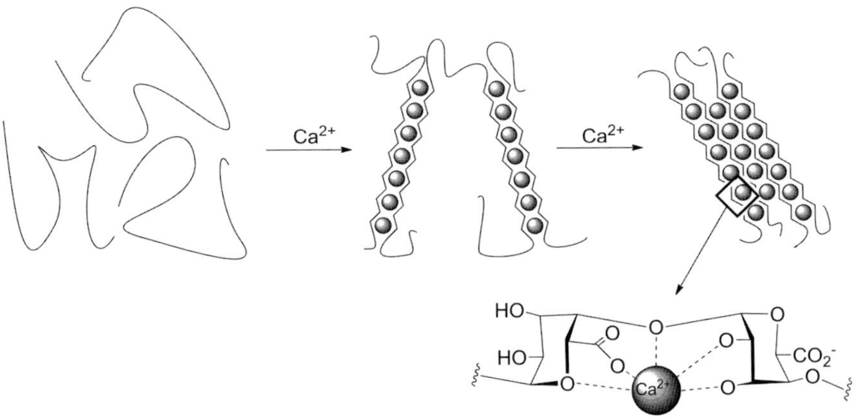

Scheme 7.2 Scheme of the egg box model; complexation of calcium via four G-units.

rheological measurements. Poly-L-guluronate chain segments show substantial enhancement (approximately 50%) of circular dichroism ellipticity in the presence of excess K^+, with smaller changes for other univalent cations: $Li^+ < Na^+ < K^+ > Rb^+ > Cs^+ > NH_4^+$ [13].

The only commercial available derivative of alginate is propylene glycol alginate, produced by esterification of the uronic acids with propylenoxide. Propylene glycol alginate is mostly utilized in food industry as stabilizer, thickener, and emulsifier. Other food additives are sodium alginate, ammonium alginate, calcium alginate (CA), and potassium alginate. All these alginate types with different cations are water soluble. In order to achieve solubility of alginate in polar organic solvents, it is necessary to exchange these cations by quaternary ammonium salts with lipophilic alkyl chains. Basically two strategies were pursued to modify the monomeric structure. On the one hand, the carboxyl group is attacked by a strong nucleophile, generally using an active ester as precursor. The second strategy uses a ring-opening of the carbohydrate by cracking the bonds between C-3 and C-4. Aqueous sodium periodate breaks vicinal diols generating two aldehydes [14]. A wide range of reactions in dimethyl sulfoxide or *N,N*-dimethylformamide are described, where different ester or amides could be synthesized [15]. Furthermore radical photo crosslinkable alginate has been synthesized by Jeon *et al.* via acrylating the carboxylic acid [16]. Several alginate derivatives have been synthesized to generate hydrogels. The generation of thermostable hydrogels can be achieved using UV radiation [17] or crosslink reagents [18, 19] for *in situ* polymerization. With regard to clinical applications, drugs or biomarker like methotrexate, doxorubicin hydrochloride, mitoxantrone dihydrochloride [20], daunomycin [21], or linear RGD-peptides [22] were attached to the alginate backbone. Afterward the modified alginates were gelled by adding calcium chloride or crosslink reagents, respectively.

Alginate with its unique material properties and characteristics has been increasingly considered as biomaterial for medical applications. Alginate has been used as excipient in tablets with modulated drug delivery. CA gels have unique intrinsic properties and exhibit biocompatibility, mucoadhesion, porosity, and ease of manipulation. Hence, much attention has recently been focused on the delivery of proteins, cell encapsulation, and tissue regeneration. Alginates play the role of an artificial extracellular matrix, especially in the area of tissue engineering, and alginate gels are widely used as wound regeneration materials [23–26].

Besides the commercially available wound dressing Kaltostat®, fibrous ropes composed of mixed calcium and sodium salts of alginic acid, new types of alginate-based dressings have been developed. Chiu *et al.* presented two new types by crosslinking alginate with ethylendiamine and polyethyleneimine, respectively. Due to the improved properties compared to Kaltostat®, the author predicts a great potential for clinical applications [25].

Diabetic foot ulcers (DFUs) are at risk of infection and impaired healing, placing patients at risk of lower extremity amputation. DFU care requires debridement and dressings. A prospective, multicentre study from Jude *et al.* compared clinical efficacy and safety of AQUACEL® hydrofiber dressings containing ionic silver (AQAg) with those of Algosteril CA dressings. When added to standard care with appropriate off-loading, AQAg silver dressings were associated with favorable clinical outcomes compared with CA dressings, specifically in ulcer depth reduction and in infected ulcers requiring antibiotic treatment. This study reports the first significant clinical effects of a primary wound dressing containing silver on DFU healing [27].

In terms of drug/protein delivery, numerous applications of CA gel beads or microspheres have been proposed. As one example, alginate nanoparticles were prepared by the controlled cation-induced gelation method and administered orally to mice. A very high drug encapsulation efficiency was achieved in alginate nanoparticles, ranging from 70% to 90%. A single oral dose resulted in therapeutic drug concentrations in the plasma for 7–11 days and in the organs (lungs, liver, and spleen) for 15 days. In comparison to free drugs (which were cleared from plasma/organs within 12–24 h), there was a significant enhancement in the relative bioavailability of encapsulated drugs. As clinical application, alginate-based nanoparticulate delivery systems have been developed for frontline antituberculosis drug carriers (e.g., for rifampicin, isoniazid, pyrazinamide, and ethambutol) [28].

Another approach for the surface modification of CA gel beads and microspheres has been the chemical crosslinking of the shell around the alginate core. The approach based on the technique of coating CA gel microspheres has also been used to produce microcapsules. This technique is very promising for the macromolecular drug delivery in biomedical and biotechnological applications [29]. Mazumder *et al.* have shown that covalently crosslinked shells can be formed around CA capsules by coating with oppositely charged polyelectrolytes containing complementary amine and acetoacetate functions. Furthermore, alginate gels were used for cell and stem cell encapsulation [30]. An approach of Dang *et al.* enables a practical route to an inexpensive and convenient process for the generation of cell-laden microcapsules without requiring any special equipment [31].

CA has been one of the most extensively investigated biopolymers for binding heavy metals from dilute aqueous solutions in order to engineer medical applications. Becker *et al.* have studied the biocompatibility and stability of CA in aneurysms *in vivo*. They depicted that CA is an effective endovascular occlusion material that filled the aneurysm and provided an effective template for tissue growth across the aneurysm neck after 30 to 90 days. The complete filling of the aneurysm with CA ensures stability, biocompatibility, and optimal healing for up to 90 days in swine [32].

Hepatocyte transplantation within porous scaffolds has being explored as a treatment strategy for end-stage liver diseases and enzyme deficiencies. The limited viability of transplanted cells relies on the vascularization of the scaffold site which is either too slow or insufficient. The approach is to enhance the scaffold vascularization before cell transplantation via sustained delivery of vascular endothelial growth factor, and by examining the liver lobes as a platform for transplanting donor hepatocytes in close proximity to the host liver. The conclusion by Kedem *et al.* has shown that sustained local delivery of vascular endothelial growth factor-induced vascularization of porous scaffolds implanted on liver lobes and improved hepatocyte engraftment [33].

Furthermore, sodium alginate is used in gastroesophageal reflux treatment [34, 35]. Dettmar *et al.* published the rapid effect onset of sodium alginate on gastroesophageal reflux compared with ranitidine and omeprazole [36]. The rate of acid and pepsin diffusion through solutions of sodium alginate was measured using *in vitro* techniques by Tang *et al.* They demonstrated that an adhesive layer of alginate present within the esophagus limits the contact of refluxed acid and pepsin with the epithelial surface [37].

In the field of nerve regeneration, Hashimoto *et al.* have developed a nerve regeneration material consisting of alginate gel crosslinked with covalent bonds. One to two weeks after surgery, regenerating axons were surrounded by common Schwann cells, forming small bundles, with some axons at the periphery being partly in direct contact with alginate. At the distal stump, numerous Schwann cells had migrated into the alginate scaffold 8–14 days after surgery. Remarkable restorations of a 50-mm gap in cat sciatic nerve were obtained after a long term by using tubular or nontubular nerve regeneration material consisting mainly of alginate gel [38].

7.3 Carrageenan

Carrageenan is a class of partially sulfated linear polysaccharides produced as main cell wall material in various red seaweeds (Rhodophyceae). The polysaccharide chain is composed of a repeating unit based on the disaccharide →3)-β-D-galactose-(1→4)-α-3,6-anhydro-D-galactose or →3)-β-D-galactose-(1→4)-α-D-galactose. Three major types can be distinguished by the number and position of the sulfate groups on the disaccharide repeating unit: κ-carrageenan (one sulfate group at position 4 of the β-D-galactose), ι-carrageenan (one sulfate group at position 4 of the β-D-galactose and one sulfate group at position 2 of the α-3,6-anhydro-D-galactose), and λ-carrageenan (one sulfate group at position 2 of the β-D-galactose and two sulfate groups at position 2 and 6 of the α-D-galactose) [39] (Scheme 7.3).

Different seaweeds produce different types of carrageenan but the biosynthesis of these commercially important polysaccharides is not completely studied yet. The most important subtype κ-carrageenan is isolated from the tropical seaweed *Kappaphycus alvarezii*, also known as *Eucheuma cottonii*. After alkali treatment, a

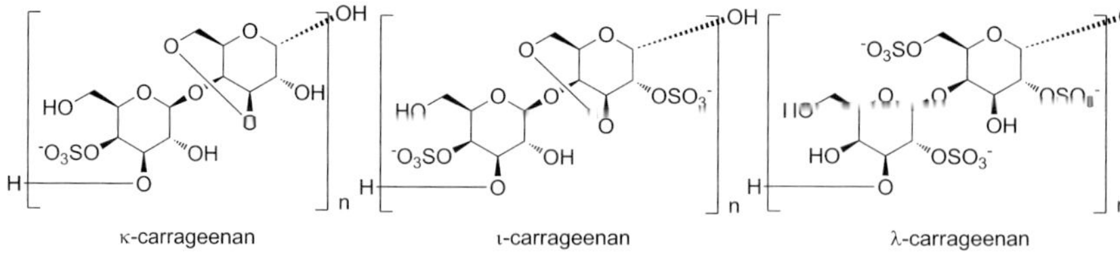

Scheme 7.3 Chemical structures of κ, ι, and λ-carrageenan.

relatively homogeneous κ-carrageenan can be obtained. *Eucheuma denticulatum* (syn. *spinosum*) is the most important source of ι-carrageenan, whereas *Gigartina pistillata* and *Chondrus crispus* mainly produce λ-carrageenan [39].

The gelling properties of the carrageenans strongly differ between the subtypes. κ-carrageenan gives strong and rigid gels, ι-carrageenan makes soft gels, and λ-carrageenan does not form gel. The gelation of a carrageenan solution is induced by cooling a hot solution that contains gel-inducing cations such as K^+ (κ-carrageenan) or Ca^{2+} (ι-carrageenan). The Na^+-form of the carrageenans does not yield a gel [39]. Detailed information on the gelling properties of the carrageenans is summarized in a recent review by Rinaudo [1].

A variety of carrageenan-degrading enzymes (carrageenase) was isolated until now. Most carrageenases are κ or ι-carrageenases, cleaving the polymeric chain of κ or ι-carrageenan in the β-glycosidic bond and yielding a di- or tetrasaccharide with a terminal 3,6-anhydrogalactose [39–41]. As the last enzyme in this context, a λ-carrageenase was cloned from *Pseudoalteromonas bacterium*, strain CL19, which was isolated from a deep-sea sediment sample. The pattern of λ-carrageenan hydrolysis shows that the enzyme is an endo-type λ-carrageenase with a tetrasaccharide of the λ-carrageenan ideal structure as the final main product. As for the other carrageenases, this enzyme also cleaves the β-1,4 linkages of its backbone structure. Remarkably, the deduced amino acid sequence shows no similarity to any reported proteins [42]. Additionally, λ-carrageenase activity was also identified and purified from the marine bacterium *Pseudoalteromonas carrageenovora* [43].

Polysaccharides are often added to dairy products to stabilize their structure, enhance viscosity, and alter textural characteristics. Also, carrageenans are used as thickener and stabilizer in the dairy industry, for example, in the production of dairy products such as processed cheese [44]. Since carrageenan is a polyanionic structure, several applications for gels with polycationic compounds such as chitosan are published. Tapia *et al.* compared chitosan–carrageenan with chitosan–alginate mixtures for the prolonged drug release. They found that the chitosan–alginate system is better than the chitosan–carrageenan system as matrix because the drug release is controlled at low percentage of the polymers in the formulation, the mean dissolution time is high, and different dissolution profiles could be obtained by changing the mode of inclusion of the polymers. In the chitosan–alginate system, the swelling behavior of the polymers controlled the drug release from the matrix. In the case of the system chitosan–carrageenan, the high capacity of carrageenan promotes the entry of water into the tablet, and therefore, the main mechanism of drug release is the disintegration instead of the swelling of the matrix [45].

In a different context, the polyelectrolyte hydrogel based on chiotosan and carrageenan was evaluated as controlled release carrier to deliver sodium diclofenac. The optimal formulation was obtained with chitosan–carrageenan as 2:1 mixture and 5% diclofenac. The controlled release of the drug was maintained under simulated gastrointestinal conditions for 8 h. Upon crosslinking with glutaric acid and glutaraldehyde, the resulting beads were found to be even more efficient and allowed the release of the drug over 24 h [46].

In a recent study, the preparation of crosslinked carrageenan beads as a controlled release delivery system was reported. Since κ-carrageenan just allowed thermoreversible gels, a protocol for an additional crosslinking using epichlorohydrin was introduced. Low epichlorohydrin concentrations led to unstable and weak beads with uneven and cracked surfaces. An optimized crosslinker concentration resulted in smooth and stable gel beads that showed great potential for the application as delivery systems in food or pharmaceutical products [47].

7.4 Cellulose and Its Derivatives

Cellulose was first described by Anselme Payen in 1838 as a residue that was obtained after aqueous extraction with ammonia and acid-treatment of plant tissues [48]. It is a carbohydrate polymer composed of β-(1→4)-linked D-glucose.

Cellulose is one of the most common polymers because it is ubiquitous in the biomass. Its chain length depends on the origin and the treatment of the polymer. The biosynthesis of cellulose has been described in numerous reviews [49]. Besides plants as polymer source, it can also be obtained from bacterial production (see Chapter 5) or from *in vitro* synthesis. Cellulose can be produced either by enzymatic polymerization of β-cellobiosyl fluoride monomers or by chemical synthesis, for example, by cationic ring-opening polymerization of glucose orthoesters. These approaches are summarized in a review from Kobayashi *et al.* [50].

The crystal structure of cellulose has been studied intensively. Two modifications of cellulose I were discovered, varying in the character of their elementary cell, which is either triclinic or monoclinic. Cellulose II is the thermodynamically most stable structure. More solid and liquid state crystal structures of cellulose and the fibrillar morphology of the polymer are summarized in the review from Klemm *et al.* [51].

Due to its numerous hydrogen bonds, cellulose is insoluble in nearly all common solvents [52]. For this reason, several cellulose solvent systems have been explored to enable its chemical modification. LiCl–dimethylacetamide mixtures as well as tetrabutylammonium fluoride in dimethyl sulfoxide or metal containing solvents, for example, cuprammonium hydroxide, have been investigated [53]. Several chemical derivatizations of cellulose can be realized in order to use cellulose as drug deliverer or for other medical applications. An overview of these modifiable functional groups is given in Scheme 7.4. For instance, cellulose can be oxidized at different positions as well as esterified or alkylated at the primary hydroxyl group. Especially the last mentioned derivatizations lead to water- and/or organic-soluble compounds, which can be used for further modifications.

The secondary alcohol groups of cellulose can be oxidized to ketones, aldehydes, or carboxylic functions depending on the reaction conditions. The product is called oxycellulose and represents an important class of biocompatible and bioresorbable polymers which is widely used in medical applications. It is known to be hemostatic, enterosorbent, and wound-healing. Furthermore, oxycellulose can be used

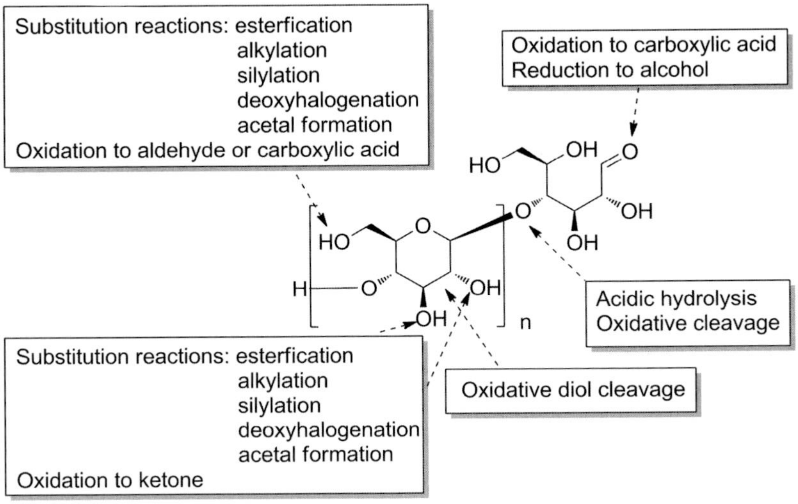

Scheme 7.4 Possible positions for chemical modification of cellulose.

as drug carrier because its carboxylic groups can be used for further derivatization, especially for the coupling of various bioactive agents such as antibiotics, antiarrhythmic drugs, and antitumor agents. By addition of these drugs to oxycellulose, their toxicity could be increased or their activity could be enhanced. For more detailed information on the synthesis and the applications of oxycellulose, see the review from Bajerová *et al.* [54]. Aldehyde-functionalized oxycellulose can be used in the field of tissue engineering. Hydrogel formation of aldehyde- and hydrazine-functionalized polysaccharides is explained in Chapter 10.

The synthesis of various cellulose esters was summarized by Seoud and Heinze [55]. They separated the functionalization process into three steps: (i) activation of the polymer by solvent, heat, or others, (ii) dissolution of the cellulose according to methods described above, and (iii) chemical derivatization. The applications of cellulose esters are multifaceted. Depending on their chemical structure, they are used as coatings for inorganic materials, laminates, optical films, and applications in the separation area such as hemodialyses and blood filtration. Several applications of cellulose esters are summarized in a review by Edgar *et al.* [56].

Another type of cellulose esters are the cellulose sulfonates, prepared from cellulose and sulfonic acid or sulfonic chloride. This class of compounds has reactive groups that can be easily substituted with nucleophilic reagents, for example, amines to yield aminocellulose, which is used as enzyme support [57].

Sodium carboxymethyl cellulose is another common cellulose derivative. This anionic, water-soluble compound is generated through etherification of the primary alcohol of cellulose. It is used as an emulsifying agent in pharmaceuticals and cosmetics [58]. Sannino *et al.* used carboxymethyl cellulose and hyaluronan hydrogels to prevent postsurgical soft tissue adhesion [59]. Both polymers were crosslinked with divinylsulfone. Rokhade and coworkers prepared semi-interpenetrating polymer network microspheres of gelatin and sodium carboxymethyl cellulose with an encapsulated anti-inflammatory agent. Glutaraldehyde served as a crosslinker in this drug release system [60].

Silylation of cellulose with chlorosilanes or silazanes leads to thermostable silyl ethers, which are more lipophilic in comparison to unmodified cellulose. Several conditions, which lead to silyl ethers with different substitution patterns, are described in a review by Klemm *et al.* [51]

Other etherified cellulose derivatives, for example, methycellulose, ethylcellulose, hydroxypropyl cellulose, and hydroxypropyl methyl cellulose, are described elsewhere [58]. Briefly, methycellulose is used in bulk laxatives, nose drops, ophthalmic preparations, and burn ointments, and ethylcellulose has a broad range of applications because it is insoluble in water but soluble in polar organic solvents.

7.5 Microbial Cellulose

Microbial cellulose (MC) belongs to the group of homopolysaccharides, which consists of only one type of monosaccharide, in the case of MC, β-D-glucose. The monomers are linked through 1→4 glycosidic bonds (Scheme 7.5).

The production of MC was first observed by A. J. Brown in 1886, who found out that cellulose was produced in resting cells from *Acetobacter xylinum* in the presence of oxygen and glucose [61]. Other bacteria which produce MC are *Agrobacterium, Acetobacter, Aerobacter, Archromobacter, Azotobacter, Rhizobium, Sarcina,* and *Salmonella*. The review from Chawla *et al.* gives an overview concerning the cultivation conditions of the different strains [62]. The fermentation process and the biosynthesis for MC are in-depth described in recent published reviews [63, 64]. Briefly, the complex process consists of three steps; namely: (i) the linear strand formation from uridine diphosphoglycose, catalyzed by cellulose synthetase, a membrane-anchored protein, (ii) the extracellular secretion of the chain, and (iii)

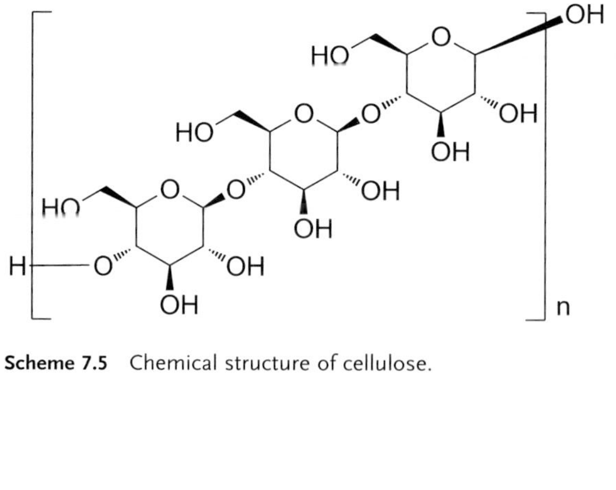

Scheme 7.5 Chemical structure of cellulose.

assembly to hierarchically composed ribbon-shaped microfibrils of approximately $80 \times 4\,nm$ [65].

Although it has the same chemical structure as plant cellulose, the MC can be obtained in higher purities and it has a higher degree of polymerization and crystallinity. The fibrils of bacterial cellulose are 100 times thinner than their plant analogs. Furthermore, it has remarkable water-holding capacity and a high tensile strength, which results from the interfibrillar hydrogen bonding. These physical properties made MC to a promising candidate for biomedical applications.

Besides the physical properties described above, MC has a lot of advantages in the wound-healing process. Due to its nanoporous structure, external bacteria cannot penetrate the wound. It is easy to sterilize, cheap, and elastic which provides a painless wound coverage and removal. The material is highly porous and allows an unhindered gas exchange [66]. Helenius *et al.* evaluated the bacterial cellulose in aspects of chronic inflammation, foreign body responses, cell ingrowth, and angiogenesis [67]. MC proved to be entirely biocompatible. This is why MC is often used as wound dressing material, which protects the wounds from infection or dehydration. The review from Czaja *et al.* summarized the application of MC in the field of wound treatment [68]. Several approaches have been made to improve and accelerate the healing process, for example, by impregnation of the MC tissue with therapeutic agents such as superoxide dismutas or poviargol [69]. Additionally, several composites with other polymers such as gelatin have been reported [70].

MC is also used as tissue material in artificial cardiovascular medicine. Therefore, it is necessary to mold the polymer into the needed shape during its synthesis. Different techniques for this demand are summarized elsewhere [51, 66]. Charpentier *et al.* published another approach in the exploration of a vascular prosthetic device. They used PETG and PCTG polyesters as backbone material, modified the surface with UV and plasma treatment and coated it with MC to reduce coagulation effects of the material [71].

Another patented system deals with the coating of endoprostheses with MC, in order to obtain biocompatible devices [72]. MC is also used as physical barrier, which separates bone cells from the surrounding tissue to prevent fibroblast cell ingrowth. This accelerates the regeneration process of the osseous cells [73, 74]. MC was tested as a 3-D scaffold for *in vitro* cell cultivation to mimic the extracellular matrix. Afterward, the overgrown tissue should be implanted into the body to replace the diseased area [75].

7.6
Chitin and Chitosan

Chitin and chitosan are structurally related aminopolysaccharides. Both polysaccharides may be regarded as derivatives of cellulose, where chitin bears an acetamido group and chitosan bears a aminogroup instead of the C-2 hydroxyl functionality (Scheme 7.6).

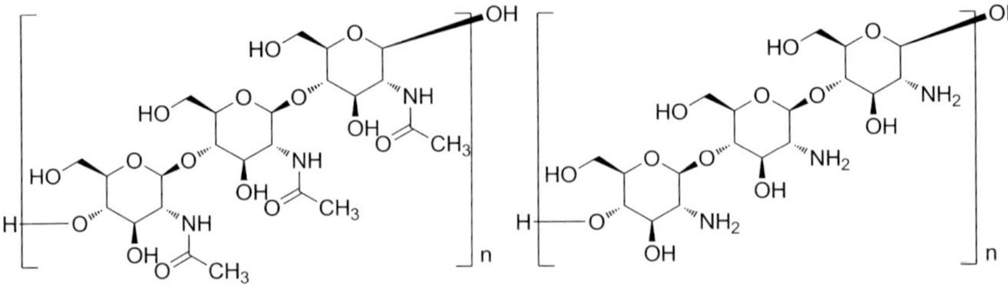

Chitin Chitosan

Scheme 7.6 Chemical structures of chitin and chitosan.

Chitin is the second most abundant biopolymer after cellulose and is found in ordered fibrils in cell walls of fungi and yeast and in the exoskeleton of crustaceans and insects. The main commercial sources of chitin are shrimp and crab shells, a waste product of the seafood production. For the production of pure chitin, the shells are deproteinized under basic conditions and subsequently demineralized under acidic conditions to remove $CaCO_3$. It is important to note that chitin shows three different crystalline structures depending on its function in nature. The outer skeletal chitin in crustaceans consists of α-chitin, squid pen consists of β-chitin, and fungi contain γ-chitin. Chitin shows a high biocompatibility, an excellent biodegradability, and a low immunogenicity. A major problem is the low solubility of chitin in water and almost all common organic solvents due to its high crystallinity [76]. *N*-deacetylation in concentrated alkali solution at high temperatures or using the enzyme chitin deacetylase (EC 3.5.1.41) leads to chitosan. The chemical *N*-deacetylation can be performed in two different ways. In a heterogeneous process, chitin is treated with 10–60% sodium hydroxide solution at 70–150 °C for up to 6 h. Chitosan prepared by this method is approximately 90% deacetylated [77]. A milder homogenous process leads to water-soluble chitosan which is 50% deacetylated by storing an alkaline solution of chitin for 77 h at room temperature [78]. The predominant thermochemical chitosan production is environmentally unsafe and hard to control, leading to broad range of products with a lower molecular mass due to partial hydrolysis of the polymeric chain. The use of chitin deacetylase (ED 3.5.1.41), which could be isolated and cloned from various fungi and insects, can circumvent some of these problems. It was shown that a 97% deacetylation of chitosan is possible using partially deacetylated chitosan as substrate. The enzymatic deacetylation of crystalline or amorphous chitin is still less effective yielding a 0.5–9.5% deacetylated product [79]. Naturally occurring chitosan is very rare and can be found together with chitin in several fungi. Since chitosan is rare, chitosan-degrading enzymes are less abundant. Lysozyme does also, in addition to its natural substrate (the glycosidic linkage of certain bacterial cell walls peptidoglycans), hydrolyze chitin and chitosans. Lysozyme is present in many tissues and secretions such as tears, saliva, and blood [80, 81]. In a detailed study, the enzymatic (lysozyme, chitinase, etc.) digestibility of various chitins and chitosans was investigated. It turned out that the digestibility of chitin by the chitinase from *Bacillus* sp. PI-7S is much higher than by lysozyme. Also β-chitin was digested more smoothly than α-chitin, and chitosan deacetylated under homogeneous conditions was hydrolyzed by lysozyme more rapidly than that under heterogeneous conditions [82].

In contrast to chitin, chitosan is highly soluble in diluted acids. The primary amino groups in chitosan are protonated below pH 6.0, resulting in a water-soluble cationic polyelectrolyte. At higher pH values, the ammonium salt gets deprotonated resulting in a neutral amino group and the polymer gets insoluble. On the other hand, this solubility transition is highly dependent on the degree of *N*-acetylation and chitosan with 50% *N*-acetylation is soluble even under alkaline conditions [78]. In addition, the anion of the acid plays an important role for the solubility of chitosan. While many acids such as acetic, citric, formic, hydrochloric, lactic, and diluted nitric acid can easily dissolve chitosan, the phosphates and sulfates of chitosan are not soluble in water [76].

Several approaches were published to solubilize chitin with and without chemical modification of the polymer. 2.77 M sodium hydroxide was reported as good solvent for chitin and the addition of urea did improve the solubility [83]. A powerful organic solvent system for chitin was first described by Austin and Rutherford. They found that lithium chloride forms a complex with the acetamide carbonyl group of chitin [84]. The resulting complex is soluble in polar organic solvents such as *N*-methyl-2-pyrrolidinone, *N,N*-dimethylacetamide, *N,N*-dimethylpropionamide, and 1,3-dimethyl-2-imidazolidinone. Chitin solutions with a concentration of 5–7% (w/v) could be obtained using these conditions [85]. Another suitable solvent system is $CaCl_2$-dihydrate saturated methanol as reported by Tamura [86]. The water content is essential and anhydrous $CaCl_2$ in methanol does not dissolve chitin at all. Two grams of α-chitin powder can be dissolved in 100 mL of $CaCl_2 \cdot (H_2O)_2$-saturated methanol but just 0.5–1 g of β-chitin is soluble under those conditions. The solubility is also affected by the degree of *N*-acetylation and the molecular weight of chitin as depicted in Figure 7.1 [87].

Another successful strategy for chitin dissolution is the synthesis of soluble chitin esters. The introduction of bulky acyl groups into the chitin chain yields chitin derivatives with improved solubility [88]. Acetylchitin is readily synthesized and spun into fibers but still polar acidic solvents such as formic acid are necessary to dissolve the material [89]. Butyrylchitin, with a larger substituent in the chain, can be synthesized using methanesulphonic acid as catalyst and solvent. This derivative is easily soluble in several organic solvents, such as acetone, methanol, ethanol, dimethylformamide, and methylene chloride [90]. A simpler method for the synthesis of highly substituted dibutyrylchitin with butyric anhydride uses 70% perchloric acid as a catalyst. Dibutyrylchitin fibers with a porous core were made by a simple method of dry spinning its 20–22% solutions in acetone. These fibers have tensile properties similar to or better than those of chitin. Alkaline hydrolysis of the butyric esters restores chitin, and even fibers with good tensile properties can be obtained by alkaline hydrolysis of dibutyrylchitin fibers in 5% sodium hydroxide at 55 °C without destroying the fiber structure [91]. The ester cleavage

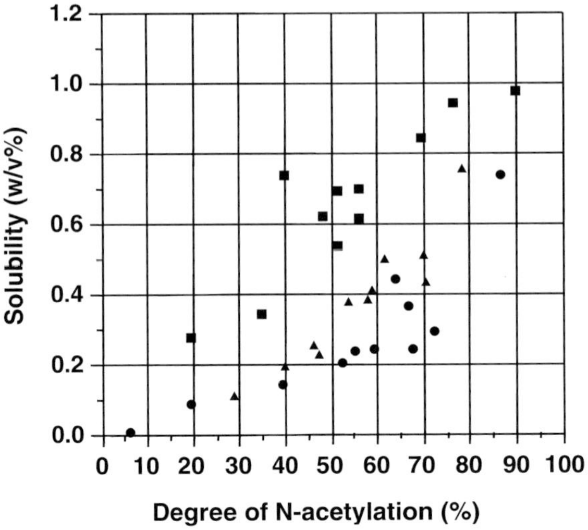

Figure 7.1 Dependence of the $CaCl_2$-dihydrate/methanol solubility of chitin with respect to the degree of acetylation and the molecular weight of chitin. Solid square, 1.2 × 104; solid triangle, 4.0 × 104; solid circle, 1.6 × 105 (with permission from Ref. [87]).

can be monitored by FTIR spectroscopy and by the weight loss of the material, which raised up to 40% for a complete hydrolysis. The restoration of the chitin structure from dibutyrylchitin fibers resulted in an increase of the degree of crystallinity and in the diameter of the fibers along with a decrease of the tensile strength [92].

The peculiar biochemical properties of chitins and chitosans remain unmatched by other polysaccharides. The major areas of application include water treatment, biomedical applications (including wound dressing and artificial skin), and personal-care products. Chitin and chitosan-based materials have unique characteristics in the area of tissue regeneration. Hemostasis is immediately obtained after application of chitin-based dressings to traumatic and surgical wounds: platelets are activated by chitin with redundant effects and superior performances compared with known hemostatic materials. To promote angiogenesis, the production of the vascular endothelial growth factor is upregulated in wound healing when macrophages are activated by chitin/chitosan. Biocompatible wound dressings derived from chitin are available in the form of hydrogels, xerogels, powders, composites, films, and scaffolds. The scaffolds are easily colonized by human cells to restore tissue defects. Chitin tubes, which can be manufactured from the tendon of the crab leg muscle or by using electrospun chitosan nonwoven, can be implanted to bridge a dissected nerve and used as alternative to autologous grafts. Chitosan is also used in cartilage tissue engineering where it provides an environment in which the chrondrocytes maintain their correct morphology and their capacity to synthesize the correct extracellular matrix. Scaffolds made of either pure β-chitin, or pure chitosan, or mixtures of both polysaccharides had the same efficiency in supporting chondrocytes. Chitosan composites can also be used in the treatment of bone defects, where it promotes growth and mineral-rich matrix deposition by osteoblasts. Especially, porous hydroxyapatite–chitin matrices have a great potential in this field of regenerative therapy [93].

7.7 Dextran

Dextran belongs to the family of homopolysaccharides, precisely to the complex, multibranched glucans. Glucans are polysaccharides which are built up by glucose monomeric units. Dextran itself is constructed by a specific form of glucose, α-D-glucopyranose. The polymeric chain consists of a substantial number of consecutive α-(1→6) linkages in their major chains, usually more than 50% of the total linkages. Further side chains result mainly from α-(1→3) and occasionally from α-(1→2) or α-(1→4) linkages. Overall the molecular weight differs from 10 to 150 kDa (Scheme 7.7).

The exact network of a specific dextran depends on its individual producing microbial strain. Dextran is produced either in *Leuconostoc mesenteroides* and other lactic acid bacteria or in certain *Gluconobacter oxydans*. The former converts sucrose into dextran with the dextransucrase enzyme, whereas the latter converts maltodextrins into dextran with the dextran dextrinase enzyme [94]. The enzyme dextransucrase catalyzes the transfer of D-glucopyranosyl moieties from sucrose to dextran, while fructose is released. Thereby dextransucrose acts substrate specific because other native saccharides like fructose, glucose, or mixtures of both are not converted. Furthermore, no adenosine triphosphate or cofactors are required [95]. According to the classification of transferases, dextransucrase is an extracellular glucosyltransferase. Until today, more than 30 sucrose glycosyltransferase genes have been sequenced and their catalytic sites have been identified. The families of glycosyltransferases and glycoside-hydrolases share related mechanistic and structural characteristics [96].

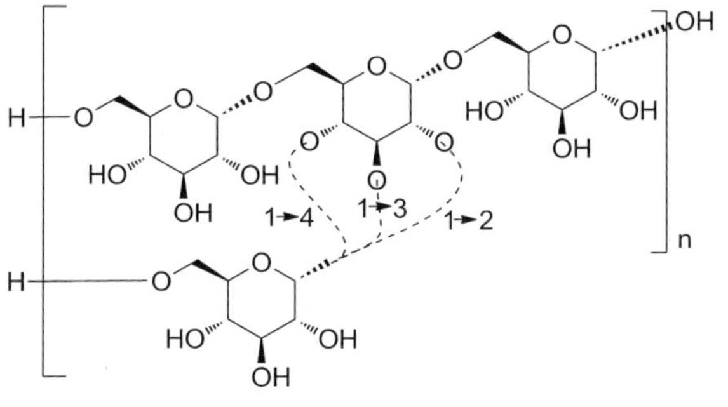

Scheme 7.7 Chemical structure of dextran with exemplary α(1→2), α(1→3), and α(1→4) linkages.

The synthesis of unbranched dextran was already published in the 1950s [97]. Nowadays, dextran can also be synthesized via cationic ring-opening polymerization of 1,6-anhydro-2,3,4-tri-*O*-allyl-β-D-glucopyranose [98]. Commercially available dextran is generally produced by dextransucrase NRRL B-512F from *L. mesenteroides*. In this process, cultures of *L. mesenteroides* were grown in sucrose-containing media with growth factors, trace minerals, and an organic nitrogen source. In former times, Naessens *et al.* have shown that forms of *G. oxydans* could be a promising alternative to *L. mesenteroides* as biocatalysts for the synthesis of dextran and oligodextrans [99].

Dextran is enzymatically degraded by dextranase into dextrose (D-glucose). Dextranases belong to the family of glycosyl-hydrolases and are subdivided into endo- and exodextranses (Figure 7.2). In organisms, these enzymes are present in human liver, intestinal mucosa, colon, spleen, and kidney. Since the first reports on *Cellvibrio fulva* dextranase in the 1940s, more than 1500 scientific papers and more than 100 patents have been issued on dextran-hydrolyzing enzymes found in a number of microbial groups, fungi being the most important commercial source of dextranase [100, 101]. Enzymatically fractionated dextran with a specific chain length and individual characteristics possesses interest in different branches of industry. It can be implemented in cosmetics, drug formulations, and vacancies, as cryoprotectants, and as stabilizers in the food industry.

In general, derivatization occurs at the hydroxyl groups in the monomeric unit. Several approaches for crosslinking dextran are published. Bis-acrylamid, epichlorhydrin, diisocyanates, phosphorus oxychloride, methacrylate, acylate, and other functions were favored. In addition, esterifications by inorganic or organic compounds have been established. In the range of inorganic ester, only the sulfates and the phosphates have gained interest. The introduction of these ionic groups leads to polyelectrolytes with an improved water solubility. In contrast to the broad

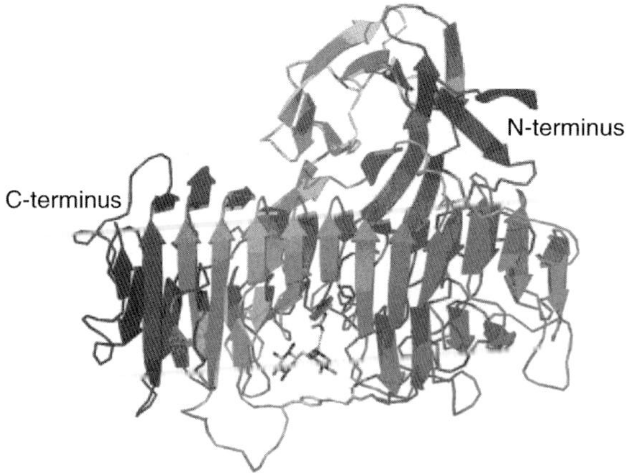

Figure 7.2 Crystal structure of endodextranase Dex49A from *Penicillium minioluteum* with isomaltose in the product-bound form (with permission from Ref. [101]).

variety of applications of 1→4- and 1→3-linked glucans after reaction with (C-2 to C-4) carboxylic acid anhydrides and chlorides, the use of dextran esters of short-chain aliphatic acids such as acetates or propionates is rather limited. Moreover, several ether derivatives were established. A detailed overview about the chemistry, biology, and application of functional polymers based on dextran is given by Heinze *et al.* [102].

For medical usage, dextran has many useful characteristics. It is highly biocompatible, slowly degraded, innocuous to the body and readily excreted from it. Hence, dextran has found application in clinic; especially, biodegradable dextran hydrogels are widely used for protein delivery [103], due to the fact that they can act as carriers for a controlled release of drugs to targeted organs mediated by slow dextranase hydrolysis. Kim *et al.* have presented the *in vitro* drug release behavior of dextran–methacrylate hydrogels using doxorubicin [104]. A recent publication has shown that injectable *in situ* crosslinking hydrogels by hydrazone formation can be used for local antifungal therapy [105]. Mostly aqueous solutions of dextran are used as blood volume expander [106]. This relies to the feature that a 6% dextran solution with an average molecular mass of about 75 kDa provides an osmotically neutral fluid comparing to the blood. Molecules with a too low molecular weight are removed rapidly from the blood circulation via the kidneys, whereas molecules with a too high molecular weight can interfere with the normal coagulation process. Furthermore, 40 kDa dextran improves the blood flow besides the plasma volume expansion, presumably mediated by reduction of blood viscosity and inhibition of erythrocyte aggregation [107].

In a further application, crosslinked dextran chains are often used as purification and/or separation material. Epichlorohydrin crosslinked dextran forms a gel which is used as release agent. A most common separation process to fractionate water-soluble substances by molecular size is Sephadex®. Elementary Sephadex, derived from SEparation PHAmarcia DEXtran, gels are prepared by crosslinking dextran to a macromolecular network of great stability [108].

Dextrans and derivatives are widely used in nutrition products, fructose syrup, and as additives in bakery, candies, and ice cream. Moreover, it has upraising potential uses as emulsifying and thickening agents, high-viscosity gums, explosives, deflocculants in paper industry, oil drilling muds, soil conditioners, and surgical sutures [109]. An interesting approach is the introduction of dextranases for the degradation and removal of dental caries. Dental plaque consists of ~20% glucans; consequently, dextranses could inhibit the synthesis of insoluble glucans [110].

7.8 Gellan

Gellan is a linear anionic bacterial exopolysaccharide with a repeating unit based on the tetrasaccharide → 3)-D-glucopyranosyl-β-(1→4)-D-glucuronopyranosyl-β-(1→4)-D-glucopyranosyl-β-(1→4)-L-rhamnopyranosyl-α-(1→3). In its native form,

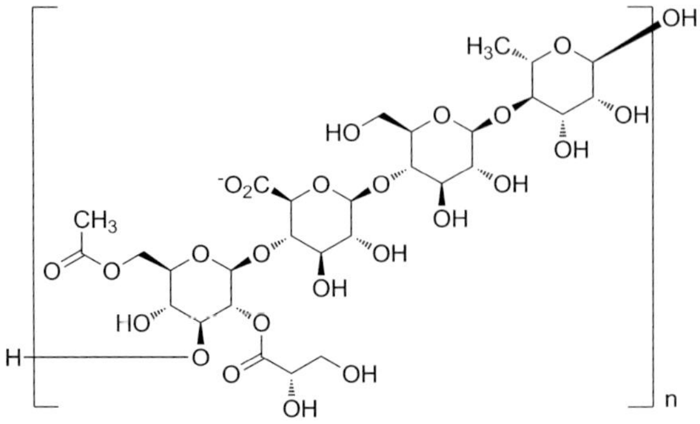

Scheme 7.8 Chemical structure of high-acyl gellan.

gellan bears an acetyl group at position 6 and an L-glyceryl group at position 2 of the glucose β-(1→4) linked to the glucuronic acid. It consists of about 50,000 sugar units and is normally deacylated before use [111] (Scheme 7.8).

Gellan gum is produced via submerged fermentation using *Sphingomonas paucimobilis* ATCC 31461. Studies on the nutritional requirements for optimal exopolysaccharide production in a salt-based synthetic medium reveal soluble starch (20 g/L) as the best carbon source and tryptone (0.5 w/v%) as the best nitrogen source. A maximum of 35.7 g/L gellan is produced under optimized conditions. After heating the crude fermentation broth at pH 10, deacetylated gellan can also be isolated by precipitation [112].

Under suitable conditions, the native gellan gum forms soft, elastic, transparent, and flexible gels. In contrast, the deacylated gellan forms stiff and brittle gels in the presence of many cations such as Ca^{2+}, Mg^{2+}, Na^+, K^+, and H^+ [113]. Physical properties such as the setting temperature, strength, and firmness of the gel depend on the type of cation used for the gelation, as well as its ionic strength. For example, potassium ions set the gels at a lower temperature than calcium ions. In contrast, calcium ions can yield gels of the same strength and quality as potassium, but at a much lower concentration of 1/40th that of the monovalent ions. The X-ray-based solid phase tertiary structure of deacylated gellan was reported by Chandrasekaran *et al.* to consist of two identical left-handed, threefold double helices [114]. The molecular origin for rheological characteristics of native gellan gum is discussed by Tako *et al.* The ^{1}H-nuclear magnetic resonance spectra of gellan in its native and deacylated form show that the L-rhamnosyl residues of native gellan involved a small number of $^4C_1$-pyranose conformations and a large number of $^1C_4$-pyranose conformations, whereas for deacylated polymer, almost all of the residues were involved in $^4C_1$-pyranose conformation [111]. Such conformational changes of the L-rhamnosyl residues contribute to the gel formation for deacetylated rhamsan gum which is effectively a gentiobiosylated form of gellan [115].

In a broad screen, Gellan was found to be degradable by a number of bacterial strains identified as *Bacillus* sp. Several Gram-negative bacteria secrete extracellular eliminase-type enzymes (lyases) which are cleaving the sequence β-D-glucopyranosyl-β-(1→4)-D-glucuronopyranosyl in the tetrasaccharide repeat unit of gellan. In most of the bacterial isolates, the lyases are predominantly endoenzymes. The enzymes are highly specific toward gellan and rhamsan but do not degrade most of the other bacterial exopolysaccharides, which are structurally related to gellan [116, 117]. Hashimoto *et al.* cloned a 140 kDa gellan lyase from *Bacillus* sp GL1. The recombinant enzyme is most active at pH 7.5, 50 °C and stable below 45 °C. The recombinant lyase is active on gellan, especially in the deacylated form, but is inert against gellan-related polysaccharides such as S-88, welan, rhamsan, and S-198 [118].

A thermostable gellan lyase, with a residual activity of 100% after 24 h incubation at 60 °C and a half-life time of 60 min at 70 °C, was isolated form a thermophilic strain *Geobacillus stearothermophilus*. The strain produces the thermostable gellan lyase extracellularly during exponential phase and the enzyme is not present in culture liquid without gellan. The enzyme has an optimal activity at 75 °C in a very large pH area between 4 and 8.5 [119].

The easiest and most common modification of gellan is the complete deacylation. The native gellan gum is dissolved in 0.2 M potassium hydroxide and stirred at room temperature for 12 h under atmosphere of nitrogen. After neutralization with 0.1 M hydrochloric acid and filtration through Celite 545, the deacylated gellan can be precipitated in the presence of 0.05% KCl by the addition of two volumes ethanol [111].

In a recent study, acrylate and maleate esters were synthesized. The esterfication is possible using homogenous (acrylic acid in water) or heterogeneous (acryloyl chloride or maleic anhydride in *N,N*-dimethylformamide or aceton) reaction conditions. These macromonomeres can be polymerized under mild conditions and lead to biodegradable thermo- and pH-stimulable hydrogels with adjustable crosslink density [120].

Self-crosslinking of aqueous gellan with 1-ethyl-3-[3-(dimethylammino)propyl]-carbodiimide leads to thermally stable hydrogels. Based on X-ray data, the structure of the gel is proposed to be a bundle of 48 gellan double helices aligned in parallel. The rigid bundles formed by associated gellan double helices constitute the junction zones which sustain the overall gel structure displaying solid-like properties. Relatively large cavities supported by rigid bundles inside the gels absorb water quickly as indicated by kinetic water uptake data [121].

Gellan was approved for use in food in Japan in 1988 and in the United States in 1992. Today gellan is used in food products, which require a highly gelled structure such as meat and vegetable aspic, jams, and jellies. As an example, reduced calorie jams can be prepared with only 0.15% of clarified gellan, and a matrix containing 0.7% gellan does not melt on sterilization. In addition, gellan is used to provide body and mouth-feel as a substitute of gelatin and to speed up the set time of jellies as a substitute of starch.

Additionally, gellan is used as solid culture media for growth of microorganisms and plants, as matrix in gel electrophoresis and to immobilize cells. Gellan has some promising properties in the area of controlled drug release. Sustained delivery of paracetamol from gels formed *in situ* in the stomach is similar to commercially available suspension. Also, gellan-based ophthalmic solutions are reported to have longer residence time in tear fluid than saline solution. In both applications, the solution of gellan mixed with drugs *in situ* forms a gel at acidic pH in the stomach or in the presence of ions in the lachrymal fluid [122].

In a first study to evaluate gellan as scaffold material for tissue engineering, gellan could be ionically crosslinked on mixing with α-modified minimum essential medium resulting in the formation of a self-supporting hydrogel. By adding a suspension of rat bone marrow cells in α-modified minimum essential medium to 1% gellan solution, it is possible to immobilize cells within the three-dimensional gellan matrix that remain viable for up to 21 days in culture. This simple approach to cell immobilization within three-dimensional constructs poses a low risk to a cell population immobilized within a gellan matrix and thus indicates the potential of gellan for use as a tissue engineering scaffold [123].

7.9 Guar Gum

Guar gum belongs to the group of galactomannans and is a neutral polysaccharide with a main chain composed of β-(1→4)-linked mannose units and a molecular weight of approximately 1×10^6 Da. On average, every second mannose is substituted at position 6 with a α-D-galactopyranose. Analysis of the products obtained by enzymatic digestion indicates that in guar galactomannan, the D-galactosyl groups are arranged mainly in pairs and triplets [124] (Scheme 7.9).

Guar gum is isolated from the seed of *Cyamopsis tetragonoloba,* which belongs to the Luguminosoe family. Guar gum forms crosslinked hydrogels on treatment with borax or calcium ions. However, its aqueous solution is nonionic and hydrocolloidal and is not affected by ionic strength or pH. In order to yield a monodisperse guar gum fraction, a purification procedure modified by Wientjes *et al.* can

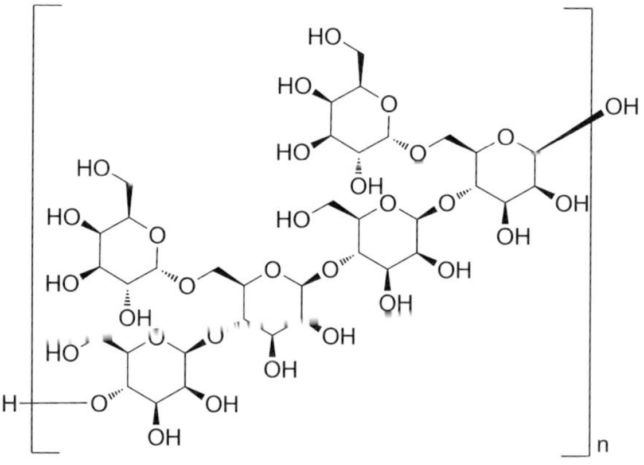

Scheme 7.9 Chemical structure of guar gum.

be applied: Crude guar gum (10 g) is treated with 200 mL boiling, aqueous 80% ethanol for 10 min. The obtained slurry is collected on a glass filter (no. 3) and washed successively with ethanol, acetone, and diethyl ether. The solid is added to 1 l of water, homogenized with a blender, and centrifuged at 2300 g for 15 min. The supernatant is precipitated in two volumes of cold acetone, redissolved in hot water, and ultracentrifuged at 82,000 g for 1.5 h at room temperature. Finally, the supernatant is precipitated with two volumes of ethanol and precipitate is collected on a glass filter (no. 4), washed with ethanol, acetone, and diethyl ether before freeze-drying [125].

Three kinds of bonds in guar gum are susceptible to enzymatic hydrolysis: the endo- and exo-β-1,4 linkages on the D-mannose backbone and the α-1,6 linkage between the mannose backbone and the galactose side chain. The enzymes that cleave these bonds are, respectively, endo- and exo-β-mannanase and α-galactosidase [126]. The 1,4-β-D-mannosidic linkages in galactomannans can be hydrolyzed by β-D-mannanases, a class of enzymes which is produced by plants, bacteria, and fungi. The efficiency of this reaction depends both on the degree of polymerization and galactose substitution levels [127]. Galactomannans with a galactose content of up to 32% are hydrolyzed with no significant change of the K_m or relative V_{max} values. However, if the galactose content approaches 34–38%, the K_m values doubles, and the relative V_{max} values decreases by 10–20%. Typically, 6% of the mannosidic linkages in guar gum are hydrolyzed [128]. Since galactomannans as guar gum are used in hydraulic fracturing of oil and gas wells, it is necessary to employ thermostable enzymes for the enzymatic degradation. Commonly applied enzymatic breakers are mixtures of hemicellulases produced by *Aspergillus niger*. This enzyme preparation is moderately thermostable with temperature optima of approximately 65 °C and it has shown to be effective for galactomannan hydrolysis and viscosity reduction. McChutchen *et al.* isolated and characterized a α-galactosidase and a β-mannanase produced by the hyperthermophilic bacterium *Thermotoga neapolitana* 5068. The purified α-galactosidase had a temperature optimum of 100–105 °C with a half-life of 130 min at 90 °C and 3 min at 100 °C. The purified β-mannanase was found to have a temperature optimum of 91 °C with a half-life of 13 h at that temperature and 35 min at 100 °C [129].

To yield derivatives with adjusted material properties, guar gum can be carboxymethylated by the reaction with the sodium salt of monochloroacetic acid in presence of sodium hydroxide. Using homogenous reaction conditions, various degrees of substitution can be synthesized. Aqueous solutions of carboxymethylated guar gum have higher viscosities compared to unmodified guar gum [130]. Another important derivative is hydroxypropyl guar gum, a hydrophobic polymer obtained from the native biopolymer via an irreversible nucleophilic substitution, using propylene oxide in the presence of an alkaline catalyst. When guar gum is modified to hydroxypropyl guar, the added hydroxypropyl groups sterically block the hydrogen bonding sites on the guar backbone and reduce the hydrogen-bonding attractions between guar molecules. In comparison to guar gum, this derivative has an improved viscosity [131].

Several protocols are published describing the synthesis of grafted guar gum [132]. Nayak and Singh described the ceric-ammonium-nitrate-initiated graft copolymerization of polyacrylamide onto hydroxypropyl guar gum by solution polymerization technique. Six grades of graft copolymers have been synthesized by varying catalyst and monomer concentrations. The percentage of grafting increases with increasing catalyst concentration and decreases with monomer concentration taking other parameters constant [133].

Native and modified guar gum is widely used in petroleum industry as additives in aqueous fracturing fluids and in drilling shallow wells. These applications utilize the properties to increase viscosity, reduce fluid loss, and decrease fluid friction. Additionally, several applications were published using guar gum in the field of controlled drug release and drug delivery. Soppirnath and Aminabhavi prepared a graft copolymer of guar gum with acrylamide, which was crosslinked with glutaraldehyde to form the hydrogel microspheres by the water-in-oil emulsification method. The microspheres were loaded with two antihypertensive drugs, verapamil hydrochloride (water-soluble) and nifedipine (water-insoluble) to investigate their controlled release characteristics. The drugs could be incorporated either during crosslinking by dissolving it in the reaction medium or after crosslinking by the soaking technique [134]. The synthesis of acryloyl guar gum and its hydrogel materials for use in the slow release of L-DOPA and L-tyrosine is described by Thakur *et al.* The material obtained has good properties as release devices for transdermal applications for the treatment of diseases like vitiligo and Parkinson's disease. The hydrogels exhibit unique swelling behavior, and respond well to the physiological stimuli such as pH and the ionic strength. A high loading capacity of L-tyrosine and L-DOPA and a slow release behavior – especially at pH 7.4 – was achieved with these hydrogel materials [135].

Tiwari *et al.* reported the synthesis of biodegradable hydrogels-based photopolymerized guar gum-methacrylate macromonomers for *in situ* fabrication of tissue engineering scaffolds. Depending on the reaction conditions, the hydrogels exhibit equilibrium swelling ratios between 22% and 63%. The degree of β-D-mannanases-induced biodegradation of the hydrogels decreased linearly with increasing gel content and the degree of methacrylation of the respective macromonomers [136].

7.10
Hyaluronic Acid (Hyaluronan)

Hyaluronic acid (HyA) is a linear, high molecular weight polysaccharide composed of β-(1→4)- and β-(1→3)-linked *N*-acetyl-D-glucosamine (GlcNAc) and D-glucuronic acid (GlcA) (Scheme 7.10). This polysaccharide was discovered in 1934 by Meyer and Palmer in the vitreous humor of cattle eyes [137]. Eponymous for the polymer was the Greek word "halos," which stands for glass, in combination with its component "uronic acid." In 1986, HyA was renamed hyaluronan due to the inter-

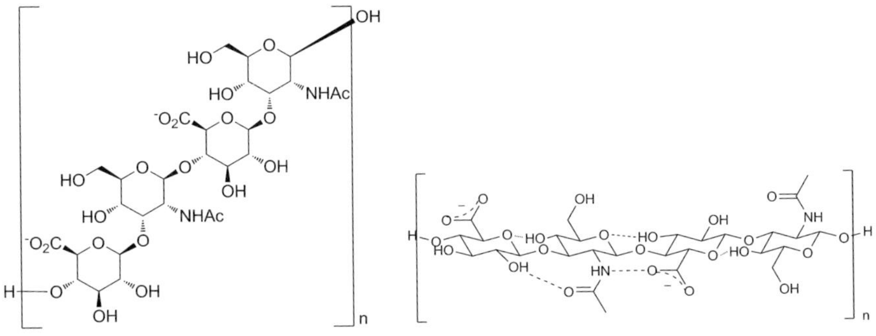

Scheme 7.10 Chemical structure of hyaluran; the dotted lines represent proposed hydrogen bonds.

national nomenclature of the IUPAC commission [138]. This name pays attention to the fact that hyaluronan has a polyanionic structure with a range of corresponding cations, for example, Na^+,K^+, and H^+.

Scheme 7.10 also shows the possible hydrogen bonds (dotted lines) of a HyA tetrasaccharide in dimethyl sulfoxide. If the polymer is solved in water, the solvent molecules are bonded between the carboxylic group of D-glucuronic acid and the amide nitrogen of *N*-acetyl-D-glucosamine. As a consequence, HyA can store large amounts of water and is highly viscous in aqueous solutions. By dissolving the HyA in alkaline solutions, a drastic decrease in the viscosity can be observed due to the deprotonation of the hydroxyl groups, which are involved in the hydrogen bonding in neutral solutions. The secondary structure of HyA was reviewed by Lapčik and Hargittai [139, 140].

Hyaluronan is synthesized by hyaluronan synthetases and can mainly be found in connective tissue, for example, in vitreous fluid of eyes, in synovial fluid, or in chicken combs [141]. The lowest concentrations were detected in blood serum. Nowadays, most of the commercially available HyA for synthetic applications is produced by bacteria, for example, by *Streptococcus equi* and *S. zooepidemicus* yielding about 4 g of HyA from 1 l, cultivation broth [139]. Unfortunately, this fermentative HyA production method bears the risk of mutations of the bacterial strains and the coproduction of toxins [142]. The analysis of HyA with electrophoretic and chromatographic techniques regarding its molecular size is reviewed by Kakehi *et al.* [143].

The biological function of HyA in the body of verbrates is dependent on the chain length of the polymer. High-molecular-weight polymers (4×10^2 to 2×10^4 kDa) are space-filling, antiangiogenic, immunosuppressive, and impede the cell differentiation, while smaller polymers play a role in ovulation, embryogenesis, protection of epithelial integrity, and wound healing as well as regeneration processes. Small oligosaccharides are inflammatory, immunostimulatory, angiogenic, and can be antiapoptotic [144].

HyA is degraded by hyaluronidases (Hyals), a very heterogeneous class of enzymes concerning their mode of action and their optimal working conditions [145]. Besides HyA, the Hyals also accept chondroitin and chondroitin sulfates as substrates. Karl Meyer separated the Hyals in three classes and his classification is still up to date [146, 147].

1) Vertebrate endo-β-*N*-acetyl-hexonaminidases, which hydrolyzes the substrate.
2) Bacterial Hyals (β-eliminases), which eliminates the glycosic linkage through introduction of a double bond.
3) A group of little explored endo-β-glucuronidases from leeches, whose mode of action resembles the verbrate Hyals.

In a vertebrate body, HyA has a very high turnover rate of about 5 g a day resulting in a half-life in the bloodstream of 2–5 min. HyA is not only degraded by Hyals but also in a catabolic pathway initiated by reactive oxygen species such as hydroxyl radicals [148]. The excretion of HyA is described in the review from Lebel [149].

The clinic applications of HyA were summarized and classified by Balazs in 2004 [142, 150].

1) **Viscosurgery:** HyA is used to protect damaged tissues and provide space during ophthalmological surgeries. Because HyA is one of the major components in the vitreous fluid of the eye, due to its viscoelastic properties, it has to be replaced, when it is lost during the extraction or replacement of damaged lenses. These surgeries are necessary after a cateract, for example, caused by diabetes mellitus. The first product on the marked for this application was Healon (noninflammatory fraction of Na-hyaluronate), produced by Biotrics, Inc., Arlington, MA and later by Pharmacia, Uppsala in Sweden.

2) **Viscoprotection:** HyA can protect healthy or wounded tissue from dryness. Healon is used to moisten the eyes pupils, although HyA is not present in tears. The advantage is that the polymer is not removed while blinking.

3) **Viscosupplementation:** HyA can be applied intra-articular, if the synovial fluid is damaged by osteoarthritis, which leads to stiff and painful joints. This disease affects about 10% of the world's population and is particular common for older people. By injecting high-molecular-weight HyA in the joint fluid, viscoelasticity increases and the production of endogenous HyA is stimulated. This is necessary because the content and molecular weight of naturally produced HyA is reduced in the patient. Additionally, inflammatory mediators are inhibited and cartilage degradation is decreased. For this application, several products containing HyA are on the marked [147].

4) **Viscoaugmentation:** HyA also fills spaces in skin, sphincter muscles, vocal, and pharyngeal tracts, and this is why, it is used in the therapy of injured tissue in the field of otolaryngology. Its shock absorption, wound healing, and osmotic functions play an especially important role in the vocal folds, due to the constant trauma caused by vibratory actions of phonation [150].

5) **Viscoseparation:** HyA is used to inhibit adhesion between two surgically traumatized tissue surfaces.

The problem with HyA containing products for the application described before is the short half-life time of the polymer, which leads to numerous repetitions of the medical treatment. Therefore, several attempts have been made to crosslink the HyA single strands to create hydrogels with very high molecular masses. This delays the degradation of the polymer.

The group from Šoltés *et al.* synthesized hydrogels for nonsurgical, viscosupplementatious applications [151]. Their patented system is the injection of a cocktail of two HyA derivates, namely, *n*-acetylurea-HyA and β-cyclodextrin-HyA, which spontaneously associate to hydrogels. To ensure that the hydrogel is not formed before the injection, a drug molecule is added which blocks the association process until it is degraded in the body. A new synthetic route to these HyA derivates was proposed by Charlot *et al.* [152]. Other hydrogels for viscoaugmentation applications were described by Shu *et al.*, who synthesized thiolated HyA derivatives and formed hydrogels through disulfide crosslinking or by coupling to α,β-unsaturated esters and amides of polyethylene crosslinkers [153]. The polymerization time could be adjusted from 10 min up to several days, depending on the chemical structure of the crosslinker.

Pulpitt *et al.* and Jia *et al.* used another crosslinking strategy based on hydrazide- and aldehyde-modified HyA [154, 155]. The hydrazide is generated through coupling of adipic dihydrazide to the carboxylic groups of HyA. The aldehyde-modified HyA can be obtained by oxidation of the diol moiety with sodium periodate. These two functional groups react spontaneously to hydrogels via hydrazone formation. Another hydrogel formation mechanism is described by Kurisawa *et al.* who synthesized an injectable HyA–tyramine conjugates which polymerized via enzyme-induced oxidative coupling [156].

HyA can be used as drug carrier to improve the solubility of pharmaceuticals or to decrease their toxicity. A review from Bettolo *et al.* describes different HyA–paclitaxel bioconjugates, which overcome the initial mentioned problems [157]. Paclitaxel is a well-known antitumor agent used for the treatment of breast and ovarian cancer. A controlled release in the tumor region can be assumed due to the overexpression of the HyA CD 44 receptor in various cancer cell lines. Approaches toward HyA as carrier substrate for carboranes in boron neutron capture therapy have been made by Di Meo *et al.* [158]. This therapy includes the delivery of ^{10}B to the tumor tissue and irradiation thereof with small doses of thermal/epithermal neutrons. This leads to ablation of the malignant cells.

There are numerous applications and chemical modification strategies to use HyA in the field of drug delivery: (i) Motokawa *et al.* developed a strategy of sustained release of erythropoietin through noncovalent encapsulation in HyA hydrogels for anemia treatment synthesized with the aldehyde/hydrazine method [159]. (ii) Kinetics of the drug release of esterified HyA–steroid conjugates for the treatment of inflammatory joint diseases was studied with NMR [160]. (iii) HyA and polyglutamate block polymers were synthesized using "click chemistry" [161].

Nanovesicles were prepared thereof and cytotoxic agents such as doxorubicin were internalized for controlled drug release. (iv) Sorbi *et al.* introduced the coupling of methotrexate to the primary alcohol of HyA for the therapy of autoimmune diseases such as rheumatoid arthritis [162].

7.11 Pullulan

Pullulan is a water soluble, neutral polysaccharide consisting of maltotriose monomeric units interconnected by α-(1→4) glycosidic bonds, whereas maltrotriose sequences are combined to yield a polymeric chain by α-(1→6) glycosidic linkages. The structural formula corresponds to [α-D-Glc$_p$-(1→4)-α-D-Glc$_p$-(1→4)-α-D-Glc$_p$(1→6)]$_n$. The assembled polysaccharide strain may display some minor structural abnormalities like maltotetraose; however, chemical properties are not affected. The regular alteration of the glycosidic bonds results in two distinctive properties: structural flexibility and enhanced solubility. The first description of pullulan has been published by Wallenfels *et al.* in the 1960s, based on the analyzed fermentation medium of *Aureobasidium pullulans* [163] (Scheme 7.11).

Because of the coexistence of α-(1→4)- and α-(1→6)-bonds in the chain, the overall structure is an intermediate between amylose and dextran. Native water-soluble pullulan is produced extracellularly by several strains such as *A. pullulans, Pullularia pullulans,* or *Dematium pullulans*. However, the bacterial cultures generate two different exopolysaccharides. Besides pullulan, a water-insoluble jelly-like β-(1→3)-glucan is also produced [164]. The detailed mechanism of the biosynthesis is not known yet, accordingly there is no explanation of the β-glucan formation. Some indications adumbrate the dependency on the individual genetic type of *A. pullulans* [165]. Local peculiarities of the biosynthesis with various *A. pullulans* strains have been investigated by Kondratyeva *et al* [166]. Pullulan can be synthesized from sucrose by cell-free enzymes of *A. pullulans* when both ATP and UDPG are added to a reaction mixture. In the case of glucose, mannose, galactose, fructose, or other carbon sources, the pathway is not clear [167]. The average molecular

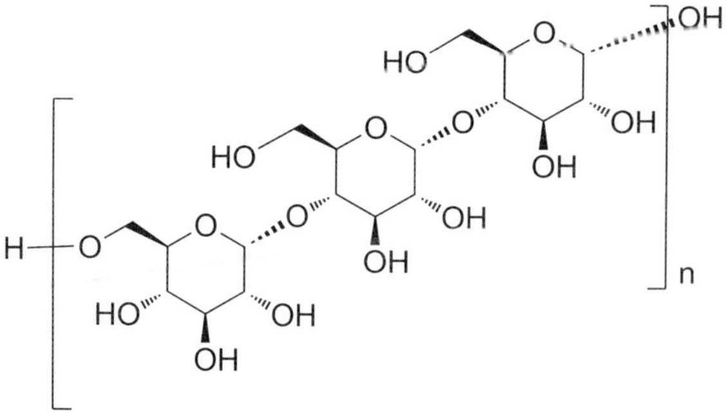

Scheme 7.11 Chemical structure of pullulan.

weight of pullulan varies in very broad ranges, from hundreds to thousands of kilodaltons, depending on the culture strain, pH, cultivation techniques, and substrates used [168]. As a subsumption of characteristics, pullulan is water soluble, nonhygroscopic in nature, moldable and spinnable, nontoxic, edible, and biodegradable. Due to these attributes, it is a good adhesive and binder and has capacity to form fibers, compression moldings, and strong oxygen-impermeable films. An actual survey about patents, new inventions related to production, cosmetical and pharmaceutical applications, as well as chemical derivatizations like esterification, etherification, hydrogenation, and carboxylation has been published by Singh *et al.* [169].

Pullulanase (pullulan-6-glucanohydrolase), which is a debranching enzyme, belongs to the family of glycosyl hydrolases. It is widely distributed among animals, plants, fungi, and bacteria and has the ability to hydrolyze α-glucosidic linkages. Complete hydrolysis of pullulan by both enzymes (1–6)-α-D- and (1–4)-α-D-pullulanase leads to isopanose as the main product. In some cases, partial hydrolysis yields isomaltose, maltose, or panose. The precise enzymatic mechanism of degradation and the resulting final product differs in each case [170] Thermal decomposition of the polymer chain occurs at temperatures about 250–280 °C.

Analogous to the broad family of polysaccharides based on glucose, chemical modifications can merely be done at one of the hydroxyl groups in the monomeric unit. As a result of the 1→4 linkages in the maltotriose units, the primary hydroxyl functions at the C-6 remain free. This leads to a higher reactivity of the C-6 alcohol compared to the C-2, C-3, and or C-4 position. Suitable modifications to carbonation, succinoylation, or carbamylation occur at the C-6 position. Nevertheless, the synthesis of carbomethylpullulan (CMP) by carboxymethylation proceeds predominantly at the C-2 position, similar to dextran. In order to establish a homogeneous sulfatation of pullulan, the reactivity of the different polysaccharide carbon atoms has been determined by Mähner *et al* [171]. They revealed the reactivity in the order C-6 > C-3 > C-2 > C-4. This result has been confirmed by Alban *et al.* who identified that the sulphatation occurs from C-6 to C-4, irrespective of the molecular weight, the production and the degree of substitution [172]. Furthermore, alkyl building blocks like chloroalkyl chlorides, chloroacetyl chlorides, or cholesteryl groups were attached to the polymeric chain. They were used to crosslink chains or form hydrogels [173]. Remarkable are syntheses with perfluoroalkyl carboxylic acids or amines in order to get alkylfluoronated compounds [174]. Consequently, pullulan and its derivatives have broad potential for food, pharmaceutical, and industrial applications.

In order to establish medical applications, the biological activity of pullulan was determined. The analyses have shown that pullulan has no mutagenic, carcinogenic, or toxicological activities, but exhibits a great affinity toward the liver and is effectively endocytosed by the parenchymal liver cells. This affinity toward liver cells qualifies pullulan as a drug delivery polymer for hepatitis therapy [175]. According to the great analogy to dextran, pullulan was also attempted for use as a blood-plasma substitute. In contrast to the assumptions, pullulan possess a short half-life in the blood, considered to the great affinity toward liver. CMP is pronounced as an auspicious drug carrier. The introduced carboxylic acid moiety induces a negative charge to the polysaccharide which results in a prolonged retention of the macromolecule within the organism [176]. CMP was found to be selectively absorbed by the spleen and lymph nodes. This fact allows the development of a new application spectrum of pullulan as immunodepressant conjugate [177]. Up to now, several CMP silalyl Lewis^X conjugates are applied to target inflammation sites, or CMP-doxorubicin derivatives were synthesized and evaluated for their antitumor activity. Analogous to the well-known anticoagulant heparin or dextran-sulfate, pullulan-sulfate has been engineered, where the gained activity is almost comparable with that of heparin. Through combination of highly hydrophilic polysaccharide chains with native or synthetic hydrophobic moieties, new promising novel biologically active compounds can be created. In the form of nanoparticles, they can find application as drug-delivery system and for the thermal and colloidal stabilization of proteins. These compact nanoparticles are formed by intermolecular aggregation of hydrophobized polysaccharides, whereas the strength of the self-association can be regulated by the length of the hydrophobic moiety [174]. The already mentioned notable fluoro derivatives are utilized as fluorosurfactant or oxygen carrier. Fluorosurfactants are necessary for targeting delivery of drugs or as contrasting agents. The latter unique ability to dissolve oxygen 20 times higher than that of blood plasma enables the development of injectable forms of perfluorocarbon-based oxygen carriers, which have no toxic or reactivity-related side effects and can be removed easily form the organism via excretion [178].

7.12
Scleroglucan

Scleroglucan is a branched homopolysaccharide, which only consists of D-glucose moieties. The polymer has a main chain of β-(1→3)-linked D-glucopyranose units. Every third glucose is substituted at position 6 with a single β-D-glucopyranose (Scheme 7.12).

The polymer is produced extracellularly by the heterotrophic filamentous fungi of the genus *Sclerotium*, for example, *Sclerotium glucanicum*, *Sclerotium rolfsii*, and *Sclerotium delphinii*. For industrial fermentation processes, mainly *S. glucanicum* and *S. rolfsii* were used. In the plant pathogenic fungi *Sclerotium* sp., scleroglucan enables the attachment of the parasite to the plant surfaces and protects the fungi against desiccation. Under optimal fermentation conditions, a maximum yield of 8.5–10 g/L scleroglucan can be obtained using *S. glucanicum*. In contrast, high sucrose fermentation of *S. rolfsii* results in an isolated yield of up to 21 g/L. Depending on the strain and the fermentation conditions, the molecular weight of scleroglucan ranges between 1.3×10^5 and 6.0×10^6 g mol^{-1}. Schizophyllan, a polysaccharide isolated from *Schizophyllum* sp., has the same molecular composition as scleroglucan but is reported with a molecular weight of 6.12×10^6 g mol^{-1} [179].

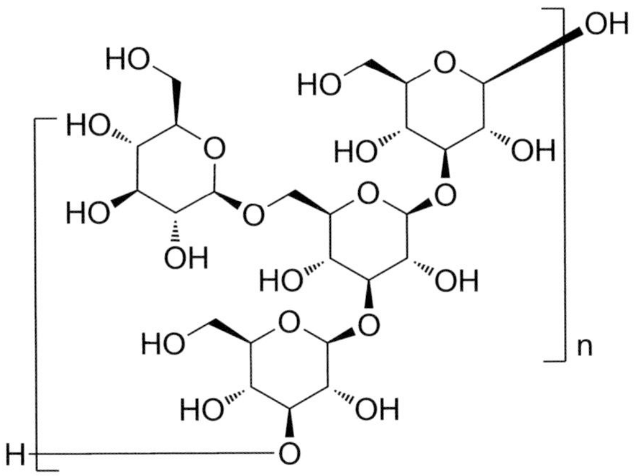

Scheme 7.12 Chemical structure of scleroglucan.

Scleroglucan readily dissolves in water and forms rigid, right-handed, triple-helical structure. The interstrand aggregation of the helices is disturbed by the side glucose groups, which protrude from the helix. The scleroglucan helix structure is thermostable, but instable at pH higher than 12.5 or toward organic solvents such as dimethyl sulfoxide. The helical structure of scleroglucan in solution is responsible for rheological properties. It forms viscous pseudoplastic solutions, which tolerate several salts, high temperature (10–90 °C), and broad pH range (pH 1 to 11) without major changes in the viscosity. Retaining more than 90% of its viscosity after 500 days at 90 °C in seawater, scleroglucan was the most stable compound among 140 other polymers tested for use in polymer flooding in oil reservoirs [179, 180].

Scleroglucan is compatible with many electrolytes such as sodium chloride (5%), sodium sulfate (5%), calcium chloride (20%), and disodium hydrogen phosphate (10%) without major changes in the rheological properties [179]. In contrast, a hydrogel gel can be obtained by addition of borax (0.1 mol borax for 1 mol repeating unit of scleroglucan) to the aqueous polymer solution. The final polymer concentration in the stable hydrogel is 0.7% (w/v) [181, 182]. Oxidation of the glucose side chain in scleroglucan using sodium periodate results in the formation of dialdehyde groups (scleraldehyde) and the production of formic acid. During the initial periodate oxidation, an increase in the storage modulus associated with the formation of gels can be observed. The mechanism of gellation can be explained by the formation of intermolecular hemiacetal linkages. These gels are to be physically stable, whereas aldehyde reduction (using sodium borohydride) or oxidation to carboxyl (using sodium chlorite) dissolved the gels [183]. The corresponding polymeric carboxylic acid after sodium chlorite oxidation is also known as sclerox and was studied in detail [184]. Scleraldehyde with a low degree of oxidation (10% and 20%), prepared by a controlled oxidation of scleroglucan, is reported to retain essentially a triple-stranded helical conformation, while the triple-stranded chains separate in single chains with increasing the degree of oxidation (40% and 100%).

The hydrogel prepared from scleraldehyde with a low degree of oxidation by crosslinking with diamines can be represented by a network composed of randomly oriented triple helices interlinked at the sites where the aldehyde groups are present [185].

The first industrial application of scleroglucan was in oil recovery, where it showed better pH and temperature stability than xanthan. In watered-out reservoirs, where seawater pressure is no longer sufficient to recover the oil, the addition of scleroglucan to improve the viscosity of the feed water can improve the process significantly. Additionally, scleroglucan lubricates the drill and controls the backpressures created during drilling [179]. In the field of pharmaceutics, numerous studies are published on applications of scleroglucan both in its native form and as derivatives. Hydrogels obtained by different crosslinking agents are suitable for a release modulation from various dosage forms. Sustained release and environment-controlled delivery systems that can be obtained from crosslinked scleroglucan represent a challenging field of applications [186]. There are numerous additional applications using the superb rheological properties and the high stability of scleroglucan ranging from food and cosmetics to paints and ceramic glazes. In most of these cases, scleroglucan is a competitor of xanthan [187].

7.13 Xanthan

Xanthan is a polysaccharide with a cellulose backbone of β-(1→4) linked D-glucose. Every second glucose is substituted at position 3 with a side chain consisting of β-D-mannose-(1→4)-β-D-glucuronic acid-(1→2)-α-D-mannose. The terminal mannose moiety is partially substituted with a pyruvate, coupled as an acetal at positions 4 and 6. The internal mannose usually bears an acetate group at position 6. The degree of pyruvate substitution varies between 30% and 40%, whereas 60–70% of the internal mannose units are acetylated. The molecular weight of xanthan is about 1,000,000 $g\,mol^{-1}$ (Scheme 7.13).

Because of the glucuronic acid moieties and the partial pyruvate substituents in the side chain, xanthan is an ionic polysaccharide and its polyelectrolytic character in water was studied in detail [188].

The polymer is produced by the fermentation with the bacteria *Xanthomonas campestris* [189], *Xanthomonas phaseoli* [190], and *Xanthomonas juglandis* [191] and other *Xanthomonas* species with an annual production of approximately 30,000 tons. The pyruvyl and acetyl content of xanthan isolates depends on the fermentation conditions and bacterial strain [192]. As shown for *X. campestris*, the production, composition, and viscosity of the xanthan synthesized by this strain are influenced by the fermentation time and nutrient exhaustion in batch culture and by the dilution rate in continuous culture. The specific rate of xanthan synthesis is maximal during exponential growth, although some xanthan is also formed during the stationary phase [189]. Under optimized conditions, a production of up to 22 g/L xanthan can be reached in a stirred tank fermentor [193]. As the product

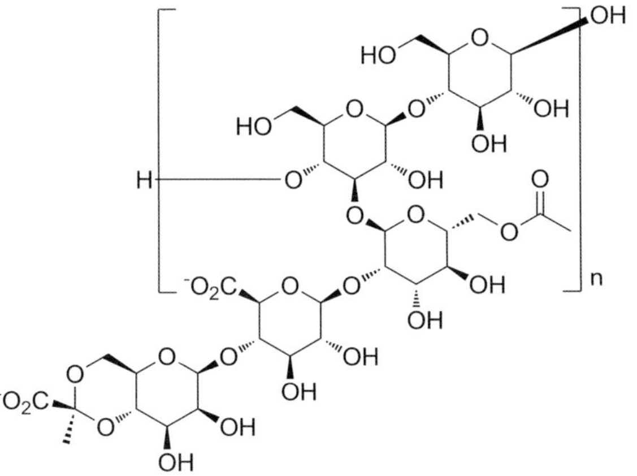

Scheme 7.13 Chemical structure of xanthan.

accumulates, the broth viscosity increases dramatically and the broth rheology presents serious problems to mixing, heat transfer, and oxygen input. This limiting factor of the process efficiency can be overcome by water-in-oil cultivation technologies, where the xanthan produced stays within the aqueous droplets dispersed in an immiscible organic phase (e.g., vegetable oil or *n*-hexadecane) and the effective viscosity of the overall system is kept low. Xanthan concentrations of more than 200 g/L culture medium can be reached by this approach [194].

Xanthan is known as a highly stable polysaccharide, which is not easily degraded by microorganisms. Still some microorganisms such as *Bacillus* sp. 13–4, *Paenibacillus alginolyticus* XL-1, *Cellulomonas* sp. LX, *Microbacterium* sp. XT11, and *Paenibacillus* sp. XD were reported to participate in depolymerization of xanthan [195]. The enzymatic basis of xanthan degradation was studied by Nankai *et al.* who analyzed the structures of xanthan depolymerization product by electrospray mass spectrometry and identified the enzymes involved in the process. In *Bacillus* sp., GL1 Xanthan is depolymerized to constituent monosaccharides by two extracellular and three intracellular enzymes. In the initial step, the terminal pyruvate-substituted mannose is cleaved by a xanthan lyase and the polymeric backbone is subsequently depolymerized by a β-D-glucanase (endoxanthanase) [196]. Recently, a novel endoxanthanase catalyzing the hydrolysis of the main chain of xanthan with an intact side chain was isolated from *Microbacterium* sp. strain XT11 and characterized. This enzyme may be used in the treatment of *Xanthomonas* infectious disease or for the biodegradation of xanthan injected into underground oil [197].

Xanthan is easily soluble in cold and hot water. The viscosity of the xanthan solution is nearly independent of temperatures up to 90 °C and of the pH value. Also higher salt concentrations have no big influence on the viscosity. The polymer has relatively good stability during sterilization and the solutions are very stable toward chemical, enzymatic, or bacterial degradation allowing a wide range of applications. Also xanthan can be crosslinked, gelled, and associated with other biopolymers such as alginates, carrageenans, galactomannans, gelatin, glucomannans, or pectins.

The practical use of xanthan gum is mainly due to its ability to induce high viscosity at low polymer concentration in an aqueous environment. It also has a unique shear thinning behavior and is less prone to degradation compared with other polysaccharides. Major markets of this biopolymer exist in food, ceramic glazing, and petroleum drilling industries. In 1969, FDA allowed xanthan for the general use in foods. Typical food applications of xanthan gum are salad dressings, sauces, gravies, dairy products, desserts, low-calorie foods, and convenience foods in general. Xanthan gum is also used in cleaners, coatings, polishes, and agricultural flowables. Furthermore, it is used as an agent in many toothpastes and cosmetic preparations [198].

The main pharmaceutical application of xanthan is the usage as tablet excipient to modulate the rate of drug delivery and drug release [4]. In a recent study, crosslinked starch–xanthan hydrogels are synthesized and the new hydrogels has a good film forming ability. The equilibrium swelling ratio, swelling rate, gel mesh size, and drug permeability of the starch–xanthan hydrogels increase with increasing crosslinker and xanthan content. The mesh sizes of the hydrogels are 2.84–6.74 nm at pH 7.4 depending on the gel composition. This mesh size range is large enough to transport most drugs from small molecules to polypeptide and proteins. The hydrogel exhibits selective permeability depending on drug charges, allowing the design of controlled release formulations of ionizable drugs [199]. No crosslinker is necessary when polyionic hydrogels are formed through interaction of the polyanion xanthan with a polycation such as chitosan. This hydrogel is porous, has a fibrous structure, good hygroscopic qualities, and is capable of immobilizing bioactive substances such as drugs or enzymes. The channels present in the fibrillar gels have a pore size between 0.1 and 1 μm, whereas the fibrils have a diameter of 0.1 μm. The polyionic hydrogel has the advantage of creating an ionic microsystem which favors the stabilization of a protein polymer by interacting with the free acid and base functions. Thus, upon coimmobilization of protease and xylanase in the xanthan-based hydrogel, the protease acitivity is increased up to 85% [200].

7.14 Summary

Polysaccharides are ubiquitous naturally occurring renewable polymers, which can be isolated from various sources such as plants, animals, and microorganisms. In nature, polysaccharides were used for energy storage, as scaffold, as signaling, and shielding element. Polysaccharides exhibit an enormous structural diversity due to different monosaccharide composition, different linkage types and patterns, and a wide spectrum of molecular weights. As a direct consequence, their physical, chemical, and biological properties are well dispersed. The next level of complexity is introduced by chemical modifications such as chain decorations and crosslinkages. Therefore, polysaccharides have a wide range of potential applications ranging from tissue engineering and regenerative medicine over food additives to explosives, deflocculants in paper industry, and oil drilling muds.

Degradability or even biodegradability is hard to define in this context and in some cases even partial depolymerization is considered as degradation. Several polysaccharides are readily degradable in vertebrates such as HyA and chitin. Several tissue engineering applications were reported for this class of polymers, and fine-tuning the degradation kinetics is an important field of research. Other polymers such as gellan, xanthan alginate, and cellulose are mainly degraded by bacteria using the polymer as energy source. As one example, all organisms known to degrade cellulose efficiently produce a battery of highly diverse enzymes with different specificities, which act together in synergism. These polysaccharides were mainly used for technical or food applications due to their rheological and material properties or as degradable excipient in tablets pharmaceutical formulations. Belonging to the bacterial degradable polymers, alginate with its unique material properties and characteristics has also been increasingly considered as biomaterial for medical applications. CA gels have unique intrinsic properties and exhibit biocompatibility, mucoadhesion, porosity, and ease of manipulation. Hence, much attention has recently been focused on alginate-based materials for protein delivery, cell encapsulation, and tissue regeneration.

Acknowledgments

Financial support was provided by funding from the Deutsche Forschungsgemeinschaft (DFG, German Research Foundation) for the Cluster of Excellence REBIRTH (From Regenerative Biology to Reconstructive Therapy).

In Memoriam

Professor Severian Dumitriu was a highly esteemed scientist and teacher and a wonderful human being. His legacy will live on within the polymeric research community through his significant contributions (over 180 scientific papers and book chapters). He devoted his life to polymeric biomaterials field progress, promoting scientific excellence and training future generations of scientists. When he started on this chapter no one could ever have imagined that he could not finalize it. Professor Dumitriu's family is grateful to the co-authors and book editors for completion and final review of this chapter.

References

1 Rinaudo, M. (2008) *Polym. Int.*, **57**, 397–430.

2 Mano, J.F., Silva, G.A., Azevedo, H.S., Malafaya, P.B., Sousa, R.A., Silva, S.S.,

Boesel, L.F., Oliveira, J.M., Santos, T.C., Marques, A.P., Neves, N.M., and Reis, R.L. (2007) *J. R. Soc. Interface*, **4**, 999–1030.
3 Liu, Z., Jiao, Y., Wang, Y., Zhou, C., and Zhang, Z. (2008) *Adv. Drug Deliv. Rev.*, **60**, 1650–1662.
4 Coviello, T., Matricardi, P., Marianecci, C., and Alhaique, F. (2007) *J. Control. Release*, **119**, 5–24.
5 Remminghorst, U. and Rehm, B.H.A. (2006) *Biotechnol. Lett.*, **28**, 1701–1712.
6 De Vos, P., De Haan, B., and Van Schilfgaarde, R. (1997) *Biomaterials*, **18**, 273–278.
7 Haug, A., Larsen, B., and Smidsrød, O. (1967) *Acta Chem. Scand.*, **21**, 691–704.
8 Sutherland, I.W. (1995) *FEMS Microbiol. Rev.*, **16**, 323–347.
9 Gimmestad, M., Ertesvåg, H., Heggeset, T.M.B., Aarstad, O., Svanem, B.I.G., and Valla, S. (2009) *J. Bacteriol.*, **191**, 4845–4853.
10 Sikorski, P., Mo, F., Skjåk-Bræk, G., and Stokke, B.T. (2007) *Biomacromolecules*, **8**, 2098–2103.
11 Thom, D., Grant, G.T., Morris, E.R., and Rees, D.A. (1982) *Carbohydr. Res.*, **100**, 29–42.
12 Thu, B., Smidsrød, O., and Skjåk-Bræk, G. (1996) *Immobilized Cells; Basics and Applications*, Elsevier Science, Amsterdam.
13 Seale, R., Morris, E.R., and Rees, D.A. (1982) *Carbohydr. Res.*, **110**, 101–122.
14 Gomez, C.G., Rinaudo, M., and Villar, M.A. (2007) *Carbohydr. Polym.*, **67**, 296–304.
15 Soon-Shiong, P., Desai, N.P., Sandford, P.A., Heintz, R.A., and So-Jomihardjo, S. (1993) Crosslinkable polysaccharides, polycations and lipids useful for encapsulation and drug release, Patent WO 93/09176.
16 Jeon, O., Bouhadir, K.H., Mansour, J.M., and Alsberg, E. (2009) *Biomaterials*, **30**, 2724–2734.
17 Baroli, D. (2006) *J. Chem. Technol. Biotechnol.*, **81**, 491–499.
18 Augst, A.D., Kong, H.J., and Mooney, D.J. (2006) *Macromol. Biosci.*, **6**, 623–633.
19 Ossipov, D.A., Piskounova, S., and Hilborn, J. (2008) *Macromolecules*, **41**, 3971–3982.
20 Bouhadir, K.H., Alsberg, E., and Mooney, D.J. (2001) *Biomaterials*, **22**, 2625–2633.
21 Bouhadir, K.H., Kruger, G.M., Lee, K.Y., and Mooney, D.J. (2000) *J. Pharm. Sci.*, **89**, 910–919.
22 Rowley, J.A., Madlambayan, G., and Mooney, D.J. (1999) *Biomaterials*, **20**, 45–53.
23 Fonder, M.A., Lazarus, G.S., Cowan, D.A., Aronson-Cook, B., Kohli, A.R., and Mamelak, A.J. (2008) *J. Am. Acad. Dermatol.*, **58**, 185–206.
24 Terrill, P.J., Goh, R.C.W., and Bailey, M.J. (2007) *J. Wound Care*, **16**, 433–438.
25 Chiu, C.-T., Lee, J.-S., Chu, C.-S., Chang, Y.-P., and Wang, Y.-J. (2008) *J. Mater. Sci. Mater. Med.*, **19**, 2503–2513.
26 Poucke, S.V., Jorens, P.G., Peeters, R., Jacobs, W., de Beeck, B.O., Lambert, J., and Beaucourt, L. (2004) *Int. Wound J.*, **1**, 207–213.
27 Jude, E.B., Apelqvist, J., Spraul, M., Martini, J., and the Silver Dressing Study Group (2007) *Diabet. Med.*, **24**, 280–288.
28 Ahmad, Z., Pandey, R., Sharma, S., and Khuller, G.K. (2006) *Indian J. Chest. Dis. Allied Sci.*, **48**, 171–176.
29 Mazumder, M.A.J., Shen, F., Burke, N.A.D., Potter, M.A., and Stöver, H.D.H. (2008) *Biomacromolecules*, **9**, 2292–2300.
30 Paul, A., Ge, Y., Prakash, S., and Shum-Tim, D. (2009) *Regenerative Med.*, **4**, 733–745.
31 Dang, T.T., Xu Q., Bratlie K.M., O'Sullivan, E.S., Chen X.Y., Langer, R., and Anderson, D.G. (2009) *Biomaterials*, **30**, 6896–6902.
32 Becker T.A., Preul M.C., Bichard W.D., Kipke D.R., and McDouqall C.G. (2007) *Neurosurgery*, **60**, 1119–1127
33 Kedem, A., Perets, A., Gamlieli-Bonshtein, I., Dvir-Ginzberg, M., Mizrahi, S., and Cohen, S. (2005) *Tissue Eng.*, **11**, 715–722.
34 Bretagne, J.F., Richard-Molard, B., Honnorat, C., Caekaert, A., and Barthélemy, P. (2006) *Presse Med.*, **35**, 23–31.
35 Cresi, F., Savino, F., Marinaccio, C., and Silvestro, L. (2006) *Arch. Dis. Child.*, **91**, 93.

36 Dettmar, P.W., Sykes, J., Little, S.L., and Bryan, J. (2006) *Int. J. Clin. Pract.*, **60**, 275–283.

37 Tang, M., Dettmar, P., and Batchelor, H. (2005) *Int. J. Pharm.*, **292**, 169–177.

38 Hashimoto, T., Suzuki, Y., Suzuki, K., Nakashima, T., Tanihara, M., and Ide, C. (2005) *J. Mater. Sci. Mater. Med.*, **16**, 503–509.

39 De Ruiter, G.A., and Rudolph, B. (1997) *Trends Food Sci. Technol.*, **8**, 389–395.

40 Collén, P.N., Lemoine, M., Daniellou, R., Guégan, J.-P., Paoletti, S., and Helbert, W. (2009) *Biomacromolecules*, **10**, 1757–1767.

41 Michel, G., Chantalat, L., Fanchon, E., Henrissat, B., Kloareg, B., and Dideberg, O. (2001) *J. Biol. Chem.*, **276**, 40202–40209.

42 Ohta, Y. and Hatada, Y. (2006) *J. Biochem.*, **140**, 475–481.

43 Guibet, M., Kervarec, N., Génicot, S., Chevolot, Y., and Helbert, W. (2006) *Carbohydr. Res.*, **341**, 1859–1869.

44 Černíková, M., Buňka, F., Pavlínek, V., Březina, P., Hrabě, J., and Valášek, P. (2008) *Food Hydrocolloids*, **22**, 1054–1061.

45 Tapia, C., Escobar, Z., Costa, E., Sapag-Hagar, J., Valenzuela, F., Basualto, C., Gai, M.N., and Yazdani-Pedram, M. (2004) *Eur. J. Pharm. Biopharm.*, **57**, 65–75.

46 Piyakulawat, P., Praphairaksit, N., Chantarasiri, N., and Muangsin, N. (2007) *AAPS PharmSciTech*, **8**, 120–130.

47 Keppeler, S., Ellis, A., and Jacquier, J.C. (2009) *Carbohydr. Polym.*, **78**, 973–977.

48 (a) Payen, A. and Hebd, C.R. (1838) *Seances Acad. Sci.*, **7**, 1052; (b) Payen, A. and Hebd, C.R. (1838) *Seances Acad. Sci.*, **7**, 1125.

49 (a) Finaev, D. (2007) *Biol. Plant.*, **51**, 407–413; (b) Bessueille, L. and Bulone, V. (2008) *Plant Biotechnol.*, **25**, 315–322.

50 Kobayashi, S., Sakamoto, J., and Kimura, S. (2001) *Prog. Polym. Sci.*, **26**, 1525–1560.

51 Klemm, D., Heublein, B., Fink, H.P., and Bohn, A. (2005) *Angew. Chem. Int. Ed.*, **44**, 3358–3393.

52 Bochek, A.M. (2003) *Russ. J. Appl. Chem.*, **76**, 1711–1719.

53 (a) Potthast, A., Rosenau, T., Buchner, R., Röder, T., Ebner, G., Bruglachner, H., Sixta, H., and Kosma, P. (2002) *Cellulose*, **9**, 41–53; (b) Ciacco, G.T., Liebert, T.F., Frollini, E., and Heinze, T.J. (2003) *Cellulose*, **10**, 125–132; (c) Saalwächter, K., Buchard, W., Klüfers, P., Kettenbach, G., Mayer, P., Klemm, D., and Dugarmaa, S. (2000) *Macromolecules*, **33**, 4094–4107.

54 Bajerová, M., Krejčová, K., Rabišková, M., Gajdziok, J., and Masteiková, R. (2009) *Adv. Polym. Tech.*, **28**, 199–208.

55 El Seoud, O.A. and Heinze, T. (2005) *Adv. Polym. Sci.*, **186**, 103–149.

56 Edgar, K.J., Buchanan, C.M., Debenham, J.S., Rundquist, P.A., Seiler, B.D., Shelton, M.C., and Tindall, D. (2001) *Prog. Polym. Sci.*, **26**, 1605–1688.

57 (a) Becher, J., Liebegott, H., Berlin, P., and Klemm, D. (2004) *Cellulose*, **11**, 119–126; (b) Tiller, J., Klemm, D., and Berlin, P. (2001) *Des. Monomers Polym.*, **4**, 315–328.

58 Kamel, S., Ali, N., Jahangir, K., Shah, S.M., and El-Gendy, A.A. (2008) *Express Polym. Lett.*, **2**, 758–778.

59 Sannino, A., Madaghiele, M., Conversano, F., Mele, G., Maffezzoli, A., Netti, P.A., Ambrosio, L., and Nicolais, L. (2004) *Biomacromolecules*, **5**, 92–96.

60 Rokhade, A.P., Agnihotri, S.A., Patil, S.A., Mallikarjuna, N.N., Kulkarni, P.V., and Aminabhavi, T.M. (2006) *Carbohydr. Polym.*, **65**, 243–252.

61 Brown, A.J. (1886) *J. Chem. Soc. Trans.*, **49**, 432–439.

62 Chawla, P.R., Bajaj, I.B., Survase, S.A., and Singhal, R.S. (2009) *Food Technol. Biotechnol.*, **47**, 107–124.

63 Shoda, M. and Sugano, M. (2005) *Biotechnol. Bioprocess Eng.*, **10**, 1–8.

64 Ross, P., Mayer, R., and Benziman, R. (1991) *Microbiol. Rev.*, **55**, 35–58.

65 Iguchi, M., Yamanaka, S., and Budhiono, A. (2000) *J. Mater. Sci.*, **35**, 261–270.

66 Czaja, W.K., Young, D.J., Kawecki, M., and Brown, R.M. Jr. (2007) *Biomacromolecules*, **8**, 1–12.

67 Helenius, G., Bäckdahl, H., Bodin, A., Nannmark, U., Gatenholm, P., and Risberg, B. (2006) *J. Biomed. Mater. Res.*, **76A**, 431–438.

68 Czaja, W., Krystynowicz, A., Bielecki, S., and Brown, R.M. Jr. (2006) *Biomaterials*, **27**, 145–151.
69 Legeza, V.I., Galenko-Yaroshevskii, V.P., Zinov'ev, E.V., Paramonov, B.A., Kreichman, G.S., Turkovskii, I.I., Gumenyuk, E.S., Karnovich, A.G., and Khripunov, A.K. (2004) *Bull. Exp. Biol. Med.*, **138**, 311–315.
70 Yasuda, K., Gong, J.P., Katsuyama, Y., Nakayama, A., Tanabe, Y., Kondo, E., Ueno, M., and Osada, Y. (2005) *Biomaterials*, **26**, 4468–4475.
71 Charpentier, P.A., Maguire, A., and Wan, W. (2006) *Appl. Surf. Sci.*, **252**, 6360–6367.
72 Loures, B.R. (2004) Endoprosthesis process to obtain and methods used, Patent WO 2004/0455448 AI.
73 Novaes, A.B. Jr., Novaes, A.B., Grisi, M.F.M., Soares, U.N., and Gaberra, F. (1993) *Braz. Dent. J.*, **4**, 65–71.
74 de Macedo, N.L., da Silva Matuda, F., de Macedo, L.G.S., Monteiro, A.S.F., Valera, M.C., and Carvalho, Y.R. (2004) *Braz. Dent. Oral Sci.*, **3**, 395–400.
75 (a) Watanabe, K., Eto, Y., Takano, S., Nakamori, S., Shibai, H., and Yamanaka, S. (1993) *Cytotechnology*, **13** 107–114; (b) Märtson, M., Viljanto, J., Laippala, P., and Saukko, P. (1998) *Eur. Surg. Res.*, **30**, 419–425.
76 (a) Kumar, M.N.V. (2008) *React. Funct. Polym.*, **46**, 1–27; (b) Pillai, C.K.S., Paul, W., and Sharma, C.P. (2009) *Prog. Polym. Sci.*, **34**, 641–678.
77 Chang, K.L.B., Tsai, G., Lee, J., and Fu, W.-R. (1997) *Carbohydr. Res.*, **303**, 327–332.
78 Sannan, T., Kurita, K., and Iwakura, Y. (1976) *Makromol. Chem.*, **177**, 3589–3600.
79 Tsigos, I., Martinou, A., Kafetzopoulos, D., and Bouriotis, V. (2000) *Trends Biotechnol.*, **18**, 305–312.
80 Berger, L.R. and Weiser, R.S. (1957) *Biochim. Biophvs. Acta*, **26**, 517–521.
81 Amano, K.L. and Lto, E. (1978) *Eur. J. Biochem.*, **85**, 97–104.
82 Shigemasa, Y., Saito, K., Sashiwa, H., and Saimoto, H. (1994) *Int. J. Biol. Macromol.*, **16**, 43–49.
83 Hu, X., Du, Y., Tang, Y., Wang, Q., Feng, T., Yang, J., and Kennedy, J.F. (2007) *Carbohydr. Polym.*, **70**, 451–458.
84 Austin, P.R., Brine, C.J., Castle, J.E., and Zikakis, J.P. (1981) *Science*, **212**, 749–753.
85 Austin, P.R. (1988) *Methods Enzymol.*, **161**, 403–407.
86 Tamura, H. (2006) *Polym. Prep. Jpn.*, **55**, 1862.
87 Tamura, H., Nagahama, H., and Tokura, S. (2006) *Cellulose*, **13**, 357–364.
88 Somorin, O., Nishi, N., Tokura, S., and Noguchi, J. (1979) *Polym. J.*, **11**, 391–396.
89 Tokura, S., Nishi, N., Somorin, O., and Noguchi, J. (1980) *Polym. J.*, **12**, 695–700.
90 Kaifu, K., Nishi, N., Komai, T., Tokura, S., and Somorin, O. (1981) *Polym. J.*, **11**, 241–245.
91 Szosland, L. and East, C.C. (1995) *J. Appl. Polym. Sci.*, **58**, 2459–2466.
92 Urbanczyk, G., Lipp-Symonowicz, B., Szosland, I., Jeziorny, A., Urbaniak-Domagala, W., Dorau, K., Wrzosek, H., Sztajnowski, S., Kowalska, S., and Sztajnert, E. (1997) *J. Appl. Polym. Sci.*, **65**, 807–819.
93 Muzzarelli, R.A.A. (2009) *Carbohydr. Polymer*, **76**, 167–182.
94 Naessens, M., Cerdobbel, A., Soetaert, W., and Vandamme, E.J. (2005) *J. Chem. Technol. Biotechnol.*, **80**, 845–860.
95 Robyt, J.F. (1985) *Encyclopaedia of Polymer Science*, vol. 4 (ed. J.I. Kroschwitz), John Wiley & Sons, Inc., New York, pp. 753–767.
96 Remaud-Simeon, M., Willemot, R.-M., Sarçabal, P., de Montalk, G.P., and Monsan, P. (2000) *J. Mol. Catal., B Enzym.*, **10**, 117–128.
97 Hehre, E.J. (1956) *J. Biol. Chem.*, **222**, 739–750.
98 Kakuchi, T., Kusuno, A., Miura, M., and Kaga, H. (2000) *Macromol. Rapid Commun.*, **21**, 1003–1006.
99 Naessens, M., Cerdobbel, A., Soetraet, W., and Vandamme, E.J. (2005) *J. Ind. Microbiol. Biotechnol.*, **32**, 323–334.
100 Khalikova, E., Susi, P., and Korpela, T. (2005) *Microbiol. Mol. Biol. Rev.*, **69**, 306–325.
101 Larsson, A.M., Andersson, R., Ståhlberg, J., Kenne, L., and Jones, T.A. (2003) *Structure*, **11**, 1111–1121.

102 Heinze, T., Liebert, T., Heublein, B., and Hornig, S. (2006) *Adv. Polym. Sci.*, **205**, 199–291.
103 Van Tomme, S.R., and Hennink, W.E. (2007) *Expert Rev. Med. Devices*, **4**, 147–164.
104 Kim, S.-H. and Chu, C.-C. (2000) *J. Biomater. Appl.*, **15**, 23–46.
105 Hudson, S.P., Langer, R., Fink, G.R., and Kohane, D.S. (2010) *Biomaterials*, **31**, 1444–1452.
106 De Belder, A.N. (1996) *Polysaccharides in Medicinal Applications* (ed. S. Dumitriu), Marcel Dekker, New York, pp. 505–523.
107 Leathers, T.D. (2002) *Biopolymers Vol. 5, Polysaccharides I: Polysaccharides from Prokaryotes* (eds E.J. Vandamme, S. De Baets, and A. Steinbüchel), Wiley-VCH Verlag GmbH, Weinheim, pp. 299–321.
108 Porath, J. and Flodin, P. (1987) *Chromatographia*, **23**, 365–369.
109 Debelder, A.N. (1993) *Industrial Gums: Polysaccharides and Their Derivatives*, 3rd edn (eds R.L. Whistler and J.N. BeMiller), Academic Press, New York, pp. 513–542.
110 Lobene, R.R. (1979) *J. Dent. Res.*, **58**, 2381–2388.
111 Tako, M., Teruya, T., Tamaki, Y., and Konishi, T. (2009) *Colloid. Polym. Sci.*, **287**, 1445–1454.
112 Nampoothiri, K.M., Singhania, R.R., Sabarinath, C., and Pandey, A. (2003) *Process Biochem.*, **38**, 1513–1519.
113 Cascone, M.G., Barbani, N., Cristallini, C., Giusti, P., Ciardelli, G., and Lazzeri, L. (2001) *J. Biomater. Sci. Polymer Edn.*, **12**, 267–281.
114 Chandrasekaran, R., Pulgjaner, L.C., Joyce, K.L., and Arnott, S. (1988) *Carbohydr. Res.*, **181**, 23–40.
115 Tako, M. and Tamaki, H. (2005) *Polym. J.*, **37**, 498–505.
116 Kennedy, L. and Sutherland, I.W. (1994) *Microbiology*, **140**, 3007–3013.
117 Jung, Y.-J., Park, C.S., Lee, H.G., and Cha, J. (2006) *J. Microbiol. Biotechnol.*, **16**, 1868–1873.
118 Hashimoto, W., Sato, N., Kimura, S., and Murata, K. (1998) *Arch. Biochem. Biophys.*, **354**, 31–39.
119 Derekova, A., Sjøholm, C., Mandeva, R., Michailova, L., and Kambourova, M. (2006) *Extremophiles*, **10**, 321–326.
120 Hamcerencu, M., Desbrieres, J., Khoukh, A., Popa, M., and Riess, G. (2008) *Carbohydr. Polym.*, **71**, 92–100.
121 Dentini, M., Desideri, P., Crescenzi, V., Yuguchi, Y., Urakawa, H., and Kajiwara, K. (2001) *Macromolecules*, **34**, 1449–1453.
122 Bajaj, I.B., Survase, S.A., Saudagar, P.S., and Singhal, R.S. (2007) *Food Technol. Biotechnol.*, **45**, 341–354.
123 Smith, A.M., Shelton, R.M., Perrie, Y., and Harris, J.J. (2007) *J. Biomater. Appl.*, **22**, 241–254.
124 McCleary, B.V., Clark, A.H., Dea, I.C.M., and Rees, D.A. (1985) *Carbohydr. Res.*, **139**, 237–260.
125 Wientjes, R.H.W., Duits, M.H.G., Jongschaap, R.J.J., and Mellema, J. (2000) *Macromolecules*, **33**, 9594–9605.
126 Cheng, Y. and Prud'homme, R.K. (2000) *Biomacromolecules*, **1**, 782–788.
127 McCleary, B.V. and Matheson, N.K. (1983) *Carbohydr. Res.*, **119**, 191–219.
128 McCleary, B.V. (1979) *Carbohydr. Res*, **71**, 205–230.
129 McCutchen, C.M., Duffaud, G.D., Leduc, P., Petersen, A.R.H., Tayal, A., Khan, S.A., and Kelly, R.M. (1996) *Biotechnol. Bioeng.*, **52**, 332–339.
130 Pal, S. (2009) *J. Appl. Polym. Sci.*, **111**, 2630–2636.
131 (a) Lapasin, R., De Lorenzi, L., Pricl, S., and Torriano, G. (1995) *Carbohydr. Polym.*, **28**, 195–202; (b) Cheng, Y., Brown, K.M., and Prud'homme, R.K. (2002) *Biomacromolecules*, **3**, 456–461.
132 Singh, V., Tiwari, A., Tripathi, D.N., and Sanghi, R. (2004) *Carbohydr. Polym.*, **58**, 1–6.
133 Nayak, B.R. and Singh, R.P. (2001) *Eur. Polym. J.*, **37**, 1655–1666.
134 Soppirnath, K.S. and Aminabhavi, T.M. (2002) *Eur. J. Pharm. Biopharm.*, **53**, 87–98.
135 Thakur, S., Chauhana, G.S., and Ahn, J.-H. (2009) *Carbohydr. Polym.*, **76**, 513–520.
136 Tiwari, A., Grailer, J.J., Pilla, S., Steeber, D.A., and Gong, S. (2009) *Acta Biomater.*, **5**, 3441–3452.
137 Meyer, K. and Palmer, J.W. (1934) *J. Biol. Chem.*, **107**, 629–634.
138 Balazs, E.A., Laurent, T.C., and Jeanloz, R.W. (1986) *Biochem. J.*, **235**, 903.

139 Lapčik, L. Jr. and Lapčik, L. (1998) *Chem. Rev.*, **98**, 2663–2684.

140 Hargittai, I. and Hargittai, M. (2008) *Struct. Chem.*, **19**, 697–717.

141 Weigel, P.H., Hascall, V.C., and Tammi, M. (1997) *J. Biol. Chem.*, **272**, 13997–14000.

142 Kogan, G., Šoltés, L., Stern, R., and Gemeiner, P. (2007) *Biotechnol. Lett.*, **29**, 17–25.

143 Kakehi, K., Kinoshita, M., and Yasueda, S. (2003) *J. Chromatogr. B*, **797**, 347–355.

144 Stern, R., Asari, A.A., and Sugahara, K.N. (2006) *Eur. J. Cell Biol.*, **85**, 699–715.

145 Stern, R. and Jedrzejas, M.J. (2006) *Chem. Rev.*, **106**, 818–839.

146 Meyer, K. and Rapport, M.M. (1952) *Adv. Enzymol.*, **13**, 199–236.

147 Volpi, N., Schiller, J., Stern, R., and Šoltés, L. (2009) *Curr. Med. Chem.*, **16**, 1718–1745.

148 Stern, R., Kogan, G., Jedrjas, M.J., and Šoltés, L. (2007) *Biotechnol. Adv.*, **25**, 537–557.

149 Lebel, L. (1991) *Adv. Drug Deliv. Rev.*, **7**, 221–235.

150 Balazs, E.A. (2004) *Chemistry and Biology of Hyaluronan*, Elsevier, Amsterdam.

151 (a) Šoltés, L., Mendichi, R., Kogan, G., and Mach, M. (2004) *Chem. Biodivers.*, **1**, 468–472; (b) Šoltés, L. and Mendichi, R. (2003) *Biomed. Chromatogr.*, **17**, 376–384.

152 Charlot, A., Heyraud, A., Guenot, P., Rinaudo, M., and Auzély-Velty, R. (2006) *Biomacromolecules*, **7**, 907–913.

153 (a) Shu, X.Z., Liu, Y., Palumbo, F.S., Luo, Y., and Prestwich, G.D. (2004) *Biomaterials*, **25**, 1339–1348; (b) Shu, X.Z., Liu, Y., Roberts, M.C., and Prestwich, G.D. (2002) *Biomacromolecules*, **3**, 1304–1311; (c) Shu, X.Z., Liu, Y., Palumbo, F., and Prestwich, G.D. (2003) *Biomaterials*, **24**, 3825–3834; (d) Shu, X.Z., Ahmad, S., Liu, Y., and Prestwich, G.D. (2006) *J. Biomed. Mater. Res. A*, **79A**, 902–912.

154 Bulpitt, P. and Aeschlimann, D. (1999) *J. Biomed. Mater. Res.*, **47**, 152–169.

155 Jia, X., Colombo, G., Padera, R., Langer, R., and Kohane, D.S. (2004) *Biomaterials*, **25**, 4797–4804.

156 Kurisawa, M., Chung, J.E., Yang, Y.Y., Gao, S.J., and Uyama, H. (2005) *Chem. Commun.*, **34**, 4312–4314.

157 Leonelli, F., Bella, A.L., Migneco, L.M., and Bettolo, R.M. (2008) *Molecules*, **13**, 360–378.

158 Di Meo, C., Panza, L., Capitani, D., Mannina, L., Banzato, A., Rondina, M., Renier, D., Rosato, A., and Crescenzi, V. (2007) *Biomacromolecules*, **8**, 552–559.

159 Motokawa, K., Hahn, S.K., Nakamura, T., Miyamoto, H., and Shimoboji, T. (2006) *J. Biomed. Mater. Res.*, **78A**, 459–465.

160 Taglienti, A., Sequi, P., and Valentini, M. (2009) *Carbohydr. Res.*, **344**, 245–249.

161 Upadhyay, K.K., Le Meins, J.-F., Misra, A., Voisin, P., Bouchaud, V., Ibarboure, E., Schatz, C., and Lecommandoux, S. (2009) *Biomacromolecules*, **10**, 2802–2808.

162 Sorbi, C., Bergamin, M., Bosi, S., Dinon, F., Aroulmoji, V., Khan, R., Murano, E., and Norbeo, S. (2009) *Carbohydr. Res.*, **344**, 91–97.

163 Bender, H., Lehmann, J., and Wallenfels, K. (1959) *Biochim. Biophys. Acta*, **36**, 310–317.

164 Kataoka-Shirasugi, N., Ikuta, J., Kuroshima, A., and Misaki, A. (1994) *Biosci. Biotech. Biochem.*, **58**, 2145–2151.

165 Simon, L., Caye-Vaugien, C., and Bouchonneau, M. (1993) *J. Gen. Microbiol.*, **139**, 979–985.

166 Kondratyeva, T.F. (1981) *Uspechi Microbiology*, **16**, 175–192.

167 Hayashi, S., Hayashi, T., Takasaki, Y., and Imada, K. (1994) *J. Ind. Microbiol.*, **13**, 5–9.

168 Shingel, K.I. (2004) *Carbohydr. Res.*, **339**, 447–460.

169 Singh, R.S., Saini, G.K., and Kennedy, J.F. (2008) *Carbohydr. Polym.*, **73**, 515–531.

170 Singh, R.S., Saini, G.K., and Kennedy, J.F. (2010) *Carbohydr. Polym.*, **80**, 401–407.

171 Mähner, C., Lechner, M.D., and Nordmeier, E. (2001) *Carbohydr. Res.*, **331**, 203–208.

172 Alban, S., Schauerte, A., and Franz, G. (2002) *Carbohydr. Polym.*, **47**, 267–276.

173 Mocanu, G., Mihai, D., Dulong, V., Picton, L., Lecerf, D. (2011) *Carbohyd. Polym.*, **84**, 276–281.

174 Glinel, K., Huguet, J., and Muller, G. (1999) *Polymer*, **40**, 7071–7081.

175 Tabata, Y., Matsui, Y., Uno, K., Sokawa, Y., and Ikada, Y. (1999) *J. Interferon Cytokine Res.*, **19**, 287–292.

176 Yamaoka, T., Tabata, Y., and Ikada, Y. (1993) *Drug Deliv.*, **1**, 75–82.

177 Masuda, K., Sakagami, M., Horie, K., Nogusa, H., Hamana, H., and Hirano, K. (2001) *Pharm. Res.*, **18**, 217–223.

178 Riess, J.G. (1998) *Blood Substitutes: Methods, Products and Clinical Trials* (ed. T.M.S. Chang), Karger Landes Systems, Basel, pp. 101–126.

179 Survase, S.A., Saudagar, P.S., Bajaj, I.B., and Singhal, R.S. (2007) *Food Technol. Biotechnol.*, **45**, 107–118.

180 Brigand, G. (1993) Scleroglucan, in *Industrial Gums* (eds. R.L. Whistler and J.N. BeMiller), Academic Press, New York, USA, pp. 461–472.

181 Coviello, T., Grassi, M., Lapasin, R., Marino, A., and Alhaique, F. (2003) *Biomaterials*, **24**, 2789–2798.

182 Grassi, M., Lapasin, R., Coviello, T., Matricardi, P., Di Meo, C., and Alhaique, F. (2009) *Carbohydr. Polym.*, **78**, 377–383.

183 Christensen, B.E., Aasprong, E., and Stokke, B.T. (2001) *Carbohydr. Polym.*, **46**, 241–248.

184 Crescenzi, V., Gamini, A., Paradossi, G., and Torri, G. (1983) *Carbohydr. Polym.*, **3**, 273–286.

185 Maeda, H., Rambone, G., Coviello, T., Yuguchi, Y., Urakawa, H., Alhaique, F., and Kajiwara, K. (2001) *Int. J. Biol. Macromol.*, **28**, 351–358.

186 Coviello, T., Palleschi, A., Grassi, M., Matricardi, P., Bocchinfuso, G., and Alhaique, F. (2005) *Molecules*, **10**, 6–33.

187 Giavasis, I., Harvey, L.M., and McNeil, B. (2002) Scleroglucan, in *Biopolymers, Polysaccharides II*, vol. 6 (eds S. De Baets, E.J. Vandamme, and A. Steinbuchel), Wiley-VCH Verlag GmbH, Weinheim, p. 37.

188 Rinaudo, M. and Milas, M. (1982) *Carbohydr. Polym.*, **2**, 264–269.

189 Tait, M.I., Sutherland, I.W., and Clarke-Sturman, A.J. (1986) *J. Gen. Microbiol.*, **132**, 1483–1492.

190 Lilly, G.V., Wilson, A.H., and Leach, J.G. (1958) *Appl. Microbiol.*, **6**, 105–108.

191 Roseiro, J.C., Amaral Collaco, M.T., Esgalhado, M.E., and Emery, A.N. (1992) *Process Biochem.*, **27**, 167–175.

192 Rottava, I., Batesini, G., Silva, M.F., Lerin, L., de Oliveira, D., Padilha, F.F., Toniazzo, G., Mossi, A., Cansian, R.L., Luccio, M.D., and Treichel, H. (2009) *Carbohydr. Polym.*, **77**, 65–71.

193 Pons, A., Dussap, C.G., and Gros, J.B. (1990) *Bioprocess Eng.*, **5**, 107–114.

194 Kuttuva, S.G., Restrepo, A.S., and Ju, L.-K. (2004) *Appl. Microbiol. Biotechnol.*, **64**, 340–345.

195 Muchová, M., Růžička, J., Julinová, M., Doležalová, M., Houser, J., Koutný, M., and Bunková, L. (2009) *Water Sci. Tech.*, **60**, 965–973.

196 Nankai, H., Hashimoto, W., Miki, H., Kawai, S., and Murata, K. (1999) *Appl. Environ. Microbiol.*, **65**, 2520–2526.

197 Li, B., Guo, J., Chen, W., Chen, X., Chen, L., Liu, Z., and Li, X. (2009) *Appl. Biochem. Biotechnol.*, **159**, 24–32.

198 Katzbauer, B. (1998) *Polym. Degrad. Stab.*, **59**, 81–84.

199 Shalviri, A., Liu, Q., Abdekhodaie, M.J., and Wu, X.Y. (2009) *Carbohydr. Polym.*, **79**, 898–907.

200 Dumitriu, S. and Chornet, E. (1998) *Adv. Drug Deliv. Rev.*, **31**, 223–246.

8
Biodegradable Shape-Memory Polymers

Marc Behl, Jörg Zotzmann, Michael Schroeter, and Andreas Lendlein

8.1
Introduction

Shape-memory polymers (SMPs) can change their shape in a predefined way on demand when exposed to a suitable stimulus. They are able to change their shape as soon as the stimulus activates molecular switching moieties. At present, most investigated SMPs are thermosensitive, which means that the shape-memory effect (SME) is triggered by heat. They change their shape once the material softens as a result of exceeding a certain switching temperature (T_{switch}).

SMPs are mainly applied in the biomedical field in implants, surgical instruments, extracorporal devices, wound covers, as well as in controlled drug release devices. Prominent examples of applications in everyday life are heat shrinkable tubing and films [1, 2], which are used for insulating electronic wiring or for packing [3]. Here, mainly covalently crosslinked polyethylene is used. Shape-memory polyurethanes (SMPU) have been designed and synthesized [4], which are used in textiles as smart fabrics [5, 6]. Further rapidly developing application fields include self-deployable sun sails in spacecraft or space structure applications [7], intelligent medical devices [8], or implants for minimal invasive surgery (MIS) [9, 10]. In this chapter, (bio)degradable SMPs will be presented, and their synthesis and applications are introduced.

SMPs belong to the group of "actively moving" polymers [11]. Most SMPs investigated so far are dual-shape polymers. Recently, triple-shape and multi-shape materials having the capability of two or even more subsequent movements were developed [12–19]. Triple-shape materials can change their shape from a temporary shape A to a possible second temporary shape B and finally to a permanent shape C. The temporary shapes are obtained by mechanical deformation of the material's permanent shape C at a temperature $T > T_{switch}$ and subsequent fixation of these deformations at lower temperatures ($T < T_{switch}$). The synthesis and the processing of the material determine the permanent shape C. In SMPs reported so far, heat or light has been used as a stimulus to trigger the SME [4, 20–23]. Indirect actuation of the SME has also been realized by irradiation with

Handbook of Biodegradable Polymers: Synthesis, Characterization and Applications, First Edition. Edited by Andreas Lendlein, Adam Sisson.
© 2011 Wiley-VCH Verlag GmbH & Co. KGaA. Published 2011 by Wiley-VCH Verlag GmbH & Co. KGaA.

infrared- [24] or UV-light [25], application of electric field [26, 27] or alternating magnetic fields [28–31], or lowering of T_{switch} below ambient temperature by plasticizers such as water [32]. The SME results from a combination of a suitable molecular architecture and a programming procedure. Therefore, intrinsic material properties such as thermal or mechanical properties can be adjusted to the needs of specific applications by small variations of molecular parameters, such as monomer ratio or main chain bonds. This approach of adjusting material properties enables the design of polymer systems. Furthermore, this approach enables the creation of multifunctional materials, which is an actual trend in polymer science. Multifunctionality is the targeted combination of material functions, which are not linked with each other [33]. Multifunctional SMP can be realized as multimaterial systems, for example, by the incorporation of particles in polymer matrices, in which each material contributes a certain function, or as one component systems by the integration of suitable functional groups or building blocks [34]. Promising approaches can be the combination of biofunctionality, hydrolytic degradability, and shape-memory functionality. Such multifunctional SMPs have a high potential for applications in the biomedical field such as MIS (see Section 8.4) [35]. In contrast to metal implants or nondegradable polymers, bioresorbable SMPs are advantageous as they do not require an additional surgery for implant removal. In addition, bulky implants created from bioresorbable SMPs and having a T_{switch} between room temperature and body temperature could be inserted to the application site through a small incision in a compressed or elongated temporary shape. As soon as the implant is placed in the body, it assumes body temperature and changes into its bulky application-relevant shape. Other promising biomedical applications include intelligent degradable suture materials, which tighten a wound with a predefined stress, stimuli-sensitive matrices for drug delivery applications, or active scaffolds for regenerative therapies.

The required bioresorbable SMPs can be realized by the introduction of hydrolyzable bonds as weak links in the polymer chain enabling the degradation of these polymers in the presence of water, which may be supported by enzymes. Figure 8.1 shows hydrolysable bonds used in degradable polymers, in order of their stability.

Biodegradable, synthetic polymers may have advantages compared to polymers from natural sources. They can be tailored to meet the specific requirements of certain applications, such as thermal and mechanical properties. In addition, the processability of synthetic polymers, for example, by extrusion or injection molding

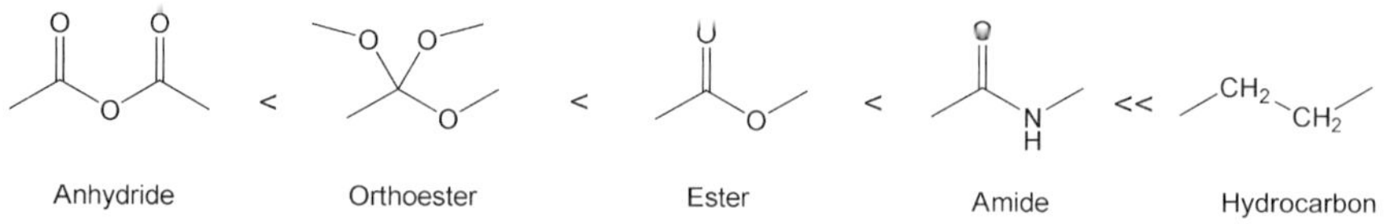

Figure 8.1 Relative stability of chemical bonds against hydrolysis occuring in common, synthetic polymers.

is much easier as they display in general a higher thermal stability as natural polymers. The tailoring of the polymer chain length of synthetic polymers enables polymers to form domains with a more defined domain size. When certain precautions are considered, a higher purity can be obtained, as a contamination with certain cell fragments can be avoided, which originate from the original source and can act as endotoxins. Consequently, polymers from natural sources require a high effort of purification, which potentially results in higher costs for such materials.

8.2
General Concept of SMPs

As the SME results from the combination of the polymer's molecular architecture/ morphology and a specific programming procedure, it can be understood as a functionalization of the polymer. The shape-memory creation procedure (SMCP), which is also called programming, and the recovery of the original shape due to the SME are schematically shown in Figure 8.2.

Suitable polymeric materials that are capable of an SME provide a polymer network architecture consisting of netpoints, chain segments, and molecular switches, with the latter being sensitive to an external stimulus, which is heat in the case of the thermally induced SME.

The permanent shape of the SMP is determined by the netpoints, which are interconnected by the chain segments. The netpoints can be of chemical (covalent bonds) or physical (intermolecular interactions) nature. Covalent bonds can be formed by the application of a suitable crosslinking chemistry, while netpoints provided by intermolecular interactions require a morphology consisting of at least two segregated domains, for example, a crystalline and an amorphous phase. In such multiphase polymers, the polymer chain segments form domains. The domains that are related to the highest thermal transition temperature (T_{perm}) are called hard domains and are acting as physical netpoints.

In the course of SMCP when the temporary shape is created, a deformation is applied to the polymer sample. This deformation requires a sufficient elastic

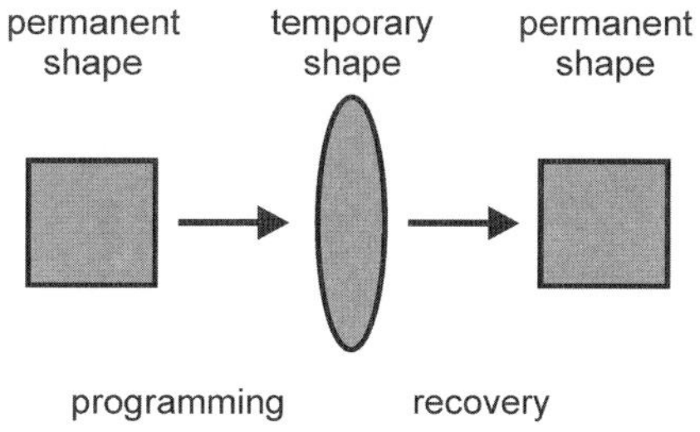

Figure 8.2 Schematic representation of the shape-memory effect. Taken from [4]. Copyright Wiley-VCH Verlag GmbH & Co. KGaA. Reproduced with permission.

deformability of the polymer network and can be reached by the chain segments, which must be capable to enable a certain orientation. The extent of the deformability increases with growing length and flexibility of these chain segments. In the polymer networks before applying the deformation, the majority of the polymer chain segments display a random coil conformation, which is the entropically favored orientation. The stimuli-triggered recoiling of the polymer chain segments, which is entropically driven, enables the recovery of the permanent shape.

The reversible fixation of the temporary shape is achieved by stimuli-sensitive switches that form additional reversible crosslinks, which can be established and cleaved on demand, and prevent in this way the recoiling of the polymer chain segments. Similarly to the permanent netpoints, these additional crosslinks can be established by the formation of chemical (covalent) reversible bonds, by the intermolecular interactions of side groups, or by the solidification of domains formed by these polymer chain segments when being cooled below their correlated thermal transition temperatures T_{trans}. Therefore, these polymer chain segments are also named switching segments; the associated domains are called switching domains. Such thermosensitive SMP can be classified according to the thermal transitions related to the solidification of the polymer chain segments. T_{trans} can be a glass transition temperature (T_g), a melting transition temperature (T_m), or a liquid crystalline transition. In all cases, heating of the SMP above the thermal transition causes a regain of flexibility of the vitrified or crystallized switching domains so that the elastic state is reached again. In thermoplastic SMP, only the domains associated to the polymer chain segments with the second highest T_{trans} are acting as switching domains.

Figure 8.3 displays a schematic representation of the molecular mechanism of the thermally induced SME, a thermoplastic SMP with $T_{trans} = T_m$, and covalent polymer networks with $T_{trans} = T_m$ (Figure 8.3b) or $T_{trans} = T_g$ (Figure 8.3c).

The SME can be quantified in cyclic, stimuli-specific tests under strain or stress control. Many degradable SMPs are triggered by heat as stimulus; consequently, the SME is determined in cyclic, thermomechanical tests. In these tests, the strain fixity rate (R_f), the strain recovery rate (R_r), and the switching temperature (T_{switch}) are determined. A single cycle includes the SMCP (programming) and the recovery of its permanent shape. The strain-controlled test consists of four steps: (1) heating of the sample to a temperature T_{high} above T_{trans} and deformation of the sample to a certain extension (ε_m) at a defined strain rate for a fixed period of time, (2) cooling to a temperature T_{low} with a certain cooling rate (β_c) while ε_m is kept constant, (3) unloading of the sample to $\sigma = 0\,\text{MPa}$ T_{low}, (4) heating of the test specimen to T_{high} while keeping the strain constant, and (5) start of the next cycle by going back to (1). In this test, the strain applied to the sample is controlled while the developing stress is recorded. In stress-controlled cyclic tests, steps (1) and (2) are adapted by keeping the stress σ constant at a maximum stress σ_m instead of keeping the sample at ε_m. The recovery step (4) is carried out by keeping $\sigma = 0\,\text{MPa}$ (Figure 8.4). In this test protocol, the deformation of the sample is monitored while the stress is controlled. T_{high} and T_{low} are adjusted to $T_{trans} \pm (20–30)$

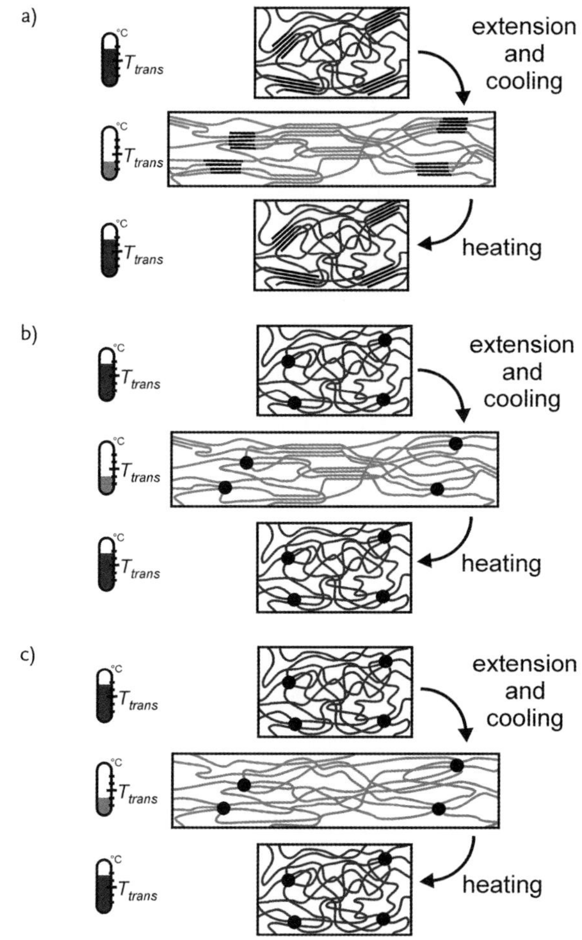

Figure 8.3 Schematic representation of the molecular mechanism of the thermally induced shape-memory effect: (a) physically crosslinked polymer network with phase-segregated domains having a crystalline or semicrystalline switching phase, (b) covalently crosslinked polymer network with crystalline or semicrystalline switching phase, and (c) covalently crosslinked polymer network with amorphous switching phase. Taken from [4]. Copyright Wiley-VCH Verlag GmbH & Co. KGaA. Reproduced with permission.

K of the examined polymer network. These cyclic, thermomechanical tests are typically performed five times. Figure 8.4b represents a three-dimensional diagram of a stress-controlled procedure. While the first cycle is used for erasing the thermal history of the polymer sample, cycles 2–5 are used for quantification of the shape-memory effect. In such a measurement, the sample is deformed at T_{high} to a maximum strain ε_m resulting in tensile stress σ_m (maximum stress) (1). The stretched specimen is then cooled to a temperature T_{low}, which is below T_{trans} (2). Several different effects of the sample behavior have to be considered, such as the

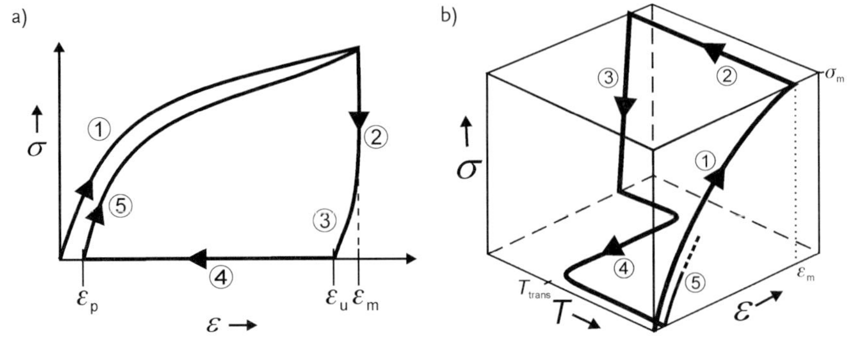

Figure 8.4 (a) ε-σ diagram of the strain-controlled programming and stress-free recovery of the shape-memory effect: (1) deformation of the sample to a maximum deformation ε_m at T_{high}; (2) cooling to T_{low} while σ_m is kept constant; (3) unloading to zero stress; (4) clamp distance is driven back to original starting distance, heating up to T_{high} while keeping $\sigma = 0$ MPa; (5) start of second cycle; (b) ε–T–σ diagram of the strain-controlled programming and stress-free recovery of the shape-memory effect: (1) stretching to ε_m at T_{high}; (2) cooling to T_{low} with constant cooling rate while σ_m is kept constant; (3) clamp distance is reduced until the stress-free state $\sigma = 0$ MPa is reached; (4) heating to T_{high} with a constant heating rate; and (5) start of the second cycle [4]. Taken from [4]. Copyright Wiley-VCH Verlag GmbH & Co. KGaA. Reproduced with permission.

strain-controlled	stress-controlled
Shape fixity ratio (R_f)	
$$R_f(N) = \frac{\varepsilon_u(N)}{\varepsilon_m} \quad (8.1)$$	$$R_f(N) = \frac{\varepsilon_u(N)}{\varepsilon_l(N)} \quad (8.3)$$
Shape recovery ratio (R_r)	
$$R_r(N) = \frac{\varepsilon_m - \varepsilon_p(N)}{\varepsilon_m - \varepsilon_p(N-1)} \quad (8.2)$$	$$R_r(N) = \frac{\varepsilon_l(N) - \varepsilon_p(N)}{\varepsilon_l(N) - \varepsilon_p(N-1)} \quad (8.4)$$

Figure 8.5 Equations for the determination of R_f and R_r from cyclic, thermomechanical measurements.

change of the expansion coefficient in the vitrified or viscoelastic state or changes in volume of the sample due to crystallization effects for $T_{trans} = T_m$ [4]. After cooling, the stress is released ($\sigma = 0$ MPa) leading to the elongation ε_u (3). Finally, the sample is heated again to T_{high} and the permanent shape ε_p is recovered (4).

From these cyclic, thermomechanical tests, the values of R_r and R_f at a given strain ε_m can be determined according to the four equations in Figure 8.5.

In a strain-controlled protocol, R_f is given by the ratio of the strain in the stress-free state after the retraction of the tensile stress in the Nth cycle $\varepsilon_u(N)$ and the maximum strain ε_m (Eq. (8.1), Figure 8.5). R_f describes the ability to fix the mechanical deformation, which has been applied during the programming process. R_r quantifies the ability of the polymer to memorize its permanent shape and it is a measure of how far the applied strain during the programming $\varepsilon_m - \varepsilon_p(N-1)$ is recovered during the SME. For that the strain that was applied during the programming in the Nth cycle, $\varepsilon_m - \varepsilon_p(N-1)$ is compared to the change in strain during the SME $\varepsilon_m - \varepsilon_p(N)$ (Eq. (8.2), Figure 8.5). The remaining strain of the samples after two successively passed cycles in the stress-free state is given by $\varepsilon_p(N-1)$ and $\varepsilon_p(N)$. In the stress-controlled protocol, R_f is represented by the ratio of the tensile strain after unloading ε_u and the strain at σ_m after cooling of the Nth cycle $\varepsilon_l(N)$ (Eq. (8.3), Figure 8.5). In such a protocol, R_r quantifies the ability of the polymer to reverse the deformation that was applied in the programming procedure $\varepsilon_l - \varepsilon_p(N-1)$ during the following shape-memory transition. For this purpose, the strain that was applied during the programming step in the Nth cycle $\varepsilon_l(N) - \varepsilon_p(N-1)$ is compared to the change of strain that occurs with the SME $\varepsilon_l(N) - \varepsilon_p(N)$ (Eq. (8.4), Figure 8.5).

8.3 Classes of Degradable SMPs

A strategy to functionalize SMP, so that they become biodegradable is the introduction of hydrolytically cleavable bonds into such polymers (see Figure 8.1) [36]. In the design of these polymers, it has to be considered that the degradation products should be either fully metabolized or excretable as fragments. This is of exceptional importance when these SMP are intended for biomedical applications. Furthermore, such degradable polymeric (bio)materials enable the application as matrix materials for controlled drug release systems that requires the exact characterization of the polymer's erosion behavior and the drug diffusion characteristics. Degradable SMPs show two types of degradation mechanisms: surface- and bulk erosion [37]. The degradation type depends on the diffusion of water into the polymer and the reactivity of the polymer functional groups (see Figure 8.1). Amorphous and crystalline segments, especially switching segments, display different degradation behavior. Amorphous segments degrade much faster due to the easier water penetration in these areas. In contrast, the penetration of water in crystalline segments is more inhibited by the dense packing of the crystalline lamellae.

In this section, an overview about degradable materials that exhibit an SME is given. SMPs can be divided into four types (see Table 8.1).

The requirements for an implant material are determined by the specific application. The key properties of degradable biomaterials are their mechanical properties, their degradation rate and degradation behavior, as well as biocompatibility and biofunctionality. Each application requires a specific combination of these properties/functions.

In the following sections, four different types of degradable SMPs are described.

Table 8.1 Overview over the four categories of SMP.

Type of netpoints of polymer network	Switching domains	Thermal transition	Example
Covalent	Crystallizable	T_m	Polymer networks from poly(ε-caprolactone) dimethacrylate [38]
Covalent	Amorphous, not crystallizable	T_g	Polymer networks from oligo[(*rac*-lactide)-*co*-glycolide] tetrol and diisocyanate [39]
Physical	Crystallizable	T_m	Polymer networks from oligo(ε-caprolactone)diole, oligo(*p*-dioxanone)diole and diisocyanate [9]
Physical	Amorphous, not crystallizable	T_g	Poly(L,L-lactide-*co*-glycolide-*co*-trimethylene carbonate) [40]

8.3.1
Covalent Networks with Crystallizable Switching Domains, $T_{trans} = T_m$

This type of polymer network consists of chain segments of homo- or copolymers and covalent netpoints. They can be prepared by (co)polymerization/poly(co)condensation of several monomers (Figure 8.6).

As a linear polyester poly(ε-caprolactone) (PCL) is hydrolytic degradable. It can be synthesized by ring-opening polymerization (ROP) of ε-caprolactone. When diols are used as initiators, macrodiols can be obtained. Covalent polymer networks can be created from these macrodiols after subsequent functionalization with polymerizable end groups, for example, dimethacrylates. These polymer networks were shown to be hydrolytically degradable and capable of an SME [38]. By the addition of a comonomer, for example, *n*-butyl acrylate, the elasticity of such polymer networks can be increased, resulting in AB copolymer networks. At the same time, T_{trans} of the network can be adjusted from 51 °C for the PCL dimethacrylate homonetwork to 44 °C for a copolymer network having 70 wt% *n*-butyl acrylate [41]. The degradability of such AB copolymer networks could be increased by the introduction of glycolide into the macrodimethacrylates [42]. The AB copolymers were prepared from poly(ε-caprolactone-*co*-glycolide) dimethacrylate and *n*-butyl acrylate as photosets. The macrodimethacrylates had a number average molecular weight (M_n) up to 13,500 g mol^{-1} and a maximum glycolide content of 21 mol%. The polymers were semicrystalline at room temperature and displayed a T_m between 18 and 53 °C. In the polymer networks, the oligo(butyl acrylate) formed the amorphous soft segment. Degradation experiments showed good hydrolytic degradability at pH 7 and 37 °C. The presence of glycolate accelerates

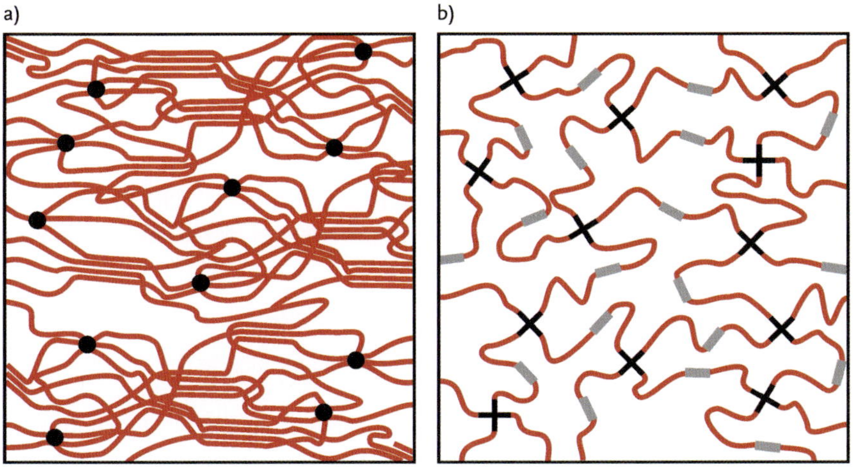

Figure 8.6 Schematic representation of covalent polymer networks. (a) Netpoints (black) consisting of acrylates or methacrylates and crystallizable switching segments (blue), for example, consisting of poly(ε-caprolactone); (b) obtained from multiarm precursors (red) acting as amorphous switching segments. The netpoints (black cross) are provided by the precursors, which were linked by small difunctional crosslinkers (gray).

the course of the hydrolytic chain scission and mass loss, and the presence of poly(*n*-butyl acrylate) segments decreases the degradation rate.

Recently, a covalent network of PCL with a percolative physical network was described [43]. Polyhedral oligosilsesquioxane (POSS) diols served as initiators for the polymerization of ε-caprolactone. The obtained oligomers were acrylated and crosslinked with a tetrathiol by photopolymerization to form a so-called double network. Here, the POSS moieties, which were located in side chains, provided a physical network, while the acrylate groups built a covalent polymer network, and the PCL chain segments contributed the switching domains. The content of POSS was varied from 22 to 47 wt% in the networks. Higher POSS content resulted in two distinct rubbery plateaus during the thermomechanical tests. T_m of the PCL moieties ranged from 39 to 47 °C and T_m of the POSS moieties from 86 to 69 °C depending on the content. Applications in tissue engineering and drug delivery were thought to be possible.

Poly[(3-hydroxybutyrate)-*co*-(3-hydroxyvalerate)], which was produced by bacteria, displayed an SME. The temporary shape was fixed by induced formation of hard domains by orientation via stretching the material [44]. The material had a very broad melting transition from approximately 37 to 115 °C and an elongation at break (ε_R) of 700%.

Recently, a stent made of an SMP from chitosan films crosslinked with an epoxy compound (ethylene glycol diglycidyl ether), which was blended with polyethylene glycol and glycerol was reported [45]. Generally, chitosan-based films are brittle because of their high crystallinity. Blending of the SMP with polyethylene glycol (M_n = 400,000 g mol^{-1}) reduced the crystallinity and enabled shape-memory properties of the material. The SME could be repeated several times and could be controlled by the hydration or dehydration of the SMP. When immersed in an aqueous buffer solution of 37 °C, the material recovered its permanent shape within 150 s. The degradability of the material was investigated in enzymatic degradation studies in lysozyme solution for 10 weeks. The material was shown to be degradable, but degradability decreased with increasing crosslinking density.

8.3.2 Covalent Networks with Amorphous Switching Domains, $T_{trans} = T_g$

In covalently crosslinked polymer networks, the general parameters for controlling the shape-memory behavior are the nature of the switching segments influencing the characteristics of the SME such as T_{switch} and the crosslink density influencing the mechanical properties.

Completely amorphous polymer networks with a thermally induced SME are described in reference [46], but were not originally developed for medical applications and are not hydrolytically degradable. Amorphous, biodegradable SMP networks could be prepared by coupling well-defined star-shaped hydroxy-telechelic polyesters with a low-molecular-weight junction unit (diisocyanate) [39]. The copolyester segments were formed by copolymerization of diglycolide and *rac*-dilactide and yielded the oligo[(*rac*-lactide)-*co*-glycolide] by ROP. The application of 1,1,1-tris(hydroxymethyl)ethane and pentaerythrite as initiators resulted in trifunctional or tetrafunctional star-shaped precursors, respectively. The mechanical properties of such polymer networks could be substantially enhanced by the introduction of an additional amorphous phase being immiscible with the first amorphous component. Incorporation of poly(propylene glycol) led to microscopic phase segregation within the amorphous networks and thus resulted in two distinct glass transitions with one T_g between −59 and −25 °C and the second T_g between 39 and 53 °C as well as good elastic properties at ambient temperature with ε_R up to 500%. The mechanical properties could be controlled by independently altering the two parameters, content and molecular weight of the poly(propylenglycol) segment [47].

The substitution of the diglycolide comonomer by other cyclic diesters in the synthesis of hydroxytelechelic copolyesters was shown to be another parameter to control T_{trans} of such amorphous polymer networks [48].

Transparent and hydrolytically degradable SMP networks with $T_{trans} = T_g$ based on acrylate chemistry could be obtained by UV polymerization of poly[(L-lactide)-*ran*-glycolide] dimethaycrylates (PLGDMA) [49]. Hydroxy telechelic poly[(L-lactide)-*ran*-glycolide]s (PLG) M_n between 1000 and 5700 g mol^{-1} were prepared by ROP from L,L-dilactide, diglycolide, and ethylene glycol as initiator using dibutyltin oxide as the catalyst. Subsequent functionalization of the PLG with methacryloyl chloride resulted in terminal methacrylate groups. T_g was shown to be almost constant at about 55 °C. Mechanical properties of these polymer networks below and above T_g differed significantly. The storage modulus E' determined by dynamic mechanical analysis at varied temperature (DMTA) was 3080 MPa at room temperature with $\varepsilon_R = 43\%$ and 8 MPa at 80 °C with $\varepsilon_R = 130\%$. The mechanical properties at temperatures higher than T_g depended on crosslinking density. A hydrolytic degradation in bulk could be expected according to the described behavior for PLG [50]. Excellent shape-memory properties with R_r close to 100% were obtained during tests under stress-control.

Phase-separated, amorphous, and degradable block copolymer networks were also prepared by photo crosslinking of the linear ABA triblock precursors poly(*rac*-lactide)-*b*-poly(propylene oxide)-*b*-poly(*rac*-lactide)dimethacrylate [51]. A polypropyleneglyocol (B-block) with $M_n = 4000\,\mathrm{g\,mol^{-1}}$ was used as a macroinitiator for poly(*rac*-lactide) synthesis, whereas poly(*rac*-lactide) blocks (A-blocks) with M_n of 2000, 4000, and 6000 $\mathrm{g\,mol^{-1}}$ were obtained. Thus, the length of the macrodimethacrylates precursors was systematically varied and SMP networks with different mechanical properties were synthesized. The T_g of the phase provided by the poly(propylene oxide) was obtained at −50 °C. An additional transition associated to the mixed phase between the phase transition resulting from the poly(propylene oxide) and the poly(*rac*-lactide) as well as the phase transition from the poly(propylene oxide) were observed, when M_n of the macro-dimethacrylate precursors was $<10\,\mathrm{g\,mol^{-1}}$. A distinct phase separation of the resulting polymer networks could be observed for macrodimethacrylate precursors with $M_n > \mathrm{g\,mol^{-1}}$. Values of ε_R from 70% to 219% could be achieved and the polymers displayed R_f values between 92% and 96% and R_r values from 87% to values over 99%, which increased with increasing poly(*rac*-lactide) content. Potential biomedical applications are intelligent implants or smart drug delivery systems.

8.3.3
Physical Networks with Crystallizable Switching Domains, $T_{trans} = T_m$

An important group of this type of SMPs is based on linear multiblock copolymers. Phase-segregated domains have to be formed by different segments being represented by different types of blocks within the linear polymer chains. A T_{trans} related to a T_m is obtained when the domains of the switching segment are crystallizable. The polymer blocks could be biodegradable polyesters or polyethers. Prominent examples for this type of multiblock copolymers are polyesterurethanes (PEUs). In such PEUs, the polyurethane segments are acting as hard segments, while the polyester segments, for example, PCL, are representing the switching segments (Figure 8.7). Thermoplastic SMP can be synthesized by direct coupling of presynthesized polymer blocks with a reactive linker, by applying the prepolymer method, or by melt blending. The application of the prepolymer method enabled the production of thermoplastic polyurethane elastomers on an industrial scale. In this process, isocyanate-terminated prepolymers are obtained by reaction of hydroxytelechelic oligoesters or -ethers with an excess of a low-molecular-weight diisocyanate. Biocompatible and simultaneously biodegradable multiblock copolymers

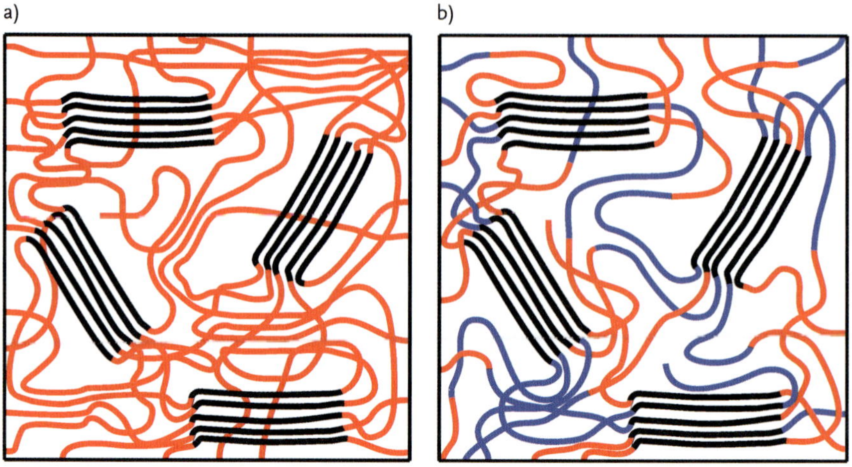

Figure 8.7 Schematic representation of physically crosslinked polymer networks: (a) polymer network with hard segments (black) and crystallizable switching segments (red); (b) multiblock copolymer with hard segment (black) and two amorphous switching segments (red, blue).

with shape-memory properties could be synthesized via co-condensation of macrodiols from poly(*p*-dioxanone) (PPDO) and PCL using an isomeric mixture of 1,6-diisocyanato-2,2,4-trimethylhexane and 1,6-diisocyanato-2,4,4-trimethylhexane as a bifunctional coupling agent [9]. The required diols were synthesized by ROP of the cyclic esters [52]. In these multiblock copolymers, named PDC, PPDO with the higher T_m was chosen as the hard segment to provide the physical crosslinks determining the permanent shape, while the crystallizable PCL is acting as switching segment. The mechanical properties strongly depend on the hard segment content. Hard segment contents of the synthesized polymers ranged from 0 to 83 wt%. The multiblock copolymers were elastic at room temperature and exhibited ε_R values up to 1000%. An increase of the PPDO amount resulted in a stiffer polymer and a decrease of the corresponding ε_R. R_f between 98% and 99.5% were determined throughout all cycles, while R_r depended on the cycle number and gradually approached values near 100%.

The same type of polymer as a blend with a poly(alkylene adipate) as mediator segment to promote the miscibility of the other segments showed also shape-memory properties, was biodegradable and is therefore a candidate for biomedical applications [53].

Phase-segregated PEUs prepared from PCL diol, ethylene glycol, and 2,4-toluene diisocyanate as linker showed shape-memory behavior and due to the PCL are biodegradable [54]. A tensile deformation of 300% was possible and R_r values between 94% and 100% could be determined. By adjusting the molecular weight of the PCL diol and the hard-to-soft ratio, the switching temperature could be adjusted to the range of 37–42 °C. In multiblock copolymers having PCL switching segments and POSS moieties in the polyurethane hard segments, a tremendous increase in the elasticity above T_{trans} was determined, which was attributed to physical crosslinks formed in the hard domains through POSS crystallization [55]. Recently, the degradation characteristics of crystalline multiblock copolymers had been investigated on SMP urethanes (SMPU) based on poly(adipate)diol ($M_w = 3500\,g\,mol^{-1}$) as switching segment and a hard segment derived from methylene-bis(4-phenylisocyanate) (MDI) and butanediol (BD). The degradation process could be divided into three phases: an induction phase, a phase of continuous degradation, and a phase of accelerated degradation. R_r remained fairly constant during phase one and decreased slowly during phase two. The increase in crystallinity in phase two was accompanied by an increase in R_f [56].

During the hydrolytic degradation of polyester segments, carboxylic acids are generated. In contrast, when some of the ester functional groups are replaced by peptide segments as in polydepsipeptides, the carboxylic acids can be directly buffered by the amino functions, which are also generated during degradation. The substitution of a polyester segment by polydepsipeptide segment in multiblock copolymers is thought to combine advantageous degradation behavior of the depsipeptide segment with the shape-memory capability of multiblock copolymers with the PCL switching segment. Thermoplastic multiblock copolymers with polydepsipeptide- and PCL segments providing shape-memory capability were synthesized via coupling of the depsipeptide oligo[3-(*R*)-isobutylmorpholine-2,5-dione]diol (PIBMD) and PCL diol ($M_n = 2900\,g\,mol^{-1}$) using TMDI [57]. In these polymer materials, the switching domains were formed by the PCL block, while the domains determining the permanent shape were formed by the polydepsipeptide segments. The shape-memory properties of such a thermoplastic multiblock copolymer with 50 wt% of PCL segments yielded in R_f and R_r values of more than 96% for all cycles, a T_{sw} around body temperature, and an ε_R value of 680%. The degradation behavior shows a mass loss of 12 wt% at 37 °C over 70 days in aqueous buffer solution.

Other physically crosslinked networks based on PCL and α-cyclodextrines (α-CD) form partial inclusion complexes using the molecular recognition of CDs. The T_m of PCL makes it suitable for clinical applications [58], but the transition temperature of the complex is around 60 °C. PCL with a molecular weight of 80,000 $g\,mol^{-1}$ was applied because polymers with lower molecular weight formed complete inclusion complexes with the α-CD resulting in a crystalline powder. The material was obtained by solvent casting at 70 °C with DMF. It was important to keep the theoretical mass proportion of the inclusion complexes between 30% and 50%, otherwise the resulting materials dissolved during purification or became too brittle. The R_r value was found to be between 95% (for 30% inclusion ratio) and 90% (for 50% inclusion ratio) and R_f values between 92% and 83% were obtained for the complexes. Biodegradability was evaluated by an enzymatic test at 37 °C.

A hyperbranched SMPU was synthesized from poly(butylenes adipate)glycol (PBAG), a hyperbranched polyester (Boltron H30), and MDI [59]. The synthesis was accomplished by a two-step process, first functionalization of PBAG with

MDI, which was followed by the addition of Boltron H30 as a chain extender. The polyurethane with 25 wt% of hard segment showed the best results in shape-memory behavior. The R_r was 96–98% for a hard segment content of 15–35 wt%. A higher content disturbed the crystallization of the PBAG soft segment.

8.3.4 Physical Networks with Amorphous Switching Domains, $T_{trans} = T_g$

In this type of SMP, the switching temperature is related to a T_g. Examples for biodegradable SMP having $T_{trans} = T_g$ are polyetherurethanes synthesized by applying the prepolymer method leading to a hard segment from MDI, BD, and poly(tetramethyleneoxide) (PTHF) or polyethylene adipate as second segment [4]. These polymers form predominantly a mixed domain acting as switching phase (Figure 8.7b). The quality of phase separation between the polyurethane segments and the polyetherurethane is determined by the molecular weight of the polytetrahydrofurandiols used as precursors. In a commercially available polyetherurethane synthesized from methylene bis(*p*-cyclohexyl isocyanate) (H12 MDI), 1,4-butanediol (BD), and PTHF diol (Tecoflex®), T_g is at 74 °C. This material is used in artificial hearts, wound dressings, and pacemaker leads [60]. Determination of shape-memory properties revealed R_f values of 100% and R_r values of 80% after the third cycle at $\varepsilon_m = 50\%$. The incorporation of silica-coated magnetic nanoparticles of iron(III)oxide core into Tecoflex® enabled the remote actuation of the thermally induced SME in alternating magnetic fields [28].

A terpolymer from L,L-dilactide, diglycolide, and trimethylene carbonate prepared by ROP with $Zr(Acac)_4$ as catalyst exhibited a single-phase amorphous material with no evidence of phase separation [61]. Usually, physically crosslinked SMP networks display at least two phases with distinct thermal transitions. Polymers prepared by random polymerization of the three monomers resulted in materials having values of M_n between 25,000 and 54,000 g mol^{-1} and PDI values between 2.0 and 2.3. R_r values from 89 to 95% at initial recovery temperatures between 38 to 42 °C enabled applications in the biomedical field. The T_g of the terpolymers changed with the composition in a predictable manner and followed a trend based on the relative amount of the three units, but stayed always close to body temperature. A further advantage is the use of a zirconium complex as initiator. This compound is significantly less toxic compared to common stannous initiator and inert in human metabolic processes.

To overcome the drawback of the low stiffness of SMP for bone regeneration applications, addition of fibers or particles is necessary to reinforce the polymers. Composites of hydroxyapatite (HAP) with poly(D,L-lactide) (PDLLA) also showed shape-memory properties which are improved compared to pure PDLLA [62]. PDLLA was synthesized by ROP and afterwards uniformly mixed with HAP particles. T_g was increasing from 54 °C for pure PDLLA to 59 °C for a 1:1 mixture. For a mixture of PDLLA/HAP 3:1 having a T_g of 53 °C, the storage modulus (E') at 23 °C was 3220 MPa and 29.6 MPa at 83 °C. The recovery rates of the materials were all above 95%. Comparable behavior was shown by a PDLLA composite with β-tricalcium phosphate composite [63].

8.4 Applications of Biodegradable SMPs

During the last few years, the research field of SMPs as materials for biomedical applications has grown rapidly. SMPs were widely investigated in specific medical instruments as highly recoverable and convenient materials for minimally invasive surgery [9], in drug delivery systems [64], vascular surgery [65], implantable devices [35], and intracranial aneurysm surgery [66]. The clinical trend toward MIS techniques sets new requirements for materials to be used as a matrix of medical devices [67]. This technique would benefit from the implantation of small objects that unfold to bulky devices of desired shape and functionality. In the following text, examples for medical devices based on biodegradable SMP will be presented and briefly discussed. Overviews for the use of SMPs for medical devices can be found in previous reviews [68, 69].

8.4.1 Surgery and Medical Devices

Depending on the respective application, implants should provide a supportive function and mechanical strength over a long period of time or only temporarily, whereas the material should then biodegrade and disappear from the site of application. So far, the suggested SMP medical devices are only based on thermosensitive SMPs. Besides triggering the shape recovery by direct exposure to heat (e.g., body temperature, flushing with warm water), indirect heating by light absorption of near-infrared dyes loaded into the material is an intensively evaluated approach particularly for intravascular applications [70].

Major targets for the evaluation of biodegradable SMPs in medical devices in the field of cardiovascular applications are stents [8, 65]. Stents are used to maintain the internal lumen of blood vessels when local flow constrictions were removed by cardiovascular intervention. Stents have to be inserted in a shape with a small outer diameter, and then subsequently expand and apply pressure against the surrounding tissue. Fully degradable stents are believed to be the future in stent technology [71]. However, so far, most studies on SMPs as stent matrices have been conducted with nondegradable SMPs. Only in some cases, partially degradable materials employing degradable polyester segments were evaluated [12].

For optimal healing after surgery with minimal scar formation, wounds have to be closed by applying a defined pressure to the wound lips. A programmed suture made from SMPs exhibiting a T_{trans} around body temperature has the capacity to exert a controllable pressure on the wound edges, holding the wound closed even after potential inflammatory swelling decayed. An example for such a suture is

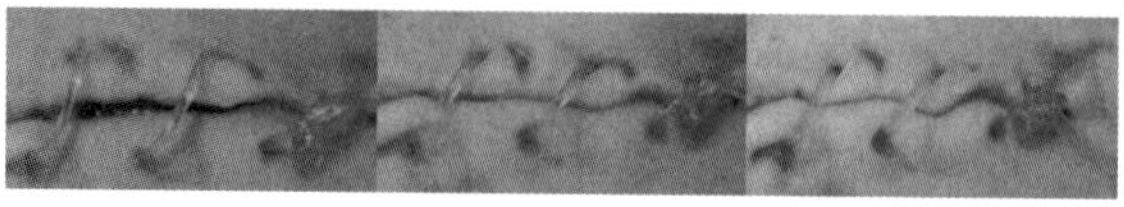

Figure 8.8 Picture series of a degradable suture from a multiblock copolymer with crystallizable switching segments used for wound closure. Taken from [9]. Reprinted with permission from AAAS, USA.

shown in Figure 8.8. The degradability of the material increases the patients' comfort as the second surgery for removal of the suture can be avoided and the risk for infections is decreased.

Other important fields of currently developed SMP applications are aneurysm treatment and clot removal devices. Aneurysms may be associated with potentially lethal rupture or uncontrolled aggregation of blood clots, which could embolize peripheral tissue when removed from the aneurysm into the blood stream. Among different treatment options, commonly used or suggested strategies that rely on endovascular implantation of medical devices include the implantation of vessel prostheses in aortic aneurysm [72, 73]. From first simulation data, SMP foams can be considered to be a promising treatment option of intracranial aneurysm, since they may reduce the risk of intraoperative rupture [74]. Preliminary *in vivo* data in dogs for aneurysms of the common carotid artery showed a successful closure, at least macroscopically in this healing vessel model [10].

For the treatment of cerebral ischemia in stroke patients, mechanical removal of intravascular blood clots has been advantageous at timepoints several hours after the onset of the stroke, where clot-dissolving standard therapies are no longer effective. A microcatheter system was introduced that is guided to the occluded vessel and passed beyond the thrombus. A nitinol wire is advanced through the catheter, deploys at the end of the catheter in a spring-like manner to a helical corkscrew-shape, and captures the clot [75]. Similar to this strategy, a corkscrew system [76] and an umbrella device [70] have been designed using commercially available covalently crosslinked thermosensitive SMP networks [77].

8.4.2
Drug Release Systems

The application of degradable polymeric biomaterials as matrix materials for pharmaceutically active agents enables stimuli-sensitive controlled drug release systems. The concept of a controlled and sustained release of drugs from biodegradable implants was developed more than 30 years ago [78]. Drugs, incorporated in a polymer, should be delivered in a controlled manner in predefined rates. In contrast to daily peroral medication, for example, with tablets, such implants should reduce the frequency of administration and provide a constant level of the desired drug in the body over an extended period of time. SMP materials as carrier for the drug would allow the implantation of bulky devices by MIS and fixation of such a device at the place of application. This has led to multifunctional materials for biomedical applications, which combine biodegradability, controlled drug release, and shape-memory capability. After implantation of a medical device, the SME can affect a fixation of the device at the site of implantation. Subsequently, the controlled release of the loaded drug is used for treating infections, reducing inflammatory responses, or, potentially, supporting regeneration processes. Finally, the degradation of the matrix can avoid a second surgery for removal of the implant.

In multifunctional SMPs that involve controlled drug release, specific processes in synthesis, processing, and programming have to be applied to enable each of the functionalities. Key for a release of bioactive molecules is their incorporation into the polymer matrix. Drug incorporation into the matrix can principally be achieved by soaking of the synthesized matrix in a drug solution and a subsequent drying step, or alternatively by mixing of defined amounts of drug with polymer network precursors and subsequent crosslinking. The latter method often allows higher drug loading, but is limited by the chemical stability of the drug under the crosslinking conditions. Drug release from polymer matrices is most often ruled either by diffusion or by degradation of the matrix. Diffusion-controlled release from a degradable matrix can, for example, be achieved for small water-soluble molecules from bulk-eroding materials [79–81]. Alternatively, the use of surface-eroding materials such as polyanhydrides [82] or poly(orthoesters) [83] allows an erosion-controlled drug release. So far, only a few systems representative for triple functional materials combining biodegradability, SME, and controlled drug release have been published. All of them belong to the bulk eroding polyesters with a diffusion-controlled release. However, type and ratio of the monomers as well as network architectures resulted in quite different capabilities of the networks.

First examples for triple functional SMPs were polymer networks from poly(ε-caprolactone-*co*-glycolide)dimethacrylate. In these semicrystalline networks, hydrophobic and hydrophilic drugs could be incorporated either by swelling of the final networks in an organic solvent saturated with the respective drug or by mixing defined amounts of the drug with the network precursors followed by irradiation (*in situ* incorporation) [84]. A semicrystalline polymer network was chosen so that the crystalline phases of the networks were used for the fixation of the temporary shape, while drug molecules should predominantly be incorporated in the amorphous phases without having a too strong influence on the melting point of the polymer crystallites or the shape fixation and recovery [85]. Large amounts of drug decreased the elongation at break of the networks, which reduced their programmability. In materials with high crystallinity in the drug-free state, no effect of drug loading on the thermomechanical properties and shape-memory functionality was observed.

Drug-loaded networks, which were shown to have high shape-fixity and shape-recovery, were subjected to hydrolytic degradation compared with drug-free samples. It was found that a diffusion-controlled release was realized before erosion of the matrix would have led to changes in the rate of drug release. Furthermore, independence of polymer functionalities could be demonstrated.

However, only limited amounts of drug could be incorporated by swelling (<1 wt%), which was theoretically high enough to ensure pharmacological efficiency of the model drugs but might be necessary to be increased for other drugs.

It was assumed that a higher drug loading by swelling can be achieved in fully amorphous SMPs such as the star-shaped PLG networks described above. In contrast to semicrystalline materials, the entire matrix is amorphous and therefore accessible for drugs [86]. Furthermore, crystallization is not required for shape fixation, thus allowing a wider variability for network composition and network architecture. Also, loading may not impact shape fixation. However, drug molecules can possibly act as softeners and reduce the T_g of amorphous SMPs, which can result in an unwanted shift in T_{switch}. Overall, SMPs as a technology platform can be envisaged to have a high potential for transfer into biomedical applications requiring biodegradable, multifunctional materials due to the generality of the underlying fundamental principles of the SME and the synthetic preparation of SMP network architectures as polymer systems, in which functions and properties can be tuned in a wide range by only small variations of their chemical structure. While basic research in SMPs is progressing rapidly, broadening the application potential of the SMP platform besides biomedical applications, the polymer system approach enables existing SMPs to be tailored to the challenging demands of the medicinal sector.

References

1 Rainer, W.C., Redding, E.M., Hitov, J.J., Sloan, A.W., and Steward, W.D. (1964) Heat-shrinkable polyethylene. US Patent, 3144398.

2 Arditti, S.J., Avedikian, S.Z., and Bernstein, B.S. (1971) Articles with Polymeric Memory US Patent, 3563973.

3 Charlesby, A. (1960) *Atomic Radiation and Polymers*, Pergamon Press, New York, pp. 198–257.

4 Lendlein, A. and Kelch, S. (2002) Shape-memory polymers. *Angew. Chem. Int. Ed.*, **41** (**12**), 2034–2057.

5 Mondal, S. and Hu, J.L. (2006) Temperature stimulating shape memory polyurethane for smart clothing. *Indian J. Fibre Text. Res.*, **31** (**1**), 66–71.

6 Hu, J. (2007) *Shape Memory Polymers and Textiles*, England Woodhead Publishing Limited, Cambridge.

7 Gall, K., Mikulas, M., Munshi, N.A., Beavers, F., and Tupper, M. (2000) Carbon fiber reinforced shape memory polymer composites. *J. Intell. Mater. Syst. Struct.*, **11** (**11**), 877–886.

8 Wache, H.M., Tartakowska, D.J., Hentrich, A., and Wagner, M.H. (2003) Development of a polymer stent with shape memory effect as a drug delivery system. *J. Mater. Sci. Mater. Med.*, **14** (**2**), 109–112.

9 Lendlein, A. and Langer, R. (2002) Biodegradable, elastic shape-memory polymers for potential biomedical applications. *Science*, **296** (**5573**), 1673–1676.

10 Metcalfe, A., Desfaits, A.C., Salazkin, I., Yahia, L., Sokolowski, W.M., and Raymond, J. (2003) Cold hibernated elastic memory foams for endovascular interventions. *Biomaterials*, **24** (**3**), 491–497.

11 Behl, M. and Lendlein, A. (2007) Actively moving polymers. *Soft Matter*, **3**, 58–67.

12 Bellin, I., Kelch, S., Langer, R., and Lendlein, A. (2006) Polymeric triple-shape materials. *Proc. Natl. Acad. Sci. USA*, **103** (**48**), 18043–18047.

13 Bellin, I., Kelch, S., and Lendlein, A. (2007) Dual-shape properties of

triple-shape polymer networks with crystallizable network segments and grafted side chains. *J. Mater. Chem.*, **17** (**28**), 2885–2891.
14 Behl, M., Bellin, I., Kelch, S., Wagermaier, W., and Lendlein, A. (2009) One-step process for creating triple-shape capability of AB polymer networks. *Adv. Funct. Mater.*, **19** (**1**), 102–108.
15 Kolesov, I.S. and Radusch, H.-J. (2008) Multiple shape-memory behavior and thermal-mechanical properties of peroxide cross-linked blends of linear and short-chain branched polyethylenes. *Express Polym. Lett.*, **2** (**7**), 461–473.
16 Pretsch, T. (2010) Triple-shape properties of a thermoresponsive poly(ester urethane). *Smart Mater. Struct.*, **19** (**1**), 015006.
17 Xie, T. (2010) Tunable polymer multi-shape memory effect. *Nature*, **464** (**7286**), 267–270.
18 Zotzmann, J., Behl, M., Feng, Y., and Lendlein, A. (2010) Copolymer networks based on poly(ω-pentadecalactone) and poly(ω-caprolactone)segments as a versatile triple-shape polymer system. *Adv. Funct. Mater.*, **20** (**20**), 3583–3594.
19 Behl, M. and Lendlein, A. (2010) Triple-shape materials. *J. Mater. Chem.*, **20** (**17**), 3335–3345.
20 Beloshenko, V.A., Varyukhin, V.N., and Voznyak, Y.V. (2005) The shape memory effect in polymers. *Russ. Chem. Rev.*, **74** (**3**), 265–283.
21 Behl, M. and Lendlein, A. (2007) Shape-memory polymers. *Mater. Today*, **10** (**4**), 20–28.
22 Mather, P.T., Luo, X.F., and Rousseau, I.A. (2009) Shape memory polymer research. *Annu. Rev. Mater. Res.*, **39**, 445–471.
23 Behl, M., Zotzmann, J., and Lendlein, A. (2010) Shape-memory polymers and shape-changing polymers. *Adv. Polym. Sci.*, **226**, 1–40.
24 Koerner, H., Price, G., Pearce, N.A., Alexander, M., and Vaia, R.A. (2004) Remotely actuated polymer nanocomposites–stress-recovery of carbon-nanotube-filled thermoplastic elastomers. *Nat. Mater.*, **3** (**2**), 115–120.
25 Lendlein, A., Jiang, H.Y., Jünger, O., and Langer, R. (2005) Light-induced shape-memory polymers. *Nature*, **434** (**7035**), 879–882.
26 Cho, J.W., Kim, J.W., Jung, Y.C., and Goo, N.S. (2005) Electroactive shape-memory polyurethane composites incorporating carbon nanotubes. *Macromol. Rapid Commun.*, **26** (**5**), 412–416.
27 Leng, J.S., Lv, H.B., Liu, Y.J., and Du, S.Y. (2007) Electroactivate shape-memory polymer filled with nanocarbon particles and short carbon fibers. *Appl. Phys. Lett.*, **91** (**14**), 144105.
28 Mohr, R., Kratz, K., Weigel, T., Lucka-Gabor, M., Moneke, M., and Lendlein, A. (2006) Initiation of shape-memory effect by inductive heating of magnetic nanoparticles in thermoplastic polymers. *Proc. Natl. Acad. Sci. USA*, **103** (**10**), 3540–3545.
29 Buckley, P.R., McKinley, G.H., Wilson, T.S., Small, W., Benett, W.J., Bearinger, J.P., McElfresh, M.W., and Maitland, D.J. (2006) Inductively heated shape memory polymer for the magnetic actuation of medical devices. *IEEE Trans. Biomed. Eng.*, **53** (**10**), 2075–2083.
30 Razzaq, M.Y., Anhalt, M., Frormann, L., and Weidenfeller, B. (2007) Thermal, electrical and magnetic studies of magnetite filled polyurethane shape memory polymers. *Mater. Sci. Eng. Struct. Mater. Prop. Microstruct. Process.*, **444** (**1-2**), 227–235.
31 Kumar, U.N., Kratz, K., Wagermaier, W., Behl, M., and Lendlein, A. (2010) Non-contact actuation of triple-shape effect in multiphase polymer network nanocomposites in alternating magnetic field. *J. Mater. Chem.*, **20** (**17**), 3404–3415.
32 Yang, B., Huang, W.M., Li, C., and Chor, J.H. (2005) Effects of moisture on the glass transition temperature of polyurethane shape memory polymer filled with nano-carbon powder. *Eur. Polym. J.*, **41** (**5**), 1123–1128.
33 Lendlein, A. and Kelch, S. (2005) Degradable, multifunctional polymeric biomaterials with shape-memory. *Mater. Sci. Forum*, **492-493**, 219–223.
34 Behl, M., Razzaq, M.Y., and Lendlein, A. (2010) Multifunctional shape-memory

polymers. *Adv. Mater.*, **22** (**31**), 3388–3410.

35 Lendlein, A. and Kelch, S. (2005) Shape-memory polymers as stimuli-sensitive implant materials. *Clin. Hemorheol. Microcirc.*, **32** (**2**), 105–116.

36 Weigel, T., Schinkel, G., and Lendlein, A. (2006) Design and preparation of polymeric scaffolds for tissue engineering. *Expert Rev. Med. Devices*, **3** (**6**), 835–851.

37 Brannon-Peppas, L. (1997) Polymers in controlled drug delivery. *Med. Plast. Biomater.*, **4**, 34–44.

38 Lendlein, A., Schmidt, A.M., Schroeter, M., and Langer, R. (2005) Shape-memory polymer networks from oligo(epsilon-caprolactone)dimethacrylates. *J. Polym. Sci. [A]*, **43** (**7**), 1369–1381.

39 Alteheld, A., Feng, Y.K., Kelch, S., and Lendlein, A. (2005) Biodegradable, amorphous copolyester–urethane networks having shape-memory properties. *Angew. Chem. Int. Ed.*, **44** (**8**), 1188–1192.

40 Zini, E., Scandola, M., Dobrzynski, P., Kasperczyk, J., and Bero, M. (2007) Shape memory behavior of novel (L-lactide-glycol ide-trimethylene carbonate) terpolymers. *Biomacromolecules*, **8** (**11**), 3661–3667.

41 Lendlein, A., Schmidt, A.M., and Langer, R. (2001) AB-polymer networks based on oligo(epsilon-caprolactone) segments showing shape-memory properties. *Proc. Natl. Acad. Sci. USA*, **98** (**3**), 842–847.

42 Kelch, S., Steuer, S., Schmidt, A.M., and Lendlein, A. (2007) Shape-memory polymer networks from oligo[(epsilon-hydroxycaproate)-*co*-glycolate] dimethacrylates and butyl acrylate with adjustable hydrolytic degradation rate. *Biomacromolecules*, **8** (**3**), 1018–1027.

43 Lee, K.M., Knight, P.T., Chung, T., and Mather, P.T. (2008) Polycaprolactone-POSS chemical/physical double networks. *Macromolecules*, **41** (**13**), 4730–4738.

44 Kim, Y.B., Chung, C.W., Kim, H.W., and Rhee, Y.H. (2005) Shape memory effect of bacterial poly[(3-hydroxybutyrate)-*co*-(3-hydroxyvalerate)]. *Macromol. Rapid Commun.*, **26** (**13**), 1070–1074.

45 Chen, M.C., Tsai, H.W., Chang, Y., Lai, W.Y., Mi, F.L., Liu, C.T., Wong, H.S., and Sung, H.W. (2007) Rapidly self-expandable polymeric stents with a shape-memory property. *Biomacromolecules*, **8** (**9**), 2774–2780.

46 Takeda, K., Akiyama, M., and Yamamizu, T. (1988) Shape-memory pore structure in porous crosslinked polystyrenes. *Angew. Makromol. Chem.*, **157**, 123–136.

47 Zotzmann, J., Alteheld, A., Behl, M., and Lendlein, A. (2009) Amorphous phase-segregated copoly(ether) esterurethane thermoset networks with oligo(propylene glycol) and oligo[(*rac*-lactide)-*co*-glycolide] segments: synthesis and characterization. *J. Mater. Sci. Mater. Med.*, **20**, 1815–1824.

48 Lendlein, A., Zotzmann, J., Feng, Y., Alteheld, A., and Kelch, S. (2009) Controlling the switching temperature of biodegradable, amorphous, shape-memory poly(*rac*-lactide)urethane networks by incorporation of different comonomers. *Biomacromolecules*, **10**, 975–982.

49 Choi, N.Y. and Lendlein, A. (2007) Degradable shape-memory polymer networks from oligo[(L-lactide)-*ran*-glycolide]dimethacrylates. *Soft Matter*, **3** (**7**), 901–909.

50 Li, S.M., Garreau, H., and Vert, M. (1990) Structure-property relationships in the case of the degradation of massive poly(α-hydroxy acids) in aqueous media. Part 2. Degradation of lactide–glycolide copolymers: PLA37.5GA25 and PLA75GA25. *J. Mater. Sci. Mater. Med.*, **1** (**3**), 131–139.

51 Choi, N.Y., Kelch, S., and Lendlein, A. (2006) Synthesis, shape-memory functionality and hydrolytical degradation studies on polymer networks from poly(*rac*-lactide)-*b*-poly(propylene oxide)-*b*-poly(*rac*-lactide) dimethacrylates. *Adv. Eng. Mater.*, **8** (**5**), 439–445.

52 Grablowitz, H. and Lendlein, A. (2007) Synthesis and characterization of,-dihydroxy-telechelic oligo(*p*-dioxanone). *J. Mater. Chem.*, **17** (**38**), 4050–4056.

53 Behl, M., Ridder, U., Feng, Y., Kelch, S., and Lendlein, A. (2009) Shape-memory capability of binary multiblock copolymer blends with hard and switching domains

provided by different components. *Soft Matter*, **5** (**3**), 676–684.

54 Ping, P., Wang, W.S., Chen, X.S., and Jing, X.B. (2005) Poly(epsilon-caprolactone) polyurethane and its shape-memory property. *Biomacromolecules*, **6** (**2**), 587–592.

55 Knight, P.T., Lee, K.M., Qin, H., and Mather, P.T. (2008) Biodegradable thermoplastic polyurethanes incorporating polyhedral oligosilsesquioxane. *Biomacromolecules*, **9** (**9**), 2458–2467.

56 Pretsch, T., Jakob, I., and Müller, W. (2009) Hydrolytic degradation and functional stability of a segmented shape memory poly(ester urethane). *Polym. Degrad. Stab.*, **94** (**1**), 61–73.

57 Feng, Y.K., Behl, M., Kelch, S., and Lendlein, A. (2009) Biodegradable multiblock copolymers based on oligodepsipeptides with shape-memory properties. *Macromol. Biosci.*, **9** (**1**), 45–54.

58 Luo, H., Liu, Y., Yu, Z., Zhang, S., and Li, B. (2008) Novel biodegradable shape memory material based on partial inclusion complex formation between α-cyclodextrin and poly(ε-caprolactone). *Biomacromolecules*, **9** (**10**), 2573–2577.

59 Cao, Q. and Liu, P. (2006) Structure and mechanical properties of shape memory polyurethane based on hyperbranched polyesters. *Polym. Bull.*, **57** (**6**), 889–899.

60 Guignot, C., Betz, N., Legendre, B., Le Moel, A., and Yagoubi, N. (2001) Degradation of segmented poly(etherurethane) Tecoflex (R) induced by electron beam irradiation: characterization and evaluation. *Nucl. Instrum. Methods Phys. Res. B*, **185**, 100–107.

61 Zini, E. and Scandola, M. (2007) Shape memory behavior of novel (L-lactide-glycolide-trimethylene carbonate) terpolymers. *Biomacromolecules*, **8**, 3661–3667.

62 Zheng, X.T., Zhou, S.B., Li, X.H., and Weng, H. (2006) Shape memory properties of poly(D,L-lactide)/hydroxyapatite composites. *Biomaterials*, **27** (**24**), 4288–4295.

63 Zheng, X., Zhou, S., Yu, X., Li, X., Feng, B., Qu, S., and Wenig, J. (2008) Effect of *in vitro* degradation of poly(D,L-lactide)/β-tricalcium composite on its shape-memory properties. *J. Biomed. Mater. Res. B*, **86B** (**1**), 170–180.

64 Ferrera, D.A. (2001) Shape-Memory Polymer Intravascular Delivery System. US Patent, 6224610..

65 Yakacki, C.M., Shandas, R., Lanning, C., Rech, B., Eckstein, A., and Gall, K. (2007) Unconstrained recovery characterization of shape-memory polymer networks for cardiovascular applications. *Biomaterials*, **28** (**14**), 2255–2263.

66 Hampikian, J.M., Heaton, B.C., Tong, F.C., Zhang, Z.Q., and Wong, C.P. (2006) Mechanical and radiographic properties of a shape memory polymer composite for intracranial aneurysm coils. *Mater. Sci. Eng. C*, **26** (**8**), 1373–1379.

67 Frost and Sullivan (2008) N39F-54-20 Medical Device Technologies Changing Healthcare.

68 Lendlein, A., Behl, M., Hiebel, B., and Wischke, C. (2010) Shape-memory polymers as technology platform for biomedical applications. *Expert Rev. Med. Devices*, **7** (**3**), 357–379.

69 Sokolowski, W., Metcalfe, A., Hayashi, S., Yahia, L., and Raymond, J. (2007) Medical applications of shape memory polymers. *Biomed. Mater.*, **2** (**1**), S23–S27.

70 Maitland, D.J., Metzger, M.F., Schumann, D., Lee, A., and Wilson, T.S. (2002) Photothermal properties of shape memory polymer micro-actuators for treating stroke. *Laser Surg. Med.*, **30** (**1**), 1–11.

71 Wykrzykowska, J.J., Onuma, Y., and Serruys, P.W. (2009) Advances in stent drug delivery: the future is in bioabsorbable stents. *Expert Opin. Drug Deliv.*, **6** (**2**), 113–126.

72 Duarte, M.P., Maldjian, C.T., and Laskowski, I. (2009) Comparison of endovascular versus open repair of abdominal aortic aneurysms: a review. *Cardiol. Rev.*, **17** (**3**), 112–114.

73 Lederle, F.A. (2009) In the clinic. Abdominal aortic aneurysm. *Ann. Intern. Med.*, **150** (**9**), ITC5-1–ITC5-15.

74 Maitland, D.J., Small, W., Ortega, J.M., Buckley, P.R., Rodriguez, J., Hartman, J., and Wilson, T.S. (2007) Prototype laser-activated shape memory polymer foam device for embolic treatment of aneurysms. *J. Biomed. Opt.*, **12** (**3**), 030504.

75 Gobin, Y.P., Starkman, S., Duckwiler, G.R., Grobelny, T., Kidwell, C.S., Jahan, R., Pile-Spellman, J., Segal, A., Vinuela, F., and Saver, J.L. (2004) MERCI 1: a phase 1 study of mechanical embolus removal in cerebral ischemia. *Stroke*, **35** (**12**), 2848–2854.

76 Metzger, M.F., Wilson, T.S., Schumann, D., Matthews, D.L., and Maitland, D.J. (2002) Mechanical properties of mechanical actuator for treating ischemic stroke. *Biomed. Microdevices*, **4** (**2**), 89–96.

77 Hayashi, S. (1993) Properties and applications of polyurethane-series shape-memory polymer. *Int. Prog. Urethanes*, **6**, 90–150.

78 Woodland, J.H. and Yolles, S. (1973) Long-acting delivery systems for narcotic antagonists. 1. *J. Med. Chem.*, **16** (**8**), 897–901.

79 Duvvuri, S., Janoria, K.G., and Mitra, A.K. (2006) Effect of polymer blending on the release of Ganciclovir from PLGA microspheres. *Pharm. Res.*, **23** (**1**), 215–223.

80 Liu, H., Finn, N., and Yates, M.Z. (2005) Encapsulation and sustained release of a model drug, Indomethacin, using CO_2-based microencapsulation. *Langmuir*, **21**, 379–385.

81 Wischke, C. and Schwendeman, S.P. (2008) Principles of encapsulating hydrophobic drugs in PLA/PLGA microparticles. *Int. J. Pharm.*, **364**, 298–327.

82 Göpferich, A. and Tessmar, J. (2002) Polyanhydride degradation and erosion. *Adv. Drug Deliv. Rev.*, **54**, 911–931.

83 Heller, J., Barr, J., Ng, S.Y., Schwach Abdellauoi, K., and Gurny, R. (2002) Poly(ortho esters): synthesis, characterization, properties and uses. *Adv. Drug Deliv. Rev.*, **54**, 1015–1039.

84 Neffe, A.T., Hanh, B.D., Steuer, S., and Lendlein, A. (2009) Polymer networks combining controlled drug release, biodegradation, and shape memory capability. *Adv. Mater.*, **21**, 3394–3398.

85 Wischke, C., Neffe, A.T., and Lendlein, A. (2010) Controlled drug release from biodegradable shape-memory polymers. *Adv. Polym. Sci.*, **226**, 177–205.

86 Wischke, C., Neffe, A.T., Steuer, S., and Lendlein, A. (2009) Evaluation of a degradable shape-memory polymer network as matrix for controlled drug release. *J. Control. Release*, **138**, 243–250.

9
Biodegradable Elastic Hydrogels for Tissue Expander Application

Thanh Huyen Tran, John Garner, Yourong Fu, Kinam Park, and Kang Moo Huh

9.1
Introduction

9.1.1
Hydrogels

Hydrogels are three-dimensional polymeric networks capable of absorbing a large amount of water or biological fluids while maintaining their basic structure [1, 2]. In the polymeric network, hydrophilic polymers are hydrated in an aqueous environment. The term "network" implies that crosslinked structures have to be present to avoid the dissolution of the hydrophilic polymer chains into the aqueous phase. Hydrogels can be classified into chemical and physical hydrogels based on the nature of crosslinking. In chemical hydrogels, the polymer chains are crosslinked by covalent bonding. If the polymer chains are crosslinked by noncovalent bonding, such networks are called physical hydrogels.

Since water molecules are the major component of the hydrogels, the mechanical strength of most hydrogels is rather low. That is, the storage moduli (G') of most hydrogels fall between several hundreds or several thousands pascals when the water content is high [3]. The poor mechanical strength and toughness after swelling are major disadvantages of using hydrogels. Therefore, the improvement of the elasticity of hydrogels is of great interest, since high elastic hydrogels are more suitable for application that bear mechanical loading, such as cartilage implant materials.

9.1.2
Elastic Hydrogels

Elastic hydrogels are hydrogels that are resilient and resistant to compression and elongation in their dried or water-swollen states. The elastic hydrogels possess the capability of withstanding cyclic mechanical strain without cracking or suffering significant permanent deformation [4]. The molecular weight of the polymers should be high enough, and the glass transition temperature (T_g) should be low

Handbook of Biodegradable Polymers: Synthesis, Characterization and Applications, First Edition. Edited by Andreas Lendlein, Adam Sisson.
© 2011 Wiley-VCH Verlag GmbH & Co. KGaA. Published 2011 by Wiley-VCH Verlag GmbH & Co. KGaA.

enough, to impart elastomeric behavior of the hydrogels [5]. Shape-memory hydrogels constitute a class of elastic hydrogels that can be elastically deformed and fixed into a temporary shape, and have ability to recover the original, permanent shape on exposure to an external stimulus such as heat or light [6].

For most biomedical applications, biodegradable elastic hydrogels are favored over nondegradable hydrogels. This is because they can be removed or eliminated by natural degradation from the applied sites in the body under relatively mild conditions, thus eliminating the need for any surgical removal processes after the system fulfills its goal. Biodegradable polymeric systems also provide flexibility in the design of delivery systems for large molecular weight drugs, such as peptides and proteins, which are not suitable for diffusion-controlled release through nondegradable polymeric systems [7]. In addition, the degradation can be utilized to control the rate of drug release and the physicochemical properties of the hydrogel systems, and thus to provide flexibility in the design of biomedical devices, such as drug–biomaterials combination products. However, proper techniques for predicting hydrogel degradation rates are critical for successful application of these degradable systems as they facilitate the design of implants with optimal degradation profiles that result in proper rates of drug release or tissue regeneration and hence maximize therapeutic effects.

9.1.3
History of Elastic Hydrogels as Biomaterials

Earlier works in elastic hydrogels were mainly focused on development of shape-memory hydrogels for fabrication of devices and implant stents. The first publication mentioning shape-memory effects in hydrogels was made by Osada *et al.* in 1995, who discovered a new phenomenon of a polymer hydrogel made by radical copolymerization of acrylic acid and *n*-stearyl acrylate having elastic memory that could be stretched to at least 1.5 times of its original length when the swollen gel is heated above 50 °C [8, 9]. Since then, biodegradable shape-memory polymers have been synthesized, including network polymers formed by crosslinking oligo(ε-caprolactone) dimethylacrylate and *N*-butylacrylate [10], a multiblock copolymer of oligo(ε-caprolactone) and oligo(*p*-dioxanone)diol [11], and polyesters of poly(propylene oxide) (PPO) with polylactide or glycolide [12]. Improvement of the stiffness and recovery force of shape-memory polymers can be achieved by the synthesis of shape-memory composites. Zheng *et al.* synthesized polylactide and hydroxyapatite composites which demonstrated better shape-memory effect than pure polylactide polymer [13].

Recently, with the increasing interest in engineering various tissues for the treatment of many types of injuries and diseases, a wide variety of biodegradable elastic hydrogels with desirable mechanical, degradation, and cytophilic properties have been developed. Elastic superporous hydrogel hybrids exhibiting mechanical resilience and a rubbery property in the fully water-swollen state have been reported by Park *et al.* These hydrogel hybrids of acrylamide (AM) and alginate could be stretched to about 2–3 times of their original lengths and could be loaded and unloaded cyclically at least 20 times. This property can potentially be exploited in the development of fast- and high-swelling elastic hydrogels for a variety of pharmaceutical, biomedical, and industrial applications [14, 15]. However, these systems lack biodegradable properties for various biomedical applications. Wen *et al.* developed biodegradable, biocompatible polyurethane-based elastic hydrogels by changing chain extenders. The hydrogels were highly elastic in its swollen state and comparable degradation and cytocompatible behaviors to polylactide. This may find the applications in both soft- and hard-tissue regeneration [16, 17]. In recent years, block copolymers of biodegradable polyesters such as poly(ε-caprolactone) (PCL), polylactides (PLAs), poly(glycolic acid) (PGA), and polylactide-*co*-glycolide (PLGA), and hydrophilic polyethylene glycol (PEG) have received considerable attention as potential biomaterials because of their combined advantages of the biodegradability of the polyesters and the biocompatibility of PEG [18–20]. The block copolymers also have some unique properties based on their amphiphilic nature. The block composition and structural characteristics can be utilized to modify various physicochemical properties such as biodegradation, permeability, swelling, elasticity, and mechanical properties [21–23]. Typical hydrogels are glassy and brittle in the dried state and it is difficult to change the shape and size of the dried state. Huh *et al.* have developed biodegradable PEG/PCL and PLGA–PEG–PLGA/PEG hydrogels showing flexible and elastic properties even in the dried state that they remain intact after repeated bending or stretching to twice the original length [24].

Further, elastic hydrogels with self-healing capacity were synthesized by hydrophobic association through micellar copolymerization of AM and a small amount of octyl phenol polyethoxy ether acrylate. These hydrogels showed high recovery even after extensive stretching and self-healing after being cut into two parts which can be used as shrinkable or thermal sensitive materials [25].

While covalently crosslinked hydrogels have the ability to control the elastic behaviors, one limiting factor is the difficulty in guaranteeing removal of impurities, such as unreacted monomers, sol fractions, nonaqueous solvents, and initiators. Feldstein *et al.* demonstrated the formation of water-absorbing, elastic, and adhesive hydrogels through hydrogen bonding of three pharmaceutical grade components poly(*N*-vinylpyrrolidone) (PVP), PEG, and poly[(methacrylic acid)-*co*-(ethyl acrylate)] [p(MAA-*co*-EA)] without introduction or formation of toxic by-products. The hydrogels are malleable under various processing conditions such as drawing, molding, and extrusion, suggesting a wide range of applications in the biomedical and cosmetic fields [26].

9.1.4
Elasticity of Hydrogel for Tissue Application

Most natural tissues, such as heart, blood vessels, skeletal muscle, tendon, and so forth, are very elastic and strong. If the biodegradable polymers are either too stiff/brittle with low elongation, or very soft with relatively low strength, the mechanical properties of these polymers are not compatible with natural tissues.

The hydrogels are good candidates for tissue applications when their elastic moduli are close to that of natural tissue components. For instance, articular cartilage contains ~70% water and bears loads up to 100 MPa, but most hydrogels, either synthetic or natural, can be easily broken indicating that they are much weaker than native cartilage tissue. The degradable elastic polyurethane hydrogels have elastic moduli ranging from 16.8 ± 3.3 to 26.6 ± 3.9 MPa, which are very close to the properties of native cartilage showing promise for soft- and hard-tissue regeneration [17].

For engineering of soft tissue, elastic hydrogel scaffolds are desirable since they are amenable to mechanical conditioning regimens that might be desirable during tissue development. Elasticity values of most of the single component hydrogels were lower than 10 kPa, while higher percentage of multicomponent hydrogels exhibited high elastic mechanical property up to 100 kPa [3]. The compressive modulus of hard tissue such as articular cartilage is in the range of 0.53–1.82 MPa [27]. In order to promote cartilage regeneration, a hydrogel scaffold must be able to exhibit mechanical integrity in the face of loading from the body, while at the same time guide appropriate cartilaginous tissue growth. A biodegradable hydrogel scaffold with elastic properties could be useful for application in cartilage treatment.

9.2 Synthesis of Elastic Hydrogels

9.2.1 Chemical Elastic Hydrogels

Chemical hydrogels are those that have covalently crosslinked networks. Thus, chemical hydrogels will not dissolve in water or other organic solvents unless covalent crosslinks are cleaved. There are generally two different methods to prepare chemical elastic hydrogels. Chemical elastic hydrogels can be prepared by polymerization of water-soluble monomers in the presence of bi- or multifunctional crosslinking agents. Chemical hydrogels can also be prepared by crosslinking water-soluble polymers using chemical reactions that involve functional groups of the polymer. Due to the high strength of the covalent linkages, the three-dimensional networks of hydrogels are permanent and the formation of crosslinks is usually irreversible.

9.2.1.1 Polymerization of Water-Soluble Monomers in the Presence of Crosslinking Agents

Polymerization of water-soluble monomers in the presence of crosslinking agents results in the formation of chemical hydrogels. Typical water-soluble monomers for the preparation of chemical elastic hydrogels include acrylic acid, AM, hydroxyethyl methacrylate, and so on. The crosslinking agents for the synthesis of elastic hydrogels are not only low-molecular-weight agents such as *N,N′*-methylenebisacrylamide but also inorganic agents such as hectorite clay.

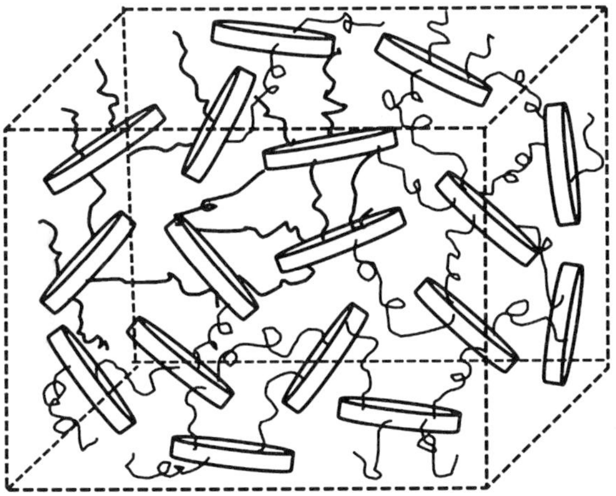

Figure 9.1 Structure of nanocomposite hydrogel using Clay-S by *in situ* polymerization.

For example, a novel highly resilient nanocomposite hydrogel with ultrahigh elongation was prepared by polymerization of monomer (AM or *N*-isopropylacrylamide (NIPAAm)) in the presence of the inorganic hectorite clay as a crosslinker (Clay-S), initiator (potassium persulfate), and accelerator (tetramethyldiamine) [28]. As shown in Figure 9.1, Clay-S forms a stable uniform dispersion in a solution that contains monomer and other reagents. Polymerization is initiated on the surfaces of the clay, and polymer chains are attached to the clay surface to form clay-brush particles, and finally, the aqueous dispersion is converted into a nanocomposite hydrogel of the uniform polymer network of Clay-S and AM, which can distribute stress evenly on each chain. The hydrogel could be elongated to 10 times of its original length and recovered to initial state.

In another approach, a hybrid of chemical and physical hydrogels was prepared from polyacrylamide and sodium alginate [14]. The copolymerization of AM monomer and *N,N′*-methylenebisacrylamide as a crosslinker and other necessary ingredients formed superporous polyacrylamide hydrogels. The crosslinking density of the hydrogel was increased by the physical crosslinking of sodium alginate with Ca^{2+}. The mechanical properties of the superporous hydrogels can be significantly increased through this interpenetrating network formation.

9.2.1.2 **Crosslinking of Water-Soluble Polymers**

Crosslinking of water-soluble polymers by the addition of bifunctional or multifunctional reagents results in chemical elastic hydrogels. Macromers are macromolecular monomers or polymers that contain two or more vinyl groups, acrylates and methacrylates being the most common. The crosslinking reactions can be catalyzed chemically, thermally, or photolytically. Photopolymerization is an increasingly common way to drive the crosslinking reaction.

Degradable polyurethane-based light-curable elastic hydrogels were synthesized from polycaprolactone diol, PEG as soft segment, lysine diisocyanate as hard segment, and 2-hydroxylethyl methacrylate as chain terminator through UV-light-initiated polymerization. The hydrogels were formed through the crosslinking of methacrylate groups in 2-hydroxylethyl methacrylate via UV light. The PCL:PEG ratios in soft segments were responsible in determining elasticity as well as the strength of the hydrogels [17].

The formation of degradable hydrogels by crosslinking macrodimethacrylates was also reported by Choi *et al.* [12]. Triblock copolymers of PLA–PPO–PLA containing polylactic acid (PLA) blocks and acrylate end groups of PPO were used to create photopolymerizable hydrogels showing shape-memory property.

Recently, the formation of elastic hydrogels from block copolymer of PEG and biodegradable polyesters has been extensively investigated. PEG is a hydrophilic polymer and its glass transition temperature is very low due to the flexible chain structure. When PEG was used as a building block for preparing hydrogels with other biodegradable polyesters such as PGA, PLA, and PCL, the hydrogels can show flexible and/or elastic properties [4, 24, 27]. PEG has two hydroxyl groups at both ends of the polymer that can be modified with a vinyl group to form a divinyl macromer. PEG acrylates are the major type of macromers for the preparation of PEG-based elastic hydrogels. For example, chemically crosslinked biodegradable elastic PEG/PCL or PLGA–PEG–PLGA/PEG hydrogels were prepared via radical crosslinking reaction of PEG-diacrylate with PCL-diacrylate or PLGA-PEG-PLGA-diacrylate in the presence of a radical initiator 2,2-azobisisobutyronitrile in a drying oven at 65 °C for 12 h [24]. Scheme 9.1 illustrates the synthetic method of PLGA-PEG-PLGA diacrylate using for crosslinking reaction with PEG-diacrylate under thermal catalyst.

9.2.2 Physical Elastic Hydrogels

Physical gels are the continuous, disordered, three-dimensional networks formed by associative forces capable of forming noncovalent crosslinks [29]. Noncovalently crosslinked hydrogels are formed when primary polymer chains contain chemical moieties capable of electrostatic, hydrogen bonding, ion dipole, or hydrophobic interaction [26]. Physical crosslinking of polymer chains can also be achieved using a variety of environment triggers (pH, temperature, and ionic strength). In physical elastic hydrogels, association of certain linear segments of long polymer molecules forms extended "junction zones." The junction zones are expected to maintain ordered structure. Although noncovalent association are reversible and weaker than chemical crosslinking, they allow solvent casting and thermal processing, and the resulting polymers often show elastic or viscoelastic properties [30].

9.2.2.1 Formation of Physical Elastic Hydrogels via Hydrogen Bonding

Examples of elastic and adhesive hydrogels via the formation of hydrogen bonding are triple blends of PVP, PEG, and p(MAA-*co*-EA). Ternary polymer blends were dissolved in ethanol under vigorous stirring, and then casted into film. The PVP/PEG/p(MAA-*co*-EA) hydrogel was formed via the stable three-dimensional

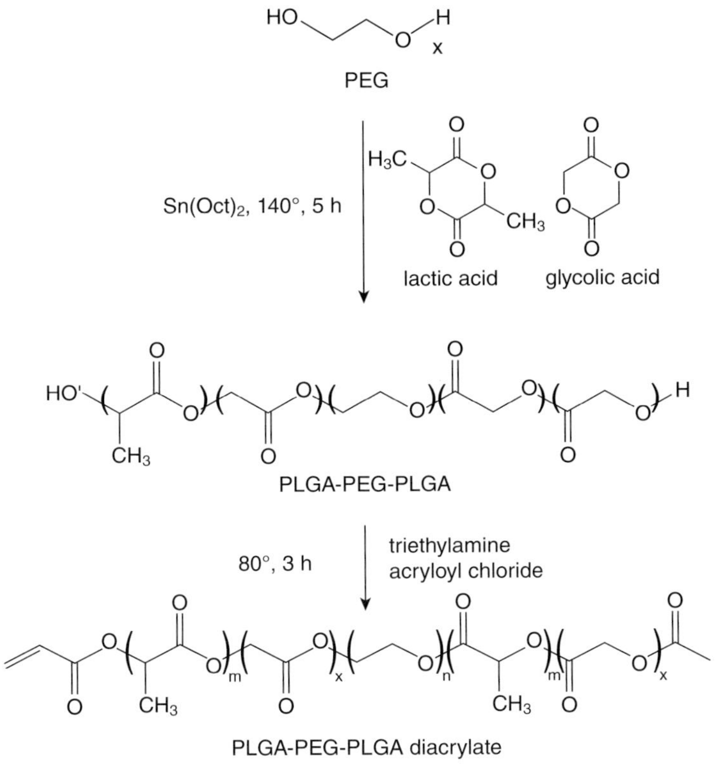

Scheme 9.1 Synthetic methods of PLGA–PEG–PLGA diacrylate.

hydrogen-bonded network in which p(MAA-*co*-EA) contains H-bond donor groups, PVP contains H-bond acceptors, and PEG contains both. The hydrogel films are malleable and retain their integrity upon hydration – a feature characteristic of covalently crosslinked hydrogels. The polymer blend films remained intact at pH 5.6 but underwent dissolution at pH 7.4 due to loss of hydrogen bonding and development of charge repulsion [26].

Hydrogen-bonding interaction can also be used to produce hydrogels by freeze-thawing. A novel double-network elastic hydrogel fabricated with PVP and PEG was prepared through a simple freezing and thawing method. PVA/PEG hydrogel structure was formed by a PVA-rich first network and a PEG-rich second component, in which hydrogen bonding existed. The two polymers were dissolved in ultrapure water and exposed to repeated cycles of freezing at −20 °C for 8 h and thawing at room temperature for 4 h. Figure 9.2 illustrates the structural formation of elastic PVA/PEG double-network hydrogels. The condensed PVA-rich phase forms microcrystals first, which bridge with one another to form a rigid and inhomogenous net backbone to support the shape of the hydrogels, and the dilute PEG-rich phase partially crystallizes among the cavities of voids of the backbone. PEG clusters in the cavities of PVA networks absorb the crack energy and relax

Figure 9.2 Schematic representation of the structural model of PVA/PEG double-network hydrogel.

the local stress either by various dissipations or by large deformation of the PEG chains. The crystalline regions of PVA essentially serve as physical crosslinks to redistribute external stresses [31].

9.2.2.2 **Formation of Physical Elastic Hydrogels via Hydrophobic Interaction**

Polymers with hydrophobic domains can crosslink in aqueous environment via reverse thermal gelation. Temperature increase promotes hydrophobic interactions resulting in the association of hydrophobic polymer chains. The physical association of hydrophobic domains holds swollen soft domains together and makes the polymers stable in water [32]. The common hydrophobic blocks which can undergo reverse thermal gelation at or near physiological temperature are PPO, PLGA, poly(*N*-isopropylacrylamide), PCL, and poly(urethane) [33].

For example, multiblock copolymers of polyethylene oxide and PCL or PLA were synthesized for the preparation of polymer films by solvent casting method. The multiblock copolymers formed thermoplastic hydrogels via hydrophobic interaction. The block copolymer films were rubbery in both dried and swollen states. The interesting property of these multiblock copolymers was that the swelling increased by increasing temperature and increased further, rather than decreasing, when the temperature was lowered to the initial temperature [30]. Other types of amphiphilic block copolymers of PCL with PLA and PGA were also synthesized to prepare elastic PCL/PLA and PCL/PGA physical hydrogels [4, 27].

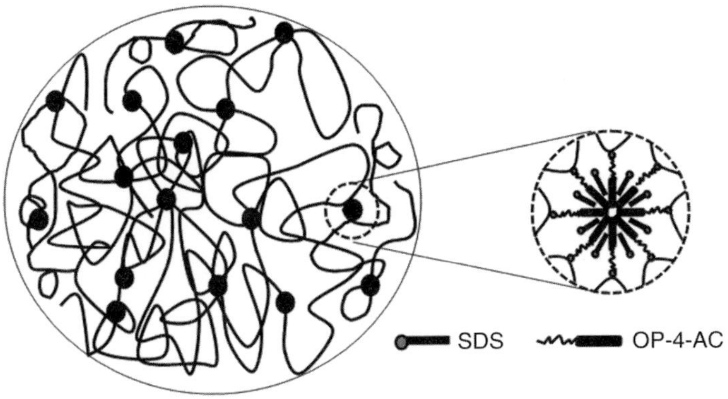

Figure 9.3 Schematic illustration of the hydrophobic association of hydrogels, which consists of associated micelles and flexible polymer chains connected by neighboring associated micelles.

In another example, a new type of physically crosslinked hydrogel via hydrophobic interaction was prepared. An elastic hydrogel with self-healing property was synthesized through micellar copolymerization of AM and a small amount of octylphenol polyethoxyether acrylate in an aqueous solution containing sodium dodecyl sulfate at 50 °C. The hydrophobically modified polyacrylamide was synthesized by the copolymerization of AM and octylphenol polyethoxyether acrylate. After polymerization, hydrophobic association of SDS and hydrophobic microblocks of hydrophobically modified polyacrylamide leads to the formation of associated micelles. These micelles act as crosslinking points, so three-dimensional polymer networks were constructed as shown in Figure 9.3 [25]. Because of the large distance between the associated micelles, all polymer chains between the crosslinking points in the hydrogels were sufficiently long and flexible.

9.3 Physical Properties of Elastic Hydrogels

Some of the most important properties of elastic hydrogels are: the gel mechanical properties, to withstand the physiological strains *in vivo* or mechanical conditioning *in vitro*; gel swelling properties to maintain cell viability; and the degradation profiles to match tissue regeneration.

9.3.1 Mechanical Property

Mechanical properties of elastic hydrogels are evaluated by the measurement of elasticity and stress relaxation. Elasticity is estimated from the tensile strength, elongation at break, and recovery after stretching. The mechanical tests are performed with hydrogel samples in a fixed cross-sectional area by pulling with a controlled, gradually increasing force until the sample changes shape or breaks. When a constant strain is applied to a rubber material, the force necessary to maintain that strain is not constant but decreases with time, this behavior is called "stress relaxation." Stress relaxation of hydrogels was determined by the following equation:

(maximum stress at a constant strain/stress at the constant strain after holding for a determined time) $\times 100$.

The tensile strength of elastic hydrogels is dependent on the crosslinking density and the flexibility of the water-soluble monomer or macromers in water. The elastic hydrogels exhibit rubberlike profiles in the stress–strain curves with very high elongation at break as shown in Figure 9.4. The recovery after stretching is usually more than 90% when applied up to the tensile strain at break.

Stress relaxation describes how polymers relieve stress under constant strain. In viscoelastic materials, stress relaxation occurs due to polymer chain rearrangement allowing permanent deformation of the materials. By this method, low values for stress relaxation indicate that polymer chain rearrangement is occurring. Example of stress relaxation of elastic hydrogel is given in Figure 9.5. The stress relaxation of all samples was more than 90%, indicating that polymer chain rearrangement is occurring only minimally. Since these hydrogels are highly crosslinked, there is little freedom for rearrangement and, as such, these materials do not deform under stress.

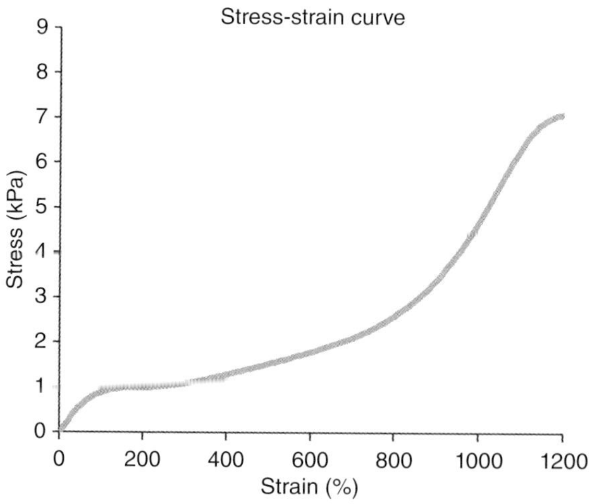

Figure 9.4 Stress–strain curve of elastic film prepared from poly(L-lactide-*co*-ε-caprolactone).

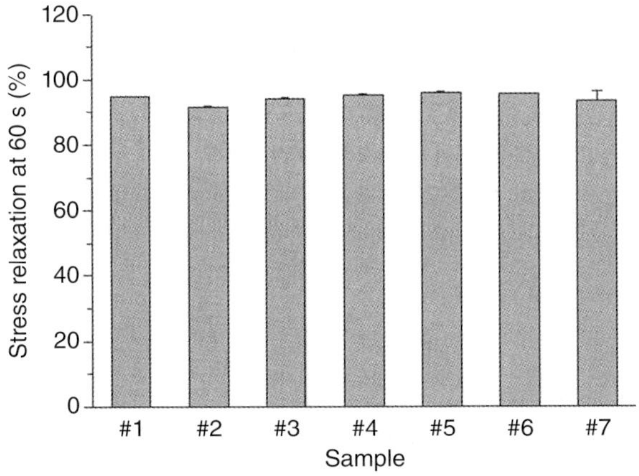

Figure 9.5 Stress relaxation of PLGA–PEG–PLGA/PEG elastic hydrogels.

9.3.2
Swelling Property

The swelling property of hydrogels is usually characterized by measuring their capacity to absorb water or aqueous solutions. The swelling ratio (R_s), which is the most commonly used parameter to express the swelling capacity of hydrogels, is defined as follows:

$$R_s = (W_s - W_d)/W_d \tag{9.1}$$

where W_s and W_d are the weights of swollen and dried hydrogels, respectively.

It is well known that the swelling ratio of hydrogels not only depends on the hydrophilic ability of the functional groups but also on the network space of the hydrogels. Generally, the hydrogel with higher network space presents higher water content. Ionization often provokes swelling due to electrostatic/osmotic repulsion of polyelectrolyte chains.

Swelling pressure is the pressure exerted on the swelling hydrogels due to the osmotic effect or the degradation of the crosslinking structure. The swelling pressure (π_{sw}) of a neutral polymer gel is determined by two opposing effects: the osmotic pressure (π_{osm}) that expands the network and the elastic pressure (π_{el}) that acts against expansion [34]:

$$\pi_{sw} = \pi_{osm} + \pi_{el} \tag{9.2}$$

where $\pi_{el} = -G'$, G' being the elastic (shear) modulus of the hydrogel.

Swelling pressure is usually measured during the degradation of hydrogels at the accelerated condition close to physiological condition, for example, in the isotonic solution of 0.154 M HCl. Up to present, no relationship between swelling pressure and swelling ratio of neutral hydrogels has been reported. However, it is known that swelling pressure gradually increases in the course of the degradation process and depends on the mechanism of the degradation. For example, the dextran gels are degraded at their backbone, the swelling pressure increases rather continuously; in the case where they are degraded at the crosslinks, it increases more discontinuously and a sudden increase occurs when the gels are completely degraded [35]. Similarly, an increase in swelling ratio of hydrogels occurs in the first period of the hydrogel degradation due to the decrease of the crosslinking density of hydrogels. As the degradation proceeded further, the swelling ratio decreases to zero since the network structure of hydrogels broke down. The swelling ratio and the swelling pressure of a hydrogel depend on internal and external factors. The internal factors are the polymer network of the hydrogel. The swelling ratio and the swelling pressure of a gel are determined by: (i) osmotic pressure, (ii) the rubber elasticity, and (iii) polymer–polymer interaction of the polymer network [36]. The external factors are parameters of the solution like the concentration and electric charge of the solute molecules or ions.

Swelling pressure of hydrogels is also measured under the osmotic condition to determine the ability of the hydrogels as a tissue expander. Osmosis is the main driving force of the volume expansion of an anhydrous gel body in the solutions of living tissue [36]. Mechanical work can be done when hydrogels transform osmosis into real pressure. In this case, the role of a gel to act as a pressure-generating device is based on the balancing of the osmotic pressure and the rubber elasticity. If the swelling pressure can overcome the resistance of the adjacent tissue, it is sufficient to dilate the surrounding tissue at the expected rate [37]. Previous research has established that a pressure close to 100 mmHg is ideal for tissue expansion [38]. As an example, the swelling pressure of PLGA–PEG–PLGA/PEG elastic hydrogels is given in Figure 9.6. All these hydrogels can create swelling pressure more than 400 mmHg which is sufficient for tissue expansion.

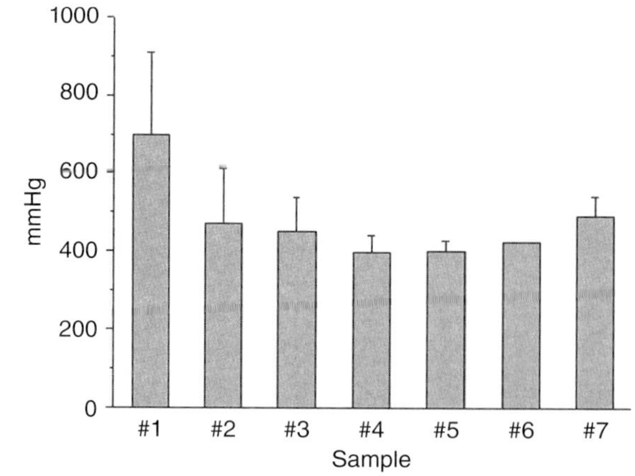

Figure 9.6 Swelling pressure of PLGA–PEG–PLGA/PEG elastic hydrogels.

9.3.3 Degradation of Biodegradable Elastic Hydrogels

Degradation of polymer hydrogel networks can occur by different mechanisms: (i) by hydrolysis of side chains or pendant groups, (ii) by cleavage of the polymeric backbone, and (iii) by cleavage of labile groups in the crosslinks [35]. Biodegradation of hydrogels occurs by either simple hydrolysis or by enzyme-catalyzed hydrolysis. *In vitro* degradation rate of biodegradable elastic hydrogels was generally evaluated by measuring the weight loss and swelling ratio of the samples in phosphate buffered saline solution at multiple time points at 37 °C with gentle shaking to mimic the *in vivo* environment. The degradation of the gel is a function of the crosslink density, as well as the hydrolytic susceptibility of chemical bonds. Hydrogels made from lower molecular weight precursors are more tightly crosslinked and thus degrade more slowly than hydrogels made from higher molecular weight precursors. Degradation of the biodegradable hydrogel network led to decreased crosslinking density, which increased the hydrogel swelling ratio. Figure 9.7 gives an example of *in vitro* degradation properties of PLGA-PEG-PLGA/PEG elastic hydrogels at different ratios of lactic acid/glycolic acid. The hydrogels showed various lag times before swelling depending on the chemical composition of the triblocks and the PLGA-PEG-PLGA/PEG block composition ratio. The hydrogels with higher content of PEG block showed lower degradation rates due to higher crosslinking density of low-molecular-weight PEG. As the degradation proceeded further, the network structure finally broke down so that the hydrogel mass was disintegrated into soluble degradation products.

9.4 Applications of Elastic Hydrogels

9.4.1 Tissue Engineering Application

Due to the high mechanical property, most biodegradable elastic hydrogels are attractive for development or regeneration of both soft- and hard tissues. For example, PEG/PCL and poly(lactide-*co*-caprolactone) (PLCL) elastic hydrogels have been investigated for use as scaffolds for cartilage regeneration [27, 39]. Chondrocyte cells were found to be dispersed evenly through the scaffold material without any further prewetting treatment, and remained viable after 3 weeks of culture. The PEG/PCL scaffold-seeded chondrocytes enhanced the gene expression of chondrogenic differentiation in a time-dependent manner. The formation of neocartilage was increased over 4 weeks after implantation in nude mice [39]. The elastic PLCL scaffolds maintained their mechanical integrity after implantation and guided cartilaginous tissue growth *in vivo* [27].

The application of cyclic mechanical strain during the smooth muscle tissue-engineering process has been found to show increased elastin and collagen

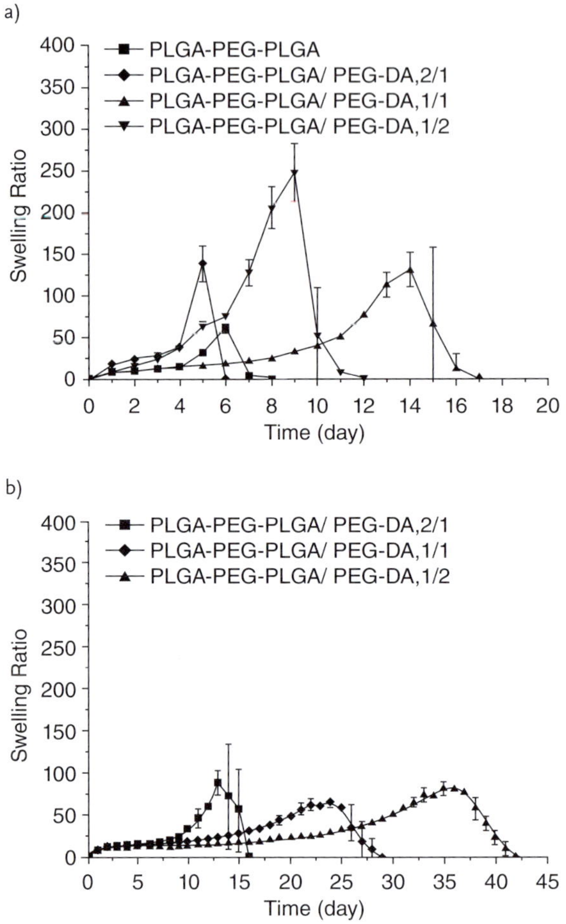

Figure 9.7 *In vitro* degradation test of PLGA–PEG–PLGA/PEG elastic hydrogels (a) LA/GA = 1 and (b) LA/GA = 4 at 37 °C.

production and tissue organization [40]. To achieve this, scaffolds must be elastic and capable of withstanding cyclic mechanical strain without cracking or suffering significant permanent deformation. Elastic biodegradable poly(glycolide-*co*-caprolactone), PLCL, and polyurethane scaffold could be used to engineer smooth-muscle-containing tissue (e.g., blood vessels and bladders) in mechanical dynamic environments [4, 5, 41]. The elastic scaffolds allowed for appropriate smooth muscle cell adhesion and subsequent tissue formation.

9.4.2
Application of Elastic Shape-Memory Hydrogels as Biodegradable Sutures

The medical application of shape-memory polymers is of great interest due to a combination of biocompatibility, tailorable transition temperature, large shape deformation and complete recovery, and elastic properties of the materials [42]. For example, the mechanical characteristics and degradability of shape-memory, multiblock copolymers can be used for the preparation of smart surgical suture. Lendlein and Langer fabricated a self-tightenable biodegradable suture from a biodegradable, elastic shape-memory polymer. The suture can be loosely connected and then heated above critical temperature to trigger the shape recovery and tighten the suture [11].

9.5 Elastic Hydrogels for Tissue Expander Applications

A material or device designed to induce skin or tissue expansion for the purpose of reconstructive and plastic surgeries has been called a tissue expander. Tissue expanders are temporary inflatable implants that are positioned under the skin to facilitate the increase of tissue dimensions for reconstruction [43]. As an example, Figure 9.8 illustrates a schematic diagram of a skin expander using a flatable balloon.

An ideal expander should have several characteristics: easily placed through a small access site, gradually enlarge over a relatively short time, well tolerated over the long term, avoid uncomfortable inflation spikes, and resistant to infection [44]. In 1982, Austad and Rose introduced a self-inflating expander that consisted of a permeable silicone membrane filled with a hypertonic saline solution [45]. However, the expansion of the silicone balloon takes too long (8–14 weeks) and induces tissue necrosis [46].

The use of hydrogels as tissue expanders in reconstructive surgery was first developed in 1992 by Downes *et al.* who exploited the osmotically driven expansion of a biocompatible poly(hydroxylethyl methacrylate) hydrogel [47]. The hydrogels are placed in their dry, contracted states, and expand gradually to their full size with over 10-fold increase in volume. Wiese verified that hydrogels are efficient materials to induce the tissue expansion using vinyl-2-pyrrolidone (VP)/methyl methacrylate (MMA) copolymeric hydrogel and demonstrating their biocompatibility and swelling pressure [46]. Once implanted, the VP/MMA absorbs body fluids that leads to gradual swelling of the device to a 250–300% in volume as shown in Figure 9.9. Wiese *et al.* also introduced the innovative self-filling device, using a hydrogel matrix consisting of MMA/VP by replacing the CH_3 groups in

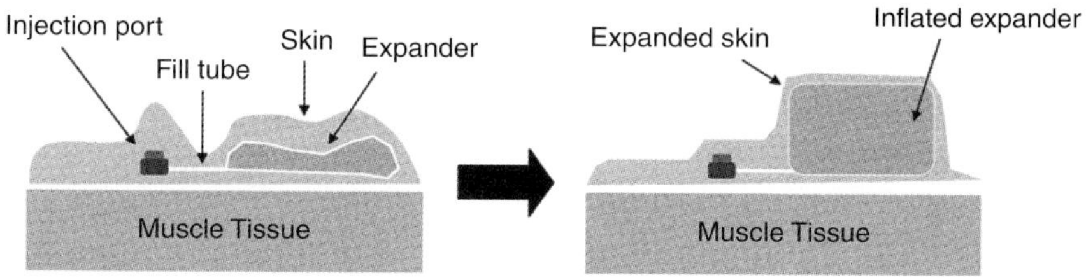

Figure 9.8 Schematic diagram of skin expander using an inflatable balloon.

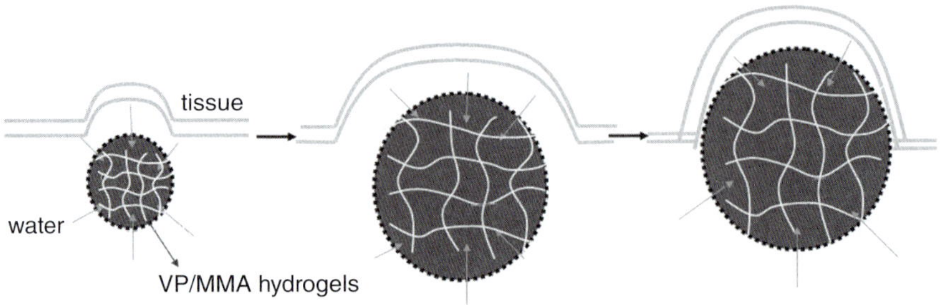

Figure 9.9 Use of VP/MMA copolymeric hydrogels for tissue expansion.

the VP/MMA hydrogel chains with COOH groups, which produced higher swelling than VP/MMA hydrogels [36]. The biocompatibility of VP/MMA hydrogel tissue expander was proved through *in vivo* test using rats; the hydrogels swell and reach their equilibrium swelling rate in 6–8 weeks by absorbing body fluids. This long *in vivo* swelling rates not only avoid tissue necrosis but also allow sufficient time for tissue growth rather than stretching of the skin [48]. Although such attempts use another material for tissue expansion, most clinically used hydrogels are still based on VP and MMA copolymers [43, 48–50].

Recently, Varga *et al.* developed thermosensitive hydrogels consisting of NIPAAm as osmotic tissue expanders. A silicate was added to improve the mechanical behavior of the polymer hydrogel. The rate of hydrogel expansion *in vivo* was highest after 2 weeks without the tissue damage. The hydrogel achieved a 25-fold increase in mass. NIPAAm polymers exhibited the most favorable viscoelastic properties, with the highest tendency to retain their preformed shape [37]. Thus, NPIAAm hydrogels allow the acquisition of more skin for reconstructive interventions. However, the current expanders lack the ability to have their shape and size changed at the time of implantation because these hydrogels are glassy and brittle in the dry state. Therefore, there has been a need to develop flexible and elastic tissue expanders so that they can be easily handled or modified appropriately according to each application. Biodegradable elastic PCL/PEG, PLA–PEG–PLA/PEG, and PLGA–PEG–PLGA/PEG hydrogels were developed for this purpose (Figure 9.10).

All the PLGA–PEG–PLGA/PEG hydrogels were flexible and elastic in dried state, and so they remain intact even after repeated bending or stretching. The hydrogels are able to generate sufficient swelling pressure (more than 400 mmHg) to expand tissue. The actual *in vivo* pressures will be substantially lower than this static condition as the skin and mucosa can stretch reducing the pressure. Furthermore, the PLGA–PEG–PLGA/PEG hydrogels exhibited the lag time before swelling; this will provide sufficient time for the wounded area to heal. The controllable degradation rate makes it possible to apply the hydrogels to various parts of body. The elastic hydrogels with self-inflating behavior, elastic, and delayed swelling properties would be useful as novel hydrogel tissue expanders (Figure 9.11).

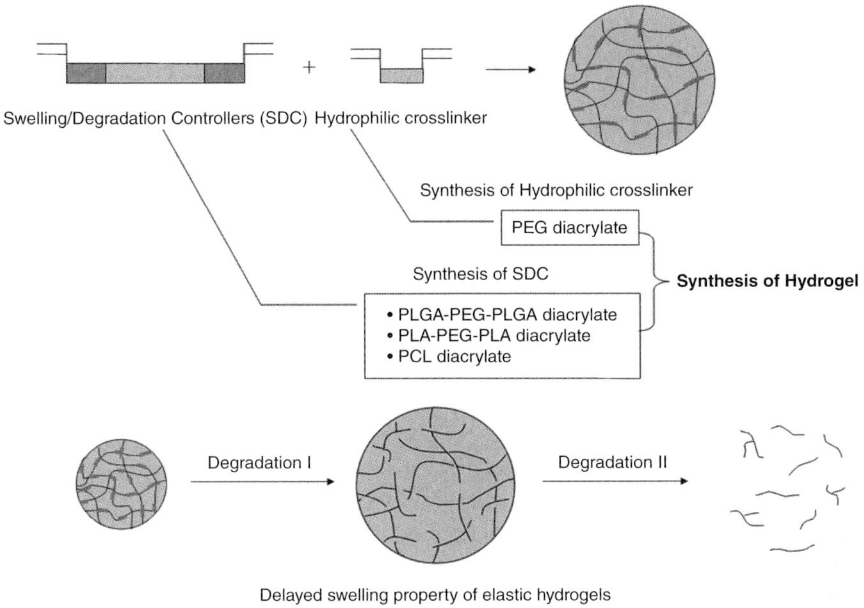

Figure 9.10 Elastic hydrogel tissue expanders with controllable swelling/degradation.

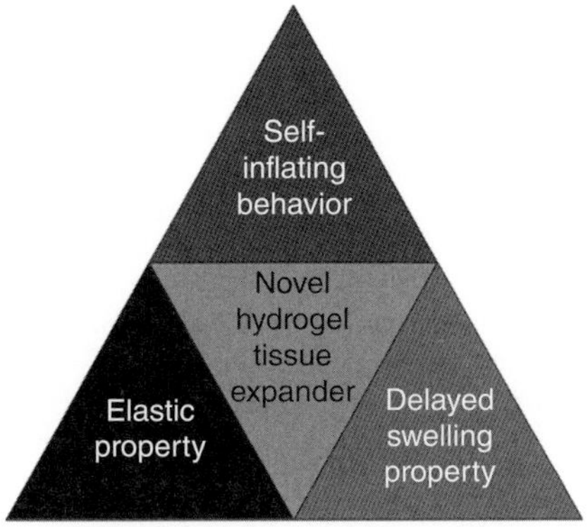

Figure 9.11 A concept of novel hydrogels for tissue expander application.

9.6 Conclusion

Although a number of attractive hydrogel systems are presently available, there are certainly novel systems with improved characteristics. A major concern in the hydrogel development is the mechanical integrity of the systems under modification and processing. Elastic hydrogels can be formed with varying polymer formulations in three-dimensional patterns. Both chemically and physically crosslinked elastic hydrogels can be rendered biodegradable through the introduction of hydrolytically sensitive groups into the networks. Due to their biocompatibility, biodegradability, and good mechanical properties, biodegradable elastic hydrogels are good candidates as biomaterials for use in medical applications, including tissue engineering. These hydrogels have been used as biodegradable sutures and scaffold materials to engineer various types of tissues in mechanical dynamic environments. Elastic hydrogels with described properties are promising expander candidates which may contribute to more effective harvesting of tissue for reconstructive interventions. However, the synthesis of biodegradable hydrogels with rubberlike elasticity and strength is not easy. Moreover, *in vivo* tests should be done to improve the clinical applicability of elastic hydrogels for tissue expansion as well as other medical applications.

References

1 Hennink, W.E. and Van Nostrum, C.F. (2002) Novel crosslinking methods to design hydrogels. *Adv. Drug Deliv. Rev.*, **54**, 13–36.

2 Kopecek, J. (2007) Hydrogel biomaterials: a smart future. *Biomaterials*, **28**, 5185–5192.

3 Yang, Z., Wang, L., Wang, J., Gao, P., and Xu, B. (2010) Phenyl groups in supramolecular nanofibers confer hydrogels with high elasticity and rapid recovery. *J. Mater. Chem.*, **20**, 2128–2132.

4 Lee, S.H., Kim, B.S., Kim, S.H., Choi, S.W., Jeong, S.I., Kwon, I.K., Kang, S.W., Nikolovski, J., Mooney, D.J., Han, Y.K., and Kim, Y.H. (2003) Elastic biodegradable poly(glycolide-*co*-caprolactone) scaffold for tissue engineering. *J. Biomed. Mater. Res. A*, **66** (**1**), 29–37.

5 Guan, J., Fujimoto, K.L., Sacks, M.S., and Wagner, W.R. (2005) Preparation and characterization of highly porous, biodegradable polyurethane scaffolds for soft tissue applications. *Biomaterials*, **26**, 3961–3971.

6 Chaterji, S., Kwon, I.K., and Park, K. (2007) Smart polymeric gels: redefining the limits of biomedical devices. *Prog. Polym. Sci.*, **32**, 1083–1122.

7 Kamath, K.R. and Park, K. (1993) Biodegradable hydrogels in drug delivery. *Adv. Drug Deliv. Rev.*, **11**, 59–84.

8 Osada, Y. and Matsuda, A. (1995) Shape memory in hydrogels. *Nature*, **376**, 219.

9 Matsuda, A., Sato, J., Yasunaga, H., and Osada, Y. (1994) Order–disorder transition of a hydrogel containing an *n*-alkyl acrylate. *Macromolecules*, **27**, 7695–7698.

10 Lendlein, A., Schmidt, A.M., and Langer, R. (2001) AB-polymer networks based on oligo(ε-caprolactone) segments showing shape-memory properties. *Proc. Natl. Acad. Sci. USA*, **98** (**3**), 842–847.

11 Lendlein, A. and Langer, R. (2002) Biodegradable, elastic shape-memory polymers for potential biomedical applications. *Science*, **296**, 1673–1676.

12 Choi, N.Y., Kelch, S., and Lendlein, A. (2006) Synthesis, shape-memory functionality and hydrolytical degradation studies on polymer networks from poly(*rac*-lactide)-*b*-poly(propylene oxide)-*b*-poly(*rac*-lactide) dimethacrylate. *Adv. Eng. Mater.*, **8** (**5**), 439–445.

13 Zheng, X., Zhou, S., Li, X., and Weng, J. (2006) Shape memory properties of poly(D,L-lactide)/hydroxyapatite composite. *Biomaterials*, **27**, 4288–4295.

14 Omidian, H., Rocca, J.G., and Park, K. (2006) Elastic, superporous hydrogel hybrids of polyacrylamide and sodium alginate. *Macromol. Biosci.*, **6**, 703–710.

15 Qui, Y. and Park, K. (2003) Superporous IPN hydrogels having enhanced

mechanical properties. *AAPS Pharm. Sci. Technol.*, **4** (**4**), 406–412.

16 Zhang, C., Zhang, N., and Wen, X. (2006) Improving the elasticity and cytophilicity of biodegradable polyurethane by changing chain extender. *J. Biomed. Mater. Res. Part B Appl. Biomater.*, **79** (**2**), 335–344.

17 Zhang, C., Zhang, N., and Wen, X. (2007) Synthesis and characterization of biocompatible, degradable, light-curable, polyurethane-based elastic hydrogels. *J. Biomed. Mater. Res. A*, **82** (**3**), 637–650.

18 Zhou, S., Deng, X., and Yang, H. (2003) Biodegradable poly(ε-caprolactone)-poly(ethylene glycol) block copolymers: characterization and their use as drug carriers for a drug carriers for a controlled delivery system. *Biomaterials*, **24**, 3563–3570.

19 Aamer, K.A., Sardinha, H., Bhatia, S.R., and Tew, G.N. (2004) Rheological studies of PLLA. PEO. PLLA triblock copolymer hydrogels. *Biomaterials*, **25**, 1087–1093.

20 Moth, C.G., Drumond, W.S., and Wang, S.H. (2006) Phase behavior of biodegradable amphiphilic poly(*l,l*-lactide)-*b*-poly(ethylene glycol)-*b*-poly(*l,l*-lactide). *Thermochim. Acta*, **445**, 61–66.

21 Chen, S., Pieper, R., Webster, D.C., and Singh, J. (2005) Triblock copolymers: synthesis, characterization, and delivery of a model protein. *Int. J. Pharm.*, **288**, 207–218.

22 Duvvuri, S., Janoria, K.G., and Mitra, A.K. (2005) Development of a novel formulation containing poly(*d,l*-lactide-*co*-glycolide) microspheres dispersed in PLGA. PEG. PLGA gel for sustained delivery of ganciclovir. *J. Control. Release*, **108**, 282–293.

23 Qiao, M., Chen, D., Ma, X., and Liu, Y. (2005) Injectable biodegradable temperature responsive PLGA–PEG–PLGA copolymers: synthesis and effect of copolymer composition on the drug release from the copolymer-based hydrogels. *Int. J. Pharm.*, **294**, 103–112.

24 Im, S.J., Choi, Y.M., Subramanyam, E., and Huh, K.M. (2007) Synthesis and characterization of biodegradable elastic hydrogels based on poly(ethylene glycol) and poly(ε-caprolactone) blocks. *Macromol. Res.*, **15** (**4**), 363–369.

25 Jiang, G., Liu, C., Liu, X., Zhang, G., Yang, M., Chen, Q., and Liu, F. (2010) Self-healing mechanism behavior of hydrophobic association hydrogels with high mechanical strength. *J. Macromol. Sci. Part A: Pure Appl. Chem.*, **47**, 335–342.

26 Bayramov, D.F., Singh, P., Cleary, G.W., Siegel, R.A., Chalykh, A.E., and Feldstein, M.M. (2008) Non-covalently crosslinked hydrogels displaying a unique combination of water-absorbing, elastic and adhesive properties. *Polym. Int.*, **57**, 785–790.

27 Jung, Y., Kim, S.H., You, H.J., Kim, S.H., Kim, Y.H., and Min, B.G. (2008) Application of an elastic biodegradable poly(L-lactide-*co*-ε-caprolactone) scaffold for cartilage tissue regeneration. *J. Biomater. Sci. Polym. Ed.*, **19** (**8**), 1073–1085.

28 Zhu, M., Liu, Y., Sun, B., Zhang, W., Liu, X., Yu, H., Zhang, Y., Kuckling, D., and Adler, H.J.P. (2006) A novel highly resilient nanocomposite hydrogel with low hysteresis and ultrahigh elongation. *Macromol. Rapid Commun.*, **27**, 1023–1028.

29 Park, K., Shalaby, W.S.W., and Park, H. (1993) Physical gels, in *Biodegradable Hydrogels for Drug Delivery*. Technomic Publishing Co., Inc., Lancaster, PA, pp. 99–129.

30 Bae, Y.H., Huh, K.M., Kim, Y., and Park, K.H. (2000) Biodegradable amphiphilic multiblock copolymers and their implication for biomedical applications. *J. Control. Release*, **64**, 3–13.

31 Zhang, X., Guo, X., Yang, S., Tan, S., Li, X., Dai, H., Yu, X., Zhang, X., Weng, N., Jian, B., and Xu, J. (2009) Double-network hydrogel with high mechanical strength prepared from two biocompatible polymers. *J. Appl. Polym. Sci.*, **112**, 3063–3070.

32 Huh, K.M. and Bae, Y.H. (1999) Synthesis and characterization of poly(ethylene glycol)/poly(L-lactic acid) alternating multiblock copolymers. *Polymer*, **40**, 6147–6155.

33 Hoare, T.R. and Kohane, D.S. (2008) Hydrogels in drug delivery: progress and challenges. *Polymer*, **49**, 1993–2007.

34 Van Thienen, T.G., Horkay, F., Braeckmans, K., Stubbe, B.G., Demeester, J., and De Smedt, S.C. (2007) Influence of free chains on the swelling pressure of PEG-HEMA and dex-HEMA hydrogels. *Int. J. Pharm.*, **337**, 31–39.

35 Stubbe, B.G., Hennink, W.E., De Smedt, S.C., and Demeester, J. (2004) Swelling pressure of hydrogels that degrade through different mechanisms. *Macromolecules*, **37**, 8739–8744.

36 Wiese, K.G., Heinemann, D.E.H., Ostermeier, D., and Peters, J.H. (2001) Biomaterial properties and biocompatibility in cell culture of a novel self-inflating hydrogel tissue expander. *J. Biomed. Mater. Res.*, **54** (2), 179–188.

37 Varga, J., Janovak, L., Varga, E., Eros, G., Dekany, I., and Kemeny, L. (2009) Acrylamide, acrylic acid and *N*-ispropylacrylamide hydrogels as osmotic tissue expander. *Skin Pharmacol. Physiol.*, **22**, 305–312.

38 Min, Z., Svensson, H., and Svedman, P. (1988) On expander pressure and skin blood flow during tissue expansion in the pig. *Ann. Plast. Surg.*, **21**, 134–139.

39 Park, J.S., Woo, D.G., Sun, B.K., Chung, H.M., Im, S.J., Choi, Y.M., Park, K., Huh, K.M., and Park, K.H. (2007) *In vitro* and *in vivo* test of PEG/PCL-based hydrogel scaffold for cell delivery application. *J. Control. Release*, **124**, 51–59.

40 Kim, B.S., Nikolovski, J., Bonadio, J., and Mooney, D.J. (1999) Cyclic mechanical strain regulates the development of engineered smooth muscle tissue. *Nat. Biotechnol.*, **17**, 979–983.

41 Kim, S.H., Kwon, J.H., Chung, M.S., Chung, E., Jung, Y., Kim, S.H., and Kim, Y.H. (2006) Fabrication of a new tubular fibrous PLCL scaffold for vascular tissue engineering. *J. Biomater. Sci. Polym. Ed.*, **17** (12), 1359–1374.

42 Liu, C., Qin, H., and Mather, P.T. (2007) Review of progress in shape-memory polymers. *J. Mater. Chem.*, **17**, 1543–1558.

43 Kobus, K.F. (2007) Cleft palate repair with the use of osmotic expanders: a preliminary report. *J. Plast. Reconstr. Aesthet. Surg.*, **60**, 414–421.

44 Mazzoli, R.A., Raymond, W.R., Ainbinder, D.J., and Hansen, E.A. (2004) Use of self-expanding, hydrophilic osmotic expanders (hydrogel) in the reconstruction of congenital clinical anophthalmos. *Curr. Opin. Ophthalmol.*, **15**, 426–431.

45 Austad, E.D. and Rose, G.L. (1982) A self-inflating tissue expander. *Plast. Reconstr. Surg.*, **70**, 588–593.

46 Wiese, K.G. (1993) Osmotically induced tissue expansion with hydrogels: a new dimension in tissue expansion: a preliminary report. *J. Craniomaxillofac. Surg.*, **21**, 309–313.

47 Downes, R., Lavin, M., and Collin, R. (1992) Hydrophilic expanders for the congenital anophthalmic socket. *Adv. Ophthalmic Plast. Reconstr. Surg.*, **9**, 57–61.

48 Swan, M.C., Tim, E.E., Goodacre, J., Czernuszka, T., and David, G.B. (2008) Cleft palate repair with the use of osmotic expanders: a response. *J. Plast. Reconstr. Aesthet. Surg.*, **61**, 220–221.

49 Ronert, M.A., Hofheinz, H., Manassa, E., Asgarouladi, H., and Olbrisch, R.R. (2004) The beginning of a new era in tissue expansion: self-filling osmotic tissue expander – four-year clinical experience. *Plast. Reconstr. Surg.*, **114** (5), 1025–1031.

50 Von See, C., Gellrich, N.C., Jachmann, U., Bormann, H., and Rucker, M. (2010) Bone augmentation after soft-tissue expansion using hydrogel expanders: effects on microcirculation and osseointegration. *Clin. Oral Implants Res.*, **21** (8), 842–847.

10
Biodegradable Dendrimers and Dendritic Polymers

Jayant Khandare and Sanjay Kumar

10.1
Introduction

The concept of using a polymer as a carrier for drug delivery system originated from the hypothesis that macromolecules could be used to improve the solubility and half-life of small molecule drugs [1, 2]. Later, it was observed that macromolecules functionalized with a drug in the form of prodrug impart added advantage by increasing accumulation in tumor tissues due to the leaky vasculature, now a concept recognized as *enhanced permeation and retention* effect [3, 4]. It has been clearly demonstrated that the macromolecular carriers have immense potential to enhance pharmacokinetics, leading to enhance the efficacy of small molecule drugs. Several carrier systems have been studied (viz., linear polymers, micellar assemblies, liposomes, polymersomes, and dendrimers) and are observed to have most of the properties required for ideal drug carrier [5]. Thus, it is not surprising that the ideal drug carrier would facilitate long blood circulation time, high accumulation in tumor tissue, high drug loading, lower toxicity, and simplicity in preparation. Within the milieu of nanocarriers, dendrimers represent a fascinating platform because of their nanosize, monodisperisty, and degree of branching to facilitate the multiple attachments of both drugs and solubilizing groups [6].

Dendrimers are excellent candidates for providing a well-defined molecular architecture, which is a result of a stepwise synthetic procedure consisting of coupling and activation steps [7]. They consist of branched, wedge-like structures called dendrons that are attached to a multivalent core, and emerge readily toward the periphery. The architecture and synthetic routes result in highly defined dendritic structure with polydispersity index near 1.00, as opposed to the much higher polydispersity of linear or hyperbranched structures [5]. The flexibility to tailor both the core and surface of these systems create them innovative *nanovehicle*, since different groups can be provided so as to optimize the properties of drug carrier. For instance, the functional periphery is one of the intriguing properties of dendritic architecture with extensive number of end groups that may be modified to afford dendrimers with tailored chemical and physical properties [8, 9]. The general

Handbook of Biodegradable Polymers: Synthesis, Characterization and Applications, First Edition. Edited by Andreas Lendlein, Adam Sisson.
© 2011 Wiley-VCH Verlag GmbH & Co. KGaA. Published 2011 by Wiley-VCH Verlag GmbH & Co. KGaA.

methods of synthesizing dendrimers are classified into (i) convergent and (ii) divergent approaches. The synthesis process involves repetitive coupling and activation steps, which makes it difficult to obtain dendrimers in high yield, at reasonable cost. These barriers have definitely limited the application of dendrimers primarily in biomedical field [7].

Dendrimers differentiate themselves largely from hyperbranched polymers in terms of their controlled size and shapes as well as narrow polydispersity [9]. Conversely, in linear polymers, the influence of end groups on physical properties such as solubility and thermal behavior is negligible at infinite molecular weight. However, in dendritic polymers, the situation is quite different. The fraction of end groups approaches a final and constant high value at infinite molecular weight, and therefore, the nature of the end groups is expected to strongly influence both the solution and the thermal properties of a dendrimer [10]. An explosion of interest has been fueled due to chemicophysical properties in dendritic macromolecules to be versatile nanomaterials, such as peripheral reactive end groups, viscosity, or thermal behavior, and differ significantly from those of linear polymers [11]. Till date, a variety of hyperbranched dendrimers and their polymeric architectures (e.g., polyglycerol (PG) dendrimers) have been implicated for diverse applications in the form of drug encapsulation, catalysis, and polymerization initiators [12–14].

This chapter highlights an overview on biodegradable dendrimers. More specifically, design of biodegradable dendritic architectures has been discussed keeping focus on challenges in designing such dendrimers; their relation of biodegradability and biocompatibility, and its biological implications.

Tomalia and Newkome *et al.* introduced well-defined and highly branched dendrimers [5, 15], and almost a decade later, the first form of biodegradable dendrimer was simultaneously published by various groups [16–18]. Groot *et al.* reported a biodegradable form of dendrimers that have been built to completely and rapidly dissociate into separate building blocks upon a single triggering event in the dendritic core [17]. These dendrimers collapse into their separate monomeric building blocks after single (chemical or biological) activation step that triggers a cascade of self-elimination reactions, thereby releasing the entire end groups from the periphery of the "exploding" cascade-release dendrimer. Thus, such multiple-releasing dendritic systems have been termed as "cascade-releasing dendrimers." The degrading dendritic system possesses two major advantages over the conventional dendrimers: (i) multiple covalently bound drug molecules can be site-specifically released from the targeting moiety by a single cleaving step, and (ii) they are selectively as well as completely degraded and therefore can be easily drained from the body [17].

Fascinatingly, Suzlai *et al.* demonstrated that the linear dendrimer could undergo *self-fragmentation* through a cascade of cleavage reactions initiated by a single triggering event [18]. The degradation of dendrimer cleavage eventually leads to two subsequent fragmentations per subunit, or geometric dendrimer disassembly. Overall, the concept of "dendritic amplification" was disclosed, in which an initial stimulus triggers the efficient disassembly of a dendrimer resulting in the ampli-

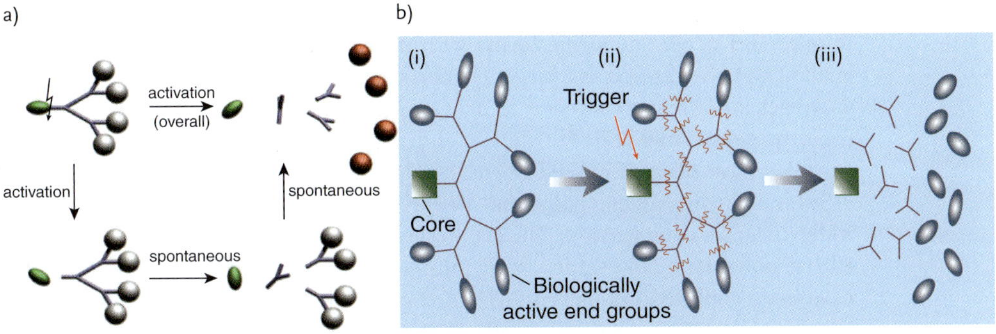

Figure 10.1 (a) Single activation of a second-generation cascade-releasing dendrimer triggers a cascade of self-eliminations and induces release of all end groups. Covalently bound end groups are depicted in gray, branched self-elimination linkers in blue, and the specified in green. The released end groups are depicted in red [17]. (b) Schematic of simultaneous release of biologically active end groups from trigger-tuned dendrimer: (i) dendrimer consists of two-dimensional part of a sphere, (ii) dendrimer is triggered with a specific signal so that the dendrimer scaffold falls apart in a chain of reactions, and (iii) the net result is observed with release of all molecules, including the end groups. In the experiments of de Groot *et al.*, the end groups represented are the anticancer drug paclitaxel [17, 19].

fication of a certain property or quality of a system due to the large increase in molecular species (dendrimer fragments) [18].

Degradable dendritic architectures mainly consist of the following classes:

1) dendrimers with degradable backbones (pH labile, enzymatic hydrolysis, etc.),
2) dendritic cores with cleavable shells (pH environment), and
3) cleavable dendritic prodrug forms.

Typically, the dendritic skeleton can be degraded or hydrolyzed based on environmental or external stimuli, for example, pH, hydrolysis, or by enzymatic degradation. Meijer and van Genderen reported that the dendrimer skeleton can be constructed in such a way that it can disintegrate into known molecular fragments once the disintegration process has been initiated (Figure 10.1a and b) [17, 19]. The dendrimers scaffold can fall apart in several steps in a chain reaction, releasing all of its constituent molecules by a single trigger. This has been demonstrated by de Groot *et al.* to achieve the release of the anticancer drug paclitaxel. Furthermore, the by-products of dendrimers degradation have proven to be noncytotoxic except for the drug paclitaxel itself [17]. The simultaneous release of biologically active end groups from a trigger-tuned dendrimer is represented in Figure 10.1. With single activation of a second generation, cascade-releasing dendrimer can trigger a cascade of self-eliminations and induces release of all end groups (Figure 10.1a). On the other hand, other forms of dendrimers can be triggered by a specific signal, and the dendrimer scaffold can fall apart in a chain of reactions. Notably, the first reaction activates the dendrimer's core, thereby initiating a cascade of "elimination" reactions leading to release of drug molecules (Figure 10.1b). The dendritic forms with many identical units mean that amplification can be achieved as a kind of explosion. However, there could be a possible drawback since if such a system is activated at the wrong time or place, the result could be devastating [17]. The details of design and synthesis of such degradable scaffolds have been discussed in the text below.

Several biodegradable polymers, dendrimers, and their prodrugs have been widely used as drug carriers [20, 21]. Recently, dendrimer carriers based on polyethers, polyesters, polyamides, melanamines or triazines, and several polyamides have been explored extensively [13, 22, 23]. Other forms, for example, dendritic polyglycerols (dPGs) are structurally defined, consisting of an aliphatic polyether backbone, and possessing multiple functional end groups [14, 24]. Since dPGs are synthesized in a controlled manner to obtain definite molecular weight and narrow molecular polydispersity, they have been evaluated for a variety of biomedical applications [25]. Hyperbranched PG analogs have similar properties as perfect dendritic structures with the added advantage of defined mono- and multifunctionalization [13, 14]. Additionally, Sisson *et al.* demonstrated PGs functionalized by emulsification method to create larger micogel structures emphasized for drug delivery [26]. Among plethora of dendritic carriers, polyester dendrimers represent an attractive class of nanomaterials due to their biodegradability trait; however, the synthesis of these nanocarriers is challenging because of the hydrolytic susceptibility of the ester bond [27, 28]. In contrast, polyamide- and polyamine-based dendrimers could withstand much wider selection of synthetic manipulations, but they do not degrade as easily in the body and thus they may be more prone to long-term accumulation *in vivo*.

Grinstaff recently described biodendrimers comprising biocompatible monomers [21] using natural metabolites, chemical intermediates, and monomers of medical-grade linear polymers. Interestingly, these dendritic macromolecules (e.g., poly(glycerol-succinic acid) dendrimer) (PGLSA) are foreseen to degrade *in vivo* (Scheme 10.1). Furthermore, these dendrimers have been tuned for degradation rate and the degradation mechanism for future *in vivo* applications.

10.2
Challenges for Designing Biodegradable Dendrimers

Biological applications of dendrimers have paved far ahead, comparatively over to its newer forms of core designs-exhibiting biodegradability. As a consequence to obtain a universal biodegradable, yet highly aqueous soluble and unimolecular dendrtic carrier capable of achieving high drug pay loading remains to be an unmet challenge. The greater aspect is to limit the early hydrolysis of the polymeric chains at the core compared to the periphery. Therefore, the prime objective remains to design biodegradable dendrimers having precise branching, molecular weight, monodispersity, and stable multiple functional appendages for covalent attachment of the bioactives.

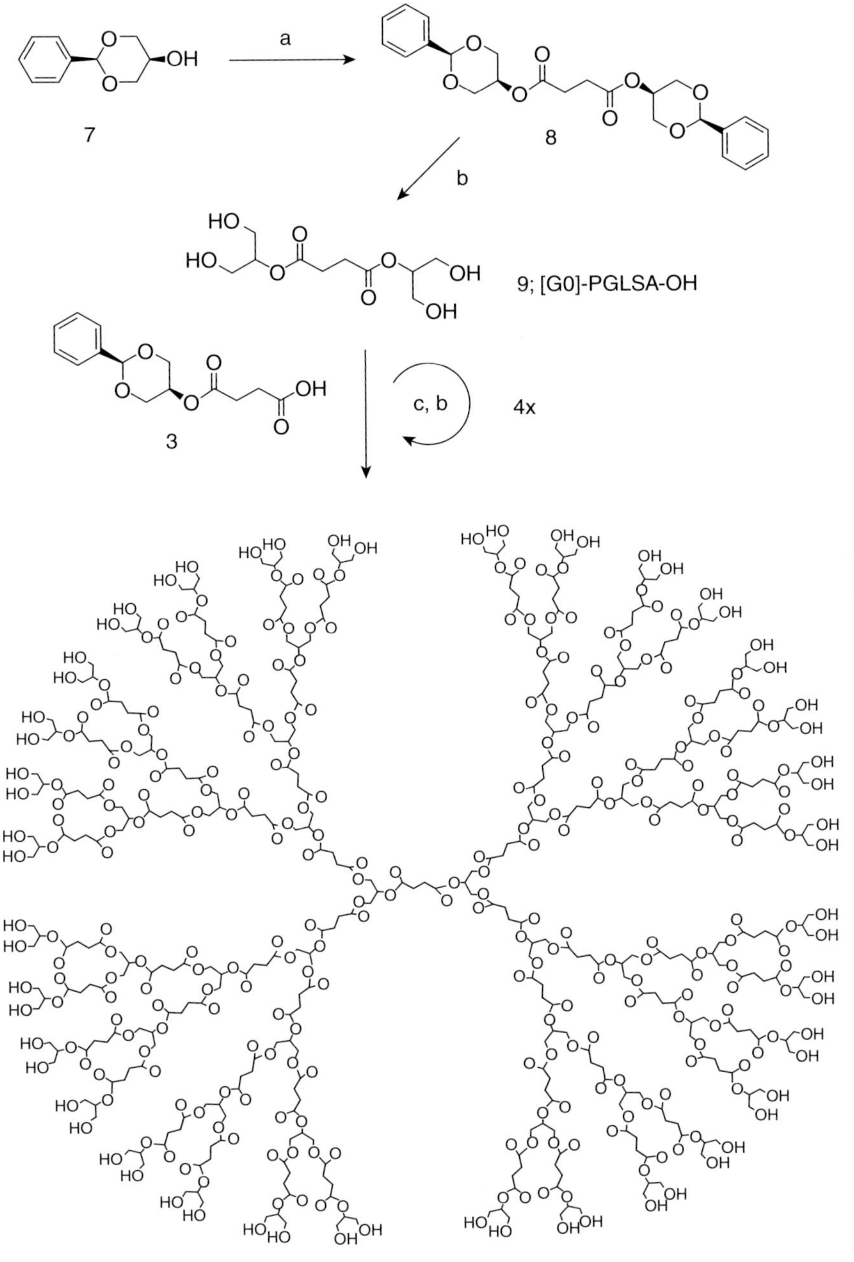

10; [G4]-PGLSA-OH

Scheme 10.1 Divergent synthetic method for G4-PGLSA-OH biodendrimer (**10**): (a) succinic acid, DPTS, DCC, CH_2Cl_2, 25 °C, 14 h; (b) 50 psi H_2, $Pd(OH)_2/C$, THF, 25 °C, 10 h; (c) 3, DPTS, DCC, THF, 25 °C, 14 h. 3 (2-(*cis*-1,3-*O*-benzylidene-glycerol)succinic acid mono ester) *cis*-1,3-*O*-benzylideneglycerol (**7**), 4-(dimethylamino)pyridinium 4-toluenesulfonate (DPTS) (**8**) [21].

It has been realized that the hydrolysis rate of polyester dendrimers dramatically depends on the hydrophobicity of the monomer, repeating units, steric environment, and the reactivity of the functional groups located within the dendrimer. Independently of one another, teams led by de Groot, Shabat, and McGrath have explored a much more advanced concept – the simultaneous release of all of dendrimer's functional groups by a single chemical trigger [16–18]. All three researchers presented that the dendrimer skeleton can be constructed to disintegrate into the known molecular fragments, once the disintegration process has been initiated. Now they have been variously termed as "cascade-release dendrimers," "dendrimer disassembly," and colorfully "self-immolative dendrimers" (SIDs), effective to perform chemical amplification reactions. Triggered by a specific chemical signal, the dendrimer scaffold can fall apart in several steps in a chain reaction, releasing all of the constituent molecules [16–18].

Szalai *et al.* [18] have reported a small dendrimer that can be disassembled geometrically by a single chemical trigger leading to two subsequent fragmentations in each subunit and completely reducing the polymer back to its monomers. The authors described dendrimers with 2,4-bis(hydroxymethyl)phenol repeat units capable of geometric disassembly of the corresponding anionic phenoxide species having labile vinylogous hemiacetals. With removal of the trigger group from 2,4-bis-(hydroxymethyl)phenol-based dendrimer subunit resulted in the formation of an *o,p*-bis(benzyl ether)phenoxide. The phenoxide – a bis(vinylogous hemiacetal) anion – cleaves to liberate alkoxide and *p*-quinone methide, which are trapped by an appropriate nucleophile under the reaction conditions, consistent with the electrophilic nature of quinine methides. The resulting phenoxide further cleaves to liberate a second equivalent of alkoxide and *o*-quinone methide, in turn trapped by the nucleophile to yield a fully cleaved phenoxide. The authors suggest that if alkoxide was analogous in structure to phenoxide, then the subsequent cleavages could occur, resulting in a geometric fragmentation through a dendrimer. Such unique dendrons are build with a core of 2,4-bis(hydroxymethyl)phenol units. The removal of a carbocation creates a phenoxide that could be cleaved and liberates two alkoxide groups in the presence of a suitable nucleophile. Small dendrimers with nitrophenoxy reporter groups and a single "trigger" group exhibit that second-generation dendrimers can be disassembled in under a minute time. If such process can be extended to higher generation dendrimers, it could be widely used to release drug molecules, in a complex form between the arms of the dendrimer vehicle [29].

The focus on biodegradable dendrimers could offer numerous advantages in biology compared to its nondegradable counterparts. Toward this direction, different biodegradable dendritic architectures have been designed. For example, SIDs have been designed possessing the capability to release all of their tail units through a self-immolative chain fragmentation. The trigger is initiated by a single cleavage event at the dendrimer's core [29]. The authors have hypothesized that by incorporation of drug molecules as tail units and an enzyme substrate as the trigger, multiprodrug units can be generated that could be activated on a single enzymatic cleavage. Such kind of biodegradable dendritic forms can be used to achieve targeted drug delivery. Another key challenge with polymeric and dendritic prodrug forms has been to achieve the complete elimination of these macromolecules from the body. More precisely, SIDs are reported to be excreted easily from the body due to their complete biodegradability [29]. Furthermore, the advantage of cleavage effect in SIDs with tumor-associated enzyme or a targeted one could be amplified and therefore may increase the number of active drug molecules in targeted tumor tissues.

The conventional method has been to attach covalently bioactive molecules to dendritic scaffolds by controlling the loading and release of active species. Chemical conjugation to a dendritic scaffold allows covalent attachment of different kinds of active molecules (imaging agents, drugs, targeting moieties, or biocompatible molecules) in a controlled ratio [14, 21, 23]. The loading as well as the release can be tuned by incorporating cleavable bonds that can be degraded under specific conditions present at the site of action (endogeneous stimuli, e.g., acidic pH, overexpression of specific enzymes, or reductive conditions as well as exogeneous stimuli, e.g., light, salt concentration, or electrochemical potential). In a recent report, Calderon *et al.* reported the use of the thiolated PG scaffold for conjugation to maleimide-bearing prodrugs of doxorubicin (DOX) or methotrexate (MTX) which incorporate either a self-immolative para-aminobenzyloxycarbonyl spacer coupled to dipeptide Phe–Lys or the tripeptide D-Ala–Phe–Lys as the protease substrate [30]. Both prodrugs were cleaved by cathepsin B, an enzyme overexpressed by several solid tumors, to release DOX or an MTX lysine derivate. An effective cleavage of PG–Phe–Lys–DOX and PG–D-Ala–Phe–Lys–Lys–MTX and release of DOX and MTX–lysine in the presence of the enzyme was observed.

Another challenge in dendritic or polymeric platforms is to tune the pharmacokinteics and extend the ability of a macromolecule to carry multiple copies of bioactive compounds [31]. This can be achieved by designing PEGylated dendrimers, which can circumvent the synthetic and biological limitations [27]. The polymeric architecture can be designed to avoid the destructive side reactions during dendrimer preparation while maintaining the biodegradability. Here, in this chapter, we highlight dendrimers with biodegradable characteristic in the presence of a suitable environment (e.g., pH). Chemical synthetic approaches have been discussed in detail, limited for their biodegradation and their biological implications.

10.2.1
Is Biodegradation a Critical Measure of Biocompatibility?

In the past, many polymers have been proven clinically safe. For example, PEG and PLGA polymers are being routinely used in delivering anticancer bioactives [23]. However, newer polymeric forms, which are currently being used in the biomedical field, are inherently heterogeneous in their structures, wherein the individual molecules have different chain lengths, due to their intrinsic polydispersed nature [8]. Therefore, their biodegradation profile is a crucial measure since the heterogeneous traits can substantially increase undesired effects on the biological activities, since it is not clear which part of the polymers with heterogeneous molecular weights is predominantly responsible for producing the undesired effect [32]. In order to minimize the heterogeneity, novel synthetic methods have to be employed for the preparation of polymers, and dendrimers for overcoming this heterogeneity, with the potential advantages of unimolecular homogeneity and defined chemical structures [33].

There have been numerous limitations to use poly(amidoamine) (PAMAM) dendrimers for biomedical applications due to their nonbiodegrading traits. Nevertheless, these polymers have shown to be biocompatible and can be easily prepared with various surface functionalities, such as $-NH_2$, $-COOH$, and $-OH$ groups, and are commercially available up to generation 10 (G10) [7]. Even though most applications of PAMAM are studied *in vitro*, a wide range of biomedical applications has been proposed in the fields of gene delivery [34], anticancer chemotherapy [35], diagnostics [36], and drug delivery [37, 38]. The cytotoxicity of PAMAM dendrimers is difficult to generalize and depends on their surface functionality, dose, and the generation of the dendrimers; however, the nonbiodegradable nature of PAMAM is one of the reasons for its toxicity [39]. Toward this end, more insights were recently described by Khandare *et al.* with respect to the structure–biocompatibility relationship of dPG derivatives possessing neutral, cationic, and anionic charges [40]. *In vitro* toxicity for various forms of dPGs was reported and compared with PAMAM dendrimers, polyethyleneimine (PEI), dextran, and linear polyethylene glycol (PEG) using human hematopoietic cell line U-937. It has been reported that dPGs possess greater cell compatibility similar to linear PEG polymers and dextran, and is therefore suitable for developing sysmetic formulation in therapeutics [40].

Polymeric and dendritic carrier systems are expected to possess suitable physicochemical properties for improved bioavailability, cellular dynamics, and targetability [23]. This is particularly true if the polymeric architectures have high surface charge, molecular weight, and a tendency to interact with biomacromolecules in blood due to their surface properties [40]. Most of the hyperbranched polymeric architectures consisting of bioactive therapeutic agents are administered by a systemic route. Therefore, their fate in blood and interactions with the plasma proteins and immune response are very critical to establish the overall biocompatibility. Studies in this direction have established the molecular and physiological interactions of the dendritic polymers with plasma components [41].

Conclusively, biodegradable dendrimers and its other architectures ideally should possess the following traits: (i) nontoxic, (ii) nonimmunogenic, and (iii) preferably be biocompatible and biodegradable. In this last instance, one of the potential virtues of dendrimers other than biodegradability comes under the heading of "multivalency" – the enhanced effect that stems from lots of identical molecules being present at the same time and place. Such simultaneous combination of multivalency and biodegradability with precision architectures can make dendrimers a greater versatile platform with many interesting biomedical applications, not least for the drug delivery [42].

10.3
Design of Self-Immolative Biodegradable Dendrimers

Polymeric forms of prodrugs have been designed and synthesized for achieving targetability in malignant tissues, due to overexpression of specific molecular receptive targets [43, 44]. The release of the free drug by a specific enzyme is very crucial for the cleavage of a prodrug-protecting group. Although many dendritic prodrugs have been designed to target the cancer, only few biodegradable approaches have been explored till date [16–19, 27, 45]. Toward this end, SIDs have been lately synthesized, which may open new opportunities for targeted drug delivery. In contrast to conventional dendrimers, SIDs are fully degradable and can be excreted easily from the body [29]. Since the dendrimer are multi-immolative, this effect may increase the number of active drug molecules in targeted tumor tissues. SID dendritic building units are conceptualized on 2,6-bis-(hydroxymethyl)-*p*-cresol (7), which has three functional groups (Scheme 10.2).

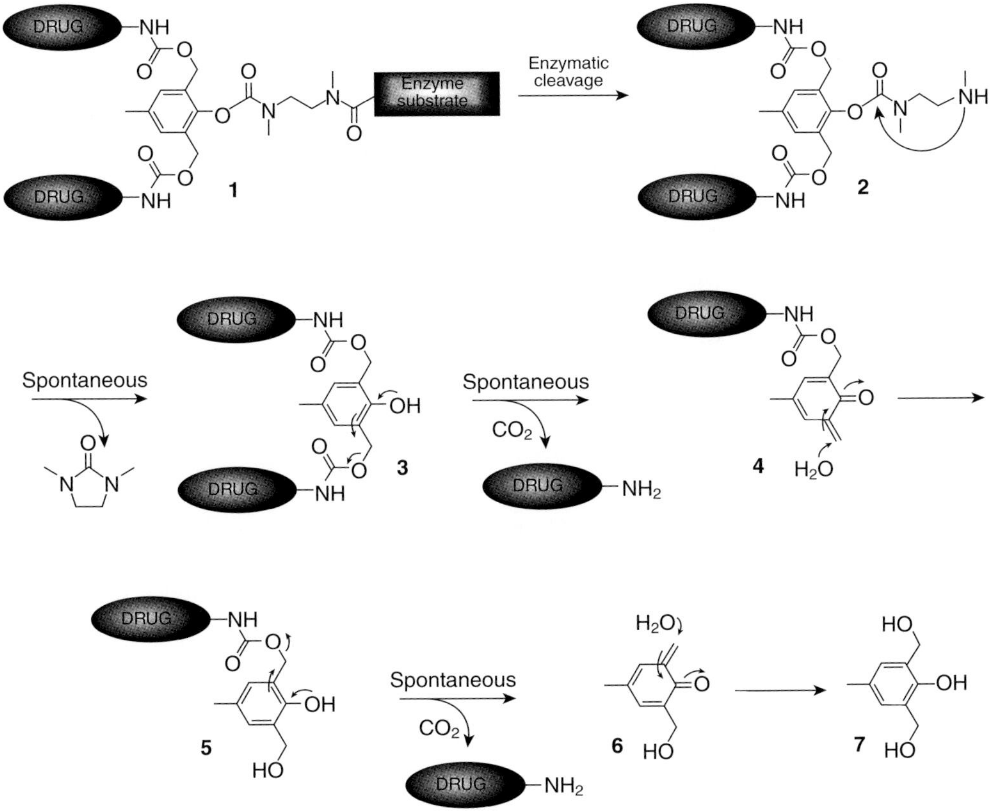

Scheme 10.2 Mechanism of dimeric prodrug activation by a single enzymatic cleavage [29].

Two hydroxybenzyl groups were attached through a carbamate linkage to drug molecules, and a phenol functionality was conjugated to a trigger by using *N,N*-dimethylethylenediamine (compound **1**) as a short spacer molecule. The cleavage of the trigger is initiative for the self-immolative reaction, starting with a spontaneous cyclization of amine intermediate **2**, to form an *N,N′*-dimethylurea derivative. On the other hand, the generated phenol **3** undergoes a 1,4-quinone methide rearrangement followed by a spontaneous decarboxylation to liberate one of the drug molecules. Similarly, the quinone methide species **4** is rapidly trapped by a water molecule to form a phenol (compound **5**), which further undergoes an 1,4-quinone methide rearrangement to liberate the second drug entity. Furthermore, the quinone methide-generated species **6** is once again trapped by a water molecule to form **7**. Thus, compound **7** is reacted with **2** equivalent of (TBS)Cl to afford phenol **8**, which is acylated with *p*-nitrophenyl (PNP) chloroformate to form carbonate **9** (Scheme 10.3). The latter is reacted with mono-Boc-*N,N′*-dimethylethylenediamine to generate compound **10**, which is deprotected in the presence of Amberlyst-15 to give diol **11**. Later, the deprotection with trifluroacetic acid (TFA) afforded an amine salt, which is reacted *in situ* with linker I (activated form of antibody 38C2 substrate) to generate compound **12** [29].

Thereafter, the latter was reacted with **2** equivalent of DOX to obtain a prodrug **14**. Acylation of diol **11** with **2** equivalent of PNP chloroformate resulted in compound dicarbonate **15**, which is reacted with **2** equivalent of camptothecin amine units to give compound **16** (Scheme 10.4). Deprotection with TFA resulted in an amine salt, which is reacted *in situ* with linker II to yield prodrug **17**. The authors selected the anticancer drug DOX and catalytic antibody 38C2 [46] as the activating enzyme. Antibody 38C2 catalyzes a sequence of retro-aldol retro-Michael cleavage reactions, using substrates that are not recognized by human enzymes.

Prodrugs of this kind can demonstrate slight toxicity increased over activation of monomeric prodrugs. Both monomeric and dimeric prodrugs showed chemical stability in the cell medium. *In vitro* and *in vivo* efficacy of the dendritic conjugates was demonstrated by activating several prodrugs. Figures 10.2 and 10.3 represent *in vitro* activity of these polymers and have been detailed in later section.

10.3.1
Clevable Shells – Multivalent PEGylated Dendrimer for Prolonged Circulation

The unique structural properties of dendrimers increasingly entice scientists to use them for many biomedical applications [9–11, 14, 19, 47]. In particular, biodegradable and disassembled dendritic molecules have been attracting growing attention [16–19]. Toward this direction, anticancer prodrugs of DOX PEGylated dendrimers have been designed for the selective activation in malignant tissues by a specific enzyme, which is targeted or secreted near tumor cells [48]. In recent studies, a family of polyestercore dendrimers based on a 2,2-bis(hydroxymethyl) propanoic acid (bis-HMPA) monomer unit, functionalized in the form of shells with eight 5 kDa PEG chains [27], was shown to be biocompatible and capable of high drug loading while facilitating high tumor accumulation through its long

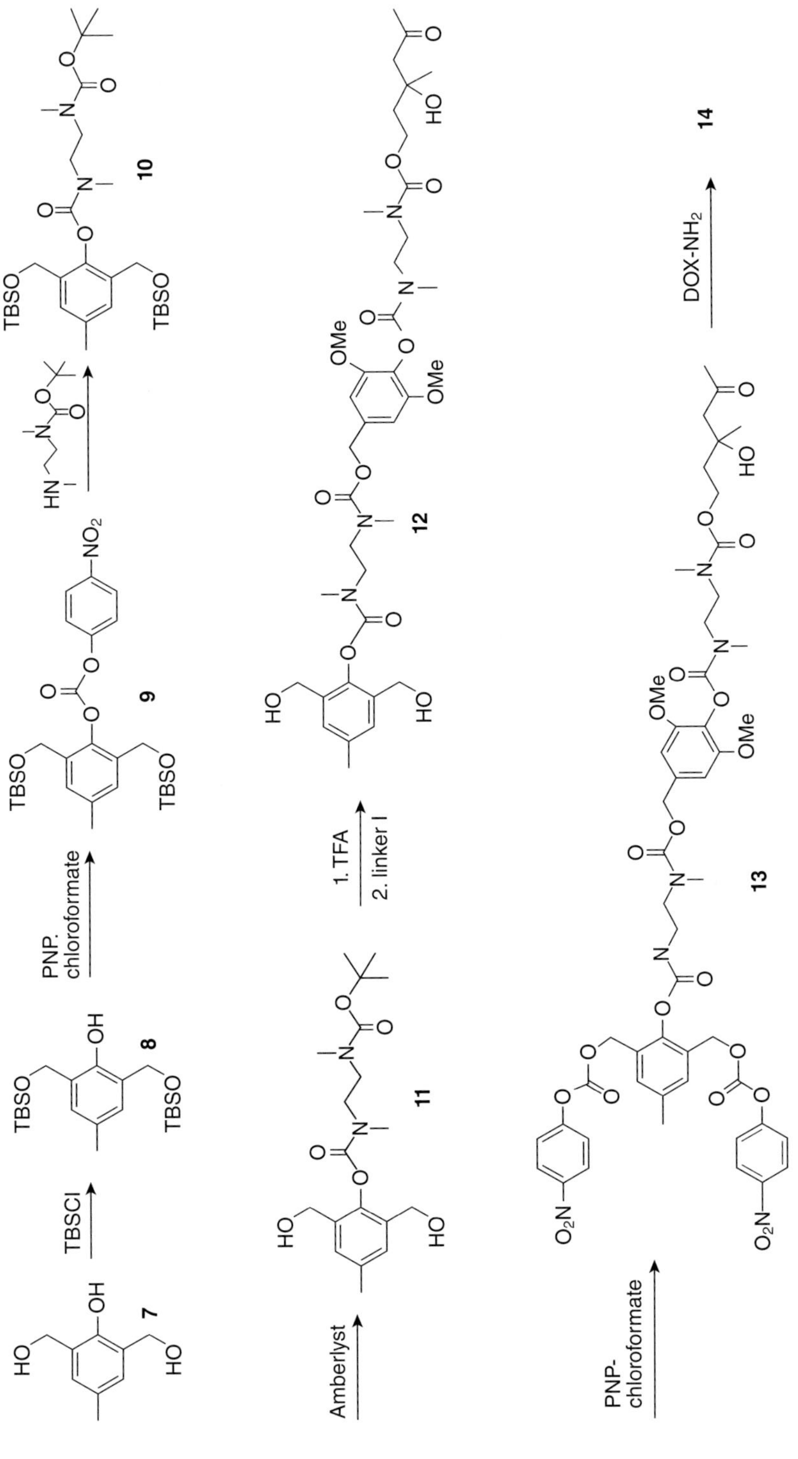

Scheme 10.3 Synthesis of the prodrug of DOX with a trigger which can be activated with catalytic antibody 38C2 [29].

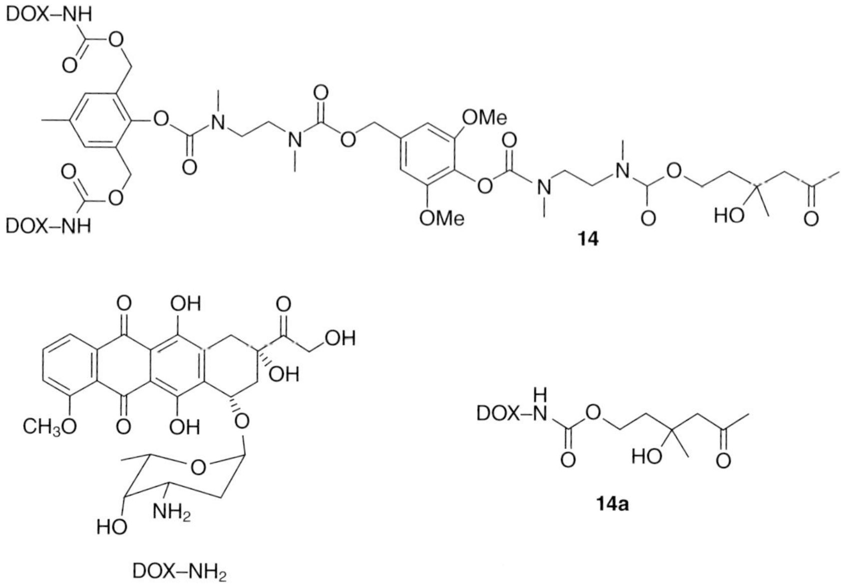

Scheme 10.4 Chemical structure of DOX prodrugs **14** and **14a** [29].

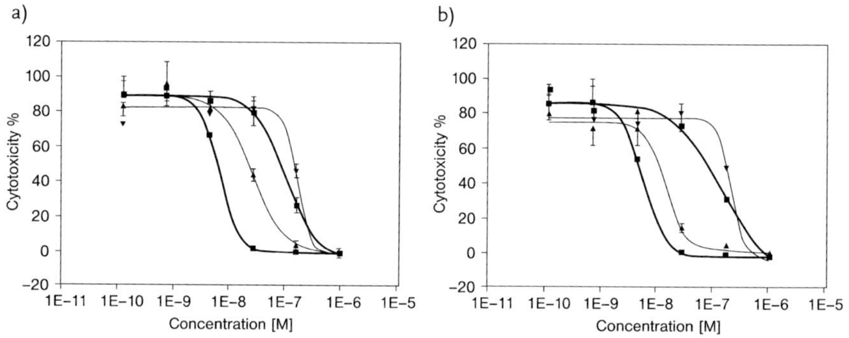

Figure 10.2 Growth inhibition assay of the human Molt-3 leukemia cell line, with addition of prodrugs in the presence and absence of catalytic antibody 38C2 (cells were incubated for 72 h): (a) (9) DOX, (b) pro-DOX **14**a, (2) pro-DOX **14a** + 1 μM 38C2, (1) solvent control; (b) (9) DOX, (b) pro-DOX **14**, (2) pro-DOX **14** + 1 μM 38C2, (1) solvent control [29].

circulation half-life. Polyester dendrimers based on bis(HMPA) monomer units have attracted a lot of attention as they are nonimmunogenic, biodegradable, and nontoxic in nature. Scheme 10.5 describes the synthesis of a core-functionalized PEGylated dendrimer [27]. In brief, the tetrafunctional pentaerythritol core **1** was tailored by benzylidene-protected bis(HMPA) monomer **2** to yield generation **1** dendrimer **3**. The protecting groups were removed by hydrogenolysis, and periph-

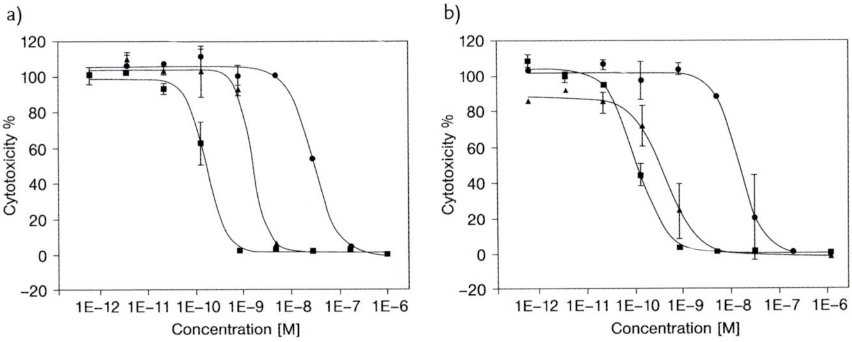

Figure 10.3 Growth inhibition assay of the human Molt-3 leukemia cell line, with prodrugs in the presence and absence of catalytic antibody 38C2 (cells were incubated for 72 h): (A) (9) CPT, (b) pro-CPT **17a**, (2) pro-CPT **17a** + 1 μM 38C2; (B) (9) CPT, (b) pro-CPT **17**, (2) pro-CPT **17** + 1 μM 38C2 [29].

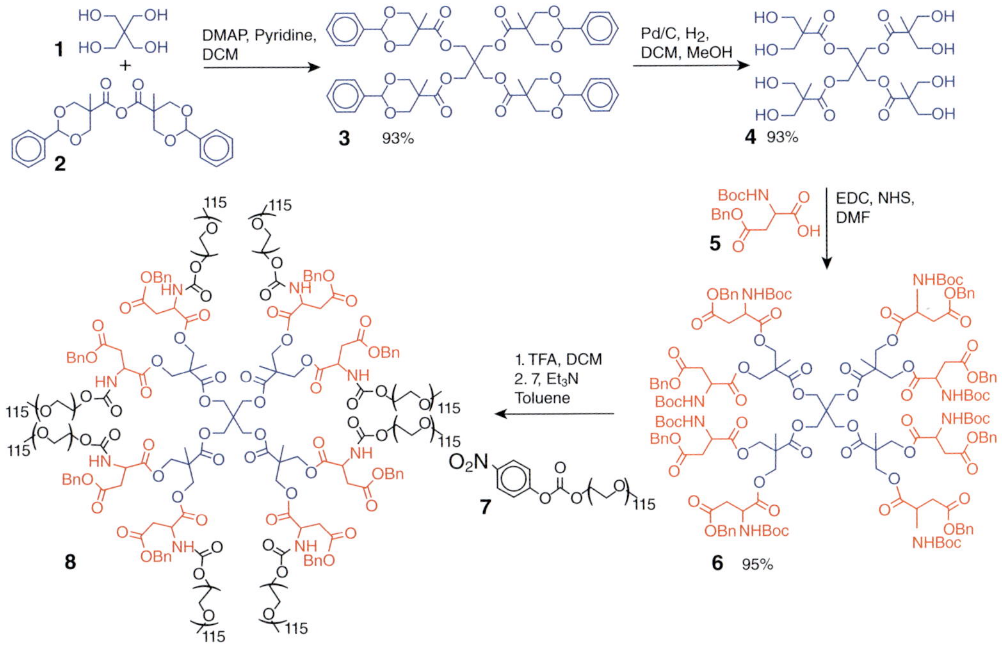

Scheme 10.5 Synthesis of symmetrically PEGylated dendrimer [27].

eral hydroxyl groups (as shown in Scheme 10.5) were functionalized using orthogonally protected aspartic acid to obtain compound **6**. Amino groups were subsequently deprotected and PEGylation was carried out with 5 kDa PEG electrophiles to obtain dendrimer **8**.

The protecting groups in benzyl ester **8** were removed by hydrogenolysis and dendrimer **9** was afforded using carboxylic acids moieties which is further envisioned for attaching drug molecules of choice. The functionalization of this dendrimer with *t*-butyl carbazate or glutamic acid derivative **10** was not successful as degradation of the dendrimer was occurring during the reaction. For further insight with respect to the degradation pathway, the authors designed dendrimer **11** and further functionalized with aspartic acid chain ends. Scaffold **11** was used since the progress of its reaction was monitored by MALDI-TOF over to PEGylated dendrimer **9** (Scheme 10.6).

Due to degradation side reaction, only a small amount of the target moiety was only reported, thereby leading to the appearance of lower molecular weight products as a result of intramolecular cyclization reactions as proposed in Scheme 10.7. This kind of cyclization reactions with benzyl ester-protected aspartic acid residues are reported in the literature [49]. Earlier, pentachlorophenol (PCP) has been used to decrease the formation of the aminosuccinyl by-product by inhibiting amide deprotonation. While, in buffered conditions, the primary amines are favorable to react with PNP carbonates and other electrophiles. It has been reported that the use of PCP was beneficial since it allows the functionalization of the carboxylic acid side chains of dendrimer **9** with protected nucleophile **10** to give dendrimer **12** (Scheme 10.6). Furthermore, DOX hydrazone conjugate **14** was synthesized by removing Boc groups from the hydrazide linkers in **12** and by condensation of the resulting amines with the ketone group of DOX **13** (Scheme 10.6). The degradation of polyester architecture was evaluated in physiological conditions. Therefore, **12** was incubated in phosphate-buffered saline buffer at 37 °C and the molecular weight with time was monitored by SEC. Due to rapid degradation, alternative dendrimer scaffolds based on robust polyamide core were explored.

10.3.1.1 **Polylysine-Core Biodegradable Dendrimer Prodrug**

Compared to polyester dendrimers, polyamide dendrimers are less susceptible to hydrolysis; however, due to increased stability, it may affect *in vivo* pharmacokinetics Recently, Fox *et al.* reported PEGylated polylysine consisting of camptothecin with 100% survival in transgenic mice using HT-29 human colon carcinoma for a period of 70 days (Scheme 10.8) [50]. It is to be noted that the very slow or incomplete degradation of the polymeric carrier's by-product may lead to toxicity [51]. Delivery of DOX using polylysine carrier is reported [50] in the form of dendrimer **18** having protected hydrazide molecules (Scheme 10.8).

As shown in Scheme 10.8, lysine dendrimer **15** was used as the starting material. Its peripheral amines were acylated with PNP-Asp(Bn)Boc to afford dendrimer **17**. The authors pointed out that the PCP additive is critical for conjugating aspartic acid to G2 lysine periphery; if not, a five-membered amino succinyl by-product may be formed via amidolysis of the benzyl ester-protected side chain. By deprotecting of amino groups of the aspartate and PEGylation with PEG-*p*-nitrophenyl, carbonate yields **18**. However, coupling *t*-butyl carbazate to the deprotected side chain of carboxylic acid terminal moieties (**19**) leads to degradation by-products such as **20** (Scheme 10.8). Furthermore, the drug was conjugated using PEGylated ester amide dendrimers (**25–27**) as represented in Scheme 10.10.

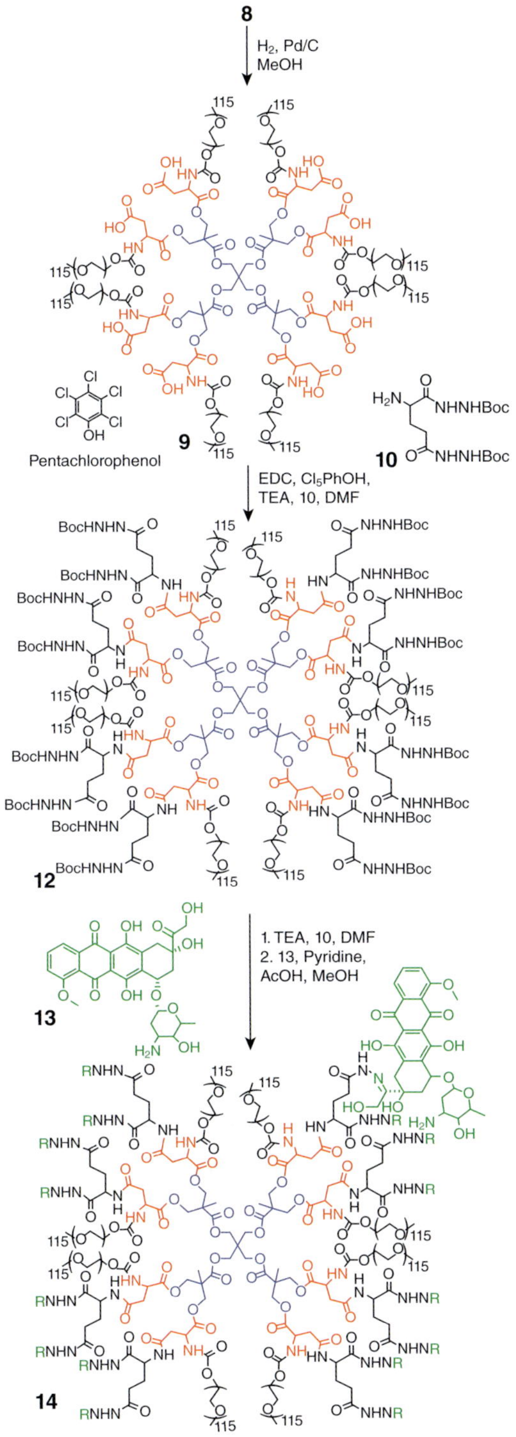

Scheme 10.6 Linker attachment and conjugation of DOX [27].

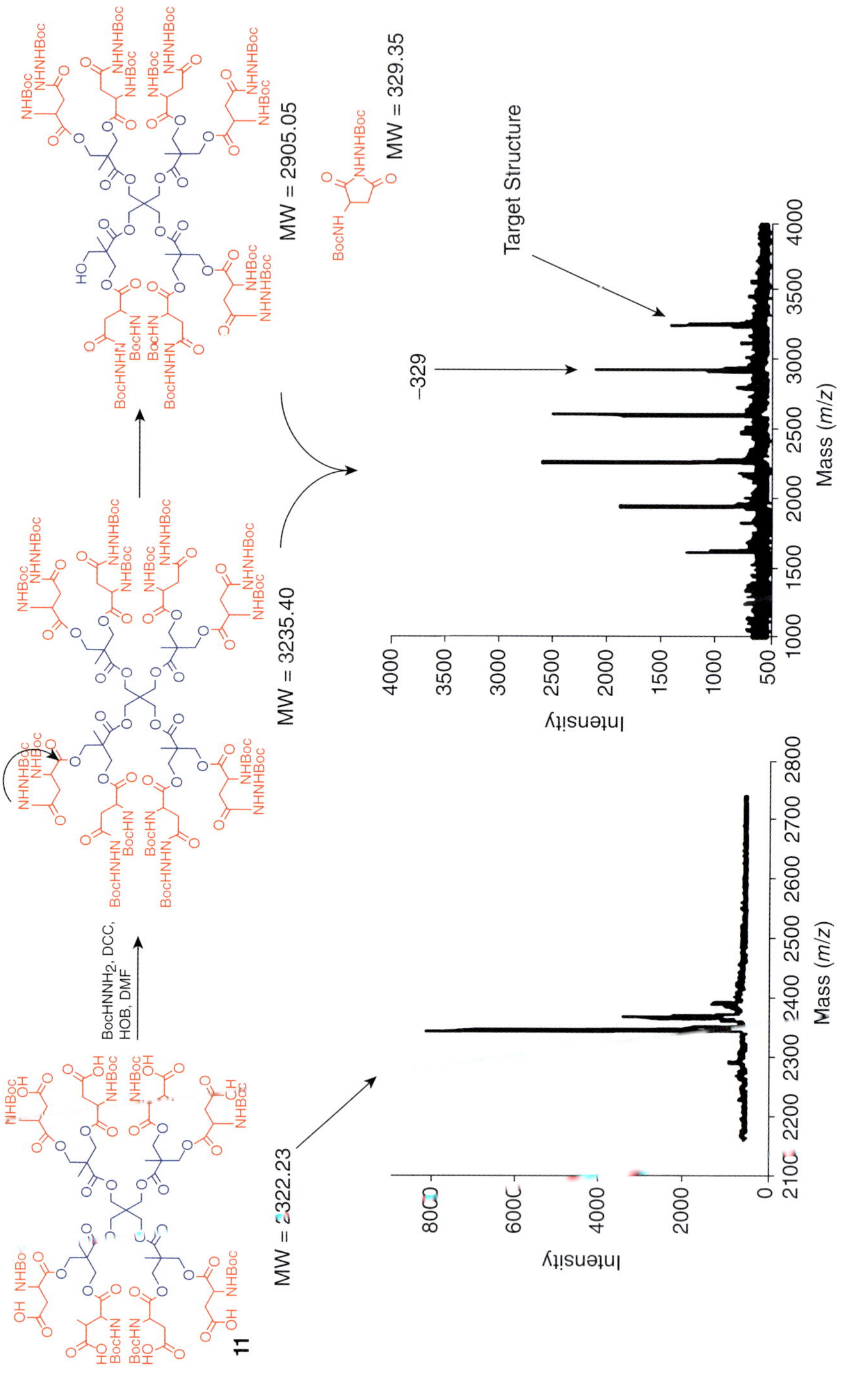

Scheme 10.7 Proposed degradation pathway for polyester dendrimer [27].

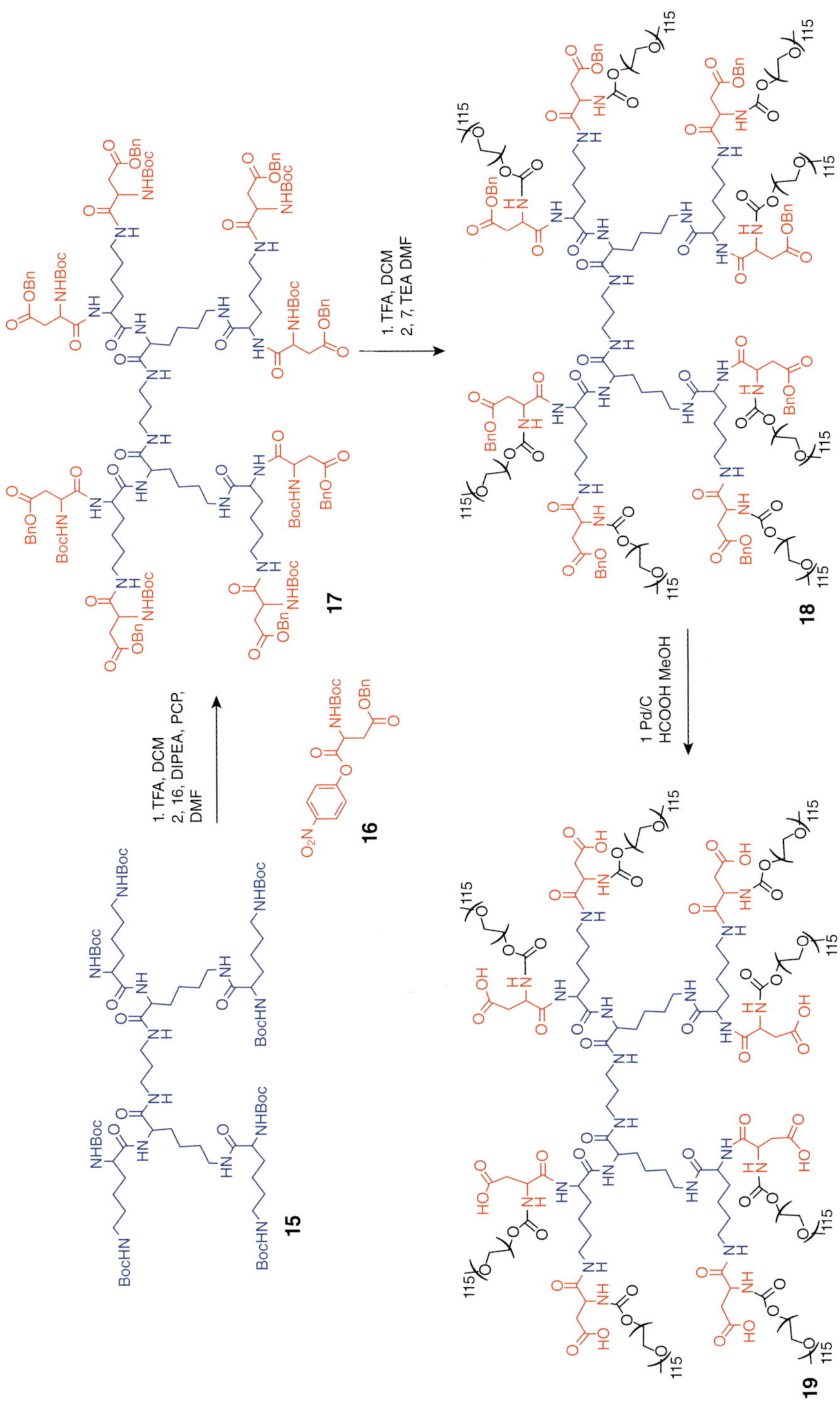

Scheme 10.8 PEGylated polylysine synthesis [27].

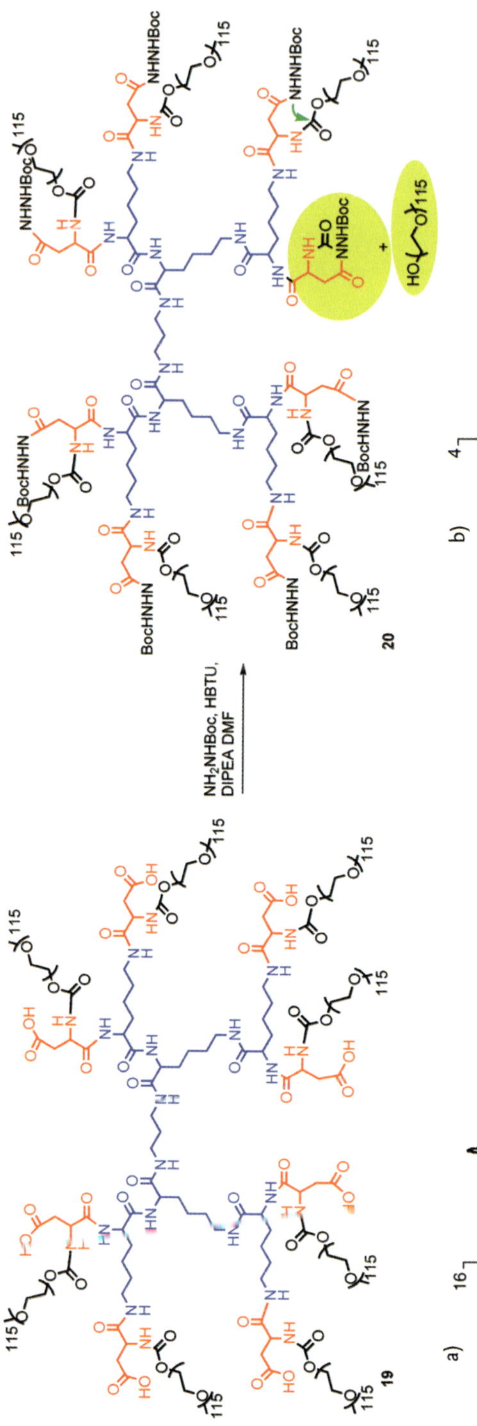

Scheme 10.9 PEGylated polylysine degradation: (a) SEC of compound **19** and (b) SEC of reaction mixture with by-product 20 [27].

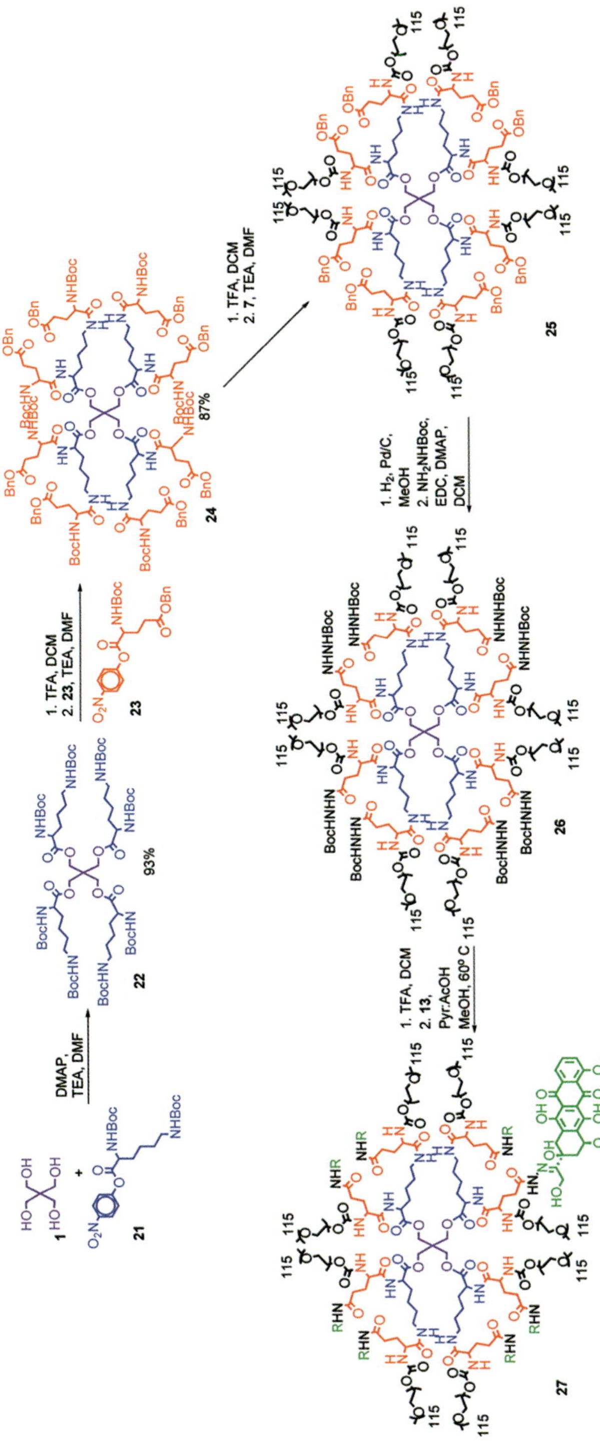

Scheme 10.10 Synthesis of drug-loaded PEGylated ester amide dendrimer [27].

Table 10.1 *In vitro* efficacies of ester amide dendrimer conjugate and controls against balb/C mice with C26 carcinoma [27].

Treatment group	No. mice	Dose (mg/kg)	Mean TGD[a] (%)	Median survival time (days)	TRD[b]	LTS
PBS[c]	10			20	0	0
Doxil	10	20	245[b]	60[b]	2	8
27	10	20	229[b]	60[b]	0	9
27	10	15	175[b]	60[b]	0	6
27	10	10	74[c]	33[b]	0	1

a) TGD, tumor growth delay, calculated from time of growth to 400 mm^3.
b) TRD, treatment-related death; LTS, long-term survivors. Compared to PBS, $P \leq 0.0001$.
c) Compared to PBS, $P = 0.004$. Compound **27** is represented in Scheme 10.10.

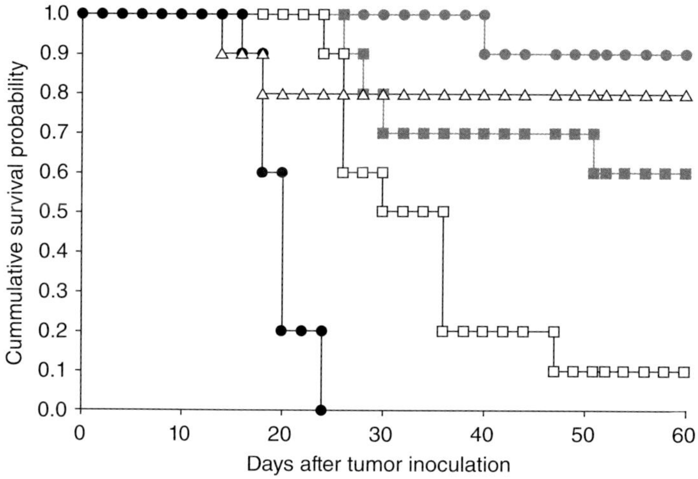

Figure 10.4 Representative survival probability versus time in Balb/C mice bearing subcutaneous C26 colon carcinoma after a single injection of PEGylated poly(ester amide) Dox conjugate or control. After tumor generation, mice were treated for 8 days. [**27** (20 mg Dox equivalent/kg); 9, **27** (15 mg Dox equivalent/kg); 0, **27** (10 mg Dox equivalent/kg); Δ, Doxil (20 mg Dox equivalent/kg); •, PBS [27].]

The *in vitro* evaluation and *in vivo* tumor efficacy of drug-loaded PEGylated ester amide dendrimer have been described in details in Section 1.4 (Table 10.1 and Figure 10.4)

10.4 Biological Implications of Biodegradable Dendrimers

After years of seeking to synthesize polymers in the form of long linear chains, the attention has been focused on highly defined polymeric topologies, especially in the form of highly branched dendritic architectures [52]. The interesting application of these intriguing and well-defined materials seems to be branching out. Especially, they offer water solubility due to enriched hydrophilic groups at the surface, ester linkages to make them biodegradable, and appropriate amounts of drug can be added so that a uniform supply of the drug is delivered when the polymer degrades. Dendrimers find implications in biology for delivering drugs, bioactives (e.g., SiRNA and peptides), diagnostics (dyes), and in targeted delivery systems (use of LHRH peptide, folic acid, antibodies, etc.) [14, 21, 22, 37, 43, 44].

Among many other crucial characteristics of cellular internalization of dendritic architectures, the ability of dendrimers to cross cell membrane is of much interest particularly for their application in drug and gene delivery. A recent study has demonstrated that dendrimers are capable to enter cells by endocytosis, but the intracellular pathway following their internalization remains controversial [53]. The intracellular trafficking property of PAMAM dendrimers was observed using confocal fluorescence microscopy with high spatial and temporal resolution in living HeLa cells. Macromolecules of different chemical functionality (neutral, cationic, and lipidated), size (from G2 up to G6), and surface charge are investigated and their internalization properties correlated with the molecular structure. So far, not many strategies have been reported to synthesize biodegradable dendrimers and among them very few suggest their biomedical applications. Section 10.5 highlights biological perspectives of biodegradable dendrimers.

Interestingly, SIDs have been developed and are introduced as a potential platform for a multiprodrug (synthesis section – Schemes 10.2–10.4 [29]). The reported dendrimers release all of their tail units, through a self-immolative chain fragmentation, which is initiated by a single cleavage at the dendrimer's core. Incorporation of drug molecules as the tail units and an enzyme substrate as the trigger generates a multiprodrug unit, which can be activated with a single enzymatic cleavage. The authors evaluated bioactivation of the dendritic prodrugs by cell-growth inhibition assay with the Molt-3 leukemia cell line in the presence and the absence of antibody 38C2 (Figures 10.2 and 10.3). The dendritic unit can be considered as a platform to develop a heterodimeric prodrug approach for remarkable increase in toxicity with its bioactivation.

When catalytic antibody 38C2 was incubated with the prodrugs, it was observed that the IC_{50} of the dimeric prodrug had shifted closer to the IC_{50} of the free DOX (Figure 10.2). The IC_{50} values of the monomeric and the dimeric prodrugs were found to be almost the same, and the prodrugs were about 200-fold less toxic than free CPT (Figure 10.3). On the other hand, catalytic antibody 38C2 was added, and both the prodrugs were activated. However, while the activity of the monomeric prodrug had shifted to a 10-fold difference from that of free CPT, the dimeric prodrug was shown to be about four times more active upon addition of 38C2, meaning that more toxicity was achieved using the dimeric prodrug and 38C2 in comparison to monomeric prodrug and the same concentration of antibody. For **27** compound administered at 15 and 10 mg DOX/kg, the treatment groups had 175% and 74% tumor growth inhibition with a medium survival time of 60 and 33 days, respectively. A dramatic 75% decrease in subcutaneous tumor size has been observed in mice that received a combination of intratumoral injections of antibody 38C2 and systemic treatments with an etoposide prodrug [54].

Tansey *et al.* described synthesis and characterization of branched poly(L-glutamic acid) (PG) containing multiple PG chains centered on a PAMAM dendrimer or PEI cores [55]. The branched PG polymers were obtained by ring-opening polymerization of benzyl ester of L-glutamic acid *N*-carboxyanhydride using PAMAM or PEI as the initiator. The polymers were degradable in the presence of the lysosomal enzyme cathepsin B, albeit more slowly than linear PG. Unlike conventional linear PG, each branched PG possessed multiple terminal amino groups. This made it possible to attach multiple targeting moieties selectively to the termini of branched PG. Conjugation of monofunctional or heterodifunctional PEG to the chain ends of branched PG demonstrated in the presence of side-chain carboxyl groups. Furthermore, folic acid, a model-targeting moiety, and the near-infrared dye indocyanine green, a model diagnostic agent, were successfully conjugated to the terminal amino groups and the side-chain carboxyl groups of branched PG, respectively. The resulting conjugate had reduced nonspecific interaction and bound selectively to tumor cells expressing folate receptors. Thus, branched PG may be useful as a polymeric carrier for targeted drug delivery.

Degradable dendrimer architectures can be conjugated with linearly branched polymers (PEG) to improve biological application due to the enhanced pharmacokinetic ability [27]. For example, van der Poll reported efficient synthesis of a robust and biodegradable PEGylated dendrimer based on a polyester–polyamide hybrid core (synthesis section, Schemes 10.8–10.10 [27]). The architecture has been designed to avoid destructive side reactions during dendrimer preparation while maintaining biodegradability. Dendrimer functionalized with DOX was also prepared from commercial starting materials in nine, high-yielding linear steps. Both the dendrimer and Doxil were evaluated in parallel using equimolar dosage in the treatment of C26 murine colon carcinoma, leading to statistically equivalent results with the most mice tumor-free at the end of the 60-day experiment. The attractive features of this dendritic drug carrier are its simple synthesis, biodegradability, and capability to deliver high payload of drugs.

Similarly, many glycol-dendrimers have been reported for applications in biology. For example, glycopeptide dendrimers containing ω-amino acids (Gly, β-Ala, γ-abu, and ε-aminohexanoic acid) are of interest for immunological studies [56]. Interestingly, biodegradable forms of dendrimers have been used as pH-sensing biodegradable near-infrared nanoprobe capable of providing complementary information through both fluorescence lifetime measurements and signal amplification in acidic environments *in vivo* [57]. Such tools may find extensive role in drug delivery systems, and such noninvasive approach may shed light on the kinetics of such drug delivery strategies *in vivo* in a cost-effective and more accurate manner.

It has been realized that in many cases the prepared dendrimeric structures are more a result of an intellectual capability to prepare some unusual compounds with new cores, branches, etc., than an exact approach based on the knowledge of size, shape, polarity, and other parameters that must be fulfilled to satisfy the strict demands of the given receptor [58].

10.5 Future Perspectives of Biodegradable Dendrimers

Now, it is possible to precisely manipulate dendrimers for their molecular weight and chemical composition, thereby to allow predictable tuning of its biocompatibility and pharmacokinetics [40].

Degradable polymeric systems have wide perspective in several established and emerging technologies such as controlled-release systems for drug delivery and photoresist methodology for microlithography among other applications [59–61] As the demand increases, the higher levels of control over the structure, properties, and performance of degradable materials would be desired [18, 20]. So far few groups have addressed the strategies for the controlled degradation of dendritic structures or the implications of the development of such systems. Researchers have now shown the perspectives of biodegradable dendrimers and for sure such forms will find immense implications in biology.

Advances in realizing the role of molecular weight and architecture on the *in vivo* behavior of dendrimers, together with recent progress in the design of biodegradable chemistries, have enabled the application of these branched polymers as antiviral drugs, tissue repair scaffolds, and targeted carriers of chemotherapeutics. It is expected that the products could reach the market soon and therefore the field must address the long-term human and environmental health consequences of dendrimer exposure *in vivo*.

The synergy due to biodegradation, multivalency, and size in nanoscale has a range of options to impart chemical "smartness" along their molecular scaffold to achieve environment-sensitive modalities; such materials are envisioned to revolutionize the existing therapeutic practices [14]. Therefore, biodegradable dendritic architectures are expected to lead to new strategies for *nanomedicine* as well as *regenerative medicine.*

10.6 Concluding Remarks

The diverse dendritic structures with multiple functional groups at the periphery for chemical modifications render dendrimers to tune biological properties. The potential virtue of dendrimers comes under the heading of "multivalancy": the enhanced effect that stems from lots of identical molecules being present at the same time and place [14, 19]. Dendrimers have shown to enter into the cells remarkably easily, with a potential to deliver drugs at the targeted site. Furthermore, there has been great emphasis to achieve the release of a drug at various pH environments. However, most demanding aspect of dendrimers is to construct them into self-disintegrating forms termed as SIDs [16, 17, 19]. Therefore, the combination of multivalancy and self-destruction characteristics of biodegradable dendrimers will have increasing interest for biomedical implications.

References

1 Kopecek, J. (1977) Soluble biomedical polymers. *Polym. Med.*, **7**, 191–221.

2 Bader, H., Ringsdorf, H., and Schmidt, B. (1984) Water-soluble polymers in medicine. *Angew. Makromol. Chem.*, **123**, 457–485.

3 Seymour, L.W. (1992) Passive tumor targeting of soluble macromolecules and drug conjugates. *Crit. Rev. Ther. Drug*, **9**, 135–187.

4 Matsumura, Y. and Maeda, H. (1986) A new concept for macromolecular therapeutics in cancer-chemotherapy-mechanism of tumoritropic accumulation of proteins and the antitumor agent SMANCS. *Cancer Res.*, **46**, 6387–6392.

5 Tomalia, D.A., *et al.* (1985) A new class of polymers: starburst-dendritic macromolecules. *Polym. J.*, **17**, 117–132.

6 Grayson, S.M. and Fréchet, J.M.J. (2001) Convergent dendrons and dendrimers: from synthesis to applications. *Chem. Rev.*, **101**, 3819–3868.

7 Esfand, R. and Tomalia, D.A. (2001) Poly(amidoamine) (PAMAM) dendrimers: from biomimicry to drug delivery and biomedical applications. *Drug Discov. Today*, **6**, 427–436.

8 Lee, J-S., Huh, J., Ahn, C-H., Lee, M., and Park, T.G. (2006) Synthesis of novel biodegradable cationic dendrimers. *Macromol. Rapid Commun.*, **27**, 1608–1614.

9 Fréchet, J.M.J. and Tomalia, D. (2001) Introduction to the dendritic state editors, in *Dendrimers and Other Dendritic Polymers* (eds J.M.J. Fréchet and D. Tomalia), John Wiley & Sons, Ltd, Chichester, UK, pp. 1–44.

10 Ihre, H.R., Padilla De Jesus, O.L., Szoka, F.C. Jr., and Frechet, J.M. (2002) Polyester dendritic systems for drug delivery applications: design, synthesis, and characterization. *Bioconjug. Chem.*, **13**, 443–452.

11 Jevprasesphant, R., Jalal, R., Attwood, D., McKeown, N.B., and D'Emanuele, A. (2003) The influence of surface modification on the cytotoxicity of PAMAM dendrimers. *Int. J. Pharm.*, **252**, 263–266.

12 Quadir, M.A., Radowski, M.R., Kratz, F., Licha, K., Hauff, P., and Haag, R. (2008) Dendritic multishell architectures for drug and dye transport. *J. Control. Release*, **132**, 289–294.

13 Haag, R., Sunder, A., and Stumb'e, J.F. (2000) An approach to glycerol dendrimers and pseudo-dendritic polyglycerols. *J. Am. Chem. Soc.*, **122**, 2954–2955.

14 Caldero'n, M., Quadir, M.A., Sharma, S.K., and Haag, R. (2010) Dendritic polyglycerols for biomedical applications. *Adv. Mater.*, **22**, 190–218.

15 Newkome, G.R., Moorefield, C.N., and Vögtle, F. (1996) *Dendritic Macromolecules: Concepts, Synthesis, Perspectives*, Wiley-VCH, Weinheim, Germany.

16 Amir, R.J., Pessah, N., Shamis, M., and Shabat, D. (2003) Self-immolative dendrimers. *Angew. Chem. Int. Ed.*, (**42**), 4494–4499.

17 de Groot, F.M., Albrecht, C., Koekkoek, R., Beusker, P.H., and Scheeren, H.W. (2003) "Cascade-release dendrimers" liberate all end groups upon a single triggering event in the dendritic core. *Angew. Chem. Int. Ed.*, (**42**), 4490–4494.

18 Szalai, M.L., Kevwitch, R.M., and McGrath, D.V. (2003) Geometric disassembly of dendrimers: dendritic amplification. *J. Am. Chem. Soc.*, **125**, 15688–15689.

19 Meijer, E.W. and van Genderen, M.H.P. (2003) Dendrimers set to self-destruct. *Nature*, **426**, 128–129.

20 Li, S., Szalai, M.L., Kevwitch, R.M., and McGrath, D.M. (2003) Dendrimer disassembly by benzyl ether depolymerization. *J. Am. Chem. Soc.*, **125**, 10516–10517.

21 Grinstaff, M. (2002) Biodendrimers: new polymeric biomaterials for tissue engineering. *Chem. Eur. J.*, **8**, 2838–2846.

22 Haag, R. (2004) Supramolecular drug-delivery systems based on polymeric core–shell architectures. *Angew. Chem.*, **116**, 280–284.

23 Duncan, R. (2006) Polymer conjugates as anticancer nanomedicines. *Nat. Rev. Cancer*, **6**, 688–701.

24 Sunder, A., Hanselmann, R., Frey, H., and Mulhaupt, R. (1999) Controlled synthesis of hyperbranched polyglycerols by ring-opening multibranching polymerization. *Macromolecules*, **32**, 4240–4246.

25 Stiriba, S.E., Frey, H., and Haag, R. (2002) Dendritische polymere fur medizinische anwendungen: auf dem weg zum einsatz in diagnostik und therapie. *Angew. Chem. Int. Ed. Eng.*, **114**, 1385–1390.

26 Sisson, A.L., Steinhilber, D., Rossow, T., Welker, P., Licha, K., and Haag, R. (2009) Biocompatible functionalized polyglycerol microgels with cell penetrating properties. *Angew. Chem. Int. Ed. Engl.*, **48**, 7540–7545.

27 van der Poll, D.G., Kieler-Ferguson, H.M., Floyd, W.C., Guillaudeu, S.J., Jerger, K., Szoka, F.C., and Fréchet, J.M. (2010) Design, synthesis, and biological evaluation of a robust, biodegradable dendrimer. *Bioconjug. Chem.*, **21**, 764–773.

28 Szuromi, P.D. (2003) Triggering polymer destruction. *Science*, **302**, 1863.

29 Shamis, M., Lode, H.N., and Shabat, D. (2004) Bioactivation of self-immolative dendritic prodrugs by catalytic antibody 38C2. *J. Am. Chem. Soc.*, **126**, 1726–1731.

30 Calderón, M., Graeser, R., Kratz, F., and Haag, R. (2009) Development of enzymatically cleavable prodrugs derived from dendritic polyglycerol. *Bioorg. Med. Chem. Lett.*, **14**, 3725–3728.

31 Khandare, J.J., Jayant, S., Singh, A., Chandna, P., Wang, Y., Vorsa, N., and Minko, T. (2006) Dendrimer versus linear conjugate: influence of polymeric architecture on the delivery and anticancer effect of paclitaxel. *Bioconjug. Chem.*, **17**, 1464–1472.

32 Svenson, S. (ed.) (2004) *"Carrier-Based Drug Delivery", ACS Symposium Series*, American Chemical Society, Washington, DC, p. 879.

33 Svenson, S. and Tomalia, D.A. (2005) Dendrimers in biomedical applications – reflections on the field. *Adv. Drug Deliv. Rev.*, **57**, 2106–2129.

34 Kukowska-Latallo, J.F., Bielinska, A.U., Johnson, J., Spindler, R., Tomalia, D., and Baker, A.J.R. Jr. (1996) Efficient transfer of genetic material into mammalian cells using starburst polyamidoamine dendrimers. *Proc. Natl. Acad. Sci. USA*, **93**, 4897–4902.

35 Malik, N., Evagorou, E.G., and Duncan, R. (1999) Dendrimer-platinate: a novel approach to cancer chemotherapy. *Anticancer Drugs*, **10**, 767–776.

36 Kobayashi, H. and Brechbiel, M.W. (2005) Nano-sized MRI contrast agents with dendrimer cores. *Adv. Drug Deliv. Rev.*, **57**, 2271–2286.

37 Patri, A.K., Majoros, I.J., and Baker, J.R. Jr. (2002) Dendritic polymer macromolecular carriers for drug delivery. *Curr. Opin. Chem. Biol.*, **6**, 466–471.

38 Khandare, J., Kolhe, P., Pillai, O., Kannan, S., Lieh-Lai, M., and Kannan, R.M. (2005) Synthesis, cellular transport, and activity of polyamidoamine dendrimer–methylprednisolone conjugates. *Bioconjug. Chem.*, **16**, 330–337.

39 Lee, C.C., MacKay, J.A., Fréchet, J.M.J., and Szoka, F.C. (2005) Designing dendrimers for biological applications. *Nat. Biotechnol.*, **23**, 1517–1526.

40 Khandare, J., Mohr, A., Calderon, M., Welker, P., Licha, K., and Haag, R. (2010) Design, synthesis, and cytotoxicity evaluation of dendritic polyglycerol architectures. *Biomaterials*, **31**, 4268–4277.

41 Malik, N., Wiwattanapatapee, R., Klopsch, R., Lorenz, K., Frey, H., Weener, J.W., *et al.* (2000) Dendrimers: relationship between structure and biocompatibility *in vitro*, and preliminary studies on the biodistribution of 125i-labelled polyamidoamine dendrimers *in vivo*. *J. Control. Release*, **65**, 133–148.

42 Khandare, J. and Minko, T. (2006) Polymer–drug conjugates: progress in polymeric prodrugs. *Prog. Polym. Sci.*, **31**, 359–397.

43 Dharap, S.S., Wang, Y., Chandna, P., Khandare, J.J., Qiu, B., Gunaseelan, S., Stein, S., Farmanfarmaian, A., and Minko, T. (2005) *Proc. Natl. Acad. Sci. USA*, **102**, 12962–12967.

44 Zhang, Y., Thomas, T.P., Desai, A., Zong, H., Leroueil, P.R., Majoros, I.J., and Baker, J.R. Jr. (2010) Targeted dendrimeric anticancer prodrug: a methotrexate-folic acid-poly(amidoamine) conjugate and a novel, rapid, "one pot" synthetic approach. *Bioconjug. Chem.*, **21**, 489–495.

45 de Groot, F.M., Damen, E.W., and Scheeren, H.W. (2001) Anticancer prodrugs for application in monotherapy: targeting hypoxia, tumor-associated enzymes, and receptors. *Curr. Med. Chem.*, **8**, 1093–1122.

46 Wagner, J., Lerner, R.A., and Barbas, C.F. III (1995) Efficient aldolase catalytic antibodies that use the enamine mechanism of natural enzymes. *Science*, **270**, 1797–1800.

47 Jensen, A.W., Maru, B.S., Zhang, X., Mohanty, D.K., Fahlman, B.D., Swanson, D.R., and Tomalia, D.A. (2005) Preparation of fullerene–shell dendrimer–core nanoconjugates. *Nano Lett.*, **5**, 1171–1173.

48 Padilla De Jesus, O.L., Ihre, H.R., Gagne, L., Frechet, J.M., and Szoka, F.C. Jr. (2002) Polyester dendritic systems for drug delivery applications: *in vitro* and *in vivo* evaluation. *Bioconjug. Chem.*, **13**, 453–461.

49 Martinez, J. and Bodanszky, M. (1978) Side reactions in peptide-synthesis, suppression of the formation of aminosuccinyl peptides with additives. *Int. J. Pept. Protein Res.*, **12**, 277–283.

50 Fox, M.E., Guillaudeu, S., Fre′chet, J.M.J., Jerger, K., Macaraeg, N., and Szoka, F.C. (2009) Synthesis and *in vivo* antitumor efficacy of PEGylated poly(l-lysine) dendrimer–camptothecin conjugates. *Mol. Pharmaceut.*, **6**, 1562–1572.

51 Kaminskas, L.M., Boyd, B.J., Karellas, P., Krippner, G.Y., Lessene, R., Kelly, B., and Porter, C.J.H. (2008) The impact of molecular weight and PEG chain length on the systemic pharmacokinetics of PEGylated poly L-lysine dendrimers. *Mol. Pharmaceut.*, **5**, 449–463.

52 Hodge, P. (1993) Polymer science branches out. *Nature*, **362**, 18–19.

53 Albertazzi, L., Serresi, M., Albanese, A., and Beltram, F. (2010) Dendrimer internalization and intracellular trafficking in living cells. *Mol. Pharmaceutics*, **7**, 680–688.

54 Shabat, D., Lode, H., Pertl, U., Reisfeld, R.A., Rader, C., Lerner, R.A., and Barbas, C.F. III (2001) *In vivo* activity in a catalytic antibody-prodrug system: antibody catalyzed etoposide prodrug activation for selective chemotherapy. *Proc. Natl. Acad. Sci. USA*, **98**, 7528–7533.

55 Tansey, W., Ke, S., Cao, X.-Y., Pasuelo, M.J., Wallace, S., and Li, C. (2004) Synthesis and characterization of branched poly(L-glutamic acid) as a biodegradable drug carrier. *J. Control. Release*, **94**, 39–51.

56 Sato, K., Hada, N., and Takeda, T. (2006) Syntheses of new peptidic glycoclusters derived from *β*-alanine: di- and trimerized glycoclusters and glycocluster-custers. *Carbohydr. Res.*, **341**, 836–845.

57 Almutairi, A., Guillaudeu, S.J., Berezin, M.Y., Achilefu, S., and Fréchet, J.M.J. (2008) Biodegradable pH-sensing dendritic nanoprobes for near-infrared fluorescence lifetime and intensity imaging. *J. Am. Chem. Soc.*, **130**, 444–445.

58 Niederhafner, P., Sebest′ik, J., and Jězek, J. (2008) Glycopeptide dendrimers. Part II. *J. Pept. Sci.*, **14**, 44–65.

59 Sawada, H. (1985) *Encyclopedia of Polymer Science and Engineering*, vol. 4 (eds H.F. Mark, N.M. Bikales, C.G. Overberger, G. Menges, and J.I. Kroschwitz), John Wiley & Sons, Inc., New York, pp. 719–745.

60 Seebach, D., Herrmann, G.F., Lengweiler, U.D., Bachmann, B.M., and Amrein, W. (1996) Enzyme-triggered disassembly of dendrimer-based amphiphilic nanocontainers. *Angew. Chem. Int. Ed. Engl.*, **35**, 2795–2797.

61 Tully, D.C., Wilder, K., Fréchet, J.M.J., Trimble, A.R., and Quate, C.F. (1999) Dendrimer-based self assembled monolayers as resists for scanning probe lithography. *Adv. Mater.*, **11**, 314–318.

11
Analytical Methods for Monitoring Biodegradation Processes of Environmentally Degradable Polymers

Maarten van der Zee

11.1
Introduction

This chapter presents an overview of the current knowledge on experimental methods for monitoring the biodegradability of polymeric materials. The focus is, in particular, on the biodegradation of materials under environmental conditions. Examples of *in vivo* degradation of polymers used in biomedical applications are not covered in detail but have been extensively reviewed elsewhere, e.g., [1–3]. Nevertheless, it is good to realize that the same principles of the methods for monitoring biodegradability of environmental polymers are also used for the evaluation of the degradation behavior of biomedical polymers.

A number of different aspects of assessing the potential, the rate, and the degree of biodegradation of polymeric materials are discussed. The mechanisms of polymer degradation and erosion receive attention and factors affecting enzymatic and nonenzymatic degradation are briefly addressed. Particular attention is given to the various ways for measuring biodegradation, including complete mineralization to gasses (such as carbon dioxide and methane), water, and possibly microbial biomass. Finally, some general conclusions are presented with respect to measuring biodegradability of polymeric materials.

11.2
Some Background

There is a worldwide research effort to develop biodegradable polymers for agricultural applications or as a waste management option for polymers in the environment. Until the end of the 20th century, most of the efforts were synthesis

Handbook of Biodegradable Polymers: Synthesis, Characterization and Applications, First Edition. Edited by Andreas Lendlein, Adam Sisson.
© 2011 Wiley-VCH Verlag GmbH & Co. KGaA. Published 2011 by Wiley-VCH Verlag GmbH & Co. KGaA.

oriented, and not much attention was paid to the identification of environmental requirements for, and testing of, biodegradable polymers. Consequently, many unsubstantiated claims to biodegradability were made, and this has damaged the general acceptance.

An important factor is that the term biodegradation has not been applied consistently. In the medical field of sutures, bone reconstruction, and drug delivery, the term biodegradation has been used to indicate degradation into macromolecules that stay in the body but migrate (e.g., UHMW polyethylene from joint prostheses), or hydrolysis into low-molecular-weight molecules that are excreted from the body (bioresorption), or dissolving without modification of the molecular weight (bioabsorption) [4, 5]. On the other hand, for environmentally degradable plastics, the term biodegradation may mean fragmentation, loss of mechanical properties, or sometimes degradation through the action of living organisms [6]. Deterioration or loss in physical integrity is also often mistaken for biodegradation [7]. Nevertheless, it is essential to have a universally acceptable definition of biodegradability to avoid confusion as to where biodegradable polymers can be used in agriculture or fit into the overall plan of polymer waste management. Many groups and organizations have endeavored to clearly define the terms "degradation," "biodegradation," and "biodegradability." But there are several reasons why establishing a single definition among the international communities has not been straightforward, including:

1) the variability of an intended definition given the different environments in which the material is to be introduced and its related impact on those environments,
2) the differences of opinion with respect to the scientific approach or reference points used for determining biodegradability,
3) the divergence of opinion concerning the policy implications of various definitions, and
4) challenges posed by language differences around the world.

As a result, many different definitions have officially been adopted, depending on the background of the defining organization and their particular interests. However, of more practical importance are the criteria for calling a material "biodegradable." A demonstrated potential of a material to biodegrade does not say anything about the time frame in which this occurs, nor the ultimate degree of degradation. The complexity of this issue is illustrated by the following common examples.

Low-density polyethylene has been shown to biodegrade slowly to carbon dioxide (0.35% in 2.5 years) [8], and according to some definitions can thus be called a biodegradable polymer. However, the degradation process is so slow in comparison with the application rate that accumulation in the environment will occur. The same applies for polyolefin–starch blends which rapidly loose strength, disintegrate, and visually disappear if exposed to microorganisms [9–11]. This is due to utilization of the starch component, but the polyolefin fraction will nevertheless persist in the environment. Can these materials be called "biodegradable"?

11.3 Defining Biodegradability

In 1992, an international workshop on biodegradability was organized to bring together experts from around the world to achieve areas of agreement on definitions, standards, and testing methodologies. Participants came from manufacturers, legislative authorities, testing laboratories, environmentalists, and standardization organizations in Europe, United States, and Japan. Since this fruitful meeting, there is a general agreement concerning the following key points [12].

1) For all practical purposes of applying a definition, material manufactured to be biodegradable must relate to a specific disposal pathway such as composting, sewage treatment, denitrification, and anaerobic sludge treatment.
2) The rate of degradation of a material manufactured to be biodegradable has to be consistent with the disposal method and other components of the pathway into which it is introduced, such that accumulation is controlled.
3) The ultimate end products of aerobic biodegradation of a material manufactured to be biodegradable are CO_2, water, and minerals and that the intermediate products include biomass and humic materials. (Anaerobic biodegradation was discussed in less detail by the participants.)
4) Materials must biodegrade safely and not negatively impact the disposal process or the use of the end product of the disposal.

As a result, specified periods of time, specific disposal pathways, and standard test methodologies were incorporated into definitions. Standardization organizations such as CEN, ISO, and ASTM were consequently encouraged to rapidly develop standard biodegradation tests so these could be determined. Society further demanded nondebatable criteria for the evaluation of the suitability of polymeric materials for disposal in specific waste streams such as composting or anaerobic digestion. Biodegradability is usually just one of the essential criteria, besides ecotoxicity, effects on waste treatment processes, etc.

In the following sections, biodegradation of polymeric materials is looked upon form the chemical perspective. The chemistry of the key degradation process is represented by Eq. (11.1) and (11.2), where $C_{polymer}$ represents either a polymer or a fragment from any of the degradation processes defined earlier. For simplicity here, the polymer or fragment is considered to be composed only of carbon, hydrogen, and oxygen; other elements may, of course, be incorporated in the polymer, and these would appear in an oxidized or reduced form after biodegradation depending on whether the conditions are aerobic or anaerobic, respectively.

Aerobic biodegradation:

$$C_{polymer} + O_2 \rightarrow CO_2 + H_2O + C_{residue} + C_{biomass} \tag{11.1}$$

Anaerobic biodegradation:

$$C_{polymer} \rightarrow CO_2 + CH_4 + H_2O + C_{residue} + C_{biomass} \tag{11.2}$$

Complete biodegradation occurs when no residue remains, and complete mineralization is established when the original substrate, $C_{polymer}$ in this example, is completely converted into gaseous products and salts. However, mineralization is a very slow process under natural conditions because some of the polymer undergoing biodegradation will initially be turned into biomass [13, 14]. Therefore, complete biodegradation, and not mineralization, is the measurable goal when assessing removal from the environment.

11.4 Mechanisms of Polymer Degradation

When working with biodegradable materials, the obvious question is why some polymers biodegrade and others do not. To understand this, one needs to know about the mechanisms through which polymeric materials are biodegraded. Although biodegradation is usually defined as degradation caused by biological activity (especially enzymatic action), it will usually occur simultaneously with – and is sometimes even initiated by – abiotic degradation such as photodegradation and simple hydrolysis. The following paragraphs give a brief introduction about the most important mechanisms of polymer degradation.

11.4.1 Nonbiological Degradation of Polymers

A great number of polymers is subject to hydrolysis, such as polyesters, polyanhydrides, polyamides, polycarbonates, polyurethanes, polyureas, polyacetals, and polyorthoesters. Different mechanisms of hydrolysis have been extensively reviewed not only for backbone hydrolysis but also for the hydrolysis of pendant groups [15–17]. The necessary elements for a wide range of catalysis, such as acids and bases, cations, nucleophiles and micellar, and phase transfer agents are usually present in most environments. In contrast to enzymatic degradation, where a material is degraded gradually from the surface inward (primarily because macromolecular enzymes cannot diffuse into the interior of the material), chemical hydrolysis of a solid material can take place throughout its cross section except for few hydrophobic polymers.

Important features affecting chemical polymer degradation and erosion include (i) the type of chemical bond, (ii) the pH, (iii) the temperature, (iv) the copolymer composition, and (v) water uptake (hydrophilicity). These features will not be discussed here, but have been covered in detail by Göpferich [4].

11.4.2
Biological Degradation of Polymers

Polymers represent major constituents of the living cells which are most important for the metabolism (enzyme proteins and storage compounds), the genetic information (nucleic acids), and the structure (cell wall constituents and proteins) of cells [18]. These polymers have to be degraded inside cells in order to be available for environmental changes and to other organisms upon cell lysis. It is therefore not surprising that organisms, during many millions of years of adaptation, have developed various mechanisms to degrade naturally occurring polymers. For the many different new synthetic polymers that have found their way into the environment only in the last 70 years, however, these mechanisms may not as yet have been developed.

There are many different degradation mechanisms that combine synergistically in nature to degrade polymers. Microbiological degradation can take place through the action of enzymes or by-products (such as acids and peroxides) secreted by microorganisms (bacteria, yeasts, fungi, etc.). Also macroorganisms can eat and, sometimes, digest polymers and cause mechanical, chemical, or enzymatic aging [19, 20].

Two key steps occur in the microbial polymer degradation process: first, a depolymerization or chain cleavage step, and second, mineralization. The first step normally occurs outside the organism due to the size of the polymer chain and the insoluble nature of many of the polymers. Extracellular enzymes are responsible for this step, acting either endo (random cleavage on the internal linkages of the polymer chains) or exo (sequential cleavage on the terminal monomer units in the main chain).

Once sufficiently small-size oligomeric or monomeric fragments are formed, they are transported into the cell where they are mineralized. At this stage, the cell usually derives metabolic energy from the mineralization process. The products of this process, apart from ATP, are gasses (e.g., CO_2, CH_4, N_2, and H_2), water, salts and minerals, and biomass. Many variations of this general view of the biodegradation process can occur, depending on the polymer, the organisms, and the environment. Nevertheless, there will always be, at one stage or another, the involvement of enzymes.

11.5
Measuring Biodegradation of Polymers

As can be imagined from the various mechanisms described above, biodegradation does not only depend on the chemistry of the polymer but also on the presence of the biological systems involved in the process. When investigating the

	(1) ***aquatic***	(2) ***high solids***
a) ***aerobic***	aerobic wastewater treatment plants surface waters, e.g., lakes and rivers marine environments	surface soils organic waste composting plants littering
b) ***anaerobic***	anaerobic wastewater treatment plants rumen of herbivores	deep sea sediments anaerobic sludge anaerobic digestion/ biogasification landfill

Figure 11.1 Schematic classification of different biodegradation environments for polymers.

biodegradability of a material, the effect of the environment cannot be neglected. Microbial activity and hence biodegradation is influenced by

1) the presence of microorganisms
2) the availability of oxygen
3) the amount of available water
4) the temperature
5) the chemical environment (pH, electrolytes, etc.).

In order to simplify the overall picture, the environments in which biodegradation occurs are basically divided in two environments: (a) aerobic (with oxygen available) and (b) anaerobic (no oxygen present). These two in turn can be subdivided into (1) aquatic and (2) high-solids environments. Figure 11.1 schematically presents the different environments, with examples in which biodegradation may occur [21, 22].

The high-solids environments will be the most relevant for measuring environmental biodegradation of polymeric materials, since they represent the conditions during biological municipal solid waste treatment, such as composting or anaerobic digestion (biogasification). However, possible applications of biodegradable materials other than in packaging and consumer products, for example, in fishing nets at sea, or undesirable exposure in the environment due to littering, explain the necessity of aquatic biodegradation tests.

Numerous ways for the experimental assessment of polymer biodegradability have been described in the scientific literature. Because of slightly different definitions or interpretations of the term "biodegradability," the different approaches are therefore not equivalent in terms of information they provide or the practical significance. Since the typical exposure to environment involves incubation of a polymer substrate with microorganisms or enzymes, only a limited number of measurements are possible: those pertaining to the substrates, to the microorganisms, or to the reaction products. Four common approaches available for studying biodegradation processes have been reviewed in detail by Andrady [13, 14]:

1) monitoring accumulation of biomass
2) monitoring the depletion of substrates
3) monitoring reaction products
4) monitoring changes in substrate properties.

In the following sections, different test methods for the assessment of polymer biodegradability are presented. Measurements are usually based on one of the four approaches given above, but combinations also occur. Before choosing an assay to simulate environmental effects in an accelerated manner, it is critical to consider the closeness of fit that the assay will provide between substrate, microorganisms, or enzymes, and the application or environment in which biodegradation should take place [23].

11.5.1 Enzyme Assays

11.5.1.1 Principle

In enzyme assays, the polymer substrate is added to a buffered or pH-controlled system, containing one or several types of purified enzymes. These assays are very useful in examining the kinetics of depolymerization, or oligomer or monomer release from a polymer chain under different assay conditions. The method is very rapid (minutes to hours) and can give quantitative information. However, mineralization rates cannot be determined with enzyme assays.

11.5.1.2 Applications

The type of enzyme to be used, and quantification of degradation, will depend on the polymer being screened. For example, Mochizuki *et al.* [24] studied the effects of draw ratio of polycaprolactone fibers on enzymatic hydrolysis by lipase. Degradability of PCL fibers was monitored by dissolved organic carbon (DOC) formation and weight loss. Similar systems with lipases have been used for studying the hydrolysis of broad ranges of aliphatic polyesters [25–30], copolyesters with aromatic segments [26, 31–33], and copolyesteramides [34, 35]. Other enzymes such as α-chymotrypsin and α-trypsin have also been applied for these polymers [36, 37]. Biodegradability of poly(vinyl alcohol) segments with respect to block length and stereochemical configuration has been studied using isolated poly(vinyl alcohol)-dehydrogenase [38]. Cellulolytic enzymes have been used to study the biodegradability of cellulose ester derivatives as a function of degree of substitution and the substituent size [39]. Similar work has been performed with starch esters using amylolytic enzymes such as α-amylases, β-amylases, glucoamylases, and amyloglucosidases [40]. Enzymatic methods have also been used to study the biodegradability of starch plastics or packaging materials containing cellulose [41–46].

11.5.1.3 Drawbacks

Caution must be taken in extrapolating enzyme assays as a screening tool for different polymers since the enzymes have been paired to only one polymer. The initially selected enzymes may show significantly reduced activity toward modified polymers or different materials, even though more suitable enzymes may exist in the environment. Caution must also be taken if the enzymes are not purified or appropriately stabilized or stored, since inhibitors and loss of enzyme activity can occur [23].

11.5.2 Plate Tests

11.5.2.1 Principle

Plate tests have initially been developed in order to assess the resistance of plastics to microbial degradation. Several methods have been standardized by standardization organizations such as the ASTM and the ISO [47–49]. They are now also used to see if a polymeric material will support growth [23, 50]. The principle of the method involves placing the test material on the surface of a mineral salts agar in a petri dish containing no additional carbon source. The test material and agar surface are sprayed with a standardized mixed inoculum of known bacteria and/or fungi. The test material is examined after a predetermined incubation period at constant temperature for the amount of growth on its surface and the rating is given.

11.5.2.2 Applications

Potts [51] used the method in his screening of 31 commercially available polymers for biodegradability. Other studies where the growth of either mixed or pure cultures of microorganisms is taken to be indicative for biodegradation have been reported [6]. The validity of this type of test and the use of visual assessment alone have been questioned by Seal and Pantke [52] for all plastics. They recommended that mechanical properties should be assessed to support visual observations. Microscopic examination of the surface can also give additional information.

A variation of the plate test is the "clear zone" technique [53], sometimes used to screen polymers for biodegradability. A fine suspension of polymer is placed in an agar gel as the sole carbon source, and the test inoculum is placed in wells bored in the agar. After incubation, a clear zone around the well, detected visually or instrumentally, is indicative of utilization of the polymer. The method has, for example, been used in the case of starch plastics [54], various polyesters [55–57], and polyurethanes [58].

11.5.2.3 Drawbacks

A positive result in an agar plate test indicates that an organism can grow on the substrate, but does not mean that the polymer is biodegradable, since growth may appear on contaminants, plasticizers present, oligomeric fractions still present in the polymer, and so on. Therefore, these tests should be treated with caution when extrapolating the data to field situations.

11.5.3 Respiration Tests

11.5.3.1 Principle

Aerobic microbial activity is typically characterized by the utilization of oxygen. Aerobic biodegradation requires oxygen for the oxidation of compounds to its mineral constituents such as CO_2, H_2O, SO_2, P_2O_5, etc. The amount of oxygen utilized during incubation, also called the biochemical (or biological) oxygen demand (BOD), is therefore a measure of the degree of biodegradation. Several test methods are based on measurement of the BOD, often expressed as a percentage of the theoretical oxygen demand (TOD) of the compound. The TOD, which is the theoretical amount of oxygen necessary for completely oxidizing a substrate to its mineral constituents, can be calculated by considering the elemental composition and the stoichiometry of oxidation [13, 59–62] or based on experimental determination of the chemical oxygen demand (COD) [13, 63].

11.5.3.2 Applications

The closed bottle BOD tests were designed to determine the biodegradability of detergents [61, 64]. These have stringent conditions due to the low level of inoculum (in the order of 10^5 microorganisms/L) and the limited amount of test substance that can be added (normally between 2 and 4 mg/L). These limitations originate from the practical requirement that the oxygen demand should not be more than half the maximum dissolved oxygen level in water at the temperature of the test, to avoid the generation of anaerobic conditions during incubation.

For nonsoluble materials such as polymers, less stringent conditions are necessary and alternative ways for measuring BOD were developed. Two-phase (semi) closed bottle tests provide higher oxygen content in the flasks and permit a higher inoculum level. Higher test concentrations are also possible, encouraging higher accuracy with directly weighing in of samples. The oxygen demand can alternatively be determined by periodically measuring the oxygen concentration in the aquatic phase by opening the flasks [60, 65–67], by measuring the change in volume or pressure in incubation flasks containing CO_2-absorbing agents [59, 68, 69], or by measuring the quantity of oxygen produced (electrolytically) to maintain constant gas volume/pressure in specialized respirometers [59, 62, 65, 66, 68].

11.5.3.3 Suitability

BOD tests are relatively simple to perform and sensitive, and are therefore often used as screening tests. However, the measurement of oxygen consumption is a nonspecific, indirect measure for biodegradation, and it is not suitable for determining anaerobic degradation. The requirement for test materials to be the sole carbon/energy source for microorganisms in the incubation media eliminates the use of oxygen measurements in complex natural environments.

11.5.4 Gas (CO_2 or CH_4) Evolution Tests

11.5.4.1 Principle

The evolution of carbon dioxide or methane from a substrate represents a direct parameter for mineralization. Therefore, gas evolution tests can be important tools in the determination of biodegradability of polymeric materials. A number of well-known test methods have been standardized for aerobic biodegradation, such as the (modified) Sturm test [70–75] and the laboratory-controlled composting test [76–79], as well as for anaerobic biodegradation, such as the anaerobic sludge test [80, 81] and the anaerobic digestion test [82, 83]. Although the principles of these test methods are the same, they may differ in medium composition, inoculum, the way substrates are introduced, and in the technique for measuring gas evolution.

11.5.4.2 Applications

Anaerobic tests generally follow biodegradation by measuring the increase in pressure and/or volume due to gas evolution, usually in combination with gas chromatographic analysis of the gas phase [84, 85]. Most aerobic standard tests apply continuous aeration; the exit stream of air can be directly analyzed continuously using a carbon dioxide monitor (usually infrared detectors) or titrimetrically after sorption in dilute alkali. The cumulative amount of carbon dioxide generated, expressed as a percentage of the theoretically expected value for total conversion to CO_2, is a measure for the extent of mineralization achieved. A value of 60% carbon conversion to CO_2, achieved within 28 days, is generally taken to indicate ready degradability. Taking into account that in this system there will also be incorporation of carbon into the formation of biomass (growth), the 60% value for CO_2 implies almost complete degradation. While this criterion is meant for water-soluble substrates, it is probably applicable to very finely divided moderately degradable polymeric materials as well [13]. Nevertheless, most standards for determining biodegradability of plastics consider a maximum test duration of 6 months.

Besides the continuously aerated systems, described above, several static respirometers have been described. Bartha and Yabannavar [86] describe a two-flask system; one flask, containing a mixture of soil and the substrate, is connected to another chamber holding a quantity of carbon dioxide sorbant. Care must be taken to ensure that enough oxygen is available in the flask for biodegradation. Nevertheless, this experimental setup and modified versions thereof have been successfully applied in the assessment of biodegradability of polymer films and food packaging materials [87–89]

The percentage of carbon converted to biomass instead of carbon dioxide depends on the type of polymer and the phase of degradation. Therefore, it has been suggested to regard the complete carbon balance to determine the degree of degradation [90]. This implies that besides the detection of gaseous carbon, also the amount of carbon in soluble and solid products needs to be determined. Soluble products, oligomers of different molecular size, intermediates, and proteins secreted from microbial cells can be measured as COD or as DOC. Solid products, biomass, and polymer remnants require a combination of procedures to separate and detect different fractions. The protein content of the insoluble fraction is usually determined to estimate the amount of carbon converted to biomass, using the assumptions that dry biomass consists of 50% protein, and that the carbon content of dry biomass is 50% [90–92].

11.5.4.3 Suitability

Gas evolution tests are popular test methods because they are relatively simple to perform and sensitive. A direct measure for mineralization is determined, and water-soluble or -insoluble polymers can be tested as films, powders, or objects. Furthermore, the test conditions and inoculum can be adjusted to fit the application or environment in which biodegradation should take place. Aquatic synthetic media are usually used, but also natural sea water [93, 94] or soil samples [86, 88, 89, 95] can be applied as biodegradation environments. A prerequisite for these media is that the background CO_2 evolution is limited, which excludes the application of real composting conditions. Biodegradation under composting conditions is therefore measured using an inoculum derived from matured compost with low respiration activity [76–78, 96, 97].

A drawback of using complex degradation environments such as mature compost is that simultaneous characterization of intermediate degradation products of determination of the carbon balance is difficult due to the presence of a great number of interfering compounds. To overcome this, an alternative test was developed based on an inoculated mineral bed-based matrix [98, 99].

11.5.5 Radioactively Labeled Polymers

11.5.5.1 Principle and Applications

Some materials tend to degrade very slowly under stringent test conditions without an additional source of carbon. However, if readily available sources of carbon are added, it becomes impossible to tell how much of the evolved carbon dioxide can be attributed to the decomposition of the plastic. The incorporation of radioactive ^{14}C in synthetic polymers gives a means of distinguishing between CO_2 or CH_4 produced by the metabolism of the polymer, and that generated by other carbon sources in the test environment. By comparison of the amount of radioactive $^{14}CO_2$ or $^{14}CH_4$ with the original radioactivity of the labeled polymer, it is possible to determine the percent by weight of carbon in the polymer which was mineralized during the duration of the exposure [51, 100–102]. Collection of radioactively labeled gasses or low-molecular-weight products can also provide extremely sensitive and reproducible methods to assess the degradation of polymers with low susceptibility to enzymes, such as polyethylene [8, 103] and cellulose acetates [104, 105].

11.5.5.2 Drawbacks

Problems with handling the radioactively labeled materials and their disposal are issues on the down side to this method. In addition, in some cases, it is difficult to synthesize the target polymer with the radioactive labels in the appropriate locations, with representative molecular weights, or with representative morphological characteristics.

11.5.6 Laboratory-Scale Simulated Accelerating Environments

11.5.6.1 Principle

Biodegradation of a polymer material is usually associated with changes in the physical, chemical, and mechanical properties of the material. It is indeed these changes, rather than the chemical reactions, which make the biodegradation process so interesting from an application point of view. These useful properties might be measured as a function of the duration of exposure to a biotic medium, to follow the consequences of the biodegradation process on material properties. The biotic media can be specifically designed in a laboratory scale as to mimic natural systems but with a maximum control of variables such as temperature, pH, microbial community, mechanical agitation, and supply of oxygen. Regulating these variables improves the reproducibility and may accelerate the degradation process. Laboratory simulations can also be used for the assessment of long-term effects due to continuous dosing on the activity and the environment of the disposal system [50].

11.5.6.2 Applications

The OECD Coupled Unit test [106] simulates an activated sludge sewage treatment system, but its application for polymers would be difficult as DOC is the parameter used to assess biodegradability. Krupp and Jewell [107] described well-controlled anaerobic and aerobic aquatic bioreactors to study degradation of a range of commercially available polymer films. A relatively low loading rate of the semicontinuous reactors and a long retention time were maintained to maximize the efficiency of biodegradation. Experimental setups have also been designed to simulate marine environments [108], soil burial conditions [108–110], composting environments [111–114], and landfill conditions [115] at laboratory scale, with controlled parameters such as temperature and moisture level, and a synthetic waste, to provide a standardized basis for comparing the degradation kinetics of films.

A wide choice of material properties can be followed during the degradation process. However, it is important to select one which is relevant to the end-use of the polymer material or provides fundamental information about the degradation process. Weight loss is a parameter frequently followed because it clearly demonstrates the disintegration of a biodegradable product [116–118]. Tensile properties are also often monitored, due to the interest in the use of biodegradable plastics in packaging applications [54, 119, 120]. In those polymers where the biodegradation involves a random scission of the macromolecular chains, a decrease in the average molecular weight and a general broadening of the molecular weight distribution provide initial evidence of a breakdown process [86, 121, 122]. However, no significant changes in material characteristics may be observed in recovered material if the mechanism of biodegradation involves bioerosion, that is, enzymatic or hydrolytic cleavage at the surface. Visual examination of the surface with various microscopic techniques can also give information on the biodegradation process [123–126]. Likewise, chemical and/or physical changes in the polymer may be followed by (combinations of) specific techniques such as infrared [10, 127] or UV spectroscopy [84, 128], nuclear magnetic resonance measurements [122–129], X-ray diffractometry [130, 131], and differential scanning calorimetry [132, 133].

11.5.6.3 Drawbacks

An inherent drawback in the use of mechanical properties, weight loss, molecular weights, or any other property which relies on the macromolecular nature of the substrate is that in spite of their sensitivity, these can only address the early stages of the biodegradation process. Furthermore, these parameters can give no information on the extent of mineralization. Especially in material blends or copolymers, the hydrolysis of one component can cause significant disintegration (and thus loss of weight and tensile properties), whereas other components may persist in the environment, even in disintegrated form [13]. Blends of starch, poly(3-hydroxybutyrate) or poly(ε-caprolactone) with polyolefins are examples of such systems [11, 43, 134].

11.5.7 Natural Environments, Field Trials

Exposures in natural environments provide the best true measure of the environmental fate of a polymer, because these tests include a diversity of organisms and achieve a desirable natural closeness of fit between the substrate, microbial agent, and the environment. However, the results of that exposure are only relevant to the specific environment studied, which is likely to differ substantially from many other environments. An additional problem is the timescale for this method, since the degradation process, depending on the environment, may be very slow (months to years) [23]. Moreover, little information on the degradation process can be gained other than the real time required for weight loss or total disintegration.

Nevertheless, field trials in natural environments are still used to extrapolate results acquired in laboratory tests to biodegradation behavior under realistic outdoor conditions [123, 135, 136].

11.6 Conclusions

The overview presented above makes clear that there is no such thing as a single optimal method for determining biodegradation of polymeric materials. First of all, biodegradation of a material is not only determined by the chemical composition and corresponding physical properties; the degradation environment in which the material is exposed also affects the rate and degree of biodegradation.

Furthermore, the method or test to be used depends on what information is requested.

One should realize that biodegradability is usually not of interest by itself. It is often just one aspect of health and environmental safety issues or integrated waste management concepts. It is fairly obvious but often neglected that one should always consider why a particular polymeric material should be (or not be) biodegradable when contemplating how to assess its biodegradability. After all, it is the intended application of the material that governs the most suitable testing environment, the parameters to be measured during exposure, and the corresponding limit values. For example, investigating whether biodegradation of a plastic material designed for food packaging could facilitate undesired growth of (pathogenic) microorganisms requires a completely different approach from investigating whether its waste can be discarded via composting (i.e., whether it degrades sufficiently rapid to be compatible with existing biowaste composting facilities).

In most cases, it will not be sufficient to ascertain macroscopic changes, such as weight loss and disintegration, or growth of microorganisms, because these observations may originate from biodegradation of just one of separate components. The ultimate fate of all individual components and degradation products must be included in the investigations. This implies that it is essential that both the polymeric materials and also intermediate degradation products have to be well characterized in order to understand the degradation process. For a good number of biodegradable materials, this means that a lot of work still needs to be done.

References

1 Hayashi, T. (1994) *Prog. Polym. Sci.*, **19**, 663.

2 Williams, D.F. and Zhong, S.P. (1994) *Int. Biodeterior. Biodegradation*, **34**, 95.

3 Buchanan, F. (ed.) (2008) *Degradation Rate of Bioresorbable Materials – Prediction and Evaluation*, Woodhead Publishing Limited, Cambridge, p. 397.

4 Göpferich, A. (1996) *Biomaterials*, **17**, 103.

5 Mabilleau, G. and Albertsson (2008) Sabokbar, in *Degradation Rate of Bioresorbable Materials – Prediction and Evaluation* (ed. F. Buchanan), Woodhead Publishing Limited, Cambridge, p. 145.

6 Albertsson, A.-C. and Karlsson, S. (1990) *Degradable Materials – Perspectives, Issues and Opportunities*, (eds S.A. Barenberg, J.L. Brash, R. Narayan, and A.E. Redpath), CRC Press, Boston, p. 263.

7 Palmisano, A.C. and Pettigrew, C.A. (1992) *Bioscience*, **42**, 680.

8 Albertsson, A.-C. and Rånby, B. (1979) *J. Appl. Polym. Sci. Appl. Polym. Symp.*, **35**, 423.

9 Austin, R.G. (1990) *Degradable Materials – Perspectives, Issues and Opportunities* (eds S.A. Barenberg, J.L. Brash, R. Narayan, and A.E. Redpath), CRC Press, Boston, p. 209.

10 Goheen, S.M. and Wool, R.P. (1991) *J. Appl. Polym. Sci.*, **42**, 2691.

11 Breslin, V.T. (1993) *J. Environ. Polym. Degrad.*, **1**, 127.

12 Anonymous (1992) *Towards Common Ground – Meeting Summary of the International Workshop on Biodegradability, Annapolis, MD, USA, 20–21 October, 1992.*

13 Andrady, A.L. (1994) *J.M.S. – Rev. Macromol. Chem. Phys.*, **C34**, 25.

14 Andrady, A.L. (2000) *Handbook of Polymer Degradation*, 2nd edn (ed. S.H. Hamid), Marcel Dekker, New York, p. 441.

15 St.Pierre, T. and Chiellini, E. (1986) *Bioact. Compat. Polym.*, **1**, 467.

16 St.Pierre, T. and Chiellini, E. (1987) *Bioact. Compat. Polym.*, **2**, 4.

17 Cameron, R.E. and Kamvari-Moghaddam, A. (2008) *Degradation Rate of Bioresorbable Materials – Prediction and Evaluation* (ed. F. Buchanan), Woodhead Publishing Limited, Cambridge, p. 43.

18 Stryer, L. (1981) *Biochemistry*, 2nd edn. W.H. Freeman and Company, New York, USA.

19 Anderson, T.A., Tsao, R., and Coats, J.R. (1993) *J. Environ. Polym. Degrad.*, **1**, 301.

20 Whitney, P.J., Swaffield, C.H., and Graffam, A.J. (1993) *Int. Biodeter. Biodegrad.*, **31**, 179.

21 Van der Zee, M., Stoutjesdijk, J.H., Van der Heijden, P.A.A.W., and De Wit, D. (1995) *J. Environ. Polym. Degrad.*, **3**, 235.

22 Eggink, G., Van der Zee, M., and Sijtsma, L. (1995) *International edition of the IOP on Environmental Biotechnology*, 7–8.

23 Mayer, J.M. and Kaplan, D.L. (1993) *Biodegradable Polymers and Packaging* (eds C. Ching, D.L. Kaplan , and E.L. Thomas), Technomic Publishing, Lancaster-Basel, p. 233.

24 Mochizuki, M., Hirano, M., Kanmuri, Y., Kudo, K., and Tokiwa, Y. (1995) *J. Appl. Polym. Sci.*, **55**, 289.

25 Tokiwa, Y. and Suzuki, T. (1981) *J. Appl. Polym. Sci.*, **26**, 441.

26 Tokiwa, Y., Suzuki, T., and Takeda, K. (1986) *Agric. Biol. Chem.*, **50**, 1323.

27 Arvanitoyannis, I., Nakayama, A., Kawasaki, N., and Yamamoto, N. (1995) *Polymer*, **36**, 2271.

28 Nakayama, A., Kawasaki, N., Arvanitoyannis, I., Iyoda, J., and Yamamoto, N. (1995) *Polymer*, **36**, 1295.

29 Walter, T., Augusta, J., Müller, R.-J., Widdecke, H., and Klein, J. (1995) *Enzym. Microb. Technol.*, **17**, 218.

30 Nagata, M., Kiyotsukuri, T., Ibuki, H., Tsutsumi, N., and Sakai, W. (1996) *React. Funct. Polym.*, **30**, 165.

31 Jun, H.S., Kim, B.O., Kim, Y.C., Chang, H.N., and Woo, S.I. (1994) *J. Environ. Polym. Degrad.*, **2**, 9.

32 Chiellini, E., Corti, A., Giovannini, A., Narducci, P., Paparella, A.M., and Solaro, R. (1996) *J. Environ. Polym. Degrad.*, **4**, 37.

33 Nagata, M., Kiyotsukuri, T., Minami, S., Tsutsumi, N., and Sakai, W. (1996) *Polym. Int.*, **39**, 83.

34 Nagata, M. and Kiyotsukuri, T. (1994) *Eur. Polym. J.*, **30**, 1277.

35 Nagata, M. (1996) *Macromol. Rap. Commun.*, **17**, 583.

36 Arvanitoyannis, I., Nikolaou, E., and Yamamoto, N. (1994) *Polymer*, **35**, 4678.

37 Arvanitoyannis, I., Nikolaou, E., and Yamamoto, N. (1995) *Macromol. Chem. Phys.*, **196**, 1129.

38 Matsumura, S., Shimura, Y., Toshima, K., Tsuji, M., and Hatanaka, T. (1995) *Macromol. Chem. Phys.*, **196**, 3437.

39 Glasser, W.G., McCartney, B.K., and Samaranayake, G. (1994) *Biotechnol. Prog.*, **10**, 214.

40 Rivard, C., Moens, L., Roberts, K., Brigham, J., and Kelley, S. (1995) *Enzym. Microb. Technol.*, **17**, 848.

41 Strantz, A.A. and Zottola, E.A. (1992) *J. Food Protect.*, **55**, 736.

42 Coma, V., Couturier, Y., Pascat, B., Bureau, G., Cuq, J.L., and Guilbert, S. (1995) *Enzyme Microb. Technol.*, **17**, 524.

43 Imam, S.H., Gordon, S.H., Burgess-Cassler, A., and Greene, R.V. (1995) *J. Environ. Polym. Degrad.*, **3**, 107.

44 Imam, S.H., Gordon, S.H., Shogren, R.L., and Greene, R.V. (1995) *J. Environ. Polym. Degrad.*, **3**, 205.

45 Vikman, M., Itävaara, M., and Poutanen, K. (1995) *J.M.S. – Pure Appl. Chem.*, **A32**, 863.

46 Vikman, M., Itävaara, M., and Poutanen, K. (1995) *J. Environ. Polym. Degrad.*, **3**, 23.

47 ASTM (2009) G21-96. Standard Practice for Determining Resistance of Synthetic Polymeric Materials to Fungi, American Society for Testing and Materials (ASTM), Philadelphia, PA, USA.

48 ASTM (1996) G22-76. Standard Practice for Determining Resistance of Plastics to Bacteria, American Society for

Testing and Materials (ASTM), Philadelphia, PA, USA (withdrawn in 2002).

49 International Standard (1997) ISO 846. Plastics – Evaluation of the action of micro-organisms, International Organization for Standardization (ISO), Genève, Switzerland.

50 Seal, K.J. (1994) *Chemistry and Technology of Biodegradable Polymers* (ed. G.J.L. Griffin), Blackie Academic and Professional, London, p. 116.

51 Potts, J.E. (1978) *Aspects of Degradation and Stabilization of Polymers* (ed. H.H.G. Jellinek), Elsevier Scientific Publishing Co., Amsterdam, p. 617.

52 Seal, K.J. and Pantke, M. (1986) *Mater. Org.*, **21**, 151.

53 Delafield, F.P., Doudoroff, M., Palleroni, N.J., Lusty, C.J., and Contopoulos, R. (1965) *J. Bacteriol.*, **90**, 1455.

54 Gould, J.M., Gordon, S.H., Dexter, L.B., and Swanson, C.L. (1990) *Agricultural and Synthetic Polymers – Biodegradability and Utilization* (eds J.E. Glass and G. Swift), American Chemical Society, Washington, DC, ACS Symposium Series 433. p. 65.

55 Augusta, J., Müller, R.-J., and Widdecke, H. (1993) *Appl. Microbiol. Biotechnol.*, **39**, 673.

56 Nishida, H. and Tokiwa, Y. (1994) *Chem. Lett.*, **3**, 421.

57 Nishida, H. and Tokiwa, Y. (1994) *Chem. Lett.*, **7**, 1293.

58 Crabbe, J.R., Campbell, J.R., Thompson, L., Walz, S.L., and Schultz, W.W. (1994) *Int. Biodeter. Biodegrad.*, **33**, 103.

59 International Standard (1999) ISO 9408:1999(E). Water quality – Evaluation of ultimate aerobic biodegradability of organic compounds in aqueous medium by determination of oxygen demand in a closed respirometer, International Organization for Standardization (ISO), Genève, Switzerland.

60 International Standard (1997) ISO 10708:1997(E). Water quality – Evaluation in an aqueous medium of the ultimate aerobic biodegradability of organic compounds – Determination of biochemical oxygen demand in a two-phase closed bottle test, International Organization for Standardization (ISO), Genève, Switzerland.

61 OECD (1993) 301D. Ready Biodegradability: Closed Bottle Test, Guidelines for Testing of Chemicals, Organization for Economic Cooperation and Development (OECD), Paris, France.

62 OECD (1993) 302C. Inherent Biodegradability: Modified MITI Test (II), Guidelines for Testing of Chemicals, Organization for Economic Cooperation and Development (OECD), Paris, France.

63 International Standard (1989) ISO 6060:1989(E). Water quality – Determination of the chemical oxygen demand, International Organization for Standardization (ISO), Genève, Switzerland.

64 International Standard (1997) ISO 10707:1997(E). Water quality – Evaluation in an aqueous medium of the "ultimate" aerobic biodegradability of organic compounds – Method by analysis of biochemical oxygen demand (closed bottle test), International Organization for Standardization (ISO), Genève, Switzerland.

65 International Standard (2004) ISO 14851:2004(E). Determination of the ultimate aerobic biodegradability of plastic materials in an aqueous medium – Method by measuring the oxygen demand in a closed respirometer, International Organization for Standardization (ISO), Genève, Switzerland.

66 ASTM (2002) ASTM D5271 02 Standard Test Method for Determining the Aerobic Biodegradation of Plastic Materials in an Activated-Sludge-Wastewater-Treatment System. American Society for Testing and Materials (ASTM), Philadelphia, PA, USA (withdrawn in 2011).

67 European Standard (2003) EN 14048:2003. Packaging – Determination of the ultimate aerobic biodegradability of packaging materials in an aqueous medium – Method by measuring the oxygen demand in a closed respirometer, European Committee for Standardization (CEN), Brussels, Belgium.

68 OECD (1993) 301F. Manometric Respirometry Test, Guidelines for Testing of Chemicals, Organization for Economic Cooperation and Development (OECD), Paris, France.

69 Tilstra, L. and Johnsonbaugh, D. (1993) *J. Environ. Polym. Degrad.*, **1**, 247.

70 ASTM (1992) D5209-92. Standard Test Method for Determining the Aerobic Biodegradation of Plastic Materials in the Presence of Municipal Sewage Sludge, American Society for Testing and Materials (ASTM), Philadelphia, PA, USA (withdrawn in 2004).

71 International Standard (2000) ISO 9439:2000(E). Water quality – Evaluation of ultimate aerobic biodegradability of organic compounds in aqueous medium – Carbon dioxide evolution test, International Organization for Standardization (ISO), Genève, Switzerland.

72 International Standard (2004) ISO 14852:2004(E). Determination of the ultimate aerobic biodegradability of plastic materials in an aqueous medium – Method by analysis of evolved carbon dioxide, International Organization for Standardization (ISO), Genève, Switzerland.

73 European Standard (2003) EN 14047:2003. Packaging – Determination of the ultimate aerobic biodegradability of packaging materials in an aqueous medium – Method by analysis of evolved carbon dioxide, European Committee for Standardization (CEN), Brussels, Belgium.

74 OECD (1993) 301B. Ready Biodegradability: Modified Sturm Test, Guidelines for Testing of Chemicals, Organization for Economic Cooperation and Development (OECD), Paris, France.

75 ASTM (2009) D6691-09. Standard Test Method for Determining Aerobic Biodegradation of Plastic Materials in the Marine Environment by a Defined Microbial Consortium or Natural Sea Water Inoculum, American Society for Testing and Materials (ASTM), Philadelphia, PA, USA.

76 ASTM (2003) D5338-98(2003). Standard Test Method for Determining Aerobic Biodegradation of Plastic Materials Under Controlled Composting Conditions, American Society for Testing and Materials (ASTM), Philadelphia, PA, USA.

77 International Standard (2007) 14855-1:2007(E). Determination of the ultimate aerobic biodegradability of plastic materials under controlled composting conditions – Method by analysis of evolved carbon dioxide – Part 1: General method, International Organization for Standardization (ISO), Genève, Switzerland.

78 International Standard (2009) 14855-2:2009(E). Determination of the ultimate aerobic biodegradability of plastic materials under controlled composting conditions – Method by analysis of evolved carbon dioxide – Part 2: Gravimetric measurement of carbon dioxide evolved in a laboratory-scale test, International Organization for Standardization (ISO), Genève, Switzerland.

79 European Standard (2003) EN 14046:2003. Packaging – Evaluation of the ultimate aerobic biodegradability of packaging materials under controlled composting conditions – Method by analysis of released carbon dioxide, European Committee for Standardization (CEN), Brussels, Belgium.

80 ASTM (2007) D5210-92. Standard Test Method for Determining the Anaerobic Biodegradation of Plastic Materials in the Presence of Municipal Sewage Sludge, American Society for Testing and Materials (ASTM), Philadelphia, PA, USA.

81 International Standard (1998) ISO 11734:1998(E). Water quality – Evaluation of the “ultimate” anaerobic biodegradability of organic compounds in digested sludge – Method by measurement of the biogas production, International Organization for Standardization (ISO), Genève, Switzerland.

82 ASTM (2011) D5511-11. Standard Test Method for Determining Anaerobic Biodegradation of Plastic Materials Under High-Solids Anaerobic-Digestion

Conditions, American Society for Testing and Materials (ASTM), Philadelphia, PA, USA.
83 ASTM (2011) D5526-94(2011)e1. Standard Test Method for Determining Anaerobic Biodegradation of Plastic Materials Under Accelerated Landfill Conditions, American Society for Testing and Materials (ASTM), Philadelphia, PA, USA.
84 Day, M., Shaw, K., and Cooney, J.D. (1994) *J. Environ. Polym. Degrad.*, **2**, 121.
85 Puechner, P., Mueller, W.-R., and Bardtke, D. (1995) *J. Environ. Polym. Degrad.*, **3**, 133.
86 Bartha, R. and Yabannavar, A. (1995) *Proceedings of the Fourth International Workshop on Biodegradable Plastics and Polymers and Fourth Annual Meeting of the Bio-Environmentally Degradable Polymer Society, Durham, NH, USA, October 11–14, 1995.*
87 Andrady, A.L., Pegram, J.E., and Tropsha, Y. (1993) *J. Environ. Polym. Degrad.*, **1**, 171.
88 Yabannavar, A. and Bartha, R. (1993) *Soil Biol. Biochem.*, **25**, 1469.
89 Yabannavar, A.V. and Bartha, R. (1994) *Appl. Environ. Microbiol.*, **60**, 3608.
90 Urstadt, S., Augusta, J., Müller, R.-J., and Deckwer, W.-D. (1995) *J. Environ. Polym. Degrad.*, **3**, 121.
91 Itävaara, M. and Vikman, M. (1995) *Chemosphere*, **31**, 4359.
92 Spitzer, B., Mende, C., Menner, M., and Luck, T. (1996) *J. Environ. Polym. Degrad.*, **4**, 157.
93 Allen, A.L., Mayer, J.M., Stote, R., and Kaplan, D.L. (1994) *J. Environ. Polym. Degrad.*, **2**, 237.
94 Courtes, R., Bahlaoui, A., Rambaud, A., Deschamps, F., Sunde, E., and Dutriex, E. (1995) *Ecotoxicol. Environ. Saf.*, **31**, 142.
95 Barak, P., Coquet, Y., Halbach, T.R., and Molina, J.A.E. (1991) *J. Environ. Qual.*, **20**, 173.
96 Pagga, U., Beimborn, D.B., Boelens, J., and De Wilde, B. (1995) *Chemosphere*, **31**, 4475.
97 Pagga, U., Beimborn, D.B., and Yamamoto, M. (1996) *J. Environ. Polym. Degrad.*, **4**, 173.
98 Tosin, M., Degli-Innocenti, F., and Bastioli, C. (1998) *J. Environ. Polym. Degr.*, **6**, 79.
99 Bellia, G., Tosin, M., Floridi, G., and Degli-Innocenti, F. (1999) *Polym. Degrad. Stabil.*, **66**, 65.
100 ASTM (2007) D6340-98. Standard Test Methods for Determining Aerobic Biodegradation of Radiolabeled Plastic Materials in an Aqueous or Compost Environment, American Society for Testing and Materials (ASTM), Philadelphia, PA, USA.
101 ASTM (2001) D6692-01. Standard Test Method for Determining the Biodegradability of Radiolabeled Polymeric Plastic Materials in Seawater, American Society for Testing and Materials (ASTM), Philadelphia, PA, USA (withdrawn in 2010).
102 ASTM (2002) D6776-02. Standard Test Method for Determining Anaerobic Biodegradability of Radiolabeled Plastic Materials in a Laboratory-Scale Simulated Landfill Environment, American Society for Testing and Materials (ASTM), Philadelphia, PA, USA.
103 Albertsson, A.-C., Barenstedt, C., and Karlsson, S. (1993) *J. Environ. Polym. Degrad.*, **1**, 241.
104 Komarek, R.J., Gardner, R.M., Buchanan, C.M., and Gedon, S. (1993) *J. Appl. Polym. Sci.*, **50**, 1739.
105 Buchanan, C.M., Dorschel, D., Gardner, R.M., Komarek, R.J., Matosky, A.J., White, A.W., and Wood, M.D. (1996) *J. Environ. Polym. Degrad.*, **4**, 179.
106 OECD (1993) 303A. Simulation Test–Aerobic Sewage Treatment: Coupled Units Test. Guidelines for Testing of Chemicals, Organization for Economic Cooperation and Development (OECD), Paris, France.
107 Krupp, L.R. and Jewell, W.J. (1992) *Environ. Sci. Technol.*, **26**, 193.
108 Kaplan, D.L., Mayer, J.M., Greenberger, M., Gross, R., and McCarthy, S. (1994) *Polym. Degrad. Stab.*, **45**, 165.
109 Dale, R. and Squirrell, D.J. (1990) *Int. Biodeterior.*, **26**, 355.
110 Seal, K.J. and Pantke, M. (1990) *Mater. Org.*, **25**, 87.

111 Gardner, R.M., Buchanan, C.M., Komarek, R., Dorschel, D., Boggs, C., and White, A.W. (1994) *J. Appl. Polym. Sci.*, **52**, 1477.

112 Buchanan, C.M., Dorschel, D.D., Gardner, R.M., Komarek, R.J., and White, A.W. (1995) *J.M.S.–Pure Appl. Chem.*, **A32**, 683.

113 Gross, R.A., Gu, J.-D., Eberiel, D., and McCarthy, S.P. (1995) *J.M.S.--Pure Appl. Chem.*, **A32**, 613.

114 European Standard (2003) EN 14045:2003. Packaging–Evaluation of the disintegration of packaging materials in practical oriented tests under defined composting conditions, European Committee for Standardization (CEN), Brussels, Belgium.

115 Smith, G.P., Press, B., Eberiel, D., McCarthy, S.P., Gross, R.A., and Kaplan, D.L. (1990) *Polym. Mater. Sci. Eng.*, **63**, 862.

116 Coma, V., Couturier, Y., Pascat, B., Bureau, G., Guilbert, S., and Cuq, J.L. (1994) *Pack. Techn. Sci.*, **7**, 27.

117 Buchanan, C.M., Boggs, C.N., Dorschel, D., Gardner, R.M., Komarek, R.J., Watterson, T.L., and White, A.W. (1995) *J. Environ. Polym. Degrad.*, **3**, 1.

118 Goldberg, D. (1995) *J. Environ. Polym. Degrad.*, **3**, 61.

119 Iannotti, G., Fair, N., Tempesta, M., Neibling, H., Hsieh, F.H., and Mueller, M. (1990) *Degradable Materials–Perspectives, Issues and Opportunities* (eds S.A. Barenberg, J.L. Brash, R. Narayan, and A.E. Redpath), CRC Press, Boston, p. 425.

120 Mergaert, J., Webb, A., Anderson, C., Wouters, A., and Swings, J. (1993) *Appl. Environ. Microbiol.*, **59**, 3233.

121 Tilstra, L. and Johnsonbaugh, D. (1993) *J. Environ. Polym. Degrad.*, **1**, 257.

122 Hu, D.S.G. and Liu, H.J. (1994) *J. Appl. Polym. Sci.*, **51**, 473.

123 Greizerstein, H.B., Syracuse, J.A., and Kostyniak, P.J. (1993) *Polym. Degrad. Stab.*, **39**, 251.

124 Lopez-Llorca, L.V. and Colom Valiente, M.F. (1993) *Micron*, **24**, 457.

125 Nishida, H. and Tokiwa, Y. (1993) *J. Environ. Polym. Degrad.*, **1**, 227.

126 Bastioli, C., Cerutti, A., Guanella, I., Romano, G.C., and Tosin, M. (1995) *J. Environ. Polym. Degrad.*, **3**, 81.

127 Kay, M.J., McCabe, R.W., and Morton, L.H.G. (1993) *Int. Biodeter. Biodegrad.*, **31**, 209.

128 Allen, N.S., Edge, M., Mohammadian, M., and Jones, K. (1994) *Polym. Degrad. Stab.*, **43**, 229.

129 Löfgren, A. and Albertsson, A.-C. (1994) *J. Appl. Polym. Sci.*, **52**, 1327.

130 Albertsson, A.-C. and Karlsson, S. (1995) *Macromol. Symp.*, **98**, 797.

131 Schurz, J., Zipper, P., and Lenz, J. (1993) *J.M.S.–Pure Appl. Chem.*, **A30**, 603.

132 Albertsson, A.-C., Barenstedt, C., and Karlsson, S. (1994) *J. Appl. Polym. Sci.*, **51**, 1097.

133 Santerre, J.P., Labow, R.S., Duguay, D.G., Erfle, D., and Adams, G.A. (1994) *J. Biomed. Mat. Res.*, **28**, 1187.

134 Iwamoto, A. and Tokiwa, Y. (1994) *Polym. Degrad. Stab.*, **45**, 205.

135 Leonas, K.K., Cole, M.A., and Xiao, X.-Y. (1994) *J. Environ. Polym. Degrad.*, **2**, 253.

136 Halley, P., Rutgers, R., Coombs, S., Kettels, J., Gralton, J., Christie, G., Jenkins, M., Beh, H., Griffin, K., Jayasekara, R., and Lonergan, G. (2001) *Starch–Stärke*, **53**, 362.

12
Modeling and Simulation of Microbial Depolymerization Processes of Xenobiotic Polymers

Masaji Watanabe and Fusako Kawai

12.1
Introduction

Microbial depolymerization processes are classified into two categories, exogenous type and endogenous type. In an exogenous depolymerization process, molecules reduce their sizes by separation of monomer units from their terminals. Examples of polymers subject to exogenous depolymerization processes include polyethylene (PE). PE is structurally a long-chain alkane of normal type. The initial step of the oxidation of *n*-alkanes is hydroxylation to produce the corresponding primary (or secondary) alcohol, which is oxidized further to an aldehyde (or ketone) and then to an acid. Carboxylated *n*-alkanes are structurally analogous to fatty acids and subject to β-oxidation processes to produce depolymerized fatty acids by liberating two carbon units (acetic acid). It is also shown by gel permeation chromatography (GPC) analysis of PEwax before and after cultivation of a bacterial consortium KH-12 that small molecules are consumed faster than large ones [1].

As is seen in the previous discussion, the mechanism of PE biodegradation is based on two essential factors: the gradual weight loss of large molecules due to the β-oxidation and the direct consumption or absorption of small molecules by cells. A mathematical model based on those factors was proposed, and PE biodegradation was studied using the model [2–5]. The biodegradability of PE between the microbial consortium KH-12 and the fungus *Aspergillus* sp. AK-3 was compared [4]. The transition of weight distribution of PE over 5 weeks of cultivation was numerically simulated using the weight distribution before and after 3 weeks of cultivation, and a numerical result is compared with an experimental result [5].

Polyethylene glycol (PEG) is another example of polymer subject to exogenous depolymerization processes. PEG is depolymerized by liberating C_2 compounds, either aerobically or anaerobically [6, 7]. The mathematical techniques originally developed for the PE biodegradation was extended to cover the biodegradation of PEG. Problems were formulated to determine degradation rates based on the weight distribution of PEG with respect to molecular weight before and after the cultivation of the microbial consortium E-1 [8]. Those problems were solved

Handbook of Biodegradable Polymers: Synthesis, Characterization and Applications, First Edition. Edited by Andreas Lendlein, Adam Sisson.
© 2011 Wiley-VCH Verlag GmbH & Co. KGaA. Published 2011 by Wiley-VCH Verlag GmbH & Co. KGaA.

numerically, and the transition of the weight distribution was simulated [9, 10]. Dependence of degradation rate on time was also considered in modeling and simulation of depolymerization processes of PEG [11–13].

Unlike exogenous type depolymerization processes in which monomer units are separated from terminals of molecules, molecules are separated internally in endogenous type depolymerization processes. Hydrolysis is often involved in endogenous type epolymerization processes, while oxidation plays an essential role in exogenous type depolymerization processes. One of the characteristics of endogenous type depolymerization processes is the rapid breakdown of large molecules to produce small molecules in an early stage of depolymerization, whereas molecules lose their weight gradually throughout these processes. Polyvinyl alcohol (PVA) is an example of polymer subject to endogenous type depolymerization. PVA is a carbon-chain polymer with a hydroxyl group attached to every other carbon unit. It is degraded by random oxidation of hydroxyl groups and hydrolysis of mono/diketones. A mathematical model for endogenous depolymerization process was proposed, and enzymatic depolymerization process of PVA was studied. [14–16]. Mathematical model originally proposed for the enzymatic degradation of PVA was applied to enzymatic degradation of polylactic acid (PLA), and the degradability of PVA and PLA was compared [17]. Dependence of degradation rate on time was considered in study of depolymerization processes of PLA [18].

In the following sections, the mathematical models for exogenous type and endogenous type depolymerization processes are described. Numerical techniques to determine degradation rates and to simulate transitions of weight distribution are illustrated. Some numerical results are also introduced.

12.2 Analysis of Exogenous Depolymerization

12.2.1 Modeling of Exogenous Depolymerization

Polyolefins are regarded as linear saturated hydrocarbons, and considered chemically inert in a natural setting. However, it has been shown that PE is slowly degraded and its degradation is promoted by irradiation or oxidation. Slow degradation of PE was shown by measurement of $^{14}CO_2$ generation [19]. Linear paraffin molecules of molecular weight up to approximately 500 were utilized by several microorganisms [20]. Oxidation of *n*-alkanes up to tetratetracontane ($C_{44}H_{90}$, mass of 618) in 20 days was reported [21]. Several experiments were performed to investigate the biodegradability of PE. Commercially available PEwax was used as a sole carbon source for soil microorganisms [1]. Microbial consortium KH-12 obtained from soil samples degraded PEwax, which was confirmed by significant weight loss (30–50%). GPC analysis of PEwax showed that small molecules were consumed faster than large ones in the depolymerization processes of PE.

While experiments revealed the nature of the microbial depolymerization process of PE, it was also viewed theoretically. PE is classified structurally as hydrocarbon, and it is subject to the following metabolic pathways [22]:

1) Terminal oxidation:

$$RCH_3 \rightarrow RCH_2OH \rightarrow RCHO \rightarrow RCOOH$$

2) Diterminal oxidation:

$$H_3CRCH_3 \rightarrow CH_3RCOOH \rightarrow HOH_2CRCOOH \rightarrow$$
$$OHCRCOOH \rightarrow HOOCRCOOH$$

3) Subterminal oxidation:

$$RCH_2CH_2CH_3 \rightarrow RCH_2CH(OH)CH_3 \rightarrow RCH_2C(O)CH_3 \rightarrow$$
$$RCH_2OC(O)CH_3 \rightarrow RCH_2OH + CH_3COOH$$

A PE molecule carboxylated by one of these oxidation processes is structurally analogous to the fatty acid, and becomes subject to β-oxidation. Then a series of terminal separation of monomer units follow.

In view of the foregoing theoretical and experimental aspects of PE biodegradation, the following assumptions were made:

1) Each molecule loses its weight by a fixed amount per unit time.
2) Some molecules are directly consumed by microorganisms.
3) The consumption rate per unit time depends on the sizes of molecules.

The mathematical model (12.1) based on these assumptions was proposed, and the biodegradability of PE was studied by analyzing the model [2–5, 14]

$$\frac{dx}{dt} = -\alpha(M)x + \beta(M+L)\frac{M}{M+L}y \quad (\alpha(M) = \rho(M) + \beta(M)) \tag{12.1}$$

where variables t and M represent the cultivation time and the molecular weight, respectively. The variable x equals $w(t,M)$ which denotes the total weight of M molecules (the PE molecules with molecular weight M) present at time t. The parameter L represents the amount of the weight loss due to the terminal separation, and the variable y is given by $y = w(t, M+L)$, that is, the total weight of $(M+L)$-molecules present at time t. The functions $\rho(M)$ and $\beta(M)$ represent the direct consumption rate and the weight conversion rate from the class of M-molecules to the class of $(M-L)$-molecules, respectively. The first term of the right-hand side of Eq. (12.1) is the total weight loss in the class of M-molecules due to the direct consumption and the β-oxidation, and the second term represents the weight conversion from the class of $(M+L)$-molecules to the class of M-molecules due to the β-oxidation.

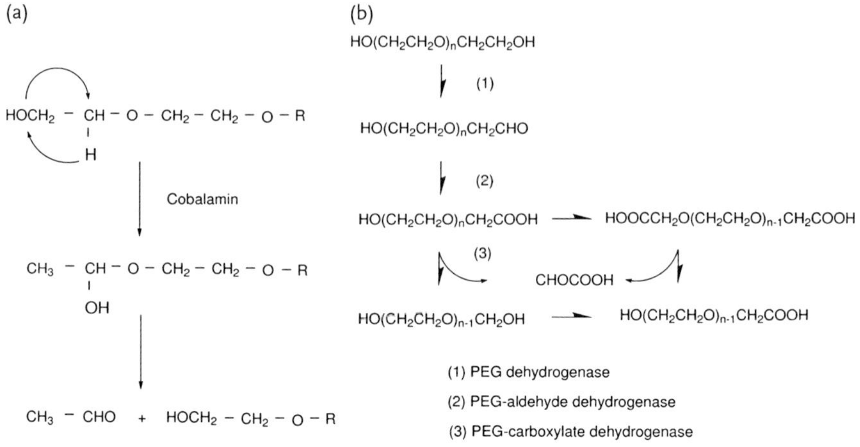

Figure 12.1 Anaerobic metabolism (a) and aerobic metabolism (b) of PEG.

The mathematical model (12.1) was originally proposed for the PE biodegradation. However, it can be viewed as a general biodegradation model for exogenous depolymerization processes, which covers not only the PE biodegradation but also other polymers such as PEG. A PEG molecule is first oxidized at its terminal, and then an ether bond is separated (Figure 12.1) [6, 7]. This process corresponds to β -oxidation for PE, and we call it oxidation because oxidation is involved throughout the depolymerization process [6, 7]. Note that $L = 44$ (CH_2CH_2O) in the exogenous depolymerization of PEG, whereas $L = 28$ (CH_2CH_2) in the β-oxidation of PE.

Equation (12.1) forms an initial value problem together with the initial condition

$$w(0, M) = f(M) \tag{12.2}$$

where $f(M)$ represents the initial weight distribution. Given the total consumption rate $\alpha(M)$ and the oxidation rate $\beta(M)$, the solution of the initial value problem is a function $w(t, M)$ that satisfies Eq. (12.1) and the initial condition (12.2). Given the initial condition (12.2) and an additional final condition at $t = T > 0$

$$w(T, M) = g(M) \tag{12.3}$$

Equation (12.1) forms an inverse problem together with the conditions (12.2) and (12.3). It is a problem to determine the degradation rates $\alpha(M)$ and $\beta(M)$ for which the solution $w(t, M)$ of the initial value problems (12.1) and (12.2) also satisfies the final condition (12.3). It has been shown that the following condition is a sufficient condition for a unique positive total degradation rate $\alpha(M)$ to exist, given the β -oxidation rate $\beta(M + L)$ and the weight distribution $w(M + L)$ [4, 5]:

$$0 < g(M) < f(M) + \frac{M\beta(M + L)}{M + L}\int_0^T w(s, M + L)ds \tag{12.4}$$

Polymer molecules must penetrate through membranes into cells in order to become subject to direct consumption. The rate of the penetration decreases, as the molecular size increases. Therefore, the rate of direct consumption must also decrease as molecular size increases. In addition, there must be a limit of penetration with respect to molecular size. It follows that $M_\rho > 0$ such that $\rho(M) = 0$ for $M > M_\rho$. Note that

$$\alpha(M) = \beta(M) \quad \text{for} \quad M > M_\rho \tag{12.5}$$

since $\alpha(M) = \rho(M) + \beta(M)$. The weight distribution of PEG with respect to the molecular weight M introduced in the following sections is given in the range $3.1 \leq \log M \leq 4.2$. The molecular weight in this range should be greater than M_ρ.

12.2.2 Biodegradation of PEG

Polyethers are utilized for constituents in a number of products including lubricants, antifreeze agents, inks, cosmetics, etc. They are also used as raw materials to synthesize detergents or polyurethanes. Those polymers are either water soluble or oily liquid, and eventually discharged into the environment [6]. Since they are not tractable to incineration or recycling, their biodegradability is an important factor of environmental protection against their undesirable accumulation [7]. Polyethers include PEG, polypropylene glycol, and polytetramethylene glycol, and they are polymers whose chemical structures are represented by the expression $HO(R{-}O)_nH$, for example, PEG: R = CH_2CH_2, polypropylene glycol: R = CH_3CHCH_2, polytetramethylene glycol: R = $(CH_2)_4$ [23].

PEG is produced in the largest quantity among polyethers. Its major part is consumed in production of nonionic surfactants. Metabolism of PEG has been well documented. PEG is depolymerized by liberating C_2 compounds, either aerobically or anaerobically [6, 7] (Figure 12.1).

12.3 Materials and Methods

12.3.1 Chemicals

All reagents used were of reagent grade.

12.3.2 Microorganisms and Cultivation

Microbial consortium E-1 was used as a PEG degrader, which was cultivated as described previously. The culture was centrifuged to remove cells and the resultant supernatant was subjected for HPLC analysis.

12.3.3 HPLC analysis

Molecular weights of PEG before and after cultivation were measured by a Tosoh HPLC ccp&8020 equipped with Tosoh TSK-GEL G2500 PW (7.5 $\phi \times 300$ mm) with 0.3 M sodium nitrate at 1.0 mL/min at room temperature. Detection was done with an RI detector (Tosoh RI-8020) (Figure 12.2). The molecular weights were calculated with authentic PEG standards (Figure 12.3). Figure 12.4 shows HPLC profiles of PEG before and after cultivation of microbial consortium E-1 based on the HPLC outputs and the PEG standards.

12.3.4 Numerical Study of Exogenous Depolymerization

Mathematical model (12.1) is appropriate for the depolymerization processes under a steady microbial population. However, the change of microbial population should be taken into account over a period in which a microbial population is still in a developing stage. In such cases, the degradation rate should be time dependent in the modeling of exogenous depolymerization processes:

$$\frac{dx}{dt} = -\beta(t,M)x + \beta(t,M+L)\frac{M}{M+L}y \tag{12.6}$$

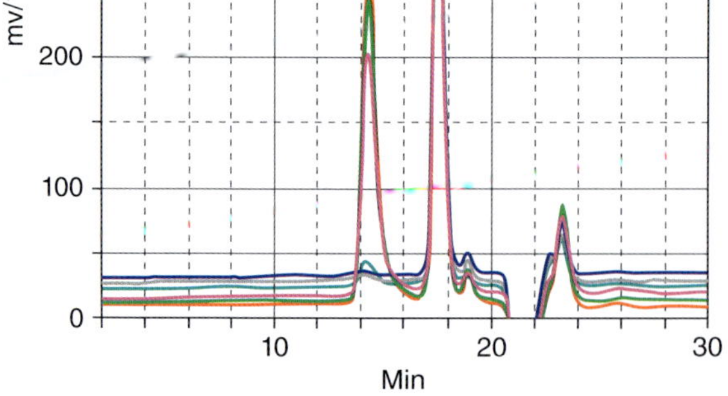

Figure 12.2 HPLC outputs of PEG before and after the cultivation of the microbial consortium E-1.

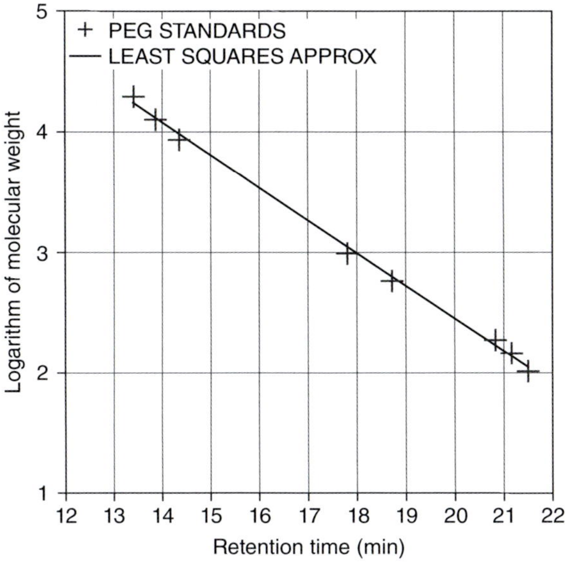

Figure 12.3 PEG standards.

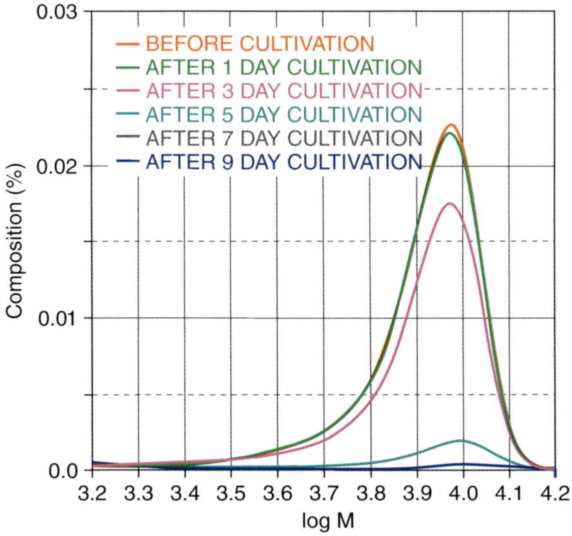

Figure 12.4 HPLC profiles of PEG before and after the cultivation of the microbial consortium E-1 [11, 12].

Solution $x = w(t,M)$ of (12.6) is associated with the initial condition (12.2). Given the degradation rate $\beta(t,M)$, Eq. (12.6) and the initial condition (12.2) form an initial value problem.

Time factors of the degradation rate such as microbial population, dissolved oxygen, or temperature affect molecules regardless of their sizes. The dependence of degradation rate on those factors is uniform over all molecules, and the degradation rate should be a product of a time-dependent part $\sigma(t)$ and a molecular dependent part $\lambda(M)$

$$\beta(t,M) = \sigma(t)\lambda(M) \tag{12.7}$$

Note that $\sigma(t)$ and $\lambda(M)$ represent the magnitude and the molecular dependence of degradability, respectively.

In order to simplify the model, let

$$\tau = \int_0^t \sigma(s)ds \tag{12.8}$$

and

$$W(\tau,M) = w(t,M), \quad X = W(\tau,M), \quad Y = W(\tau,M+L)$$

Then

$$\frac{dX}{d\tau} = \frac{dx}{dt}\frac{dt}{d\tau} = \frac{1}{\sigma(t)}\frac{dx}{dt}$$

and the exogenous depolymerization model (12.6) is converted into the equation

$$\frac{dX}{d\tau} = -\lambda(M)x + \lambda(M+L)\frac{M}{M+L}Y \tag{12.9}$$

This equation governs the transition of weight distribution $w(\tau,M)$ under the time-independent or time-averaged degradation rate $\lambda(M)$. Given the initial weight distribution $f(M)$, Eq. (12.9) forms an initial value problem together with the initial condition

$$W(0,M) = f(M) \tag{12.10}$$

Given an additional condition at $\tau = \mathrm{T}$, Eq. (12.9) forms an inverse problem together with the initial condition (12.10) and the final condition (12.11), for which the solution of the initial value problems (12.9) and (12.10) also satisfies the final condition

$$W(\mathrm{T},M) = g(M). \tag{12.11}$$

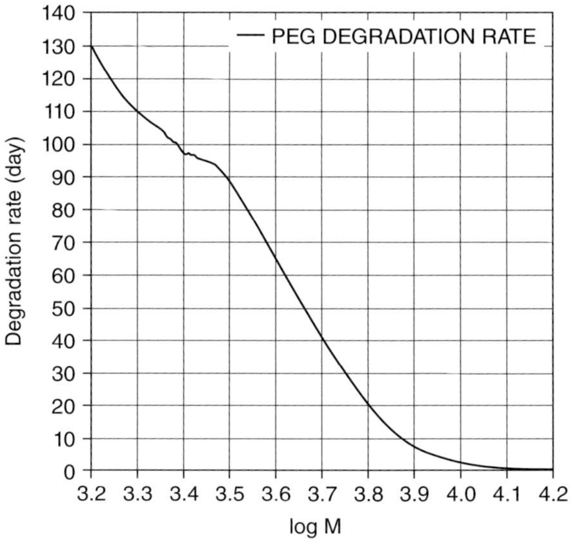

Figure 12.5 Degradation rate based on the weight distribution of PEG before and after the cultivation of the microbial consortium E-1 for 3 days [11, 12].

When the solution $W(\tau, M)$ of the initial value problem (12.9), (12.10) satisfies the condition (12.11), solution $w(t, M)$ of the initial value problems (12.6) and (12.2) satisfies the condition (12.3), where

$$\mathrm{T} = \int_0^T \sigma(s)\mathrm{d}s \tag{12.12}$$

Note that the inverse problem consisting of (12.9)–(12.11) is essentially identical to the inverse problems (12.1)–(12.3). Numerical techniques developed for the latter was applied to the former to find the degradation rate $\lambda(M)$ based on the weight distribution before and after cultivation for 3 days [12, 13] (Figure 12.5).

12.3.5
Time Factor of Degradation Rate

A microbial population grows exponentially in a developing stage, and the increase of biodegradability results from increase of microbial population. It is appropriate to assume that the time factor of the degradation rate $\sigma(t)$ is an exponential function of time

$$\sigma(t) = e^{at+b} \tag{12.13}$$

In view of Eq. (12.8)

$$\tau = \int_0^t \sigma(s)ds = \int_0^t e^{as+b}ds = \frac{e^b}{a}(e^{at} - 1) \qquad (12.14)$$

It has been shown that the parameters a and b are uniquely determined provided the weight distribution is given at $t = T_1$ and $t = T_2$, where $0 < T_1 < T_2$, and let

$$\mathrm{T}_1 = \int_0^{T_1} \sigma(s)ds \qquad (12.15)$$

$$\mathrm{T}_2 = \int_0^{T_2} \sigma(s)ds \qquad (12.16)$$

The condition (12.15) leads to

$$\sigma(t) = e^b e^{at} = \frac{a\mathrm{T}_1 e^{at}}{e^{aT_1} - 1} \qquad (12.17)$$

Now in view of (12.14),

$$\tau = \mathrm{T}_1 \frac{e^{at} - 1}{e^{aT_1} - 1} \qquad (12.18)$$

Equation (12.16) leads to

$$\mathrm{T}_2 = \mathrm{T}_1 \frac{e^{aT_2} - 1}{e^{aT_1} - 1}$$

which is equivalent to the equation

$$h(a) = 0 \qquad (12.19)$$

where

$$h(a) = \frac{e^{aT_2} - 1}{e^{aT_1} - 1} - \frac{\mathrm{T}_2}{\mathrm{T}_1}$$

It has been shown that the condition

$$\frac{T_2}{T_1} < \frac{\mathrm{T}_2}{\mathrm{T}_1} \qquad (12.20)$$

is a necessary and sufficient condition for Eq. (12.19) to have a unique positive solution [11].

In order to determine a and b, let $T_1 = \mathrm{T}_1 = 3$. The initial value problems (12.9) and (12.10) were solved numerically with the degradation rate shown in Figure 12.5 to reach the weight distribution at $\tau = 30$ (Figure 12.6). Note that Figure 12.6

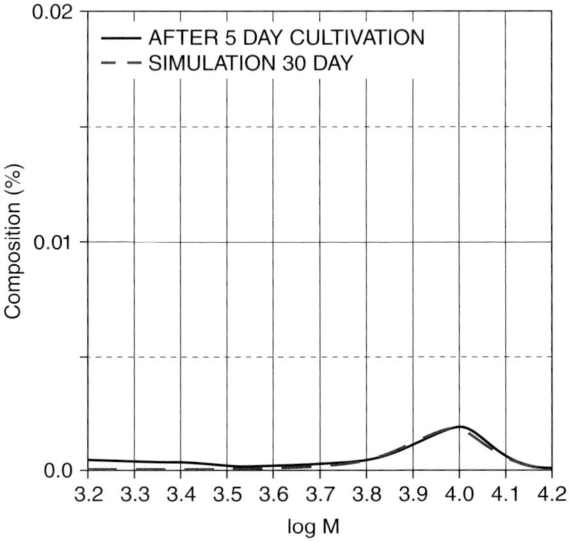

Figure 12.6 Weight distribution of PEG after cultivation for 30 days according to the time-independent model based on the initial value problems (12.9) and (12.10), and the degradation rate shown in Figure 12.5. The experimental result obtained after cultivation for 5 days is also shown [12, 13].

also shows the weight distribution after cultivation for 5 days. It is appropriate to set $T_2 = 5$ and $T_2 = 30$. Equation (12.19) was solved numerically with the Newton's method, and a numerical solution, which was approximately equal to 1.136176, was found [12].

12.3.6 Simulation with Time-Dependent Degradation Rate

Once the degradation rate $\beta(t, M) = \sigma(t)\lambda(M)$ is determined, the initial value problems (12.6) and (12.2) can be solved directly to see how the numerical results and the experimental results agree. The initial value problem was solved numerically with techniques based on previous results [3–5].

Given the initial weight distribution shown in Figure 12.4, the degradation rate $\lambda(M)$ shown in Figure 12.5, and the function $\sigma(t)$ given by Eq. (12.17) with the value of a obtained numerically, the initial value problems (12.6) and (12.2) was solved numerically with the Adams–Bashforth–Moulton predictor–corrector in PECE mode in conjunction with the Runge–Kutta method to generate approximate solutions in the first three steps [24]. Figure 12.7 shows the transition of the weight distribution for 5 days under cultivation of the microbial consortium E-1. Figure 12.8 shows the numerical result and the experimental results for the weight distribution after 1-day cultivation. Note that no information concerning the weight distribution after 1-day cultivation was used to determine the degradation

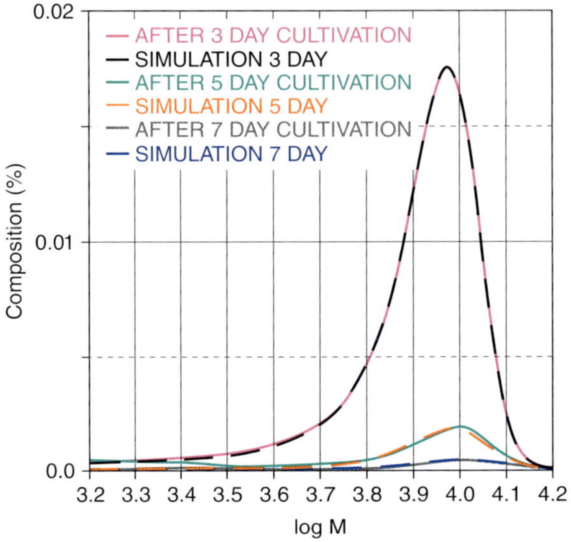

Figure 12.7 The weight distribution of PEG before and after 5-day cultivation, and the transition of the weight distribution based on the initial value problems (12.6) and (12.2) with $\sigma(t) = e^{at+b}$, $a \approx 1.136176$, $b = \ln\{a\mathrm{T}_1/(e^{aT_1} - 1)\}$, and $T_1 = \mathrm{T}_1 = 3$ [12].

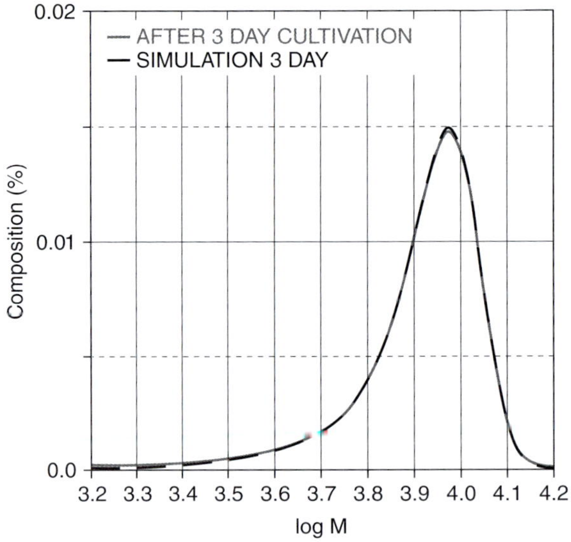

Figure 12.8 The weight distribution of PEG after 1-day cultivation and the weight distribution based on the initial value problems (12.6) and (12.2) with $\sigma(t) = e^{at+b}$, $a \approx 1.136176$, $b = \ln\{a\mathrm{T}_1/(e^{aT_1} - 1)\}$, and $T_1 = \mathrm{T}_1 = 3$ [12].

rate. Nevertheless, Figure 12.8 shows an acceptable agreement between the numerical result and the experimental result.

12.4
Analysis of Endogenous Depolymerization

12.4.1
Modeling of Endogenous Depolymerization

PVA is in general degraded in a succession of two processes: oxidation of a couple of pendant hydroxyl groups either by oxidase or by dehydrogenase followed by hydrolysis. The sequence of reactions results in a cleavage of carbon–carbon chain at a carbonyl group and an adjacent methyne group [25] (Figure 12.9). Matsumura *et al.* proposed a new metabolism of PVA by oxidation of a hydroxyl group and aldolase reaction of a monoketone structure, which results in a cleavage of carbon–carbon chain between a methyne group adjacent to a carbonyl group and an adjacent hydroxymethyne group [26] (Figure 12.9). Irrespective of metabolic pathways, PVA is in general depolymerized by oxidation and the resultant cleavage

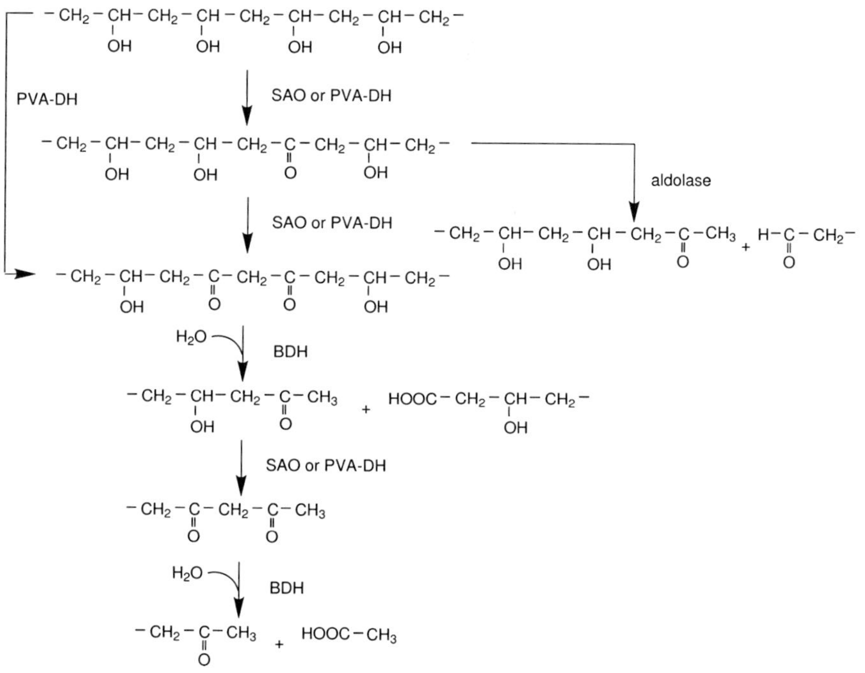

Figure 12.9 Metabolic pathways of PVA [25, 26]. SAO: secondary alcohol oxidase, PVA-DH: PVA dehydrogenase, BDH: β-diketone hydrolase.

of carbon–carbon chain between two carbonyl groups/a carbonyl group and an adjacent hydroxymethyne group, which produces smaller molecules of random sizes.

In order to mathematically model endogenous depolymerization processes of polymers such as PVA, let $w(t,M)$ be its weight distribution with respect to the molecular weight M at time t. Denote by $C(A,B)$ the class of all molecules whose molecular sizes lie between A and B. Then the total weight of $C(A,B)$ present at time t is the integral of $w(t,M)$ with respect to M over the interval $[A,B]$

$$\int_A^B w(t,M)\mathrm{d}M \tag{12.21}$$

For $K \leq M$, let $p(t,K,M)$ denote the time rate of transition from $w(t,M)$ to $w(t,K)$ due to endogenous depolymerization. The transition of the weight from $C(A,B)$ to $C(D,E)$ per unit time is the integral of $p(t,K,M)$ with respect to (K,M) over the region R

$$\iint_R p(t,K,M)\mathrm{d}M\,\mathrm{d}K$$

where

$$R = \{(K,M) \mid K \leq M,\ D \leq K \leq E,\ A \leq M \leq B\}$$

The total weight decrease in $C(A,B)$ per unit time is given by

$$\int_A^B \int_0^M p(t,K,M)\mathrm{d}K\,\mathrm{d}M \tag{12.22}$$

while the total weight increase per unit time is given by

$$\int_A^B \int_M^\infty p(t,M,K)\mathrm{d}K\,\mathrm{d}M \tag{12.23}$$

The equation

$$\frac{\mathrm{d}}{\mathrm{d}t}\int_A^B w(t,M)\mathrm{d}M = \int_A^B \frac{\partial w}{\partial t}(t,M)\mathrm{d}M$$

holds for the rate of change of the quantity (12.21), and it equals the difference between the quantities (12.22) and (12.23)

$$\int_A^B \frac{\partial w}{\partial t}(t,M)\mathrm{d}M = -\int_A^B \int_0^M p(t,K,M)\mathrm{d}K\,\mathrm{d}M + \int_A^B \int_M^\infty p(t,M,K)\mathrm{d}K\,\mathrm{d}M$$

which leads to

$$\int_A^B \left\{\frac{\partial w}{\partial t}(t,M) + \int_0^M p(t,K,M)\mathrm{d}K - \int_M^\infty p(t,M,K)\mathrm{d}K\right\}\mathrm{d}M = 0$$

Since this equation holds for an arbitrary interval $[A,B]$

$$\frac{\partial w}{\partial t}(t,M)+\int_0^M p(t,K,M)\mathrm{d}K-\int_M^\infty p(t,M,K)\mathrm{d}K=0$$

It follows that $w=w(t,M)$ satisfies Eq. (12.24) [13–16]

$$\frac{\partial w}{\partial t}(t,M)=-\int_0^M p(t,K,M)\mathrm{d}K+\int_M^\infty p(t,M,K)\mathrm{d}K \tag{12.24}$$

Let $\gamma(t,M)$ be the amount which $w(t,M)$ loses per unit time and per unit weight. The amount, which $w(t,M)$ loses per unit time, is $\gamma(t,M)w(t,M)$, which is expressed in terms of $p(t,K,M)$

$$\gamma(t,M)w(t,M)=\int_0^M p(t,K,M)\mathrm{d}K$$

This amount is distributed over weight classes over the interval $[0,M]$. For $K\in[A,B]$, let $q(K,M)$ denote the increase in $w(t,K)$ per unit weight due to the weight loss in $w(t,M)$.

Then $p(t,K,M)=\gamma(t,M)q(K,M)w(t,M)$. It follows that

$$\begin{aligned}\int_A^B \gamma(t,M)w(t,M)\mathrm{d}M &= \int_A^B\int_0^M p(t,K,M)\mathrm{d}K\,\mathrm{d}M\\ &= \int_A^B\int_0^M \gamma(t,M)q(K,M)w(t,M)\mathrm{d}K\,\mathrm{d}M\\ &= \int_0^M q(K,M)\mathrm{d}K\int_A^B \gamma(t,M)w(t,M)\mathrm{d}M\end{aligned}$$

Since this equation must hold for an arbitrary interval $[A,B]$

$$\int_0^M q(K,M)\mathrm{d}K=1 \tag{12.25}$$

Equation (12.24) leads to

$$\frac{\partial w}{\partial t}(t,M)=-\gamma(t,M)w+\int_M^\infty \gamma(t,K)q(M,K)w(t,K)\mathrm{d}K \tag{12.26}$$

Given an initial weight distribution in terms of a prescribed function $f(M)$, Eq (12.26) forms an initial value problem together with the initial condition

$$w(0,M)=f(M) \tag{12.27}$$

Given an additional weight distribution at $t=T>0$ in terms of a prescribed function $g(M)$

$$w(T,M)=g(M) \tag{12.28}$$

Equation (12.26) and the conditions (12.27) and (12.28) form an inverse problem to find the degradation rate $\gamma(t,M)$, for which the solution of the initial value problems (12.26) and (12.27) also satisfies the condition (12.28).

A time factor of degradability such as enzyme concentration or temperature affects degradation regardless of molecular sizes. Then it can be assumed that the degradation rate $\gamma(t,M)$ is a product of a function of t and a function of M:

$$\gamma(t,M) = \sigma(t)\lambda(M)$$

Let

$$\tau = \int_0^t \sigma(s)\mathrm{d}s$$

and

$$W(\tau,M) = w(t,M)$$

Then

$$\frac{\partial W}{\partial \tau} = \frac{\partial w}{\partial t}\frac{\partial t}{\partial \tau} = \frac{1}{\sigma(t)}\frac{\partial w}{\partial t}$$

It follows from Eq. (12.26) that

$$\frac{\partial W}{\partial \tau}(t,M) = -\lambda(M)W + \int_M^\infty \lambda(K)q(M,K)W(\tau,K)\mathrm{d}K \qquad (12.29)$$

Equation (12.29) is associated with initial condition

$$W(0,M) = f(M). \qquad (12.30)$$

Let

$$\mathrm{T} = \int_0^T \sigma(s)\mathrm{d}s$$

Then the inverse problem is formulated to find the degradation rate $\lambda(M)$ for which the solution of the initial value problems (12.29) and (12.20) also satisfies the condition

$$W(\mathrm{T},M) = g(M) \qquad (12.31)$$

The initial value problems (12.29) and (12.30) correspond to the initial value problems (12.26) and (12.27), and the inverse problems (12.29)–(12.31) correspond to the inverse problems (12.26)–(12.28).

Suppose that $q(K,M)$ is a product of a function of K and a function of M:

$$q(K,M) = c(K)d(M)$$

Then Eq. (12.29) becomes

$$\frac{\partial W}{\partial \tau}(t,M) = -\lambda(M)W + c(M)\int_M^\infty \lambda(K)d(K)W(\tau,K)\mathrm{d}K \qquad (12.32)$$

The condition (12.25) leads to

$$\int_0^M q(K,M)\mathrm{d}K = \int_0^M c(K)d(M)\mathrm{d}K = d(M)\int_0^M c(K)\mathrm{d}K = 1$$

In case the number of degraded molecules are uniformly distributed over the interval [0, M] [15–17]

$$c(K) = 2K \quad d(M) = \frac{1}{M^2} \qquad (12.33)$$

For these $c(K)$ and $d(M)$, Eq (12.32) is the model originally proposed for the enzymatic degradation of PVA. In case the weight of degraded molecules are uniformly distributed over the interval [0, M] [18]

$$c(K) = 1 \quad d(M) = \frac{1}{M} \qquad (12.34)$$

Differentiating both sides of Eq. (12.32) with respect to M leads to

$$\begin{aligned}\frac{\partial^2 W}{\partial M \partial \tau} &= \frac{\partial}{\partial M}\{-\lambda(M)W\} + c'(M)\int_M^\infty \lambda(K)d(K)W(\tau,K)\mathrm{d}K - \lambda(M)c(M)d(M)W \\ &= \frac{\partial}{\partial M}\{-\lambda(M)W\} + \frac{c'(M)}{c(M)}\left\{\frac{\partial W}{\partial \tau} + \lambda(M)W\right\} - \lambda(M)c(M)d(M)W\end{aligned}$$

and

$$\frac{\partial}{\partial M}\left\{\frac{\partial W}{\partial \tau} + \lambda(M)W\right\} = \frac{c'(M)}{c(M)}\left\{\frac{\partial W}{\partial \tau} + \lambda(M)W\right\} - \lambda(M)c(M)d(M)W \qquad (12.35)$$

Numerical techniques have been developed for the inverse problem to find the degradation rate $\lambda(M)$ for which the solution of the initial value problems (12.35) and (12.30) also satisfies the condition (12.31), in case $c(K)$ and $d(M)$ are given by (12.33). Those techniques can be extended to cover the general case.

12.4.2 Analysis of Enzymatic PLA Depolymerization

The experimental and analytical study of endogenous depolymerization is continued to cover degradation of PLA [17, 18]. PLA used for the experiment was poly(L-lactide). Figure 12.10 shows the GPC patterns of PLA given in terms of retention time versus voltage. The figure also shows the baselines of the GPC patterns. Figure 12.11 shows standards of molecular weight versus retention time. The figure also shows the least-squares approximation that fits the standards. Figure 12.12 shows the weight distribution of PLA before and after enzymatic degradation for 5 and 67 h based on the shift according to the baselines shown in Figure 12.10, the transformation according to the least-squares approximation shown in Figure 12.11, and scaling according to the residual amounts. An inverse problem was formulated to determine the degradation rate for which the solution of the initial value problem also satisfies the weight distribution after incubation for 5 h. PLA was solved in chloroform and emulsified by sonication for enzymatic degradation. As the time elapsed, chloroform was lost by evaporation, resulting in reduced degradation rates. In previous studies [18], the temporal change of degradability was considered, and a temporally dependent degradation rate was incorporated into the endogenous depolymerization model. The endogenous depolymerization is studied in this chapter. A mathematical model of endogenous depolymerization, which covers the previously proposed models as special cases, is introduced.

It is shown that a general model can be transformed to a form to which previously developed techniques are applicable. Techniques to solve an inverse problem

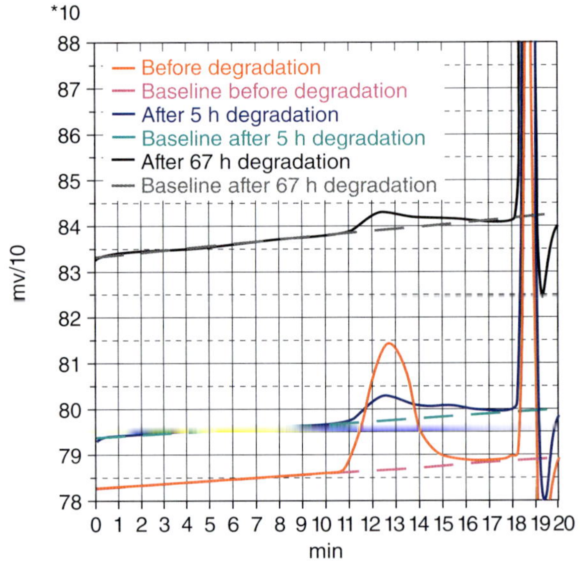

Figure 12.10 GPC patterns of PLA given in terms of retention time versus voltage [16, 17]. The baselines are also shown.

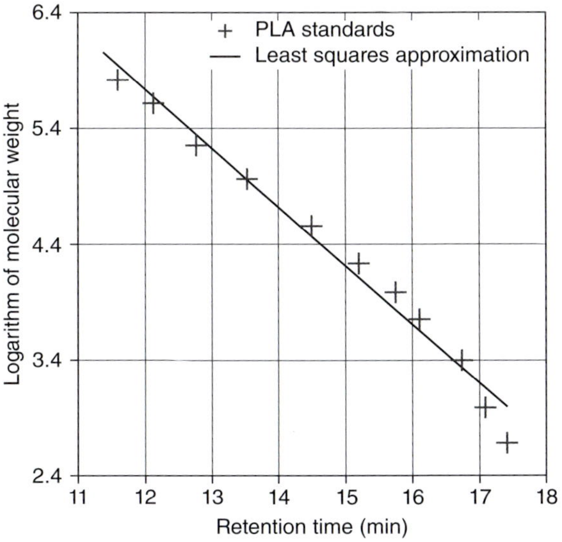

Figure 12.11 Standards of molecular weight versus retention time, and the least-squares approximation which fits the standards [17].

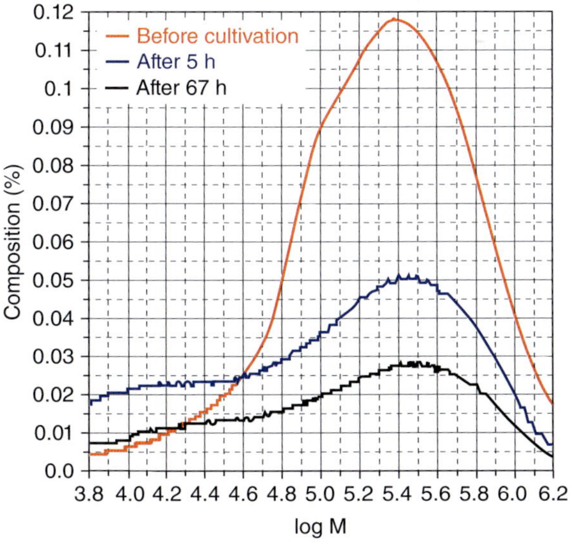

Figure 12.12 Weight distribution of PLA before and after enzymatic degradation. Residual amounts of PLA after incubation for 5 and 67 h were 40% and 27%, respectively [17, 18].

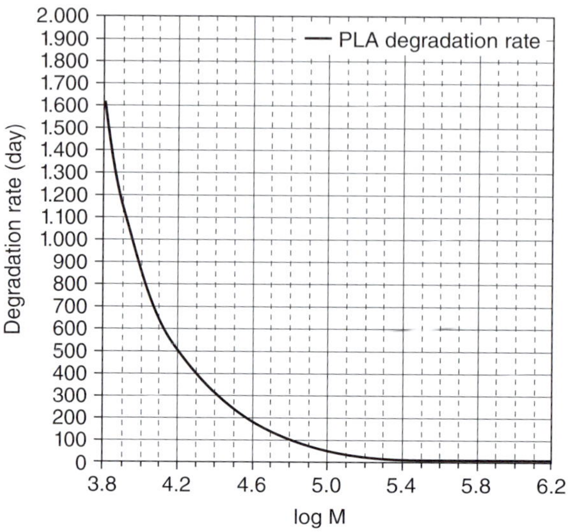

Figure 12.13 Degradation rate of PLA based on the GPC profiles obtained before and after incubation for 5 h shown in Figure 12.12. $c(K) = 1$ and $d(M) = 1/M$ [17].

to determine the degradation rate and to simulate the transition of weight distribution are described, and numerical results are introduced.

The weight distribution before incubation shown in Figure 12.12 was set as the initial condition (12.30), and the weight distribution after incubation for 5 h shown in Figure 12.12 was set as the final condition (12.30) to solve the inverse problem numerically for the function $c(K)$ and $d(M)$ given by the expressions (12.34). Figure 12.13 shows the graph of the degradation rate $\lambda(M)$. Figure 12.14 shows a result of numerical simulation for transition of weight distribution over incubation period for 10 h based on the degradation rate shown in Figure 12.13. Figure 12.15 shows the experimental result for weight distribution after incubation for 5 h and a numerical result to simulate the experimental result based on the degradation rate shown in Figure 12.13.

Chloroform used to dissolve PLA was lost by evaporation as the time elapsed. The loss of chloroform resulted in reduction of the degradation rate. Figure 12.16 shows the experimental result for weight distribution after incubation for 67 h and a numerical result for the weight distribution after incubation for 8.5 h based on the degradation rate. The figure shows that it takes only 8.5 h to reach the stage after incubation for 67 h with the average degradation rate over incubation period for the first five hours.

12.4.3 Simulation of an Endogenous Depolymerization Process of PLA

A technique to determine the time factor $\sigma(t)$ has been proposed [18]. Since the decrease of degradability was due to evaporation of chloroform, it is appropriate to assume that $\sigma(t)$ is an exponential function of time:

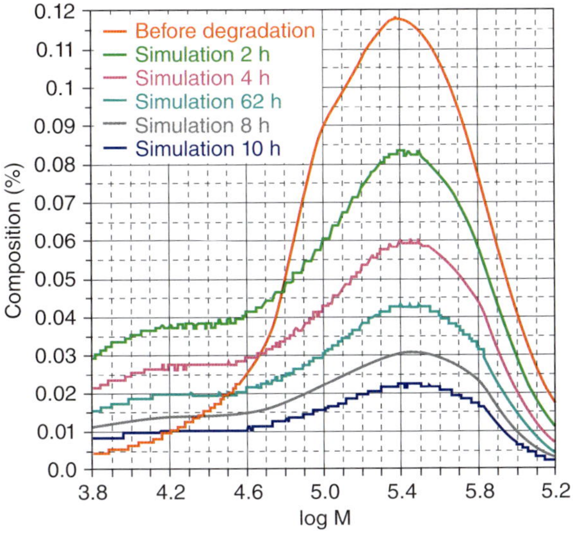

Figure 12.14 Transition of weight distribution for over incubation period for 10 h based on the degradation rate shown in Figure 12.13 [18].

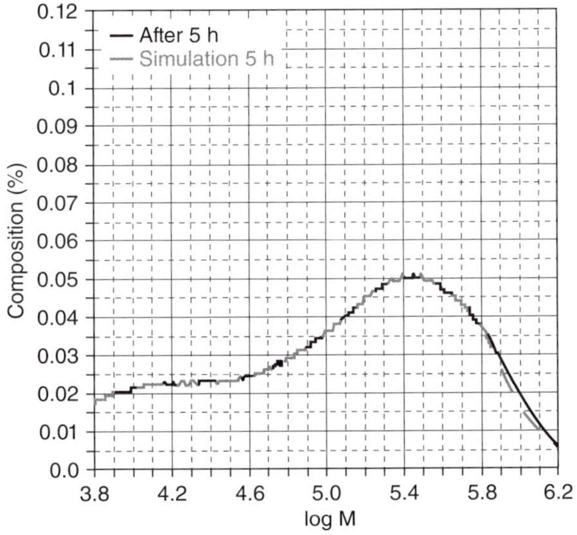

Figure 12.15 Weight distribution after incubation for 5 h and numerical result to simulate the experimental result [18].

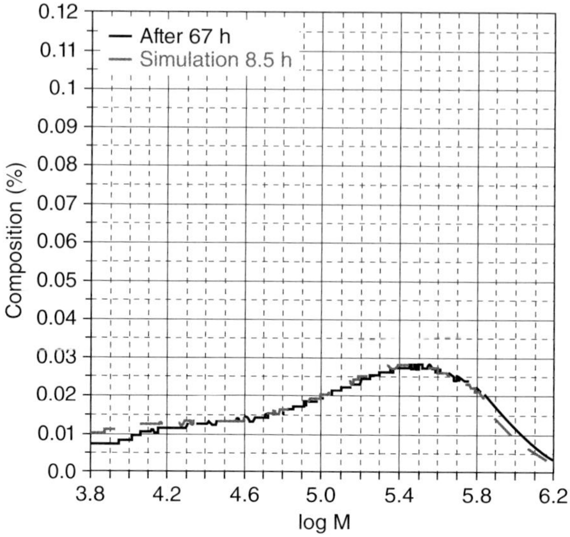

Figure 12.16 Weight distribution after 67 h of incubation and numerical result to simulate the weight distribution after incubation for 8.5 h [18].

$$\sigma(t) = e^{-at+b} \tag{12.36}$$

Then τ is given by

$$\tau = \int_0^t \sigma(s)ds = \int_0^t e^{-as+b}ds = \frac{e^b}{a}(1-e^{-at}) \tag{12.37}$$

Given T_1 and T_2 with $T_1 < T_2$, let

$$\mathrm{T}_1 = \int_0^{T_1} e^{-as+b}ds \tag{12.38}$$

$$\mathrm{T}_2 = \int_0^{T_2} e^{-as+b}ds \tag{12.39}$$

In particular, in view of the result shown in Figure 12.16, the values of parameters can be set as follows: $T_1 = \mathrm{T}_1 = 5/24$ (day), $T_2 = 67/24$ (day), and $\mathrm{T}_2 = 8.5/24$ (day).
Equations (12.38) and (12.39) lead to

$$\mathrm{T}_1 = \frac{e^b}{a}(1-e^{-aT_1})$$

It follows from Eq. (12.37) that

$$\tau = \mathrm{T}_1 \frac{1-e^{-at}}{1-e^{-aT_1}}$$

Now Eq. (12.39) leads to

$$\mathrm{T}_2 = \mathrm{T}_1 \frac{1-e^{-aT_2}}{1-e^{-aT_1}} \tag{12.40}$$

Define $h(a)$ by

$$h(a) = \frac{1-e^{aT_2}}{1-e^{aT_1}} - \frac{\mathrm{T}_2}{\mathrm{T}_1}$$

Now Eq. (12.40) is equivalent to

$$h(a) = 0 \tag{12.41}$$

Equation (12.41) was solved numerically for the values of parameters: $T_1 = \mathrm{T}_1 = 5/24$, $T_2 = 67/24$, and $\mathrm{T}_2 = 8.5/24$, and an approximate value of the solution $a \approx 4.259$ was found. Figure 12.17 shows a result of numerical simulation for the transition of weight distribution of PLA over incubation period for 67 h.

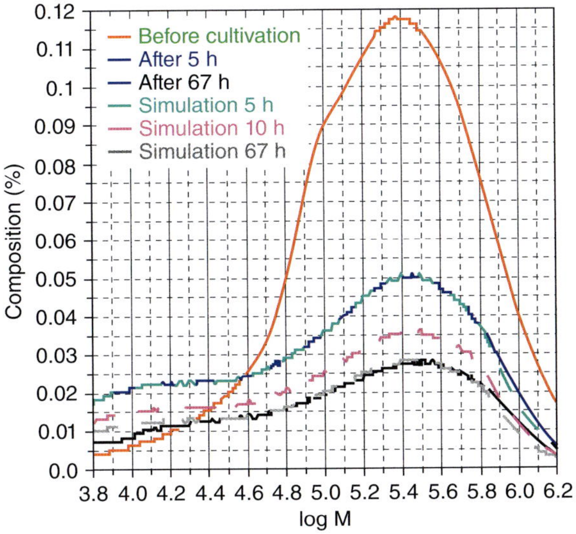

Figure 12.17 Transition of weight distribution over incubation period for 67 h [18].

12.5 Discussion

The degradation rate $\lambda(M)$ of the exogenous depolymerization model is the ratio of the total weight of M-molecules degraded per unit time. It also represents the ratio of the number of M-molecules that undergo exogenous depolymerization processes. It might be assumed to be independent of the molecular size M, for the exogenous depolymerization processes take place only at the terminals of the molecules. In practice, however, this is the case for molecules of moderate sizes. As metabolic enzymes are located in cell membranes, they take effect in periplasms, and at least one end of a molecule must penetrate through its outer membrane to become subject to an exogenous depolymerization process [7].

In the presence of sufficient number of microorganism, the number ratio of molecules in contact with microorganisms in a fixed period of time can be assumed to be independent of the molecular weight. When a molecule makes contact with a microorganism, a part of a fixed length should be taken into a cell in a fixed period. If that part happens to contain one of the terminals, the enzyme takes effect and an exogenous depolymerization process starts. The possibility for a part of a fixed length to contain a terminal becomes less when the molecule becomes large. This is the reason why the oxidation rate is a decreasing function of the molecular weight. The result shown in Figure 12.5. clearly indicates the dependence of the rate of membrane transport with respect to the molecular weight, and our mathematical analysis has revealed the role of membrane transport in exogenous depolymerization processes, which is unpredictable from experimental results alone.

The exogenous depolymerization model originally developed for PE biodegradation has been applied successfully to PEG biodegradation. The numerical results show how molecules are incorporated into cells in exogenous depolymerization processes. The validity of the result concerning the oxidation rate has been confirmed by the numerical simulation (Figures 12.7 and 12.8). This is a typical microbial depolymerization process of exogenous type, where monomer units are split from the terminals of molecules.

The only factor assumed in construction of the endogenous depolymerization model was random separation of molecules. There are other factors of weight changes in processes through the metabolic pathways of PVA as was described, but those changes should be negligible when compared to the weight shift due to random separation of molecules due to enzymatic degradation. As a mathematical model for enzymatic degradation of PVA, a linear second-order hyperbolic partial differential has been derived from the original model. Given a prescribed function that represents the degradation rate and an initial condition, it forms an initial value problem of a linear partial differential equation. On the other hand, given the initial condition and an additional final condition, it forms an inverse problem to determine the degradation rate for which the solution of the initial value problem also satisfies the final condition.

It is shown that the inverse problem can be reduced to a nonlinear ordinary differential equation whose unknown variable represents the weight flux into the class of all molecules of a fixed size from the classes of all larger molecules due to random depolymerization. The necessary and sufficient condition, under which the initial value problem of the ordinary differential equation has a solution defined locally, has been established. The theory was described by introducing a numerical result based on the ordinary differential equation. Note that one can hardly predict the existence of its solution for a given range of the independent variable. On the contrary, the inverse problem has been solved successfully for the given range of the molecular weight applying the HPLC data obtained before and after enzymatic degradation of PLA. The result of numerical simulation shows that the technique to solve the inverse problem is practically acceptable.

Acknowledgments

The authors thank Ms Y. Shimizu for the technical support. This work was supported by JSPS KAKENHI 20540118.

References

1 Kawai, F., Shibata, M., Yokoyama, S., Maeda, S., Tada, K., and Hayashi, S. (1999) Biodegradability of Scott–Gelead photodegradable polyethylene and polyethylene wax by microorganisms. *Macromol. Symp.*, **144**, 73–84.

2 Kawai, F., Watanabe, M., Shibata, M., Yokoyama, S., and Sudate, Y. (2002) Experimental analysis and numerical simulation for biodegradability of polyethylene. *Polym. Degrad. Stab.*, **76**, 129–135.

3 Watanabe, M., Kawai, F., Shibata, M., Yokoyama, S., and Sudate, Y. (2003) Computational method for analysis of polyethylene biodegradation. *J. Comput. Appl. Math.*, **161** (1), 133–144.

4 Kawai, F., Watanabe, M., Shibata, M., Yokoyama, S., Sudate, Y., and Hayashi, S. (2004) Comparative study on biodegradability of polyethylene wax by bacteria and fungi. *Polym. Degrad. Stab.*, **86**, 105–114.

5 Watanabe, M., Kawai, F., Shibata, M., Yokoyama, S., Sudate, Y., and Hayashi, S. (2004) Analytical and computational techniques for exogenous depolymerization of xenobiotic polymers. *Math. Biosci.*, **192**, 19–37.

6 Kawai, F. (2002) Microbial degradation of polyethers. *Appl. Microbiol. Biotechnol.*, **58**, 30–38.

7 Kawai, F. (1995) Breakdown of plastics and polymers by microorganism. *Adv. Biochem. Eng. Biotechnol.*, **52**, 151–194.

8 Kawai, F., and Yamanaka, H. (1986) Biodegradation of polyethylene glycol by symbiotic mixed culture (obligate mutualism). *Arch. Microbiol.*, **146**, 125–129.

9 Watanabe, M., and Kawai, F. (2005) Numerical simulation of microbial depolymerization process of exogenous type. Proceeding of 12th Computational Techniques and Applications Conference, CTAC-2004, Melbourne, Australia in September 2004, Editors: Rob May and A. J. Roberts, ANZIAM J., vol. 46(E), pp. C1188–C1204, http://anziamj.austms.org.au/ojs/index.php/ANZIAMJ/article/view/1014 (accessed 23 March 2011).

10 Watanabe, M. and Kawai, F. (2004) Analysis of biodegradability for polyethylene glycol via numerical simulation. *Environ. Res. Contr.*, **26**, 17–22. (in Japanese).

11 Watanabe, M. and Kawai, F. (2007) Mathematical study of the biodegradation

of xenobiotic polymers with experimental data introduced into analysis. Proceedings of the 7th Biennial Engineering Mathematics and Applications Conference, EMAC-2005, Melbourne. Editors: Andrew Stacey and Bill Blyth and John Shepherd and A. J. Roberts, ANZIAM J., vol. 47, pp. C665–C681, http://anziamj.austms.org.au/ojs/index/php/ANZIAMJ/article/view/1069 (accessed 23 March 2011).

12 Watanabe, M. and Kawai, F. (2009) Modeling and simulation of biodegradation of xenobiotic polymers based on experimental results, biosignals 2009. Second International Conference on Bio-inspired Systems and Signal Processing, Proceedings, Porto–Portugal, 14–17 January 2009, pp. 25–34.

13 Watanabe, M. and Kawai, F. (2009) Mathematical analysis of microbial depolymerization processes of xenobiotic polymers. Proceedings of the 14th Biennial Computational Techniques and Applications Conference, CTAC-2008, Editors: Geoffry N. Mercer and A. J. Roberts, ANZIAM J., 50 (CTAC-2008), pp. C930–C946. http://anziamj.austms.org.au/ojs/index.php/ANZIAM/article/view/1465 (accessed 23 March, 2011).

14 Watanabe, M. and Kawai, F. (2003) Analysis of polymeric biodegradability based on experimental results and numerical simulation. *Environ. Res. Contr.*, **25**, 25–32. (in Japanese).

15 Watanabe, M. and Kawai, F. (2003) Numerical simulation for enzymatic degradation of poly(vinyl alcohol). *Polym. Degrad. Stab.*, **81** (3), 393–399.

16 Watanabe, M. and Kawai, F. (2006) Mathematical modelling and computational analysis for enzymatic degradation of xenobiotic polymers. *Appl. Math. Modell.*, **30**, 1497–1514.

17 Watanabe, M., Kawai, F., Tsuboi, S., Nakatsu, S., and Ohara, H. (2007) Study on enzymatic hydrolysis of polylactic acid by endogenous depolymerizaion model. *Macromol. Theory Simul.*, **16**, 619–626. doi: 10.1002/mats.200700015.

18 Watanabe, M. and Kawai, F. (2008) Modeling and analysis of biodegradation of xenobiotic polymers based on experimental results. Proceedings of the 8th Biennial Engineering Mathematics and Applications Conference, EMAC-2007. Editors: Geoffry N. Mercer and A. J. Roberts, ANZIAM J., 49 (EMAC-2007). pp. C457–C474, http://anziamj.austms.org.au/ojs/index.php/ANZIAM/article/view/361 (accessed 23 March 2011).

19 Albertsson, A.-C., Andersson, S.O., and Karlsson, S. (1987) The mechanism of biodegradation of polyethylene. *Polym. Degrad. Stab.*, **18**, 73–87.

20 Potts, J.E., Clendinning, R.A., Ackart, W.B., and Niegisch, W.D. (1972) The biodegradability of synthetic polymers. *Polym. Preprints*, **13**, 629–634.

21 Haines, J.R. and Alexander, M. (1974) Microbial degradation of high-molecular-weight alkanes. *Appl. Microbiol.*, **28**, 1084–1085.

22 Kawai, F. (1999) Sphingomonads involved in the biodegradation of xenobiotic polymers. *J. Ind. Microbiol. Biotechnol.*, **23**, 400–407.

23 Kawai, F. (1993) Biodegradability and chemical structure of polyethers. *Kobunshi Ronbunshu*, **50** (10), 775–780.

24 Lambert, J.D. (1973) *Computational Methods in Ordinary Differential Equations*, John Wiley & Sons, Ltd, Chichester.

25 Kawai, F. (2002) Xenobiotic polymers, in *Great Development of Microorganisms* (ed. T. Imanaka), NTS. Inc., Tokyo, pp. 865–870.

26 Matsumura, S., Tomizawa, N., Toki, A., Nishikawa, K., and Toshima, K. (1999) Novel poly(vinyl alcohol)-degrading enzyme and the degradation mechanism. *Macromolecules*, **32**, 7753–7761.

13
Regenerative Medicine: Reconstruction of Tracheal and Pharyngeal Mucosal Defects in Head and Neck Surgery

Dorothee Rickert, Bernhard Hiebl, Rosemarie Fuhrmann, Friedrich Jung, Andreas Lendlein, and Ralf-Peter Franke

13.1
Introduction

13.1.1
History of Implant Materials

The 20th century can be called the era of synthetical polymers. Poly(methyl methacrylate) (PMMA) was firstly recognized as promising implant material through war-wounded pilots in World War II: Soft tissue and eye injuries induced by and containing small fractions of bursting windows of airplane cockpits (PMMA) led to minute foreign body reactions only. Szilagyi *et al.* reported first clinical experiences with *polyethylene terephthalate* as vascular arterial prostheses in 1958 [1]. In the 1960s, J. Charnley, an orthopedic surgeon from United Kingdom developed a functional and cemented total hip endoprosthesis based on steel and ultrahigh molecular weight polyethylene inlays which were cemented into the femoral bone using PMMA as "cement." Beginning at the end of the 1960s, there was a focus on the development of degradable polymeric implant materials.

Since then the availability of so-called polymer systems allows a large-scale variation of material characteristics, for example, of mechanical properties or hydrolytic degradation and thus to adapt these materials to specific local requirements in the organism [2].

13.1.2
Regenerative Medicine

Due to the shift in morbidity spectrum during the last decades and the recent demographic development in the world, the clinical medicine has to deal more and more with diseases gradually leading to a loss of function of important cell and organ systems. In many cases, these diseases cannot be cured by the currently available therapies and the patients have to remain in permanent therapy resulting in high costs.

Handbook of Biodegradable Polymers: Synthesis, Characterization and Applications, First Edition. Edited by Andreas Lendlein, Adam Sisson.
© 2011 Wiley-VCH Verlag GmbH & Co. KGaA. Published 2011 by Wiley-VCH Verlag GmbH & Co. KGaA.

Regenerative medicine is highly interdisciplinary and deals with the restitution, substitution, regeneration of nonfunctional or more or less functionally impaired cells, tissues, organs through biological replacement, for example, through tissues produced *in vitro* or through the stimulation of the body's own regeneration and/or repair processes [3, 4].

Important success in stem cell research [5, 6] and the extracorporeal tissue growth in bioreactors show the potential of regenerative medicine [7–9]. The euphoric visions to grow complete and functional organs *in vitro* right now, however, were recognized to be very premature. This is also due to a lack in basic research and the development of multifunctional implant materials [10].

13.1.3
Functionalized Implant Materials

The experience with polymer implants used in medicine led to a profile of requirements for future polymeric implant materials. The functionality of implant materials has to be broadened. They should be stimuli sensitive and, for example, change their physicochemical behavior due to external stimuli or to biological processes induced at the site of implantation. Bioactive substances like peptides, proteins, or carbohydrates might be immobilized by polymers or released from implants in a well-defined process. The most up-to-date trend in polymer sciences is the development of degradable biomaterials showing multifunctionality. This implies that specific functionalities like hydrolytic degradation, physiological and biomechanical tissue compatibilities, and shape-memory can be adjusted to regiospecific requirements at the site of implantation [11, 12].

AB-copolymer networks are an example for an implant material that can be functionalized.

These networks are produced by photocrosslinking of *n*-butyl acrylate with oligo(ε-caprolacton)dimethacrylate as macrocrosslinker [13, 14]. The incorporation of flexible polybutylacrylate segments allows, for example, the tailoring of material elasticity, which is an important determinant of the biomechanical functionality of this polymer system in the temperature range between room and body temperature. AB-copolymer networks are slowly biodegradable due to their hydrolytically cleavable polyester chain segments. Another group of multifunctional, degradable polymers are multiblock copolymer systems [15–17] containing poly(*p*-dioxanone) hard segments and crystallizable poly(ε-caprolactone) soft segments. Due to their degradability, stimuli sensitivity, biocompatibility, and functionality, these copolymer networks are termed multifunctional. Biomechanical characteristics as well as types and periods of degradation can be adjusted as well.

13.1.4
Sterilization of Polymer-Based Degradable Implant Materials

The sterilization of implant materials is a precondition for their biomedical use. Polymer-based and especially hydrolytically degradable biomaterials in general

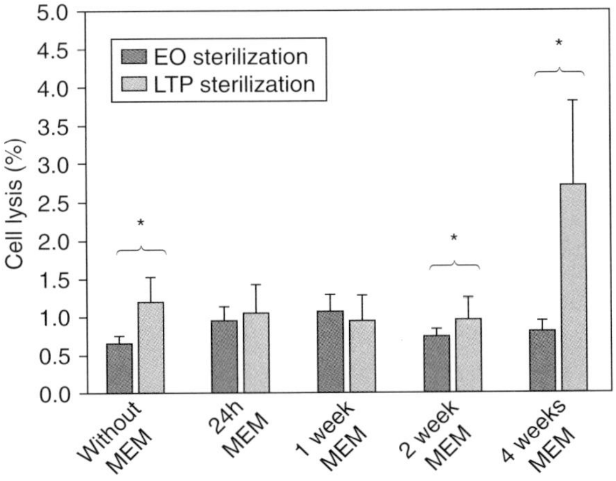

Figure 13.1 Mean rate of cell lysis after different sterilization techniques. Mean rate of cell lysis after EO and LTP sterilization of the polymer samples and different incubation time in physiological solution (MEM). Statistically significant differences of the mean rates of cell lysis were found for the differently sterilized samples without MEM incubation, as well as after 2 and 4 weeks of incubation with MEM. Abbreviations: EO = ethylene-oxide sterilization, LTP = low-temperature plasma sterilization, MEM = minimal essential medium. Reprinted with permission from [20]. Copyright 2003 Wiley Periodicals, Inc.

have a considerably lower thermal and chemical stability as ceramic or metallic materials. They are generally not sterilized with conventional sterilization methods like heat sterilization (temperatures between 160 and 190 °C) or steam sterilization (121 and 134 °C) to avoid a damage of polymers. Sterilization applying ionizing irradiation can change the chemical structure of polymers either by chain degradation or by new crosslinking of chains, so that surface characteristics as well as thermal and mechanical bulk properties can be strongly influenced [18]. A change of the chemical surface structure of implant materials can influence their biocompatibility *in vitro* and *in vivo* [19]. Since the sterilization of polymer-based biomaterials makes high demands on the sterilization method, low-temperature sterilization methods like plasma sterilization (low-temperature plasma sterilization) and sterilization with ethylene oxide are in the focus of intensive contemporary research [20–23] (Figure 13.1).

13.2 Regenerative Medicine for the Reconstruction of the Upper Aerodigestive Tract

Head and neck surgery is concerned with the reconstruction of damaged local tissues like mucosa, cartilage, bone, or skin due to congenital anomalies, progressive diseases, as well as therapeutical interventions. Fistulae of different genesis are associated with most serious complications in the head and neck area [24–26].

These fistulae cause high rates of morbidity and mortality through the development of sepsis, pneumonia, or bleeding from destruction of the carotid wall. The permanent secretion from fistulae and the cervical soft tissue defects (especially of pharyngocutaneous fistulae) is associated with a tremendous reduction of life quality of patients and their stigmatization [24]. Due to postoperative salivary fistulae in oncological patients, their irradiation may not be possible within the planned periods so that therapeutical aims cannot be reached. Contemporary therapeutical options in the treatment of pharyngocutaneous fistulae depend on the size of fistulae and on the indication of a postoperative adjuvant irradiation therapy.

13.2.1 Applications of Different Implant Materials in Tracheal Surgery

In the 1950s, a great number of experiments for the tracheal reconstruction were performed in animals using different materials like acrylresin [27], tantalum [28], stainless steel [29], polyethylene [30], nylon [31],, and teflon [32]. The great number of materials used and the short survival time of the animals demonstrated that the problem of tracheal reconstruction using implant materials could not be solved at this time. The importance of biocompatibility of implant materials and the variable requirements depending on the implantation site became obvious at the end of the 1950s. After the successful application of Dacron™ as arterial prosthesis (1958), it was realized that an appropriate material was not available for the tracheal reconstructive surgery showing the necessary elasticity, rigidity, and biocompatibility. At the end of the 1950s and the beginning of 1960s, there were first trials for the temporary application of polymeric implant materials in the tracheal reconstruction. These materials were covered with mucosa from the urinary or gall bladders to induce growth of connective tissues or bone around tracheal stents. It was called temporary application because the implant material should be removed after the newly grown cartilage or bone in the former tracheal defect zone reached a sufficient stability, so that the reconstructed tracheal tissues would not collapse. Although cartilage and bone tissues could be demonstrated histologically at the site of implantation, a sufficient tracheal stability could not be gained in any one of the animals and all animals died of respiratory insufficiency following tracheal obstruction after the removal of the differently coated implant materials [33, 34]. In the 1960s and 1970s, further materials were tested for tracheal reconstruction, for example, Marlex™ networks (polyethylene/polypropylene networks) [35], silicon rubber [36], and Marlex™ networks covered with cartilage and/or tracheal mucosa [37, 38]. These new materials also did not fulfill the comprehensive requirements for tracheal reconstruction regarding mechanical strength and adequate flexibility to avoid vascular corrosion induced by mechanical irritation. These materials lacked biocompatibility, an air- and liquid tight integration of the implant materials into the adjacent body tissues, an adequate stability against bacterial invasion, and, especially, the epithelialization of the implants with a functional tracheal epithelium [35–38].

Wenig *et al.* showed in 1987 that through application of a fibroblast collagen matrix for the tracheal reconstruction of circumscript defects, the rate of tracheal stenosis could be reduced significantly [39]. In 1989, Schauwecker *et al.* demonstrated the importance of biomechanical properties of implant materials depending on the site of implantation and that the porosity of the material surface was important for the integration of implants in surrounding tissues. These authors applied an isoelastic polyurethane prosthesis with different porosities at the luminal and abluminal surfaces for the reconstruction of 38-mm-long defects of the cervical trachea of 19 dogs. Besides end-to-end anastomosis these authors applied inverted and everted techniques of anastomosis. The mean survival time of animals in case of the inverted technique was 27.7 days, in case of the everted technique 11.3 days, and in case of the end-to-end anastomosis 19.5 days. The worst complications leading to a termination of these trials were local infections and insufficiencies of anastomosis in 12 of the animals and extensive stenoses accompanied by respiratory insufficiency in seven animals. The authors observed that polyurethane prostheses with porous surfaces developed a tight integration into surrounding tissues, but in none of the animals, the luminal prosthetic surface was inhabited by a mucociliary epithelium. The authors attributed the high rate of complications primarily to the animal model chosen because the cervical mobility in dogs was said to be much higher than in humans, pigs, or rats [40].

13.2.2
New Methods and Approaches for Tracheal Reconstruction

Key factors compromising the therapeutical success seem to be the absent regeneration of a functional mucociliary tracheal epithelium enabling the mucociliary clearance, foreign body reactions induced by implant materials, infections, and the necessity of reoperations in preoperated areas. The tissue-engineering technique was described by Langer and Vacanti in 1993 and had three key components: cells for the tissue regeneration, polymer scaffolds as a matrix to support migration, proliferation and differentiation of cells as well as regulating factors which specifically influence the cellular behavior [41]. The following demands on a tracheal prosthesis were made: It should be a flexible construct but able to endure compression which is inhabited by a functional respiratory epithelium [42]. The complete epithelialization of prostheses is thought to be the main condition to allow an adequate mucociliary clearance and to guarantee a reliable barrier against infection and invading connective tissue. There are still very few studies applying the methods of tissue engineering to produce tracheal replacements and to examine these *in vitro* and *in vivo*. Studies introduced by Vacanti *et al.* in 1994 were trend-setting where constructs based on polyglycolic acid and inhabited by bovine chondrocytes and tracheal epithelial cells were applied to close circumferential tracheal defects in rats [43]. In a consecutive study, respiratory epithelial cells were isolated and injected into cartilage cylinders grown *in vitro* [44]. Examinations of these constructs revealed mature cartilage tissues as well as epithelial structures with a submucosal connective tissue. After 3 weeks in culture, different stages of differentiation of a multilayered highly prismatic epithelium could be documented showing also some ciliary cells. In consecutive experiments, these authors developed a tracheal replacement based on chondrocytes and fibroblasts which was implanted into sheep. The tracheal replacement thus generated could not be shown to develop kinocilia within the respiratory epithelial cells and therefore was not fully functional [45].

Besides the use of different implant materials in experimental and clinical trials during the last 50 years [27–30], there were many other attempts with autologous or allogenic tissues of different origin like fasciae, skin, bone and periost, cartilage and perichondrium, muscle, esophagus, pericardium, intestine, and dura mater [46–50]. Again, high rates of complications were reported, for example, high rates of stenosis and necrosis, of anastomotic insufficiencies, and a lack of mucociliary clearance.

At the end of the 1990s and the beginning of 2000, biodegradable stents were introduced in reconstructive tracheal surgery. Lochbihler *et al.* described in 1997 for the first time the application of a resorbable intratracheal stent made of polyglactine 910 filaments copolymerized with polydioxanone for the temporary stabilization of a tracheal stenosis in rats [51]. Korpela *et al.* applied a spirally shaped and reinforced stent made of poly(L-lactide) to bridge tracheal stenoses in an animal model [52, 53]. Robey *et al.* described in 2000 the application of a biodegradable *poly[(L-lactide)-co-glycolide]* (PLGA) stent for the endotracheal stabilization of reconstructed circumscript defects in the anterior tracheal wall of rabbits using the faszia lata. Stenoses in those animals receiving intratracheal resorbable stents were significantly smaller than those in animals without stents. The high mortality rates of 17% in the implant group and 23% in the control group were mainly caused by the functionally relevant tracheal stenoses. This was the reason why the approach combining the use of autologous materials and biodegradable stents was not accepted. The authors assumed that through controlled release of growth relevant factors from the biodegradable polymeric scaffolds, the potential of this method could be enhanced so that the enhancement especially of cartilage growth would render the reconstructed tracheal segments more stabile [54].

The treatment of subglottic stenoses, especially in children, still is a high challenge in spite of all the progress in surgery. Cotton and Seid in 1980 introduced the anterior cricoid split [55]. After several modifications of this technique and bearing in mind the contraindications, more than 90% of the children can nowadays be extubated without problems. In spite of the progress, in children undergoing single-step surgical therapy to treat subglottic stenoses, it is necessary to use postoperative intubation over several days as an intratracheal splinting. An external splinting by metallic microplates in the surgical tracheal reconstruction was described first time by Zalzal and Deutch in 1991 [56]. Weisberger and Nguyen applied metallic Vitallium™ miniplates for the external splinting of cartilage transplants in the reconstructive tracheal surgery, and 10 of 13 patients (77%) were successfully extubated immediately after surgery [57]. Willner and Modlin introduced resorbable miniplates in the reconstructive tracheal surgery. These resorbable plates were fixed by sutures in the region of the tracheal defect which diminished the stability in comparison to fixation by screws [58]. Following the successful application of resorbable plates and screws made of PLGA in the pediatric craniofacial surgery [59, 60], Long *et al.* described the external fixation of rib cartilage transplants by PLGA miniplates and screws in the tracheal reconstruction of subglottic stenoses in dogs in 2001. All of the 10 animals could be extubated without problems directly postoperatively. In all of these animals, there was an adequate widening of the subglottic stenoses over the whole period of observation (up to 90 days postoperatively). Two of the animals developed necroses in the cartilage transplants but in spite of this an endoluminal epithelialization was demonstrated histologically. The eight other animals showed a complete epithelialization of the transplants [61]. Since the degradation of PLGA *in vivo* [60] clearly exceeds an observation period of 90 days like in this study, long-term results are missing concerning the resorption of PLGA in tracheal applications and also the influence of degradation products of PLGA on the mucociliary clearance.

Kojima *et al.* described the production of tissue-engineered tracheal equivalents from cylindrical pieces of cartilage and equipped with an endoluminal epithelium in 2003. Cartilage and epithelial cells were harvested from the septal cartilage of sheep and grown *in vitro*. After proliferation and cultivation *in vitro*, the cartilage cells were seeded on a polyglycolic acid matrix. To shape the construct, the cell polymer scaffold was fixed around a silicon tube and then, for cultivation under *in vivo*, conditions, implanted under the skin in the back of nude mice. Precultivated epithelial cells were suspended in a hydrogel and injected into the cartilage cylinders. After removal of the stabilizing silicon tubes, the tissue-engineered constructs were harvested after 4 weeks of implantation. The morphology of the constructs produced by tissue engineering was described to be similar to the native sheep trachea. Maturated cartilage and the generation of a pseudolayered epithelium were demonstrated histologically. Proteoglycanes and hydroxyproline contents of the constructs were comparable to native cartilage so that the authors assumed that there might be a sufficient stability of such a construct *in vivo* [62]. It is thought that such a tissue-engineered construct in comparison to the earlier applied methods might have the potential to further growth after implantation *in vivo*, which could open new perspectives for the tracheal reconstruction in children. Cartilage was harvested so far from ribs, nasal septum, and ears, and also from tracheal and joint cartilage. While Kojima *et al.* assumed that the elastic cartilage from ears might not have the ideal biomechanical properties needed to produce tracheal constructs [62], other authors were less critical in the application of elastic cartilage from ears for the tissue engineering of cartilage in tracheal reconstruction [63].

Tracheal resection with the following end-to-end anastomosis is currently the therapeutical "gold standard" in the treatment of tracheal stenoses, when less than 50% of the tracheal length in adults and less than 1/3 of the tracheal length in small children have to be removed [64, 65]. The reconstruction of longer stenoses is a therapeutical challenge not solved at the moment. The tracheal reconstruction of such long segments by transplants necessitates an adequate blood supply to avoid the necrosis of the transplants. Jaquet *et al.* examined different three-component grafts in animals to simulate the anatomical structure of the trachea composed of mucosa, cartilage, and adventitia. Transplants consisting of cartilage from the ear and oral mucosa were revascularized through the laterothoracic fascia in rabbits. The epithelialization of three-component grafts was significantly enhanced through the application of perforated mucosa (40% epithelialization of the constructs after application of perforated mucosa versus 10% epithelialization after application of nonperforated mucosa). In all of the 20 operated animals, there was a sufficient vascularization, and necroses were not detected in the transplants [66]. The authors assumed that the production of vascularized composite grafts is an option for the reconstruction of longer tracheal stenoses. A successful application of these constructs in animals and clinical studies is missing, however.

A completely different approach for the reconstruction of longer tracheal segments was chosen by other groups who applied aortal autografts for the tracheal reconstruction in pigs [67] and in sheep [68, 69]. In both animals, the implants were stabilized postoperatively by silicon stents. Immunosuppression was not applied in either of the animal models. In pig implants, an epithelialization with metaplastic epithelial cells, newly grown cartilage, and nonorganized elastic fibers were demonstrated. In sheep implants, there were initial inflammatory reactions followed by the growth of a mucociliary epithelium and the development of new cartilaginous tracheal rings [69]. In 2006, this group published results from the tracheal reconstruction of a longer segment in a human patient applying an aortal autograft. After the resection of a 7-cm-long cervical tracheal segment due to a tracheal carcinoma situated directly caudal of the cricoid cartilage and localized clearly intratracheally without regional lymph nodes or distant metastases, there was a tracheal reconstruction applying a segment of the autologous, infrarenal aorta of this 68-year-old patient. The excised aortal segment was replaced by a Dacron™ prosthesis. A chronical obstructive pulmonary disease, a peripheral arterial occlusive disease, and a myocardial infarction (17 years before the tracheal reconstruction) were known from this patient. The patient was extubated without problems 12 h postoperatively. There was an endotracheal stabilization applying a silicon stent 3 days postoperatively. An adjuvant irradiation of the whole trachea with 30 Gy was started on the 15th day postoperatively. Four weeks postoperatively, an acute dyspnea appeared in the patient due to granulation in the region of the proximal anastomosis which was treated with a further stent application proximal to the first stent. Both stents could be removed without problems 3 months later. Afterward no further granulomatous tissues could be diagnosed endoscopically at the anastomotic sites. Clinically no more states of dyspnea appeared. The patient died due to septical shock in the course of pneumonia in both lungs 6 months postoperatively. Since family members did not accept autopsy, no further details of the performance of the aorta-based tracheal construct could be revealed [70].

Although the aorta-based allogenic tracheal constructs did not perform too well in the pig, this approach in two animal models and in humans was remarkable both from clinical and from scientific perspectives. From a clinical perspective, the use of aortal segments offers a tubular structure, comparable in diameter to the trachea, which is air and fluid tight, flexible and with high mechanical strength, and is available in the afforded amount. There are problems, however, with the lack of biomechanical stability not avoiding the collapse of airways and with the missing epithelialization. From a scientific perspective, this approach allows the use of decellularized tissues, even of allogenic ones, as preformed, long-distance scaffolds in tracheal reconstruction, which enable the ingrowth and differentiation of the patient's own precursor/stem cells assumed to be needed for the regeneration of functional tissues. The application of tracheal-based allogenic constructs exploiting a decellularized donated human trachea was successfully applied by Macchiarini *et al.* in the reconstruction of a main bronchus of a 13-year-old female patient with a severe bronchio malacia. All cellular and MAC antigens are removed from the trachea which was then feeded with epithelial cells and chondrocytes developed *in vitro* from mesenchymal stem cells of the recipient. The scaffold allowed the unobstructed function of the patient's airways directly after surgery. Now almost 1 year later, the bronchoscopical findings are still regular with appropriate mechanical characteristics and a sufficient bronchociliary clearance. An immunosuppressive therapy was not necessary. The combination of autologous cells with appropriate implant scaffolds is thought to be a well applicable therapeutical option for the reconstruction of the airways [71]. A lot of efforts in basic science and clinical research have still to be spent until the growth of biomechanically loadable segmental cartilage can be engineered on demand and tissue-engineered tracheal constructs will be inhabited by fully functional epithelial cells [72].

13.2.2.1 Epithelialization of Tracheal Scaffolds

The first application in humans of an artificial trachea produced according to principles of regenerative medicine was published by Omori in 2005. A papillary carcinoma in the thyroid of a 78-year-old woman necessitated a hemithyroidectomy together with the resection of the anterior tracheal wall. The tracheal wall defect was reconstructed by a patch based on a Marlex™ net covered with collagen. Two months postoperatively, endoscopic analysis revealed the epithelialization of the scaffold. And there was also a sufficient mechanical stability in the scaffold. Two years after surgery, there were still no respiratory complications or insufficiencies. In spite of missing long-term results, the authors were convinced that new therapeutical options will be offered for the reconstructive tracheal surgery by regenerative medicine [73].

The relatively long period of 2 months needed to epithelialize the patch, which was applied in the tracheal reconstruction, points to a problem that could not get adequately solved. After application of novel polypropylene collagen scaffolds for the reconstruction of circumscript tracheal defects in dogs, the complete epithelialization of the scaffold could be demonstrated 8 months postoperatively only [74]. A fully functional tracheal epithelium is essential as a physical barrier against the extratracheal milieu, as regulator for the comprehensive metabolic functions of the airways including transport of fluids and ions and for the mucociliary clearance and the patency of the airways [75]. The early development of a complete and functionally adequate epithelialization of tracheal scaffolds is of critical importance for the biofunctionality of implants and constructs produced following the principles of tissue engineering. The research on mechanisms of regeneration and differentiation of respiratory epithelial cells in contact with tissue-engineered constructs started only recently. Before that, the research concerning the differentiation mechanisms of respiratory epithelial cells was focused on their differentiation in the embryonic phase [76] and on the development and differentiation of epithelial cells from precursor/stem cells [77]. It was shown that basal cells of the human trachea probably are precursors of respiratory epithelial cells [77, 78]. The tracheal epithelium is mainly composed of ciliary cells, goblet cells, and basal cells [79–81]. Basal cells are essential for the generation of precursor cells which are fundamental for the regeneration of epithelial damage [77, 78, 82–84].

Nomoto *et al.* seeded the scaffold material used by Omori with tracheal epithelial cells of rats *in vitro*. These epithelial cells expressed *in vitro* the cytokeratines 14 and 18 as typical intermediate filaments of epithelial cells as well as occludin, a constituent of tight junctions in epithelial cells which is a main component of the barrier against diffusion of soluble substances into the intercellular space. The cell-seeded scaffolds were applied for the reconstruction of cervical tracheal defects of 3 mm length in rats. Over the whole period of observation (30 days) *in vivo*, the artificial trachea was covered with epithelium. Partially, a single- or double-layered epithelium was found not carrying cilia, whereas other parts displayed prismatic epithelial cells with functional cilia [85]. In a further development of this technique, a thin collagen matrix (Vitrigel™) was applied for 3D growth of cells in the scaffold. This 3D matrix enhanced the growth of epithelial cells as well as the invasion of mesenchymal cells. There was a clearly accelerated regeneration of functional epithelial cells carrying cilia after tracheal reconstruction in rats using Vitrigel-coated scaffolds compared to noncoated scaffolds [86].

The importance of epithelial–mesenchymal interactions for morphogenesis, homeostasis, and regeneration of the epithelium are well known from literature since several years [87–89]. During epithelial regeneration, epithelial precursors arrived from the borders of epithelial damage to proliferate and differentiate there. Mesenchymal cells situated below the epithelium regulate epithelial growth and differentiation through generation of an appropriate biomatrix and through synthesis and release of growth relevant factors [90, 91]. Fibroblasts are also important participants in the interactions between epithelial and mesenchymal cells and strongly influence epithelial regeneration in wound healing. They are able to secrete a variety of growth factors like keratinocyte growth factor, epidermal growth factor, and hepatocyte growth factor [92, 93]. The importance of fibroblasts was shown already for epidermal wound healing [93], oral [94] and corneal epithelial regeneration [95], and also for tracheal epithelial regeneration [96]. The cocultivation of epithelial cells and tracheal fibroblasts *in vitro* induced the generation of a layered epithelium containing epithelial cells with cilia, goblet cells, and basal cells. Moreover, a basal membrane was constituted *in vitro* between epithelial cells and fibroblasts where the presence of integrin-β4 was demonstrated, which is a specific marker of basal membranes and of epithelial mucin secretion [96].

In further studies, the authors demonstrated the potential of heterotopic fibroblasts (from dermis, nasal, and oral mucosa) for tracheal epithelial regeneration. Regeneration of epithelial cells in contact with different heterotopic fibroblasts showed different characteristics in structure, development of cilia, secretion of mucins, and expression of ion and water channels, for example, aquaphorines and Na^+/K^+ ATPase. In contact with nasal fibroblasts, however, no mature and fully functional tracheal epithelium was generated *in vitro*. Dermal fibroblasts induced the generation of an epidermal like epithelium. Especially the cocultivation with fibroblasts from the oral mucosa induced the regeneration of a morphologically and functionally regular tracheal epithelium. This was comparable to the regeneration of epithelium *in vitro* after cocultivation with tracheal fibroblasts. Fibroblasts from the tracheal and the oral mucosa expressed keratinocyte growth factor, epidermal growth factor, and hepatocyte growth factor. Fibroblasts from the oral mucosa enhanced proliferation and migration of epithelial cells *in vitro* similarly to the tracheal fibroblasts. Since the explantation of oral mucosa is clearly less invasive than that of tracheal mucosa, there seems to be a very promising method available now to develop scaffolds with a functionally adequate epithelium for the tracheal reconstruction [97].

In 2008, the same group used this technique of cocultivation of epithelial cells and tracheal fibroblasts to produce a tracheal scaffold seeded with cells *in vitro* and applied the tissue-engineered scaffold for the tracheal reconstruction in rats [98]. The authors could demonstrate a fully functional epithelium *in vivo*. Besides the cocultivation of tracheal epithelial cells and fibroblasts, also the cocultivation of tracheal epithelial cells and mesenchymal stem cells for the "*in vitro*" reconstruction of a fully functional tracheal epithelium is described in the literature. The epithelium thus produced showed morphological, histological, and functional characteristics of the tracheal mucosa. The authors assumed that the cocultivation with mesenchymal stem cells could play a main role in tissue engineering in future [99].

13.2.2.2 Vascular Supply of Tracheal Constructs

A problem not adequately solved so far is the vascular supply of scaffolds and of tissue constructs developed from these scaffolds *in vivo*. In contrast to other parenchymal organs, the trachea is supplied by a network of small blood vessels which is evidently not easy to generate. Microanastomoses were not successful in animal models [100, 101] and therefore not further persecuted. It is known from the literature that after tracheal reconstruction, the capillary network present at the anastomosis proceeded in the direction of the implant only 2 cm at maximum and that this process of revascularization took several months [102]. In tracheal implants, which were longer than 3 cm, there was a lysis of the epithelium with a consecutive destruction of the basal membrane followed by the development of granulomatous tissues producing a tracheal stenosis. While bioreactors allow the growth of autologous cells [103] and functional tissues and are routinely used for the generation of osteochondral constructs, and tissue-engineered heart valves, there are very few studies showing the application of bioreactors for the generation of tracheal scaffolds. Decisive problems hindering the application of tracheal scaffolds in humans are the missing epithelialization and revascularization of the constructs. Tan *et al.* published in 2006 the concept of a so-called *in vivo* bioreactor for the generation of tracheal constructs. They proposed layered scaffolds with a porous catheter within the inner layer of the scaffold for a continuous supply of cells and nutrition media and an outer layer of the construct granting the necessary stability. In contrast to traditional bioreactors in which nutrition media mainly flow around the constructs, now a perfusion system was planned within the scaffolds similar to the blood vessel distribution *in vivo* [104]. This group seeded in a next step a phase-segregated multiblock copolymer (DegraPol™) [105] with human tracheal epithelial cells and offered a continuous supply of cells and nutrition media via a porous catheter within the scaffolds. The continuous perfusion of the tubular biodegradable scaffolds coincided with an adequate epithelialization of the constructs and an accelerated vascularization in the chorioallantois membrane assay. The authors assumed that the concept of the *in vivo* bioreactor allows a more physiological process in the reconstruction of tissues and that better initial conditions are granted for the problem so far not solved, the vascularization of tracheal scaffolds [106].

13.2.3
Regenerative Medicine for Reconstruction of Pharyngeal Defects

The reconstruction of the pharynx by degradable, multifunctional polymeric materials would be a novel therapeutical option in head and neck surgery. The use of implant materials for the reconstruction of pharyngeal defects is currently at the early beginning. Until now, there are only data concerning the use of implant materials in the area of the oral mucosa and the palate available. Hallén *et al.* injected crosslinked hyaluronic acid in rats in the dorsal pharynx wall to treat velopharyngeal insufficiency. In all animals, an early inflammatory reaction due to the hyaluronic acid was found. Six months after injection, the hyaluronic acid was still detectable at the original localization of injection and surrounded by connective tissues. Despite lacking of long-term results, the authors assumed that the injection of crosslinked hyaluronic acid is appropriate for the augmentation of a slight velopharyngeal insufficiency in humans [107]. Ophof *et al.* implanted skin substrates after cell seeding with oral keratinocytes *in vitro* into palatinal wounds in dogs as a model for closure of cleft palate by tissue-engineered constructs. In all six animals, the loss of the epithelium and a distinctive degradation of the skin substrates were detectable. The authors concluded that an adequate integration of these tissue-engineered constructs required an early and sufficient revascularization of the scaffolds *in vivo* [108]. A main focus in tissue engineering of oral mucosa is currently the use of novel dermal scaffolds and epithelial cell culture methods including 3D models. An updated review is given by Moharamzadeh *et al.* [109].

Despite numerous biomedical applications of tissue-engineered constructs in almost all medical fields, up to now there are no literature data available regarding the pharyngeal reconstruction with implant materials after tumor resection neither in animal models nor in humans. The availability of multifunctional polymeric implant materials, which can be adapted according to the anatomical, physiological, biomechanical, and surgical requirements [12, 16], facilitates the development of novel therapeutical options also in head and neck surgery. A main scientific topic of the own group is the biocompatibility testing of an elastic degradable AB-copolymer networks [13, 14] *in vitro* and *in vivo*, which seems to be appropriate for the reconstruction of pharyngeal defects due to its physicochemical characteristics.

13.3
Methods and Novel Therapeutical Options in Head and Neck Surgery

13.3.1
Primary Cell Cultures of the Upper Aerodigestive Tract

The use of cell cultures is an essential tool in nearly all biological and medical research laboratories. The biocompatibility testing should be conducted with cultures of site-specific cells depending on the biomedical application to assess the specific interaction between the biomaterial and site-specific different cells [110]. Thus, the biocompatibility testing of a polymeric material which seems to be appropriate for the reconstruction of pharyngeal defects should be conducted with primary cell cultures of the pharynx. The knowledge about the interactions between the implant materials and cells/tissues is a basic requirement for an ideal adaptation of a polymeric material according to the specific needs of the upper aerodigestive tract (ADT). In our studies, primary cell cultures of the oral cavity, the pharynx, and the esophagus were established and biochemically characterized. Immunocytological investigations showed different relative amounts of epithelial, fibroblastic, and smooth muscle cells depending on the anatomical site of explantation [111]. Relatively little is known about the mechanisms of regular and delayed wound healing of the pharyngeal epithelium. Therefore, a comprehensive characterization of primary cell cultures of the pharynx was a first step for the development and establishment of novel therapeutical options [111, 112].

13.3.2
Assessment and Regulation of Matrix Metalloproteases and Wound Healing

The amount and organization of the extracellular matrix in normal wounds is determined by a dynamic balance between overall matrix synthesis, deposition, and degradation [113]. A strictly controlled degradation of the extracellular matrix is an important process for the regular wound healing. An imbalance between degradation and synthesis of the matrix during wound healing would cause a delayed wound healing with fistulae and ulcerations in case of outbalanced degradation of the extracellular matrix or hypertrophic scars and keloids in case of outbalanced synthesis of the extracellular matrix [114].

Matrix metalloproteases (MMPs) are a class of structurally related, zinc-dependent endopeptidases that are collectively responsible for the degradation of extracellular matrix proteins. MMPs have an important function in wound healing [115, 116]. Under regular conditions *in vivo*, the expression and activation of MMPs is strictly controlled. The activity of MMPs is regulated at the level of transcription and zymogen activation and can be inhibited by specific inhibitors: the tissue inhibitors of metalloproteases TIMPs. Recently, four different TIMPs (TIMP 1–4) were identified and cloned [117]. In the literature, different MMP- and TIMP levels were reported in regular and delayed wound healing [118, 119]. The delicate balance between the activity of MMPs and TIMPs plays a key role in building a functional extracellular matrix. Up to now, little is known about the mechanisms of wound healing and MMP expression of cells of the upper ADT *in vitro* and *in vivo* [120, 121].

A comprehensive characterization of the MMP- and TIMP expressions of cells of the upper ADT is a basic requirement to develop and establish novel therapeutical options in head and neck surgery in case of delayed wound healing after surgical treatment. A main focus of the own biocompatibility testing was the analysis of the MMP- and TIMP expressions of primary cell cultures of the upper ADT after cell seeding on different modifications of the polymeric implant material to gain the knowledge for an optimal adaptation of these materials to the specific requirements of the upper ADT.

Among the primary cell cultures investigated, cells of the pharynx were seeded on the surface of a multifunctional copolymer as well as on the surface of commercially available polystyrene cell culture dishes as control. On both surfaces, cells became adherent, proliferated, and reached confluence. No statistically significant differences of the mean cell numbers were found on Day 1, 3, 6, 9, and 12 of cell growth after cell seeding [112]. The highest MMP-1-, MMP-2-, and TIMP levels were found on Day 1 of cells' growth on both surfaces. There were decreasing levels during the following time of the investigation (Figure 13.2). No statistically significant differences of the MMP- and TIMP expressions were detectable between the polymer and the control surfaces. The kinetics of MMP-2 expression were analyzed on the protein level and by RT-PCR on the mRNA level (Figure 13.2) [112]. Based on the current results, the adhesion, proliferation, and differentiation of the primary cell cultures of the pharynx were not influenced by the multifunctional copolymer.

13.3.3 Influence of Implant Topography

The integration of a material in the surrounding tissues is a basic requirement for a successful clinical application of an implant material *in vivo*. The surface characteristics of materials including their surface topography and chemical composition are of very high importance for the interaction between the material and cells and tissues [122, 123]. Until now, some cellular processes are known, which could be useful to assess the cellular behavior on implant materials. Most of this knowl-

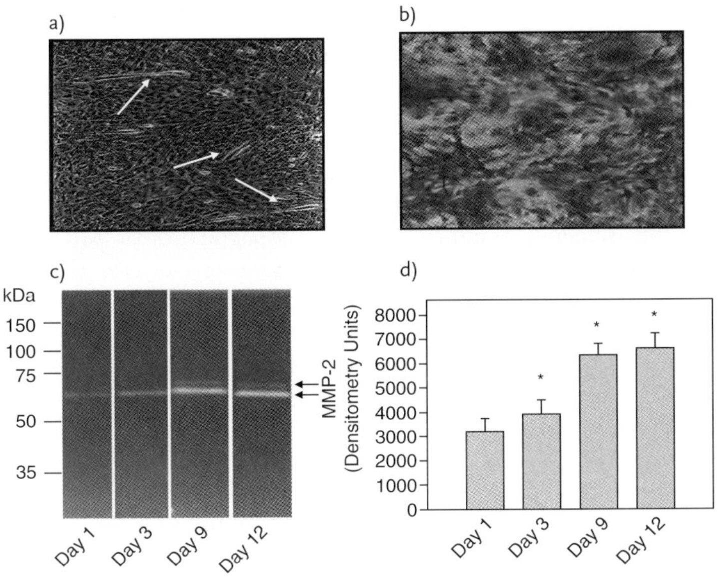

Figure 13.2 Histological findings of pharyngeal cells and results of zymography of MMP-2 of pharyngeal cells grown on a polymer surface. (a) Phase-contrast microscopy of pharyngeal cells grown on polystyrene surface of commercially available cell culture dishes is shown. Pharyngeal cells showed a confluent monolayer on the surface of 35-mm cell cultures dishes after 3 days with the typical cuboid morphology of epithelial cells. Smooth muscle cells of the pharyngeal epithelium are labeled by white arrows (magnification ×20). (b) In order to better visualize the pharyngeal cells after cell seeding on the polymer surface, Coomassie Blue staining was used. Pharyngeal cells began to form colonies after cell seeding and started to become confluent on Day 3 of cell growth (magnification ×20). (c) SDS-substrate gel electrophoresis (zymography) of primary cell cultures of the pharynx grown on the polymer surface is shown. The kinetics of appearance and activity levels of 72 kDa (MMP-2) band of pharyngeal cells are shown on Day 1, 3, 9, and 12 of cell growth. Bands are marked by arrows. The gelatinolytic activities of media conditioned by pharyngeal cells grown on the polymer surface were normalized to equal cell numbers. (d) Scanning densitometry units of the gelantinolytic bands are shown. Statistical analysis was performed to determine differences of MMP-2 levels between Day 1 and the subsequent days of cell growth. Statistically significant differences ($P \leq 0.05$) are marked by a star. Data taken from three independent experiments (values are mean ± SD). Parts c and d reprinted from [111], Copyright 2007, with permission from IOS Press.

edge is based on cell culture investigations and it is unknown if these mechanisms are also found *in vivo* [124, 125]. A fundamental requirement for a successful application of degradable implant materials for the pharyngeal reconstruction *in vivo* is a saliva-tight integration of the material in surrounding tissues to avoid salivary fistulae with destruction of neighboring soft tissue. The development of long-term degradable polymeric scaffolds for pharyngeal reconstruction has to guarantee an adequate biocompatibility and biofunctionality as well as growth of a functional tissue formation considering the specific physiological and mechanical requirements of the upper ADT. Important progress in biomaterial research of the last years was made in the improvement of cell adhesion and proliferation by the optimization of scaffold design with respect to specific requirements of the different implantation sites *in vivo* [126]. Main aspects of the research work were focused on the influence of different macroscopical and microscopical design parameters on the local differentiation of variable cells. Other aspects dealt with the controlled release of growth factors [127, 128]. Until now, relatively little is known about the influence of different surface topographies of polymeric implant materials on the gene expression and synthesis of enzymes that are directly involved in extracellular matrix remodeling [129, 130].

Our own results demonstrated the importance of the surface structure of polymeric implant materials on the cellular behavior depending on surface roughness (smooth versus rough surfaces). The cell adhesion, proliferation, as well as the kinetics of secretion and activity of MMP-1, MMP-2-, and TIMPs differed significantly depending on the type of cells and on the surface structure of the copolymer. Significantly greater average total cell numbers of oral and pharyngeal primary cells were found after cell seeding on the rough surface compared to the smooth polymer surface. Esophageal cells showed the highest cell numbers on the control (polystyrene). Oral and pharyngeal cells revealed similar kinetics of appearance and activity of MMP-1, MMP-2, and TIMPs with the highest values on Day 1, followed by a decrease of the activity levels on the rough polymer and the control surface. Oral and pharyngeal cells seeded on the smooth polymer surface displayed an opposite pattern with the lowest activity of MMP-1, MMP-2, and TIMPs on Day 1 and the highest values on Day 12. Esophageal primary cell cultures showed a comparable kinetic pattern of appearance and activities on all three different surfaces (smooth and rough polymer surface, control surface) with the lowest MMP-1-, MMP-2,- and TIMP expression on Day 1 and the highest values on Day 12 [131].

The presence or absence of the extracellular matrix or components of it govern the proliferation, differentiation, and biochemical activities of different primary cell cultures of the upper ADT. These results were confirmed by data from the literature, which also showed the influence of the surface topography on the gene expression and synthesis of the enzymes directly involved in extracellular matrix remodeling [132].

The results of these experiments suggest a specific influence of surface topography on the behavior of cells in contact with implant materials. The knowledge of the exact mechanisms of the cell–biomaterial interactions is a basic requirement for the development of an "ideal" implant material to establish cell- and tissue-optimized novel therapeutical options in head and neck surgery based on polymeric implant materials.

13.3.4
Application of New Implant Materials in Animal Models

The use of degradable implant materials in the area of the upper ADT makes high demands on the chemical, enzymatic, bacterial, and mechanical stability of a material. A premature degradation of the implant material would probably cause extensive salivary fistulae with high mortality potentially culminating in carotid artery rupture. Because of the chemical conditions in the upper ADT with changing pH values, enzymatical, bacterial, and particular mechanical load during deglutition and digestion, the reconstruction of the upper ADT by a degradable implant material requires adequate chemical, enzymatical, bacterial, and mechanical stabilities of the scaffold material. We established a standardized radical critical defect in the gastric wall of rats which was closed by an elastic long-term degradable polymeric implant. The stomach was used as a "worst-case" application site to test the stability of the implant material under extreme chemical, enzymatical, bacterial, and mechanical load. In this model, the mortality of the gastric breakdown of sutures with fistulae and local or generalized peritonitis in the follow-up is comparable to the mortality of insufficiencies and salivary fistulae of the pharynx. The implantation group included 42 animals. A primary wound closure of the gastric wall defect without biomaterial implantation was conducted in the control group ($n = 21$). Furthermore, a so-called baseline group which included animals kept under the same housing conditions without any surgical procedure was investigated ($n = 21$). The implantation periods or times of observation were 1 week, 4 weeks, and 6 months [133, 134].

Fundamental parameters investigated in this animal model were a tight closure between the polymer and surrounding tissues, the chemical and mechanical stability of the implant material, and the integration of the polymer in the surrounding tissue as well as the question of tissue regeneration after reconstruction of the defect with the polymeric implant material. Gastrointestinal complications like fistulae, perforation, or peritonitis did not occur in any of the animals. A liquid- and gas-tight anastomosis between the polymer and the adjacent stomach wall existed in all animals of the implantation group [133]. To test the impermeability between the implant material and adjacent gastric wall, the intragastric pressure was measured after maximal dilatation of the stomach by air insufflation (Figure 13.3) [133]. Neither in the implantation group nor in the control group a delayed wound healing was observed macroscopically or microscopically after 1 week, 4 weeks, and 6 months of implantation time after primary wound closure. After 1 week, a beginning regeneration of the gastric wall was detected starting from the border area of the gastric wall defect. After 4 weeks and 6 months, a regular multilayered stomach tissue as known from histology was found in the former defect zone of the gastric wall (Figure 13.4). In the control group, the defect was replaced by scar tissue [134]. Furthermore, the systemical influence of the AB-copolymer network was investigated. It is well known from literature that the peritoneum is a very sensitive compartment for inflammatory reactions in the organism depending on the biocompatibility of implant materials [135]. Incompatibilities of implant materials, a too early degradation, or the accumulation of degradation products are expected to cause local inflammatory reactions originating acute-phase reactions concomitant with the induction of gene expression of acute-phase proteins. The concentrations of the acute-phase proteins α_1-acid glycoprotein and haptoglobin, however, did not show statistically significant differences between the AB-copolymer network and the control group [136]. The analyses of the mechanisms of the integration of the implant material in the adjacent tissues as well as the mechanisms of material

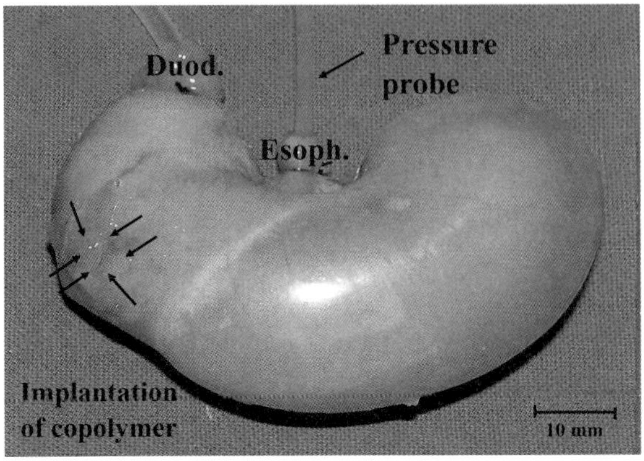

Figure 13.3 Aspect of the explanted stomach after 1 week of copolymer implantation. The polymer implantation site is marked by arrows. A flexible tube for air insufflation was inserted in the duodenum. The pressure was measured by a probe in the resected esophagus. The pressure probe is marked by an arrow. A special anatomical feature of the rat stomach becomes overt: the stringent separation between the glandular part of the stomach where the copolymer was implanted (marked by arrows) and the nonglandular part. The influence of this special anatomical feature on the biofunctionality of the polymeric material is unknown so far and needs to be investigated in another animal model. Abbreviations: Duod. = Duodenum; Esoph. = Esophagus. Reprinted by permission from [133], available at "http://www.reference-global.com/." Copyright 2006, Walter de Gruyter GmbH & Co. KG.

Figure 13.4 Macroscopical and histological findings after polymer implantation. (a) The explanted stomach is shown after 1 week of implantation. The polymer is marked by a black line. The mucosa started to overgrow the polymer from the border area (marked by stars). (b) Histological findings are shown after 1 week of implantation. The marginal area next to the defect zone showed a regular stomach epithelium marked by stars. According to the macroscopical findings, the beginning of tissue regeneration was detectable from the marginal area next to the defect zone. The polymeric material used for defect closure was removed due to the xylene and ethanol treatment and cutting of paraffin sections and was not detectable on most of the histological sections (defect closures by polymer are marked by arrows). (c) After 4 weeks of implantation time, the polymer was almost detached from the stomach and was just fixed by single sutures. The former defect was closed by regenerated tissue (marked by a star). (d) The histological findings after 2 weeks of implantation time are shown. The marginal areas next to the former defect zone are marked by stars. The former defect zone (marked by arrows) was regenerated by histological regular formated stomach epithelium. (e) After 6 months of implantation time, the polymer was completely detached from the stomach wall in all animals. (f) Histological findings after 1 month of implantation time are shown. Histologically regular formated stomach epithelium was found in the former defect zone (marked by an arrow) in all animals of the implantation group. No differences were detectable between the epithelium of the marginal area next to the former defect zone (marked by a star) and the regenerated epithelium of the former defect zone (marked by an arrow). Reprinted from [134] with permission. Copyright 2007, Georg Thieme Verlag KG, Stuttgart, New York.

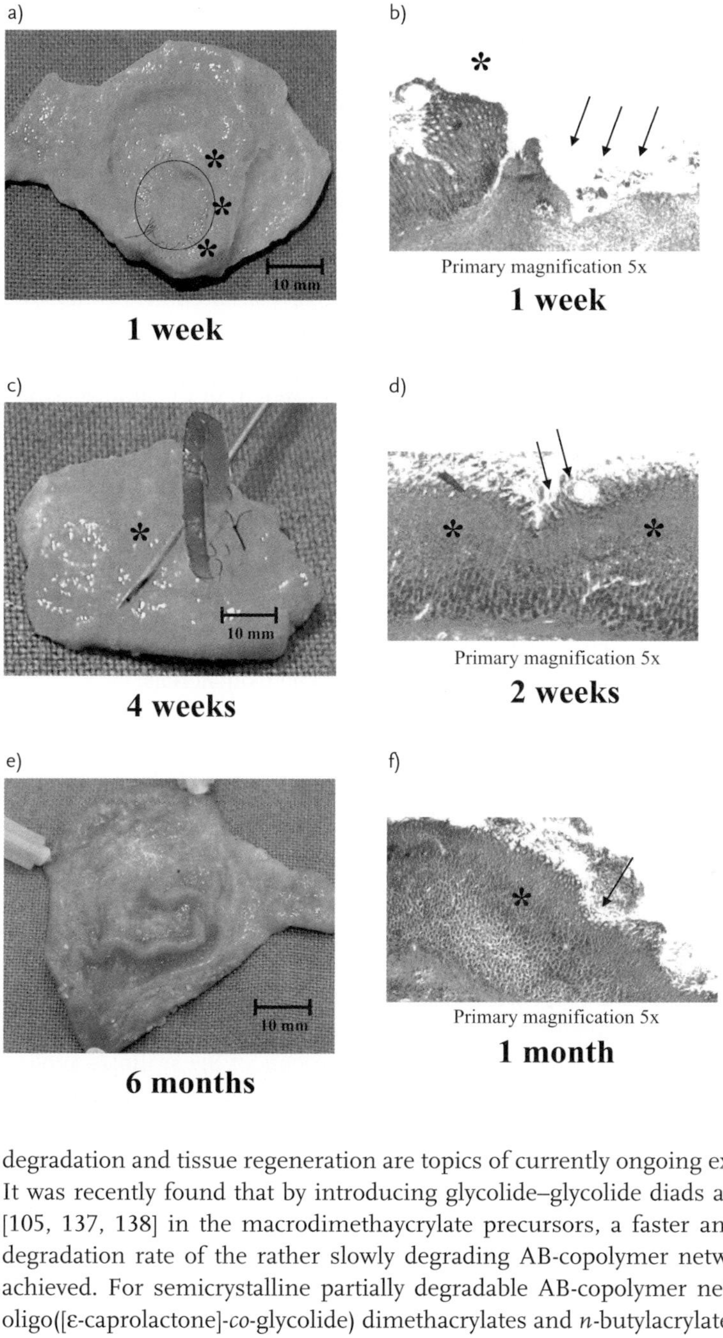

degradation and tissue regeneration are topics of currently ongoing examinations. It was recently found that by introducing glycolide–glycolide diads as weak links [105, 137, 138] in the macrodimethaycrylate precursors, a faster and adjustable degradation rate of the rather slowly degrading AB-copolymer networks can be achieved. For semicrystalline partially degradable AB-copolymer networks from oligo([ε-caprolactone]-*co*-glycolide) dimethacrylates and *n*-butylacrylate of different molar glycolide contents *in vitro* higher degradation rates of AB networks with higher χG were measured by mass loss, decrease of G, and increase of Q due to the glycolide containing ester bonds and especially glycolide–glycolide diades in the oCG, which can be considered as weak links [105, 137]. Upon cleavage of glycolide containing ester bonds, the remaining oligo(ε-caprolactone) segments regain mobility, can rearrange, and crystallize as shown by slightly increasing T_m during degradation, for example, AB-CG(21)-10. The degradation *in vivo* was only slightly accelerated compared to *in vitro* conditions in the studied time frame for glycolide-free AB-CG(0)-10 networks. This suggests that enzymes, which are known to be major contributors to the degradation of poly(ε-caprolactone) [139] could not very well access the semicrystalline poly(ε-caprolactone) segments in the bulk of the AB networks [140].

In the experiments performed with the AB-copolymer networks so far, the chemical, hydrolytical, and enzymatic stability as well as the biomechanical functionality of the polymeric implant material were shown under the extreme conditions of the stomach. The postoperative increase in weight of the animals [133], the impermeability between the implant material and adjacent tissues of the gastric wall [133], the concentrations of the acute-phase proteins α_1-acid glycoprotein and haptoglobin [136], as well as the lack of gastrointestinal complications suggest that the wound healing was not negatively influenced by the degradable AB-copolymer network during the time of investigation. On the contrary, a support of tissue regeneration by the implant material was detected. The results available so far regarding the tissue compatibility allow to regard the AB-copolymer network as a very promising implant material for the development of novel therapeutical options in head and neck surgery based on degradable biomaterials.

13.4 Vascularization of Tissue-Engineered Constructs

The vitality and functionality of tissue-engineered constructs depends on an adequate blood supply with oxygen and nutrients as well as on the removal of metabolites. Most of the tissues/organs successfully tissue engineered until now are relatively thin and/or avascular like cartilage, skin, or urinary bladder. Therefore, wound healing-driven angiogenesis in recipients is thought to be sufficient to supply the tissue-engineered constructs with oxygen and nutrients in many cases. It was suggested that the supply of blood and nutrients of the scaffolds applied for pharyngeal reconstruction could be sufficient because the used implant materials are relatively thin (<100 μm). In any case, the applied scaffolds should support angiogenesis. The investigation of the influence of polymeric implant materials on the angiogenesis is therefore an important aspect of biocompatibility testing.

In our investigations *in vitro*, we showed that bovine capillary endothelial cells (ECs) of the adrenal cortex [141] became adherent on the copolymer surface and developed confluent cell layers [142]. Also, in the chorioallantois membrane assay, no negative influence of the copolymer samples on the vascularization was detectable [142, 143]. A controlled release of angiogenic factors from vesicles on the polymer surface according to the principles of drug delivery to support angiogenesis is a scientific topic of currently ongoing investigations.

At present an adequate vascularization of the cellular colonized scaffolds *in vivo* is one of the most critical points for tissue engineering of complex and metabolic challenging organs like heart or liver. In case of parenchymal organs, the tissue-engineered microcirculation has to be connected to the recipients' circulation. The currently available techniques for the vascularization of tissue-engineered constructs can be classified in "*in vitro*" and "*in vivo*" methods. In the last years, considerable progress was made to solve the problems of building microcirculatory networks for comprehensive 3D constructs. Kunz-Schughart *et al.* developed a 3D cell culture system with cocultivation of human skin fibroblasts and ECs of the umbilical cord. They found a support of migration, vitality, and development of tubular structures of the ECs by fibroblasts. Based on such models, knowledge about the integration of capillary structures in engineered tissues can be gained [144]. Au *et al.* approached the vascularization of tissue-engineered scaffolds by cocultivation of constructs with blood vessel cells like endothelial and perivascular cells. The authors demonstrated that the co-implantation of the scaffolds with tissue-specific cells and endothelial and perivascular cells led to the development of vascular structures *in vivo*, connecting the scaffolds and recipient's circulation. The stability and adequate functionality of these vascular structures have been shown for more than one year. Based on these results, the authors assumed that this technique of co-implantation is a promising approach for the vascularization of tissue-engineered constructs [145].

On the other side, there are still numerous unsolved problems with and beyond the connection of scaffolds to the recipient's vascularization, like the maintenance or increase of vascular density with an increase of tissue or organ mass or activity, the maturation of functionally inadequate vessels, as well as the unwanted regression of vascular structures. One of the answers to these problems might be gained in future through a comprehensive knowledge about the regulation of the heterogeneous ECs in different organs. Furthermore, an extensive knowledge about the mechanisms of the molecular processes of cellular interactions between ECs, pericytes, and smooth muscle cells and between blood vessels and parenchymal cells are needed. Beyond that, the mechanical characteristics of blood vessels like permeability, elasticity, and compressibility have to be analyzed and the design of nonthrombogenic surfaces of implant materials have to be devised. A review about the current knowledge of microcirculation engineering as a basic requirement for a successful tissue engineering of parenchymal organs is given by Lokmic *et al.* [146].

13.5
Application of Stem Cells in Regenerative Medicine

Stem cells have the capacity for self-renewal and capability of differentiation to various cell lineages. Thus, they represent an important building block for regenerative medicine and tissue engineering. These cells can be broadly classified into embryonic stem cells and nonembryonic stem cells comprising adult stem cells.

Embryonic stem cells are called pluripotent and can differentiate in all cell types of the three embryonic germ layers. The adult stem cells are multipotent and the differentiation of these cells is committed to only one of the germ layers. Embryonic stem cells have a great potential but their use is limited by several ethical and scientific considerations. Limiting factors for the use of embryonic and adult stem cells next to ethical considerations [147, 148] are problems associated with extensive cell expansion *in vitro* [149], problems with *in vitro* cultivation on implant materials [150, 151], cell apoptosis following implantation [152], as well as vascularization [153].

Stem cells were already studied by Becker *et al.* in 1963 who injected bone marrow cells into irradiated mice and noticed that nodules developed in the spleens of the mice in proportion to the number of bone marrow cells injected [154, 155]. They concluded that each nodule arose from a single marrow cell. Later on, they found evidence that these cells were capable of infinite self-renewal, one of the main characteristics of stem cells.

Stem cells have been used successfully in experimental and clinical studies for bone, cartilage, spinal cord, cardiac, and bladder regeneration. A current review about the application of stem cells in the field of regenerative medicine is given by Bajada *et al.* [5].

In 2001, Vacanti *et al.* reported the successful tissue engineering of the distal phalanx and the replacement of this bone in a 36-year-old patient who suffered partial avulsion of the thumb [156]. However, only 25% of the normal strength was obtained. Quarto *et al.* reported on the use of autologous culture-expanded bone marrow stromal cells combined with porous hydroxyapatite for the reconstruction of critical sized defects (bone segmental defects 4–7 cm long) of tibia, ulna, and humerus. The results were encouraging, with good graft integration and return to functionality [157]. In 2006, Hibi *et al.* published the use of tissue engineering to augment bone formation in humans in combination with vertical distraction osteogenesis (DO) by an osteocutaneous fibular transplant for the reconstruction of the mandible after irradiation. DO is a method for the elongation of the bone that is used, among others, in the surgical reconstruction of facial skull to bridge bony defects of different genesis. To promote 3D bone formation and shorten the consolidation period, the authors applied tissue-engineered osteogenic material ("injectable bone") in a patient who was treated with vertical DO and an osteocutaneous fibular flap to reconstruct the mandible. The material which comprised autologous mesenchymal stem cells was culture-expanded and then induced to be osteogenic in character. Platelet-rich plasma was activated with thrombin and calcium chloride and infiltrated into the distracted tissue at the end of distraction and injected into a space created labially with a titanium mesh at implant placement. The reconstructed mandible was expanded from 10 to 25 mm in height despite a lacerated and opened labial periosteum in the distracted area. The authors assumed that DO assisted by tissue engineering could be the therapy of choice in future for the surgical reconstruction of bony defects [158]. Furthermore, the authors used this technique of tissue-engineered osteogenic material ("injectable bone") successfully for the osteoplastic reconstruction in cleft palate in a 9-year-old girl [159].

While stem cells are successfully in clinical use for the regeneration of articular cartilage since several years [160], the complete reconstruction of the auricle by tissue engineering is still a great challenge in head and neck surgery. The reasons are complex and especially related to the unsolved problems of scaffold design and of the differentiation induction of stem cells to produce elastic ear cartilage [161]. There are numerous other less-attended fields of research in head and neck surgery needing stem cell technology, for example, the mucosal reconstruction in the upper ADT. A first approach is the development of ciliated epithelium by cocultivation of stem cells with site-specific cells [78].

While these technologies are already in use for the reconstruction of the mucosa of the urinary tract [162] and the cornea [163] and for teeth regeneration [164], the development of mucosal reconstruction in head and neck surgery except for the salivary gland tissue [165–167] is still at the relative beginning.

13.6 Conclusion

The clinical applicability and quality of an implant material are exclusively shown in successful clinical use. The profile of demands on a material is therefore determined by the conditions *in vivo*. The chemical, enzymatical, bacterial, and mechanical conditions of the upper ADT make high demands on an implant material for the mucosal reconstruction in this area. In reconstructive surgery of the trachea, none of the different implant materials investigated by versatile methodical approaches was successfully introduced into the clinical use. For the reconstruction of pharyngeal defects based on the principles of regenerative medicine, there exist neither animal models nor a clinical application in humans until now. Based on the progress in polymer chemistry, multifunctional implant materials are available nowadays, which can selectively initiate biological processes in a physiological environment and/or change their physicochemical characteristics in reaction to external stimuli. The availability of such multifunctional implant materials and the progress in tissue engineering resulted in the establishment of novel therapeutical options in different medical fields. Applying stem cell technology, further progress is expected for the reconstruction of different tissues based on the principles of tissue engineering. To benefit from the potential of such technologies for the development and the establishment of novel therapeutical options in head and neck surgery, clinicians have to be involved in these interdisciplinary scientific projects of regenerative medicine.

References

1 Szilagyi, D.E., France, L.C., Smith, R.F., and Whitcomb, J.G. (1958) The clinical use of an elastic dacron prosthesis. *AMA Arch. Surg.*, **77**, 538–531.

2 Kohane, D.S. and Langer, R. (2008) Polymeric biomaterials in tissue engineering. *Pediatr. Res.*, **63**, 487–491.

3 Mason, C. and Dunnill, P. (2007) Lessons for the nascent regenerative medicine industry from the biotech sector. *Regen. Med.*, **2**, 753–756.
4 Mason, C. and Dunnill, P. (2008) A brief definition of regenerative medicine. *Regen. Med.*, **3**, 1–5.
5 Bajada, S., Mazakova, I., Richardson, J.B., and Ashammakhi, N. (2008) Updates on stem cells and their applications in regenerative medicine. *J. Tissue Eng. Regen. Med.*, **2**, 169–183.
6 Slater, B.J., Kwan, M.D., Gupta, D.M., Panetta, N.J., and Longaker, M.T. (2008) Mesenchymal cells for skeletal tissue engineering. *Expert Opin. Biol. Ther.*, **8**, 885–893.
7 Schulz, R.M. and Bader, A. (2007) Cartilage tissue engineering and bioreactor systems for the cultivation and stimulation of chondrocytes. *Eur. Biophys. J.*, **36**, 539–568.
8 Breymann, C., Schmidt, D., and Hoerstrup, S.P. (2006) Umbilical cord cells as a source of cardiovascular tissue engineering. *Stem. Cell Rev.*, **2**, 87–92.
9 Ott, H.C. and Taylor, D.A. (2006) From cardiac repair to cardiac regeneration–ready to translate? *Expert Opin. Biol. Ther.*, **6**, 867–878.
10 Shastri, V. and Lendlein, A. (2009) Materials in regenerative medicine. *Adv. Mater.*, **21**, 3231–3234.
11 Gall, K., Yakacki, C.M., Liu, Y., Shandas, R., Willett, N., and Anseth, K.S. (2005) Thermomechanics of the shape memory effect in polymers for biomedical applications. *J. Biomed. Mater. Res. A*, **73**, 339–348.
12 Langer, R. and Tirrell, D.A. (2004) Designing materials for biology and medicine. *Nature*, **428**, 487–492.
13 Lendlein, A. and Kelch, S. (2005) Degradable, multifunctional biomaterials with shape-memory. *Mater. Sci. Forum*, **492**, 219–223.
14 Lendlein, A., Kratz, K., and Kelch, S. (2005) Smart implant materials. *Med. Device Technol.*, **16**, 12–14.
15 Grablowitz, H. and Lendlein, A. (2007) Synthesis and characterization of alpha,omega-dihydroxy-telechelic oligo(*p*-dioxanone). *J. Mater. Chem.*, **17**, 4050–4056.
16 Lendlein, A. and Langer, R. (2002) Biodegradable, elastic shape-memory polymers for potential biomedical applications. *Science*, **296**, 1673–1676.
17 Lendlein, A. and Kelch, S. (2005) Shape-memory polymers as stimuli-sensitive implant materials. *Clin. Hemorheol. Microcirc.*, **32**, 105–116.
18 Burkoth, A.K., and Anseth, K.S. (1999) MALDI-TOF characterization of highly cross-linked, degradable polymer networks. *Macromolecules*, **32**, 1438–1444.
19 Thevenot, P., Hu, W., and Tang, L. (2008) Surface chemistry influences implant biocompatibility. *Curr. Top. Med. Chem.*, **8**, 270–280.
20 Rickert, D., Lendlein, A., Schmidt, A.M., Kelch, S., Roehlke, W., Fuhrman, R., and Franke, R.P. (2003) *In vitro* cytotoxicity testing of AB-polymer networks based on oligo(epsilon-caprolactone) segments after different sterilization techniques. *J. Biomed. Mater. Res. B*, **67**, 722–731.
21 Barron, D., Collins, M.N., Flannery, M.J., Leahy, J.J., and Birkinshaw, C. (2008) Crystal ageing in irradiated ultra high molecular weight polyethylene. *J. Mater. Sci. Mater. Med.*, **19**, 2293–2299.
22 Yakacki, C.M., Lyons, M.B., Rech, B., Gall, K., and Shandas, R. (2008) Cytotoxicity and thermomechanical behavior of biomedical shape-memory polymer networks post-sterilization. *Biomed. Mater.*, **3**, 15010.
23 An, Y.H., Alvi, F.I., Kang, Q., Laberge, M., Drews, M.J., Zhang, J., Matthews, M.A., and Arciola, C.R. (2005) Effects of sterilization on implant mechanical property and biocompatibility. *Int. J. Artif. Organs*, **28**, 1126–1137.
24 Cavalot, A.L., Gervasio, C.F., Nazionale, G., Albera, R., Bussi, M., Staffieri, A., Ferrero, V., and Cortesina, G. (2000) Pharyngocutaneous fistula as a complication of total laryngectomy: review of the literature and analysis of case records. *Otolaryngol. Head Neck Surg.*, **123**, 587–592.
25 Makitie, A.A., Irish, J., and Gullane, P.J. (2003) Pharyngocutaneous fistula. *Curr.*

Opin. Otolaryngol. Head Neck Surg., **11**, 78–84.
26 Fung, K., Teknos, T.N., Vandenberg, C.D., Lyden, T.H., Bradford, C.R., Hogikyan, N.D., Kim, J., Prince, M.E., Wolf, G.T., and Chepeha, D.B. (2007) Prevention of wound complications following salvage laryngectomy using free vascularized tissue. *Head Neck*, **29**, 425–430.
27 Longmire, W.P. (1948) Tracheal wounds and injuries, repair of large defects. *Ann. Otol. Rhinol. Laryngol.*, **572**, 875–873.
28 Ferguson, D.J., Wild, J.J., and Wangensteen, O.H. (1950) Experimental resection of the trachea. *Surgery*, **28**, 597–619.
29 Bucher, R.M., Burnett, E., and Rosenmond, G.P. (1951) Experimental reconstruction of the trachea and bronchial defects with stainless steel wire mesh. *J. Thorac. Surg.*, **21**, 572–583.
30 Craig, R.L., Holmes, G.W., and Shabart, E.J. (1953) Resection and replacement with prothesis. *J. ThoracaV. Surg.*, **25**, 384–396.
31 Holle, F. (1953) Healing conditions of tracheobronchial tree and its plastic reconstruction. Experimental study. *Arch. Klin. Chir.*, **277**, 1–35.
32 Ekestrom, S. (1956) Experimental reconstruction of intrathoracic trachea. *Acta. Chir. Scand.*, **110**, 367–372.
33 Rush, B. and Cliffton, E. (1956) Experimental reconstruction of the trachea with bladder mucosa. *Surgery*, **40**, 1105–1110.
34 Bell, J.W. (1960) Experimental repair of tracheal defects with gallbladder mucosa. *Chest*, **38**, 140–147.
35 Beal, A.C., Harrington, O.B., Greenberg, S.D., Morris, G.C., and Usher, F.C. (1962) Tracheal replacement with heavy Marlex mesh. *Arch. Surg.*, **87**, 390–396.
36 Graziano, J.L., Spinazzola, A., and Neville, W.E. (1967) Prosthetic replacement of the tracheal carina. *Ann. Thorac. Surg.*, **4**, 1–11.
37 Greenberg, S.D. and Wilms, R.K. (1962) Tracheal prosthesis: an experimental study in dogs. *Arch. Otolaryngol.*, **75**, 335–371.
38 Poticha, S.M. and Lewis, F.J. (1966) Experimental replacement of the trachea. *J. Thorac. Cardiovasc. Surg.*, **52**, 61–67.
39 Wenig, B.L., Reuter, V.C., Steinberg, B.M., and Strong, E.W. (1987) Tracheal reconstruction: *in vitro* und *in vivo* animal pilot study. *Laryngoscope*, **97**, 959–965.
40 Schauwecker, H.H., Gerlach, H., Planck, H., and Bücherl, E.S. (1989) Isoelastic polyurethane prosthesis for segmental trachea replacement in beagle dogs. *Artif. Organs*, **13**, 216–21860.
41 Langer, R. and Vacanti, J.P. (1993) Tissue engineering. *Science*, **260**, 920–92661.
42 Grillo, H.C. (2002) Tracheal replacement: a critical review. *Ann. Thorac. Surg.*, **73**, 1995–2004.
43 Vacanti, C.A., Paige, K.T., Kim, W.S., Sakata, J., Upton, J., and Vacanti, J.P. (1994) Experimental tracheal replacement using tissue engineered cartilage. *J. Pediatr. Surg.*, **29**, 201–205.
44 Sakata, J., Vacanti, C.A., Schloo, B., Healy, G.B., Langer, R., and Vacanti, J.P. (1994) Tracheal composites tissue engineered from chondrocytes, tracheal epithelial cells and synthetic degradable scaffolding. *Transplant. Proc.*, **26**, 3309–3310.
45 Kojima, K., Bonassar, L.J., Roy, A.K., Vacanti, C.A., and Cortiella, J. (2002) Autologous tissue-engineered trachea with sheep nasal chondrocytes. *J. Thorac. Cardiovasc. Surg.*, **123**, 1177–1184.
46 Fonkalsrud, E.W. and Sumida, S. (1971) Tracheal replacement with autologous esophagus for tracheal stricture. *Arch. Surg.*, **102**, 139–142.
47 Sabas, A.A., Uez, J.B., Rojas, O., Inones, A., and Aranguren, J.A. (1977) Replacement of the trachea with dura mater. Experimental work. *J. Thorac. Cardiovasc. Surg.*, **74**, 761–765.
48 Kon, M. and van den Hooff, A. (1983) Cartilage tube formation by perichondrium: a new concept for tracheal reconstruction. *Plast. Reconstr. Surg.*, **72**, 791–797.
49 Cohen, R.C., Filler, R.M., Konuma, K., Bahoric, A., Kent, G., and Smith, C.

(1985) The successful reconstruction of thoracic tracheal defects with free periostal grafts. *J. Pediatr. Surg.*, **20**, 852–858.
50 Har-El, G., Krespi, Y.P., and Goldsher, M. (1989) The combined use of muscle flaps and alloplasts for tracheal reconstruction. *Arch. Otolaryngol. Head Neck Surg.*, **115**, 1310–1131.
51 Lochbihler, H., Hoelzl, J., and Dietz, H.G. (1997) Tissue compatibility and biodegradation of new absorbable stents for tracheal stabilization: an experimental study. *J. Pediatr. Surg.*, **32**, 717–720.
52 Korpela, A., Aarnio, P., Sariola, H., Törmälä, P., and Harjula, A. (1998) Comparison of tissue reactions in the tracheal mucosa surrounding a bioabsorbable and silicone airway stents. *Ann. Thorac. Surg.*, **66**, 1772–1776.
53 Korpela, A., Aarnio, P., Sariola, H., Törmälä, P., and Harjula, A. (1999) Bioabsorbable self-inforced poly-L-lactide, metallic and silicone stents in the management of experimental tracheal stenosis. *Chest*, **115**, 490–495.
54 Robey, T.C., Välimaa, M.S., Murphy, H.S., Törmälä, P., Mooney, D.J., and Weatherly, R.A. (2000) Use of internal bioabsorbable PLGA "finger-type" stents in a rabbit tracheal reconstruction model. *Arch. Otolaryngol. Head Neck Surg.*, **126**, 985–991.
55 Cotton, R.T. and Seid, A.B. (1980) Management of the extubation problem in the premature child: anterior cricoid split as an alternative to tracheotomy. *Ann. Otol. Rhinol. Laryngol.*, **89**, 508–511.
56 Zalzal, G.H. and Deutch, E. (1991) External fixation using microplates after laryngotracheal expansion surgery. *Arch. Otolaryngol. Head Neck Surg.*, **117**, 155–159.
57 Weisberger, E.C. and Nguyen, C.T. (1996) Laryngotracheal reconstruction using a vitallium alloy miniplate. *Ann. Otol. Rhinol. Laryngol.*, **105**, 363–366.
58 Willner, A. and Modlin, S. (1995) Extraluminal laryngotracheal fixation with absorbable miniplates. *Arch. Otolaryngol. Head Neck Surg.*, **121**, 1356–1360.
59 Pietrzak, W.S., Sarver, D.R., and Verstynen, B.S. (1997) Bioabsorbable polymer science for the practicing surgeon. *J. Craniofac. Surg.*, **107**, 87–91.
60 Eppley, B.L. and Reilly, M. (1997) Degradation characteristics of PLLA-PGA bone fixation devices. *J. Craniofac. Surg.*, **8**, 116–120.
61 Long, C.M., Conlex, S.F., Kajdacsy-Balla, A., and Kerschner, J.E. (2001) Laryngotracheal reconstruction in canines. Fixation of autologous costochondral grafts using polylactic and polyglycolic acid miniplates. *Arch. Otolaryngol. Head Neck Surg.*, **127**, 570–575.
62 Kojima, K., Bonassar, L.J., Roy, A.K., Mizuno, H., Cortiella, J., and Vacanti, C.A. (2003) A composite tissue-engineered trachea using sheep nasal chondrocyte and epithel cells. *FASEB J.*, **17**, 823–828.
63 Kamil, S.H., Eavey, R.D., Vacanti, M.P., Vacanti, C.A., and Hartnick, C.J. (2004) Tissue-engineered cartilage as AF graft source for laryngotracheal reconstruction. *Arch. Otolaryngol. Head Neck Surg.*, **130**, 1048–1051.
64 George, M., Lang, F., Pasche, P., and Monnier, P. (2005) Surgical management of laryngotracheal stenosis in adults. *Eur. Arch. Otorhinolaryngol.*, **262**, 609–615.
65 Herrington, H.C., Weber, S.M., and Andersen, P.E. (2006) Modern management of laryngotracheal stenosis. *Laryngoscope*, **116**, 1553–1557.
66 Jaquet, Y., Pilloud, R., Lang, F.J.W., and Monnier, P. (2004) Prefabrication of composite grafts for long-segment tracheal reconstruction. *Arch. Otolaryngol. Head Neck Surg.*, **130**, 1185–1190.
67 Jaillard, S., Holder-Espinasse, M., Hubert, T., Copin, M.C., Duterque-Coquillaud, M., Wurtz, A., and Marquette, C.H. (2006) Tracheal replacement by allogenic aorta in the pig. *Chest*, **130**, 1397–1404.
68 Martinod, E., Seguin, A., and Holder-Espinasse, M. (2005) Tracheal regeneration following tracheal replacement with an allogenic aorta. *Ann. Thorac. Surg.*, **79**, 942–949.

69 Martinod, E., Seguin, A., Pfeuty, K., Fornes, P., Kambouchner, M., Azorin, J.F., and Carpentier, A.F. (2003) Long-term evaluation of the replacement of the trachea with an autologous aortic graft. *Ann. Thorac. Surg.*, **75**, 1572–1578.

70 Azorin, J.G., Bertin, F., and Martinod, E. (2006) Tracheal replacement with an aortic autograft. *Eur. J. Cardiothorac. Surg.*, **29**, 261–263.

71 Macchiarini, P., Jungebluth, P., Go, T., Asnaghi, M.A., Rees, L.E., Cogan, T.A., Dodson, A., Martorell, J., Bellini, S., Parnigotto, P.P., Dickinson, S.C., Hollander, A.P., Mantero, S., Conconi, M.T., and Birchall, M.A. (2008) Clinical transplantation of a tissue-engineered airway. *Lancet*, **372** (9655), 2023–2030. Erratum in: Lancet. 2009 Feb 7;373(9662):462.

72 Ernst, A. and Ashiku, S. (2006) Tracheal transplantation: are we any closer to the holy grail of airway management? *Chest*, **130**, 1299–1300.

73 Omori, K., Nakamura, T., Kanemaru, S., Asato, R., Yamashita, M., Tanaka, S., Magrufov, A., Ito, J., and Shimizu, Y. (2005) Regenerative medicine of the trachea: the first human case. *Ann. Otol. Rhinol. Laryngol.*, **114**, 429–433.

74 Yamashita, M., Kanemaru, S.I., Hirano, S., Magrufov, A., Tamaki, H., Tamura, Y., Kishimoto, M., Omori, K., Nakamura, T., and Ito, J. (2007) Tracheal regeneration after partial resection: a tissue engineering approach. *Laryngoscope*, **117**, 497–502.

75 Mall, M.A. (2008) Role of cilia, mucus, and airway surface liquid in mucociliary dysfunction: lessons from mouse models. *J. Aerosol Med. Pulm. Drug Deliv.*, **21**, 13–24.

76 Biesalski, H.K. and Nohr, D. (2003) Importance of vitamin-A for lung function and development. *Mol. Aspects Med.*, **24**, 431–440.

77 Evans, M.J., Van Winkle, L.S., Fanucchi, M.V., and Plopper, C.G. (2001) Cellular and molecular characteristics of basal cells in airway epithelium. *Exp. Lung Res.*, **27**, 401–415.

78 Hajj, R., Baranek, T., Le Naour, R., Lesimple, P., Puchelle, E., and Coraux, C. (2007) Basal cells of the human adult airway surface epithelium retain transit-amplifying cell properties. *Stem. Cells*, **25**, 139–148.

79 Ziegelaar, B.W., Aigner, J., Staudenmaier, R., Lempart, K., Mack, B., Happ, T., Sittinger, M., Endres, M., Naumann, A., Kastenbauer, E., and Rotter, N. (2002) The characterisation of human respiratory epithelial cells cultured on resorbable scaffolds: first steps towards a tissue engineered tracheal replacement. *Biomaterials*, **23**, 1425–11438.

80 Hicks, W. Jr, Hall, L. 3rd, Sigurdson, L., Stewart, C., Hard, R., Winston, J., and Lwebuga-Mukasa, J. (1997) Isolation and characterization of basal cells from human upper respiratory epithelium. *Exp. Cell Res.*, **237**, 357–363.

81 Mercer, R.R., Russell, M.L., Roggli, V.L., and Crapo, J.D. (1994) Cell number and distribution in human and rat airways. *Am. J. Respir. Cell. Mol. Biol.*, **10**, 613–624.

82 Yokoyama, T. (2004) Motor or sensor: a new aspect of primary cilia function. *Anat. Sci. Int.*, **79**, 47–54.

83 Davis, C.W. and Dickey, B.F. (2008) Regulated airway goblet cell mucin secretion. *Annu. Rev. Physiol.*, **70**, 487–512.

84 Hong, K.U., Reynolds, S.D., Watkins, S., Fuchs, E., and Stripp, B.R. (2004) *In vivo* differentiation potential of tracheal basal cells: evidence for multipotent and unipotent subpopulations. *Am. J. Physiol. Lung Cell. Mol. Physiol.*, **286**, 643–649.

85 Nomoto, Y., Suzuki, T., Yasuhiro, T., Kobayashi, K., Miyake, M., Hazama, A., Wada, I., Kanemaru, S., Nakamura, T., and Omori, K. (2006) Tissue engineering for regeneration of the tracheal epithelium. *Ann. Otol. Rhinol. Laryngol.*, **115**, 501–506.

86 Tada, Y., Suzuki, T., Takezawa, T., Nomoto, Y., Kobayashi, K., Nakamura, T., and Omori, K. (2008) Regeneration of tracheal epithelium utilizing a novel bipotential collagen scaffold. *Ann. Otol. Rhinol. Laryngol.*, **117**, 359–365.

87 El Ghalbzouri, A. and Ponec, M. (2004) Diffusible factors released by fibroblasts support epidermal morphogenesis and deposition of basement membrane components. *Wound Repair Regen.*, **12**, 359–367.

88 Xia, W., Phan, T.T., Lim, I.J., Longaker, M.T., and Yang, G.P. (2004) Complex epithelial–mesenchymal interactions modulate transforming growth factor-beta expression in keloid-derived cells. *Wound Repair Regen.*, **12**, 546–556.

89 Harrison, C.A., Dalley, A.J., and Mac Neil, S. (2005) A simple *in vitro* model for investigating epithelial/mesenchymal interactions: keratinocyte inhibition of fibroblast proliferation and fibronectin synthesis. *Wound Repair Regen.*, **13**, 543–550.

90 Imaizumi, F., Asahina, I., Moriyama, T., Ishii, M., and Omura, K. (2004) Cultured mucosal cell sheet with a double layer of keratinocytes and fibroblasts on a collagen membrane. *Tissue Eng.*, **10**, 657–664.

91 Cedidi, C.C., Wilkens, L., Berger, A., and Ingianni, G. (2007) Influence of human fibroblasts on development and quality of multilayered composite grafts in athymic nude mice. *Eur. J. Med. Res.*, **12**, 541–555.

92 Nishimura, T., Toda, S., Mitsumoto, T., Oono, S., and Sugihara, H. (1998) Effects of hepatocyte growth factor, transforming growth factor-beta1 and epidermal growth factor on bovine corneal epithelial cells under epithelial-keratinocyte interaction in reconstruction culture. *Exp. Eye Res.*, **66**, 105–116.

93 Wilson, S.E., Chen, L., Mohan, R.R., Liang, Q., and Liu, J. (1999) Expression of HGF, KGF, EGF and receptor messenger RNAs following corneal epithelial wounding. *Exp. Eye Res.*, **68**, 377–397.

94 Costea, D.E., Loro, L.L., Dimba, E.A., Vintermyr, O.K., and Johannessen, A.C. (2003) Crucial effects of fibroblasts and keratinocyte growth factor on morphogenesis of reconstituted human oral epithelium. *J. Invest. Dermatol.*, **121**, 1479–1486.

95 Daniels, J.T. and Khaw, P.T. (2000) Temporal stimulation of corneal fibroblast wound healing activity by differentiating epithelium *in vitro*. *Invest. Opthalmol. Vis. Sci.*, **41**, 3754–3762.

96 Kobayashi, K., Nomoto, Y., Suzuki, T., Tada, Y., Miyake, M., Hazama, A., Kanemaru, S., Nakamura, T., and Omori, K. (2006) Effect of fibroblasts on tracheal epithelial regeneration *in vitro*. *Tissue Eng.*, **12**, 2619–2628.

97 Kobayashi, K., Suzuki, T., Nomoto, Y., Tada, Y., Miyake, M., Hazama, A., Nakamura, T., and Omori, K. (2007) Potential of heterotopic fibroblasts as autologous transplanted cells for tracheal epithelial regeneration. *Tissue Eng.*, **13**, 2175–2184.

98 Nomoto, Y., Kobayashi, K., Tada, Y., Wada, I., Nakamura, T., and Omori, K. (2008) Effect of fibroblasts on epithelial regeneration on the surface of a bioengineered trachea. *Ann. Otol. Rhinol. Laryngol.*, **117**, 59–64.

99 Le Visage, C., Dunham, B., Flint, P., and Leong, K.W. (2004) Coculture of mesenchymal stem cells and respiratory epithelial cells to engineer a human composite respiratory mucosa. *Tissue Eng.*, **10**, 1426–1435.

100 Letang, E., Sánchez-Lloret, J., Gimferrer, J.M., Ramírez, J., and Vicens, A. (1990) Experimental reconstruction of the canine trachea with a free revascularized small bowel graft. *Ann. Thorac. Surg.*, **49**, 955–958.

101 Costantino, P.D., Nuss, D.W., Snyderman, C.H., Johnson, J.T., Friedman, C.D., Narayanan, K., and Houston, G. (1992) Experimental tracheal replacement using a revascularized jejunal autograft with an implantable Dacron mesh tube. *Ann. Otol. Rhinol. Laryngol.*, **101**, 807–814121.

102 Grillo, H.C. (2003) The history of tracheal surgery. *Chest Surg. Lin. N. Am.*, **13**, 175–189.

103 Fisher, R.J. and Peattie, R.A. (2007) Controlling tissue microenvironments: biomimetics, transport phenomena, and reacting systems. *Adv. Biochem. Eng. Biotechnol.*, **103**, 1–73.

104 Tan, Q., Steiner, R., Hoerstrup, S.P., and Weder, W. (2006) Tissue-engineered trachea: history, problems and the future. *Eur. J. Cardiothorac. Surg.*, **30**, 782–786.

105 Lendlein, A., Neuenschwander, P., and Suter, U.W. (1998) Tissue-compatible multiblock copolymers for medical applications, controllable in degradation rate and mechanical properties. *Macromol. Chem. Phys.*, **199**, 2785–2796.
106 Tan, Q., Steiner, R., Yang, L., Welti, M., Neuenschwander, P., Hillinger, S., and Weder, W. (2007) Accelerated angiogenesis by continuous medium flow with vascular endothelial growth factor inside tissue-engineered trachea. *Eur. J. Cardiothorac. Surg.*, **31**, 806–811.
107 Hallén, L. and Dahlqvist, A. (2002) Cross-linked hyaluronan for augmentation of the posterior pharyngeal wall: an experimental study in rats. *Scand. J. Plast. Reconstr. Surg. Hand Surg.*, **36**, 197–201.
108 Ophof, R., Maltha, J.C., Kuijpers-Jagtman, A.M., and Von den Hoff, J.W. (2008) Implantation of tissue-engineered mucosal substitutes in the dog palate. *Eur. J. Orthod.*, **30**, 1–9.
109 Moharamzadeh, K., Brook, I.M., Van Noort, R., Scutt, A.M., and Thornhill, M.H. (2007) Tissue-engineered oral mucosa: a review of the scientific literature. *J. Dent. Res.*, **86**, 115–124.
110 Falconnet, D., Csucs, G., Grandin, H.M., and Textor, M. (2006) Surface engineering approaches to micropattern surfaces for cell-based assays. *Biomaterials*, **27**, 3044–3063.
111 Rickert, D., Franke, R.P., Fernández, C.A., Kilroy, S., Yan, L., and Moses, M.A. (2007) Establishment and biochemical characterization of primary cells of the upper aerodigestive tract. *Clin. Hemorheol. Microcirc.*, **36**, 47–64.
112 Rickert, D., Lendlein, A., Kelch, S., Moses, M.A., and Franke, R.P. (2005) Expression of MMPs and TIMPs in primary epithelial cell cultures of the upper aerodigestive tract seeded on the surface of a novel polymeric biomaterial. *Clin. Hemorheol. Microcirc.*, **32**, 117–128.
113 Clark, R.A.F. (1995) *The Molecular and Cellular Biology of Wound Repair*, 2nd edn, Plenum Press, New York, pp. 3–50.
114 Li, J., Chen, J., and Kirsner, R. (2007) Pathophysiology of acute wound healing. *Clin. Dermatol.*, **25**, 9–18.
115 Ravanti, L. and Kähäri, V.M. (2000) Matrix metalloproteinases in wound repair (review). *Int. J. Mol. Med.*, **6**, 391–407.
116 Xue, M., Le, N.T., and Jackson, C.J. (2006) Targeting matrix metalloproteases to improve cutaneous wound healing. *Expert Opin. Ther. Targets*, **10**, 143–155.
117 Nagase, H., Visse, R., and Murphy, G. (2006) Structure and function of matrix metalloproteinases and TIMPs. *Cardiovasc. Res.*, **69**, 562–573.
118 Moses, M.A., Marikovsky, M., Harper, J.W., Vogt, P., Eriksson, E., Klagsbrun, M., and Langer, R. (1996) Temporal study of the activity of matrix metalloproteinases and their endogenous inhibitors during wound healing. *J. Cell. Biochem.*, **60**, 379–386.
119 Soo, C., Shaw, W.W., Zhang, X., Longaker, M.T., Howard, E.W., and Ting, K. (2000) Differential expression of matrix metalloproteinases and their tissue-derived inhibitors in cutaneous wound repair. *Plast. Reconstr. Surg.*, **105**, 638–647.
120 Bennett, J.H., Morgan, M.J., Whawell, S.A., Atkin, P., Roblin, P., Furness, J., and Speight, P.M. (2000) Metalloproteinase expression in normal and malignant oral keratinocytes: stimulation of MMP-2 and -9 by scatter factor. *Eur J Oral Sci*, **108**, 281–291.
121 Miyazaki, Y., Hara, A., Kato, K., Oyama, T., Yamada, Y., Mori, H., and Shibata, T. (2008) The effect of hypoxic microenvironment on matrix metalloproteinase expression in xenografts of human oral squamous cell carcinoma. *Int. J. Oncol.*, **32**, 145–151.
122 Hole, B.B., Schwarz, J.A., Gilbert, J.L., and Atkinson, B.L. (2005) A study of biologically active peptide sequences (P-15) on the surface of an ABM scaffold (PepGen P-15) using AFM and FTIR. *J. Biomed. Mater. Res. A*, **74**, 712–721.
123 Pfister, P.M., Wendlandt, M., Neuenschwander, P., and Suter, U.W. (2007) Surface-textured PEG-based hydrogels with adjustable elasticity: synthesis and characterization. *Biomaterials*, **28**, 567–575.
124 Tang, Z.G. and Hunt, J.A. (2006) The effect of PLGA doping of

polycaprolactone films on the control of osteoblast adhesion and proliferation *in vitro*. *Biomaterials*, **27**, 4409–4418.

125 Rohman, G., Pettit, J.J., Isaure, F., Cameron, N.R., and Southgate, J. (2007) Influence of the physical properties of two-dimensional polyester substrates on the growth of normal human urothelial and urinary smooth muscle cells *in vitro*. *Biomaterials*, **28**, 2264–2274.

126 Rompen, E., Domken, O., Degidi, M., Pontes, A.E., and Piattelli, A. (2006) The effect of material characteristics, of surface topography and of implant components and connections on soft tissue integration: a literature review. *Clin. Oral. Implants Res.*, **17**, 55–67.

127 Tatard, V.M., Venier-Julienne, M.C., Saulnier, P., Prechter, E., Benoit, J.P., Menei, P., and Montero-Menei, C.N. (2005) Pharmacologically active microcarriers: a tool for cell therapy. *Biomaterials*, **26**, 3727–3737.

128 Tatard, V.M., Sindji, L., Branton, J.G., Aubert-Pouëssel, A., Colleau, J., Benoit, J.P., and Montero-Menei, C.N. (2007) Pharmacologically active microcarriers releasing glial cell line–derived neurotrophic factor: survival and differentiation of embryonic dopaminergic neurons after grafting in hemiparkinsonian rats. *Biomaterials*, **28**, 1978–1988.

129 Davies, J.E. (2007) Bone bonding at natural and biomaterial surfaces. *Biomaterials*, **28**, 5058–5067.

130 Brown, R.A. and Phillips, J.B. (2007) Cell responses to biomimetic protein scaffolds used in tissue repair and engineering. Review.*Int. Rev. Cytol.*, **262**, 75–150.

131 Rickert, D., Franke, R.P., Lendlein, A., Kelch, S., and Moses, M.A. (2007) Influence of the surface structure of a multiblock copolymer on the cellular behavior of primary cell cultures of the upper aerodigestive tract *in vitro*. *J. Biomed. Mater. Res. A*, **83**, 558–569.

132 Mudera, V.C., Pleass, R., Eastwood, M., Tarnuzzer, R., Schultz, G., Khaw, P., McGrouther, D.A., and Brown, R.A. (2000) Molecular responses of human dermal fibroblasts to dual cues: contact guidance and mechanical load. *Cell Motil. Cytoskeleton*, **45**, 1–9.

133 Rickert, D., Scheithauer, M.O., Coskun, S., Lendlein, A., Kelch, S., and Franke, R.P. (2006) First results of the investigation of the stability and tissue integration of a degradable, elastomeric copolymer in an animal model. *Biomed. Tech.*, **51**, 116–124.

134 Rickert, D., Lendlein, A., Coskum, S., and Scheithauer, M.O. (2007) Polymeric biomaterials in head and neck surgery: first results of biocompatibility testing of a degradable polymer in an animal model. *Laryngorhinootologie*, **86**, 507–514.

135 Busuttil, S.J., Drumm, C., and Plow, E.F. (2005) In vivo comparison of the inflammatory response induced by different vascular biomaterials. *Vascular*, **13**, 230–235.

136 Rickert, D., Scheithauer, M.O., Coskun, S., Kelch, S., Lendlein, A., and Franke, R.P. (2007) The influence of a multifunctional, polymeric biomaterial on the concentration of acute phase proteins in an animal model. *Clin. Hemorheol. Microcirc.*, **36**, 301–311.

137 Park, T.G. (1995) Degradation of poly(lactic-*co*-glycolic acid) microspheres: effect of copolymer composition. *Biomaterials*, **16**, 1123–1130.

138 Lendlein, A., Neuenschwander, P., and Suter, U.W. (2000) Hydroxy-telechelic copolyesters with well defined sequence structure through ring-opening polymerization. *Macromol. Chem. Phys.*, **201**, 1067–1076.

139 Kulkarni, A., Reiche, J., Hartmann, J., Kratz, K., and Lendlein, A. (2008) Selective enzymatic degradation of poly (ε-caprolactone) containing multiblock copolymers. *Eur. J. Pharm. Biopharm.*, **68**, 46–56.

140 Wischke, C., Neffe, A.T., Steuer, S., Engelhardt, E., and Lendlein, A. (2010) AB-polymer networks with cooligoester-and poly(*n*-butyl acrylate)-segments as multifunctional matrix for controlled drug release. *Macromol. Biosci.*, **10**, 1063–1072.

141 Folkman, J., Haudenschild, C., and Zetter, B.R. (1979) Long-term culture of capillary endothelial cells. *Proc. Natl. Acad. Sci. USA*, **76**, 5217–5221.

142 Rickert, D., Lendlein, A., Kelch, S., and Franke, R.P. (2003) The importance of angiogenesis in the interaction between polymeric biomaterials and surrounding tissue. *Clin. Hemorheol. Microcirc.*, **28**, 175–181.

143 Rickert, D., Lendlein, A., Peters, I., Moses, M.A., and Franke, R.P. (2006) Biocompatibility testing of novel multifunctional polymeric biomaterials for tissue engineering applications in head and neck surgery: an overview. *Eur. Arch. Otorhinolaryngol.*, **263**, 215–222.

144 Kunz-Schughart, L.A., Schroeder, J.A., Wondrak, M., van Rey, F., Lehle, K., Hofstaedter, F., and Wheatley, D.N. (2006) Potential of fibroblasts to regulate the formation of three-dimensional vessel-like structures from endothelial cells *in vitro*. *Am. J. Physiol. Cell Physiol.*, **290**, 1385–1398.

145 Au, P., Tam, J., Fukumura, D., and Jain, R.K. (2007) Small blood vessel engineering. *Methods Mol. Med.*, **140**, 183–195.

146 Lokmic, Z. and Mitchell, G.M. (2008) Engineering the microcirculation. *Tissue Eng. Part B Rev.*, **14**, 87–103.

147 Shapiro, R.S. (2008) Future issues in transplantation ethics: ethical and legal controversies in xenotransplantation, stem cell, and cloning research. *Transplant. Rev.*, **22**, 210–215.

148 Kastenberg, Z.J. and Odorico, J.S. (2008) Alternative sources of pluripotency: science, ethics, and stem cells. *Transplant. Rev.*, **22**, 215–222.

149 Unger, C., Skottman, H., Blomberg, P., Dilber, M.S., and Hovatta, O. (2008) Good manufacturing practice and clinical-grade human embryonic stem cell lines. *Hum. Mol. Genet.*, **17**, 48–53.

150 Burdick, J.A. and Vunjak-Novakovic, G. (2008) Review: engineered microenvironments for controlled stem cell differentiation. *Tissue Eng. Part A*, **15**, 205–219.

151 Little, L., Healy, K.E., and Schaffer, D. (2008) Engineering biomaterials for synthetic neural stem cell microenvironments. *Chem. Rev.*, **108**, 1787–1796.

152 Vacanti, C.A. (2006) History of tissue engineering and a glimpse into its future. *Tissue Eng.*, **12**, 1137–1142.

153 Griffith, C.K., Miller, C., Sainson, R.C., Calvert, J.W., Jeon, N.L., Hughes, C.C., and George, S.C. (2005) Diffusion limits of an *in vitro* thick prevascularized tissue. *Tissue Eng.*, **11**, 257–266.

154 Schwab, A.P. and Satin, D.J. (2008) The realistic costs and benefits of translational research. *Am. J. Bioeth.*, **8**, 60–62.

155 Becker, A.J., McCulloch, E.A., and Till, J.E. (1963) Cytological demonstration of the clonal nature of spleen colonies derived from transplanted mouse marrow cells. *Nature*, **197**, 452–454.

156 Vacanti, C.A., Bonassar, L.J., Vacanti, M.P., and Shufflebarger, J. (2001) Replacement of an avulsed phalanx with tissue-engineered bone. *N. Engl. J. Med.*, **344**, 1511–1514.

157 Quarto, R., Mastrogiacomo, M., Cancedda, R., Kutepov, S.M., Mukhachev, V., Lavroukov, A., Kon, E., and Marcacci, M. (2001) Repair of large bone defects with the use of autologous bone marrow stromal cells. *N. Engl. J. Med.*, **344**, 385–386.

158 Hibi, H., Yamada, Y., Kagami, H., and Ueda, M. (2006) Distraction osteogenesis assisted by tissue engineering in an irradiated mandible: a case report. *Int. J. Oral. Maxillofac. Implants*, **21**, 141–147.

159 Hibi, H., Yamada, Y., Ueda, M., and Endo, Y. (2006) Alveolar cleft osteoplasty using tissue-engineered osteogenic material. *Int. J. Oral. Maxillofac. Surg.*, **35**, 551–555.

160 Granero-Molto, F., Weis, J.A., Longobardi, L., and Spagnoli, A. (2008) Role of mesenchymal stem cells in regenerative medicine: application to bone and cartilage repair. *Expert Opin. Biol. Ther.*, **8**, 255–268.

161 Ciorba, A. and Martini, A. (2006) Tissue engineering and cartilage regeneration for auricular reconstruction. *Int. J. Pediatr. Otorhinolaryngol.*, **70**, 1507–1515.

162 Yamzon, J.L., Kokorowski, P., and Koh, C.J. (2008) Stem cells and tissue engineering applications of the

genitourinary tract. *Pediatr. Res.*, **63**, 472–477.

163 Yang, X., Moldovan, N.I., Zhao, Q., Mi, S., Zhou, Z., Chen, D., Gao, Z., Tong, D., and Dou, Z. (2008) Reconstruction of damaged cornea by autologous transplantation of epidermal adult stem cells. *Mol. Vis.*, **14**, 1064–1070.

164 Bluteau, G., Luder, H.U., De Bari, C., and Mitsiadis, T.A. (2008) Stem cells for tooth engineering. *Eur. Cell. Mater.*, **16**, 1–9.

165 Szlávik, V., Szabó, B., Vicsek, T., Barabás, J., Bogdán, S., Gresz, V., Varga, G., O'Connell, B., and Vág, J. (2008) Differentiation of primary human submandibular gland cells cultured on basement membrane extract. *Tissue Eng. Part A*, **14**, 1915–1926.

166 Sato, A., Okumura, K., Matsumoto, S., Hattori, K., Hattori, S., Shinohara, M., and Endo, F. (2007) Isolation, tissue localization, and cellular characterization of progenitors derived from adult human salivary glands. *Cloning Stem Cells*, **9**, 191–205.

167 Lombaert, I.M., Brunsting, J.F., Wierenga, P.K., Faber, H., Stokman, M.A., Kok, T., Visser, W.H., Kampinga, H.H., de Haan, G., and Coppes, R.P. (2008) Rescue of salivary gland function after stem cell transplantation in irradiated glands. *PloS ONE*, **3**, 1–13.

14
Biodegradable Polymers as Scaffolds for Tissue Engineering

Yoshito Ikada

Abbreviations

bFGF	basic fibroblast growth factor
BMCs	bone marrow cells
BMPs	bone morphogenetic proteins
ECM	extracellular matrix
ES	embryonic stem
FDA	Federal Drug Administration
GTR	guided tissue regeneration
iPS	induced pluripotent stem
IRB	Institutional Review Board
IVC	inferior vena cava
MSCs	mesenchymal stem cells
OA	osteoarthritis
PCBM	particulate cancellous bone and marrow
PCL	poly(ε-caprolactone)
PGA	polyglycolide
PLA	polylactide
PLLA	poly-L-lactide
TCPC	total cavopulmonary connection
TEVAs	tissue-engineered vascular autografts
VEGF	vascular endothelial growth factor
3D	three-dimensional

14.1
Introduction

Tissue engineering is a new paradigm that offers new medical means for clinicians and patients who need new tissues for their defective or lost ones [1]. It has long been recognized that the limb of salamanders and newts are readily regenerated when lost. The ability to regenerate damaged human organs would constitute a

Handbook of Biodegradable Polymers: Synthesis, Characterization and Applications, First Edition. Edited by Andreas Lendlein, Adam Sisson.
© 2011 Wiley-VCH Verlag GmbH & Co. KGaA. Published 2011 by Wiley-VCH Verlag GmbH & Co. KGaA.

medical revolution. However, the regeneration of human tissues is currently almost impossible except for a few tissues including blood cells, epithelia, and bones. The reason may be disclosed when the developmental biology will have much more advanced in the near future.

Notwithstanding, it seems instructive for scientists and engineers of tissue engineering to first learn mechanisms of the natural development progress of human embryo and adults, even if the biological environments for the embryonic organogenesis are substantially different from those for the tissue regeneration in adults.

14.2 Short Overview of Regenerative Biology

Throughout the history of experimental biology, certain organisms have repeatedly attracted the attention of researchers. For instance, we cannot look at the phenomenon of limb regeneration in newts or starfish without wondering why we cannot grow back our own arms and legs [2]. The reactivation of development in postembryonic life to restore missing tissues has been a source of fascination to humans. We are beginning to find answers to the great problem of regeneration, so that we might be able to alter the human body so as to permit our own limbs, nerves, and organs to regenerate. This would mean that severed limbs could be restored, that diseased organs could be removed and regrown, and that nerve cells altered by age, disease, or trauma could once again function normally. To bring these treatments to humanity, we first have to understand how regeneration occurs in those species that have this ability. Gilbert points out three major ways by which regeneration can occur [2]. The first mechanism involves the dedifferentiation of adult structures to form an undifferentiated mass of cells that then become respecified. This type of regeneration is called epimorphosis, characteristic of regenerating limbs. The second mechanism is called morphallaxis. Here, regeneration occurs through the repatterning of existing tissues with little new growth. Such regeneration is seen in hydra. A third intermediate type of regeneration can be thought of as compensatory regeneration. Here, the cells divide, but maintain their differentiated functions. They produce cells similar to themselves and do not form a mass of undifferentiated tissue.

14.2.1 Limb Regeneration of Urodeles

When an adult salamander limb is amputated, the remaining cells are able to reconstruct a complete limb, with all its differentiated cells arranged in the proper order. It is appropriate to begin with the example of the urodele amphibian limb, simply because the adult urodele responds to amputation by regenerating a perfect replica of the original limb.

As a result of decades of research, we have considerable knowledge about the cell- and tissue-level biology of limb regeneration [3]. Of particular significance are those findings that indicate that once the regeneration cascade progresses to blastemal stages, the mechanisms controlling growth and pattern formation are the same as those in developing limbs [4]. Thus, the challenge to understanding what might be needed to induce regeneration in humans becomes focused on the developmental signals controlling the transformation of the differentiated stump into a blastema. In addition, a number of key requirements necessary for a successful regeneration response have been disclosed. These include the formation of a wound epidermis that creates a permissive environment necessary for a regeneration response, the dedifferentiation of cells at the injury site, the requirement for adequate innervation, and the need to reinitiate patterning programs involved in limb outgrowth. The absence of any one of these requirements will result in regenerative failure. If we assume that the successful induction of limb regeneration in higher vertebrates will proceed in a manner similar to urodeles, then we can anticipate that all of these requirements must be satisfied at the amputation site. These requirements for a regeneration response may be potential barriers to regeneration in higher vertebrate limbs. Urodele limb regeneration is characterized by the formation of a blastema composed of undifferentiated mesenchymal cells from which many of the different tissues of the regenerated limb develop. Similarly, regeneration of developing tissues proceeds via a blastema-like stage with the re-expression of developmentally relevant genes.

Despite the value of the urodele limb as a model for a regeneration, research progress in recent years has been relatively slow due to difficulties of bringing the power of functional analysis to bear on urodeles. It seems likely that the critical breakthroughs in regeneration research will come from the identification of the molecules that control the early events, preceding the convergence of the regeneration and development pathways. Given techniques for efficient high-throughput screening and analysis of differentially expressed genes, combined with techniques for identifying interacting molecules, urodeles will provide the opportunity to identify all the candidate genes for the control of limb regeneration. With the ability to test the function of these genes, it will be possible to identify the molecules that regulate the key steps in the process, allowing for the realization of the longed-for goal of human regeneration.

14.2.2
Wound Repair and Morphogenesis in the Embryo

Adult wound healing is notoriously imperfect and generally results in fibrosis and scar contracture with poor reconstitution of epidermal and dermal structures at the site of the healed wound, whereas embryonic wounds heal extremely well, rapidly, efficiently, and perfectly.

Adult wound closure involves active movements of both connective tissue and epidermis. The exposed connective tissue of the wound–the granulation tissue – contracts to tug the wound edges together and, as this is happening, the epidermis migrates to cover over the exposed connective tissue. The embryo also utilizes a combination of connective tissue contraction and re-epithelialization movements to close a wound, but the cellular mechanisms for both movements are quite different in embryo and adult. Another major difference between adult and embryonic tissue repair concerns the extent of inflammation during healing – at adult wound sites, there is always an extensive inflammatory response, but in the embryo, inflammation is minimal, if not nonexistent. Wound healing is an initial and critical event in any regeneration response. If wound healing occurs perfectly, that is, without scarring, then the skin (epidermal and dermal tissues) can be considered to have regenerated. Indeed, embryonic and fetal wounds heal rapidly without scarring, just as embryonic limb buds and fetal digits are able to respond to amputation by mounting a regenerative response. During a limb regenerative response, wound closure results in the formation of a specialized structure, the wound epidermis, which creates a subepidermal environment essential for regeneration. It seems likely that a similar type of subepidermal environment will be necessary for a regeneration response during healing of the skin. It seems unlikely that successful limb regeneration can occur under healing conditions that results in the deposition of scar tissue. Thus, scar-free wound healing is likely to be a necessary precondition for a successful regeneration response.

14.2.3 Regeneration in Human Fingertips

The transition from urodele limb studies to experimental attempts to induce a regenerative response in higher vertebrates has met with few successes, none resulting in a normal limb. This has led to the general conclusion that a "magic bullet" for regeneration is unlikely, but that the induction of a regeneration response will involve a coordinated effort to overcome multiple barriers to regeneration. While the regenerating urodele limb is the system of choice, alternative approaches are to study the limited regenerative responses that are known to occur in the limbs of higher vertebrates: digit tip regeneration in adult mammals. In fact, human digit tips can regenerate. Digit tip regeneration in adult primates (including humans) and rodents occurs without the formation of a blastema; instead, fibroblastic cells appear to be involved in the regeneration response. Fingertip amputations are among the most common traumas seen in hospital emergency rooms [5]. There are numerous reports that a conservative treatment consisting simply of covering the amputation wound with sterile dressings and allowing it to heal by secondary intention (i.e., without assisted wound closure) will result in the regeneration of the missing distal portion of the finger [6]. The phenomenon of fingertip regeneration in humans was initially described for children, but later shown to extend to adults. For both children and adults, regeneration of the fingertip involves the integrated regeneration of many tissues including nail matrix, nail bed, finger pulp, sensory organs, dermis, and epidermis, all of which reform to a normal or nearnormal cosmetic and physiological state through healing by secondary intention. Elongation of the distal phalangeal bone during regeneration has only been documented for children [7], but most studies lack radiographic data that allow for the assessment of bone regrowth. Animal models for digit tip regeneration in adults demonstrate distal bone growth associated with a regeneration response. There are several documented instances of regeneration of the distal phalangeal element of the toe following traumatic injury or voluntary resection to relieve hummer toe [8]. Thus, it would appear that the regenerative capabilities in human limbs include the tips of both fingers and toes.

14.2.4
The Development of Bones: Osteogenesis

The skeleton is generated through three lineages: the somites generate the axial skeleton, the lateral plate mesoderm generates the limb skeleton, and the cranial neural crest gives rise to the branchial arch and craniofacial bones and cartilage. There are two major modes of bone formation or osteogenesis, and both involve the transformation of a preexisting mesenchymal tissue into bone tissue. The direct conversion of mesenchymal tissue into bone is called intramembranous ossification. In other cases, the mesenchymal cells differentiate into cartilage, and this cartilage is later replaced by bone. The process by which a cartilage intermediate is formed and replaced by bone cells is called endochondral ossification. The cranial neural crest cells form bones through intramembranous ossification. In the skull, neural crest-derived mesenchymal cells proliferate and condense into compact nodules. As shown in Figure 14.1, some of these cells develop into capillaries; others change their shape to become osteoblasts, committed bone precursor cells. The osteoblasts secrete a collagen–proteoglycan osteoid matrix that is able to bind calcium. Upon embedding in the calcified matrix, osteoblasts become osteocytes. As calcification proceeds, bony spicules radiate out from the region

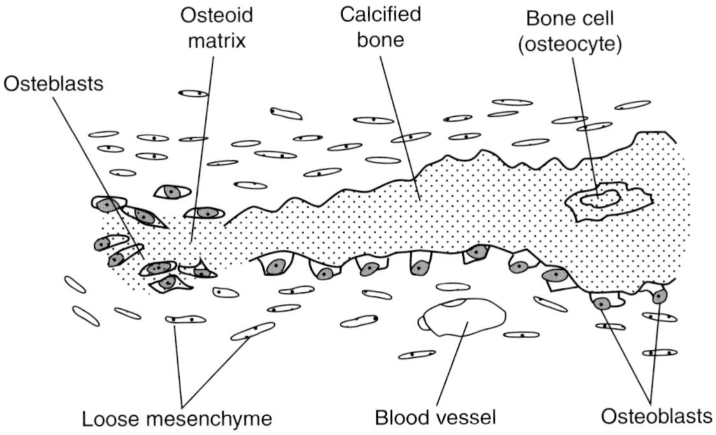

Figure 14.1 Schematic diagram of intramembranous ossification.

where ossification began. Furthermore, the entire region of calcified spicules becomes surrounded by compact mesenchymal cells that form the periosteum. The cells on the inner surface of the periosteum also become osteoblasts and deposit matrix parallel to the existing spicules. The mechanism of intramembranous ossification involves bone morphogenetic proteins (BMPs) and the activation of a transcription factor called Runx2.

Endochondral ossification involves the formation of cartilage tissue from aggregated mesenchymal cells and the subsequent replacement of cartilage tissue by bone [9]. This is the type of bone formation characteristic of the vertebrae, ribs, and limbs. The process of endochondral ossification can be divided into five stages, as shown in Figure 14.2. First, the mesenchymal cells commit to becoming cartilage cells. This commitment is caused by paracrine factors that induce the nearby mesodermal cells to express two transcription factors, which will then activate

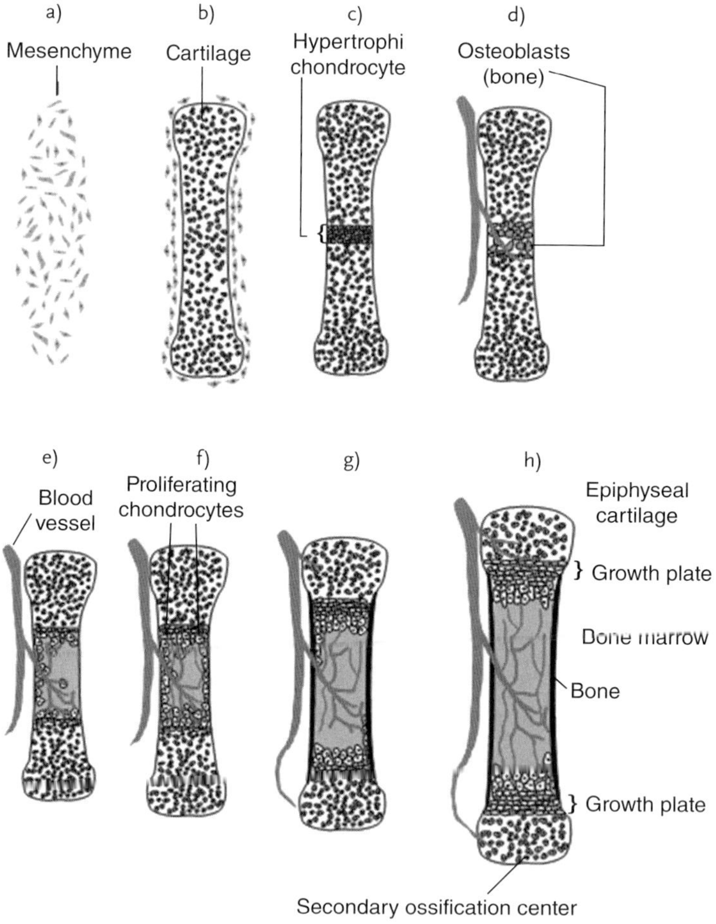

Figure 14.2 Schematic diagram of endochondral ossification.

cartilage-specific genes. During the second phase of endochondral ossification, the committed mesenchymal cells condense into compact nodules and differentiate into chondrocytes.

During the third phase of endochondral ossification, the chondrocytes proliferate rapidly to form the cartilage model for the bone. As they divide, the chondrocytes secrete a cartilage-specific extracellular matrix (ECM). In the fourth phase, the chondrocytes stop dividing and increase their volume dramatically, becoming hypertrophic chondrocytes. These large chondrocytes alter the matrix they produce (by adding collagen X and more fibronectin) to enable it to become mineralized by calcium carbonate. They also secrete the angiogenesis factor, vascular endothelial growth factor (VEGF), which can transform mesodermal mesenchymal cells into blood vessels. A number of events lead to the hypertrophy and mineralization of the chondrocytes, including an initial switch from aerobic to anaerobic respiration, which alters their cell metabolism and mitochondrial energy potential. Hypertrophic chondrocytes secrete numerous small membrane-bound vesicles into the ECM. These vesicles contain enzymes that are active in the generation of calcium and phosphate ions and initiate the mineralization process within the cartilaginous matrix. The hypertrophic chondrocytes, their metabolism and mitochondrial membranes altered, then die by apoptosis.

In the fifth phase, the blood vessels induced by VEGF invade the cartilage model. As the hypertrophic chondrocytes die, the cells that surround the cartilage model differentiate into osteoblasts. These cells express the Runx2 transcription factor, which is necessary for the development of both intramembranous and endochondral bone. The replacement of chondrocytes by bone cells is dependent on the mineralization of the ECM. This remodeling releases VEGF, and more blood vessels are made around the dying cartilage. These blood vessels bring in both osteoblasts and chondroclasts (which eat the debris of the apoptotic chondrocytes). Eventually, all the cartilage is replaced by bone. Thus, the cartilage tissue serves as a model for the bone that follows.

14.2.5 Regeneration in Liver: Compensatory Regeneration

Today, the standard assay for liver regeneration is to remove specific lobes of the liver (i.e., a partial hepatectomy), leaving the others intact. The removed lobe does not grow back, but the remaining lobes enlarge to compensate for the loss of the missing liver tissue. The amount of liver regenerated is equivalent to the amount of liver removed. The liver regenerates by the proliferation of the existing tissues. The regenerating liver cells do not fully dedifferentiate when they reenter the cell cycle. No regeneration blastema is formed. Rather, the five types of liver cells – hepatocytes, duct cells, fat-storing (Ito) cells, endothelial cells, and Kupffer macrophages – each begin dividing to produce more of themselves. Each type of cell retains its cellular identity, and the liver retains its ability to synthesize the liver-specific enzymes necessary for glucose regulation, toxin degradation, bile synthesis, albumin production, and other hepatic functions. As in the regenerating salamander limb, there is a return to some embryonic conditions in the regenerating liver. Fetal transcription factors and products are made, as are the cyclins that control cell division. But the return to the embryonic state is not as complete as in the amphibian limb.

14.3 Minimum Requirements for Tissue Engineering

14.3.1 Cells and Growth Factors

The leading player in tissue engineering is cells because it is only this living microsystem that is able to regenerate living tissues. This is different from the conventional artificial organs and tissues, where biomaterials play a pivotal role. Very recently, pluripotent stem cells such as embryonic stem (ES) and induced pluripotent stem (iPS) cells have attracted extraordinarily much attention, but these cells cannot be applied directly to tissue engineering. The cells applicable to tissue engineering should be differentiated to regenerate target tissues or will be readily differentiated depending on the environment surrounding the predifferentiated cells. The cells closely associated with tissue engineering include fibroblast, osteoblast, chondrocyte, epithelial cell, and smooth muscle cell. In addition, it should be mentioned that numerous organs contain multipotent stem cells, even in the adult. Multipotent stem cells can give rise to a limited set of adult tissue types. However, they are not as easy to use as pluripotent ES cells. First, they appear to have a relatively low rate of cell division and do not proliferate readily. Second, they are difficult to isolate, and are often fewer than one of every thousand cells in an organ. The mesenchymal stem cells (MSCs) from the bone marrow are still relatively undifferentiated, but commited to a certain lineage, and have the capability to readily differentiate based on the circumstance.

To produce a clinically applicable size of tissues by tissue engineering, we need a large number of cells, but the amount of cells that can be harvested from patients is limited. Therefore, attempts have been made to multiply the harvested cells retaining the ability to generate tissues. One of the unsolved problems in tissue engineering is to multiply the MSC keeping the undifferentiated state. If this is achieved in culture, the benefit is potentially enormous. An addition of high concentrations of basic fibroblast growth factor (bFGF) in culture has been claimed to facilitate the MSC multiplication [10], but the positive effect has not always been reported.

Cytokines greatly affect tissue engineering in terms of cell multiplication, cell differentiation, and neovascularization. A well-known example is BMPs that are able to induce ectopic bone formation without any cell addition. Similarly, bFGF encourages capillary formation without exogenous cell addition. Such vascularization is critical for nutrient supply to cells in the regeneration site. An important strategy associated with growth factors in tissue engineering is not to use a bolus dose of growth factors but to maintain the growth factor concentration at an optimal level for a certain period. For the sustained delivery of biologically active agents, carriers or delivery vehicles are generally employed, but there are few reports that have explored carriers effective in the sustained release of growth factors. Much more efforts are required to enhance the beneficial effects of growth factors on tissue engineering.

14.3.2
Favorable Environments for Tissue Regeneration

There are two modes of tissue engineering for tissue construction. One is *in vitro* (*ex vivo*) tissue engineering and the other *in vivo* (*in situ*) tissue engineering. In the beginning of tissue engineering research, many people attempted to construct living tissues outside the human body, that is, *in vitro* or *ex vivo*. Although a number of joint ventures were established to this end, most of them failed in the *in vitro* production of clinically applicable tissues on large scales. It may imply that it is difficult for us to create the artificial environment that is effective for cells to generate tissues outside the human body. Generally, a substrate to which cells attach is required for cells to survive, proliferate, and differentiate. It will be not difficult to prepare such substrates from biomaterials, but continuous supply of oxygen and nutrients to cells producing tissues is a hard task, because the supply is often disturbed by the tissues produced. The ideal route for oxygen and nutrient supply to cells is through capillaries, but sufficient capillary formation is impossible in the *in vitro* tissue engineering. This may be the reason for very limited applications of *in vitro* tissue engineering mostly to epidermal production. Tissue engineering below means the *in vivo* tissue engineering unless specified.

An essential requirement for tissue engineering is to provide cells with a favorable environment for tissue regeneration. In the case of *in vitro* tissue engineering, we, researchers, should create the environment that is the most effective for the cells in terms of tissue regeneration including cell proliferation, migration, and differentiation. In contrast to the *in vitro* tissue engineering, we do not need to create the optimal environment by ourself in the *in vivo* tissue engineering. The patient body will produce the most effective environment for the tissue regeneration by itself, if we could effectively support it.

What tissue engineers can help cells is to offer a good substrate for cell attachment, an effective barrier for preventing undesirable cells from invasion into the regeneration site, and a facility for promoting capillary formation, in other words, neovascularization. When a permissive environment optimal for cells to regenerate tissues is formed expectedly by these supplies, tissue regeneration will smoothly proceed by itself. However, a very large number of current studies on tissue engineering but a very small number of clinical trials so far imply that such an environment optimal for the *in vivo* tissue engineering can be produced only with great difficulty.

14.3.3
Need for Scaffolds

It should be noted that the natural ECM, a major component of connective tissues, is not a template or scaffold in organogenesis of embryo, but simply a product accompanying the embryogenesis. This suggests that it is not reasonable to regard a scaffold as an artificial ECM, although the current major topic in scaffold research is to mimic the natural ECM.

In discussing the rational design of scaffolds, it is necessary and pertinent to divide scaffolds into two groups (Scaffold type I and type II) on the basis of the cells to be seeded in scaffolds. Scaffold type I is used for differentiated cells including fibroblast, osteoblast, and chondrocyte, as represented in Figure 14.3. Figure 14.4 demonstrates Scaffold type II that is used for not yet fully differentiated pro-

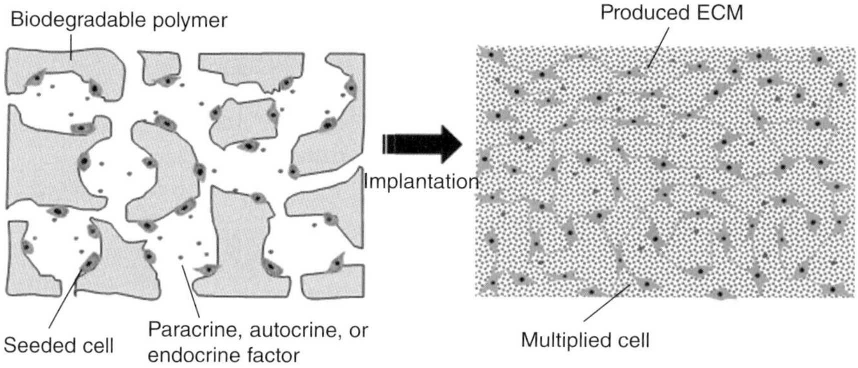

Figure 14.3 Scaffold type I for differentiated cells.

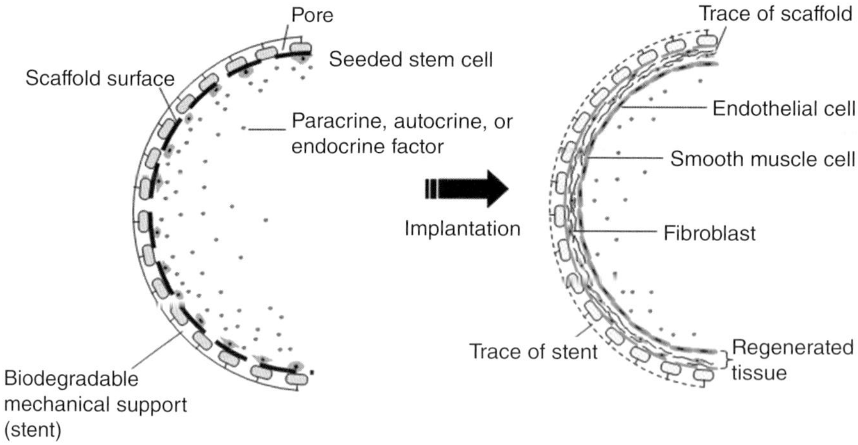

Figure 14.4 Scaffold type II for progenitor cells.

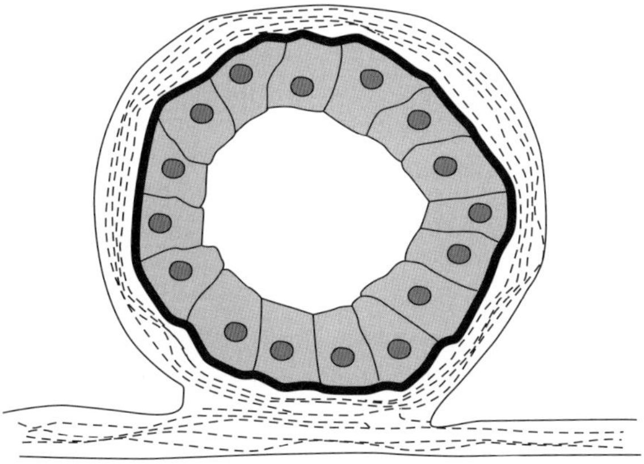

Figure 14.5 Cells assembling during organogenesis of embryo.

genitor or stem cells such as MSCs. The cells seeded in Scaffold type I produce mostly the ECM consisting of fibrous proteins, proteoglycans, and glycoproteins, constructing a connective tissue, combined with differentiated but still active cells. In this case, the 3D structure of the regenerated tissue may be regulated by the 3D structure of the scaffold.

In contrast to Scaffold type I, Scaffold type II primarily provides a perforated surface for progenitors to proliferate and differentiate into the target cells. The wall tissue of large-calibered blood vessels is exemplified in Figure 14.4. The perforated structure acts to allow oxygen and nutrient supply from the surrounding. Generally, the scaffold surface may be fabricated with a thin porous material to accommodate cells as many as possible. In the beginning of embryonic organogenesis, cells assemble into a characteristic form, as demonstrated in Figure 14.5. This cell assembling must be definitely affected by biological signaling in addition to cell–cell interactions. The biological signaling will be conveyed by endocrine factors migrated into the regeneration site from the adjacent environment as well as the autocrine and paracrine factors secreted by the seeded cells. These cytokines transform recruited precursor cells from the host into the cells producing target tissues. Similarly, the factors will dictate the fate of the cells attaching to the surface of Scaffold type II, finally resulting in regeneration of the tissue with the shape different from the scaffold.

When a large-sized, lost tissue is to be replaced by a neo tissue regenerated *in situ*, a mechanical support will be tempolarily necessary until to the full regeneration of the tissue. The tissues requiring such mechanical supports include large tubular tissues such as large-calibered blood vessels, trachea, and large tubular bones. For instance, when a partially lost aorta is to be replaced with a regenerated tissue, a tubular template should be placed in the lost site. If the tubular material is too weak in mechanical strength, it will undergo rupture before the formation of a new tissue. Loss of large bones also requires a mechanical support until to bone regeneration. A problem accompanying these events is the disturbance of tissue regeneration by the supporting materials. Such a trouble would not arise if small experimental animal models like rat are used for tissue engineering studies. Only the use of bioabsorbable materials that will be resorbed, matching with the neo tissue formation, would circumvent this problem.

14.4
Structure of Scaffolds

When a scaffold is defined as any biomaterials used to encourage tissue regeneration, it may include the substrate for cell attachment, the barrier for cells to retain the site for tissue regeneration, the guide for cells to create a tissue giving the contour of the regenerated tissue, the carrier for the sustained release of growth factors, and the mechanical support until to tissue regeneration. It is unlikely that a single biomaterial can address all these requirements, although such a simple case is often seen in studies using rats as animal model. It should be emphasized that the scaffold that is clinically applicable is practically different from that for small animals, mostly because of difference in mechanical strength. The material property necessary and common to all scaffolds is temporally controlled biodegradability.

14.4.1
Surface Structure

The most important role of scaffolds in tissue engineering is to provide an attachment site for the cells responsible to the tissue regeneration. Similar to embryonic development, multiple cells should assemble to a specified form for tissue formation. To this end, cells would bind each other through the cell–cell interactions and, in addition, cells attach to a substrate for their survival, proliferation, and differentiation. If tissue regenerates by the help of growth factors alone, a carrier, not a substrate, will be required for their sustained delivery.

Fibronectin is well known as a cell-adhesive protein and has been very often attempted to immobilize on the scaffold surface. However, chemical modification of synthetic polymer materials with entire ECM molecules or relevant peptide fragments is not always necessary for scaffolds used in tissue engineering, because fibronectin molecules are more or less present in both serum and body liquids and adsorb to the scaffold surface unless it is too hydrophilic like nonionic hydrogels or too hydrophobic like fluorinated polymers. As the fibronectin adsorption needs a certain period, scaffolds lacking immobilized fibronectin would take a longer time for cell attachment than those with immobilized fibronectin. Collagen is also cell-adhesive and hence frequently employed for the enhancement of cell attachment.

Because of poor cell adhesion, hydrogel scaffolds need surface modification, when applied in tissue engineering, but they are basically not appropriate for scaffolds, since their mechanical strength is too low to retain the environment necessary for tissue regeneration.

The surface of Scaffold type II has mostly curvature that guides the regeneration of complicatedly shaped tissues. It is this surface contour of scaffolds that determines the 3D structure of tissues with a complicated contour.

14.4.2 Porous Structure

Most of the scaffolds studied in tissue engineering have porous structure which can be created by salt leaching, freeze-drying, sintering, or other much sophisticated technologies. The pores are not independent with each other, but interconnected. There are several reasons for the porous structure. One is to make route for the transport of oxygen and nutrients to the cells in the scaffolds and of the cell waste products to the outside. For the oxygen and nutrient supply, the pore size is less important than the porosity. Another reason, especially for Scaffold type II, is to fill cells as many as possible in the limited space of scaffold. Clearly, the scaffold with interconnected pores has a higher accommodation capacity for cells than nonporous materials. In addition, it is also critical for the new ECM produced by the seeded cells to have space.

The scaffolds (type I) prepared by electrospinning have many interstices, but they are too small for cells to infiltrate, although sufficiently large for nutrient supply. To facilitate neovascularization inside a porous scaffold, the pore size should be much larger than the capillary diameter with recommended size around 200–300 μm.

14.4.3 Architecture of Scaffold

The marked difference between Scaffold type I and type II is the dimension; type I is 3D, while type II is 2D. The surface is characteristic to type II, but the thickness or depth is important for type I. The thickness of the target tissue is influenced by the thickness of Scaffold type I, while the surface of type II regulates the regeneration of tissue. Probably, the cells evenly distributed on the scaffold surface will produce a new tissue toward the outside of the scaffold or into the new environment. New tissue ingrowth into the scaffold inside will be prevented by the scaffold material still remaining without being resorbed. Even if Scaffold type II has thickness, it has nothing to do with the thickness of the regenerated tissue. In this case, the scaffold is not the mold for the tissue to be regenerated. Scaffold type II is very thin with low mechanical strength, but it is not recommended to increase the mechanical strength by making the scaffold thicker, because the presence of a large mass of biomaterial at the location of tissue regeneration will disturb the proceeding of tissue regeneration, even if the biomaterials are biodegradable. This is because it is extremely difficult to match the biodegradation rate with the tissue regeneration. An effective method for strengthening scaffolds is to make use of stenting or reinforcement. Composites made from a porous sheet and a stent or a reinforcement material are often overlooked, but they will yield a

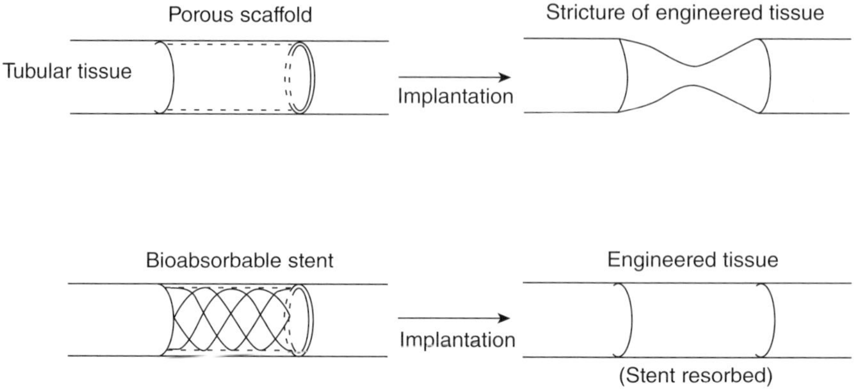

Figure 14.6 Protection of a tubular tissue by a biodegradable stent.

beneficial surface of Scaffold type II for tissue engineering. Moreover, stenting is very effective in preventing a tubular scaffold from stenosis (narrowing), as demonstrated in Figure 14.6.

14.4.4 Barrier and Guidance Structure

Any scaffolds for tissue engineering should retain and protect the environment where the tissue regeneration proceeds. The protection is generally performed by placing a barrier membrane around the permissive environment for the tissue regeneration. Well-known examples are the barrier membrane clinically used as a sheet for the guided tissue regeneration of periodontal tissues and as a tube for the regeneration of peripheral nerves. A long tube will guide the end of the extending peripheral nerve to reach the destination. The protection against invading cells and overloading from the outside will be also achieved by bulky 3D scaffolds to a certain extent.

In addition, a barrier membrane will act as a container for the bone marrow harvested from patients. The MSCs present in the bone marrow will initiate tissue regeneration in the protected environment where the bone marrow has been filled.

14.5 Biodegradable Polymers for Tissue Engineering

To fulfill the diverse needs in tissue engineering, various biodegradable materials have been exploited as scaffolds for tissue regeneration. Strictly speaking, biodegradable polymers are not identical to bioabsorbable (absorbable or resorbable) polymers, because the mechanism of polymer disappearance from the implanted site varies. Biodegradation is defined as chain scission of polymers constructing a biomaterial to shorter chains, finally to the monomer or oligomers in biological environments which contain a variety of hydrolytic enzymes, while bioabsorption simply means that the biomaterial has disappeared from the implanted location by any means. They include both enzymatic biodegradation and no chain scission of polymers. The polymer disappearance without any chain scission occurs due to the dissolution of polymer chains into the body fluid as a result of release of crosslinks that have made the polymer water-insoluble. Such crosslinking is mostly of physical type such as salt bridging with calcium ion, electrostatic interaction between cationic and anionic charges, and hydrogen bonding.

An example of salt bridging is alginate crosslinked with Ca^{2+}. If polyethylene glycol is a component of a biodegradable block copolymer, the water-soluble polyethylene glycol portion will be absorbed into the body fluid upon biodegradation of the other component. Such absorbable polymers are here included in biodegradable polymers.

A broad variety of biodegradable polymers have been studied, but those which are clinically applicable as scaffolds for tissue engineering are not many and very limited.

14.5.1 Synthetic Polymers

Traditional aliphatic poly(α-hydroxy acid)s alone have been virtually used as scaffold biomaterials for large animal experiments among numerous synthetic, biodegradable polymers. This may be due to the ease of the polyesters for fabricating porous scaffolds with different porosities and pore sizes, mechanically strong or elastomeic scaffolds, biodegradable scaffolds with different biodegradation rates, and nontoxic scaffolds with proved biosafety. Additionally, the Federal Drug Administration (FDA)/CE mark approval of several of these polyesters has motivated the application of those polymers in the tissue engineering field.

Polyglycolide (PGA) that includes here both glycolide homopolymer and glycolide-L-lactide (90:10) copolymer has the largest medical use among all commercially available, biodegradable polymers. The medical application of PGA is almost as sutures. Nonwoven fabrics fabricated from PGA fibers are also commercially available and have been used as scaffolds for tissue engineering, greatly contributing to the rapid progress of tissue engineering during the initial phase of tissue engineering research.

It is extremely difficult to prepare porous scaffolds starting from PGA powders in academic laboratories, because PGA of high molecular weights is soluble only in specific solvents like 1,1,3,3-hexafluoro-2-propanol. In contrast, copolymers of glycolide and lactide around equal monomer ratios (PGLA) are soluble in conventional organic solvents and hence have been widely used for scaffold fabrication, although the tensile strength of PGLA scaffolds is much less than PGA nonwoven fabric scaffolds. The property characteristic to PGA and PGLA is their high biodegradation rates; the mechanical strength decreases to the half within one month in the presence of water. It follows that the scaffolds prepared from these aliphatic polyesters can be applied only to the tissue engineering where tissue regeneration proceeds at relatively high rates.

In contrast to these glycolide polymers, polylactide (PLA) synthesized from L-lactide or D,L-lactide monomer undergo hydrolysis at much lower rates than glycolide polymers. It has been observed that a part or all the parts of PLA mass still remain without resorption when implanted up to one year. This implies that the application of PLA to tissue engineering may be limited to the tissues which require relatively long periods of time for regeneration.

Poly(ε-caprolactone) (PCL) has often been used for scaffold studies, probably because of its high processability (low melting point, many available organic solvents, and high strength product). However, PCL scaffolds would not be applicable to clinical tissue engineering, simply because PCL materials with high mechanical strength is virtually non-biodegradable *in vivo*. Clearly, only biodegradable polymers can be used for preclinical and clinical trials of tissue engineering.

In marked contrast to PLA and PCL, copolymers from P[LA/CL] are partially crystalline and produce strong, elastomeric scaffolds that have biodegradation rates ranging between those of PGA and PLA.

14.5.2
Biopolymers

Biodegradable biopolymers or biomacromolecules include polysaccharides, proteins (polypeptides), and nucleic acids. The most frequently used polymer for scaffold fabrication among these biopolymers is collagen, whereas nucleic acids have been scarcely used in tissue engineering studies. As collagen is ubiquitously distributed in our body, it is no wonder that many researchers have chosen collagen as a candidate material for scaffold fabrication, although mad cow disease has greatly hampered the use of collagen. Porous collagen sheet can be prepared by freeze-drying of aqueous collagen solution, followed by crosslinking with glutaraldehyde or dehydrothermal treatment. The porous structure can be controlled by changing the freeze-drying temperature. Lower freezing temperature yields scaffolds with smaller pores. Porous collagen sheets have been clinically applied to the skin tissue engineering from the 1980s, often making composite with glycosaminoglycan. Freeze-drying of aqueous solution of gelatin, denatured collagen, also produces porous gelatin sheets, but they have been applied to tissue engineering much less frequently than collagen sheets. As collagen and gelatin are able to serve as carrier of growth factors, the scaffolds fabricated from these polypeptides may have features different from others.

The low mechanical strength and high rate of degradation of biopolymers can be improved by chemical crosslinking. Fibrin glue which quickly forms upon mixing fibrinogen with thrombin has been used as scaffold preferably by surgeons. This biomedical hydrogel has low mechanical strength, but is capable of holding a large number of cells. In addition, this gel can seal another scaffold, if it has large interstices.

Chitin and chitosan are biomaterials that have been common choices for scaffold studies among polysaccharides. Scaffolds of high mechanical strength can be prepared from these crystalline biopolymers, but it should be mentioned that chitosan undergoes no appreciable biodegradation *in vivo*, at least, in rat model [11].

14.5.3
Calcium Phosphates

In contrast to organic biomaterials, inorganic biomaterials or minerals have been used for scaffolds only to a limited extent due to poor processability into highly porous structures and brittleness despite their good osteoconductivity. Among them are calcium phosphate compounds, because they are more or less biodegradable. Tetracalcium phosphate [$Ca_4O(PO_4)_2$] is resorbed more quickly than tricalcium phosphate [$Ca_3(PO_4)_2$] which has wide application as scaffold with interconnected pores. Hydroxyapatite [$Ca_{10}(PO_4)_6(OH)_2$] has been most widely used as biomaterials in orthopedic and oral surgery, but this calcium phosphate has been applied less frequently to tissue engineering than tricalcium phosphate, because hydroxyapatite is resorbed at much lower rates. However, the low resorption rate may not matter if we take it into consideration that hydroxyapatite is virtually identical to the mineral part of natural bones.

14.6
Some Examples for Clinical Application of Scaffold

14.6.1
Skin

Bilayered biomaterials composed of an inner porous collagen sheet, with or without chondroitin-6-sulfate, and an outer silicone layer have been clinically applied to skin tissue engineering [12, 13]. When placed on wounds even without any cell seeding, the collagen scaffold is replaced by a regenerated dermis-like tissue, while the silicone layer can be readily peeled off.

14.6.2
Articular Cartilage

Wakitani *et al.* applied tissue engineering to the repair of human articular cartilage defects in osteoarthritis (OA) knee joints [14]. The study group comprised 24 knees of 24 patients with osteoarthritis knee, who underwent a high tibial osteotomy. Adherent cells in bone-marrow aspirates were expanded by culture, embedded in a collagen gel scaffold, transplanted into the articular cartilage defect in the medial femoral condyle, and covered with autologous periosteum at the time of 12 high tibial osteotomies. The other 12 subjects served as cell-free controls. In the cell-transplanted group, as early as 6.3 weeks after transplantation, the defects were covered with white to pink soft tissue, in which metachromasia was partially observed. Forty-two weeks after transplantation, the defects were covered with white soft tissue, in which metachromasia was partially observed for almost all the area of the sampled tissue and hyaline cartilage-like tissue was partially observed.

14.6.3
Mandible

Autologous particulate cancellous bone and marrow (PCBM) that is rich in osteogenic progenitor cells and bone matrices has excellent properties as bone graft because it has full bone formation ability. However, PCBM does not have structural strength and the ability to hold its desired shape by itself. Kinoshita *et al.* developed a mesh manufactured from PLA that can sustain high mechanical strength for a long period of time [15]. PLA monofilaments with diameter of 0.3 or 0.6 mm were woven into a mesh. Figure 14.7 shows the PLA mesh and tray scaffold used for mandible regeneration. After preclinical studies with dogs, they started clinical studies using the PLA sheet/tray and autologous PCBM as illustrated in Figure 14.8. In eight hospitals in Japan, 62 cases underwent mandibular reconstruction between 1995 and 2001 [16]. Mesh trays were used in 28 cases and mesh sheets in six cases. The PCBM was harvested from the iliac bone of patients

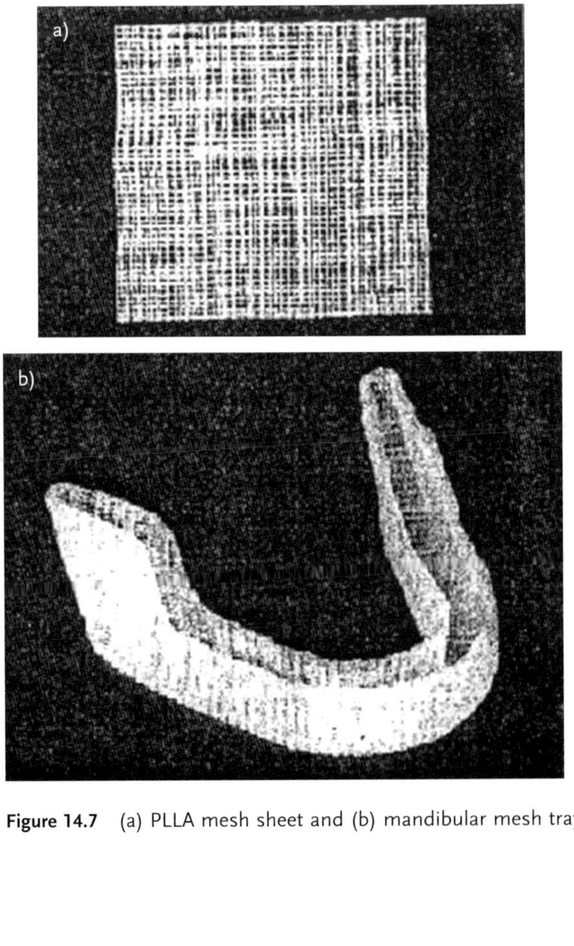

Figure 14.7 (a) PLLA mesh sheet and (b) mandibular mesh tray.

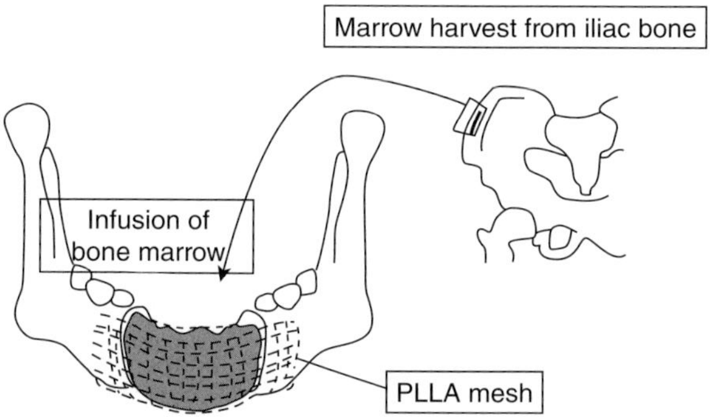

Figure 14.8 Schematic drawing of mandibular reconstruction by means of PLLA mesh and PCBM. The PLLA mesh tray was adjusted to the shape and size of the bone defect, cutting with scissors and warming at about 70 °C. The PLLA mesh tray was fixed to the residual bone with stainless steel wires and filled with PCBM taken from the iliac bone.

and 10–40 g of PCBM was transplanted to each patient. The clinical results were evaluated as *excellent* when the area of osteogenesis was over two-thirds in comparison to right after operation, based on X-ray films 6 months after surgery. Results were evaluated as *good* when osteogenesis was less than two-thirds with no reconstruction required. All other results were graded as *poor*. Forty cases were judged as *excellent*, 17 cases as *good*, and 10 cases as *poor*.

14.6.4
Vascular Tissue

Shin'oka *et al.* reconstructed peripheral pulmonary artery in a 4-year-old girl with the patient's own venous cells [17]. After that, three patients underwent tissue-engineered graft implantation with cultured autologous venous cells. However, since cell culturing was time-consuming and xenoserum had to be used, they began to use bone marrow cells (BMCs), readily available on the day of surgery, as a cell source. Matsumura *et al.* evaluated the endothelial function and mechanical strength of tissue-engineered vascular autografts (TEVAs) constructed with autologous mononuclear BMCs and a P(LA/CL) scaffold using a canine inferior vena cava (IVC) model. The mechanical strength change *in vitro* with time is shown in Figure 14.9 [18]. Figure 14.10 indicates no statistical differences in strength among IVCs of dog (shown as a control) and 6- and 12-month TEVAs. Encouraged by this successful result of the supplementary examination in the dog IVC replacement model, a 5-mL/kg specimen of bone marrow was aspirated from patients under general anesthesia before skin incision. The P(LA/CL) tube serving as a scaffold for the cells was the same as used for dogs. Twenty-three tissue-engineered conduits for extracardiac total cavopulmonary connection (TCPC) and 19 tissue-engineered patches were used for the repair of congenital heart defects. Mean follow-up after surgery was 490 ± 276 days [19]. There were no complications such as thrombosis, stenosis, or obstruction of the tissue-engineered autografts. The maximal trans-sectional area was calculated and compared with the implanted

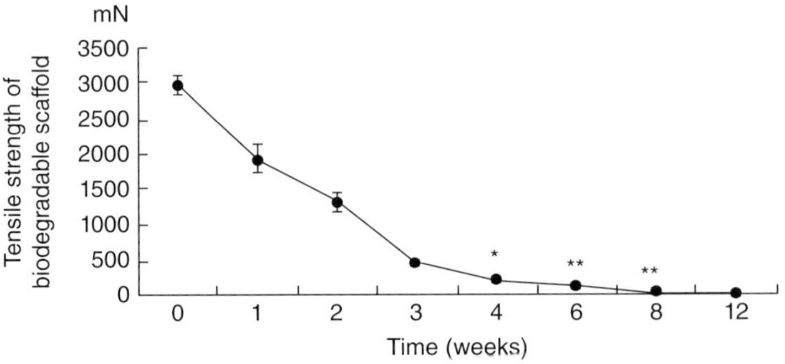

Figure 14.9 Tensile strength of biodegradable scaffold *in vitro*. The biodegradable scaffold used in this study diminished continuously within 1 month. Data represent the mean ± standard error of five samples at each time point. $*P < 0.01$, $**P < 0.001$ vs. week 0.

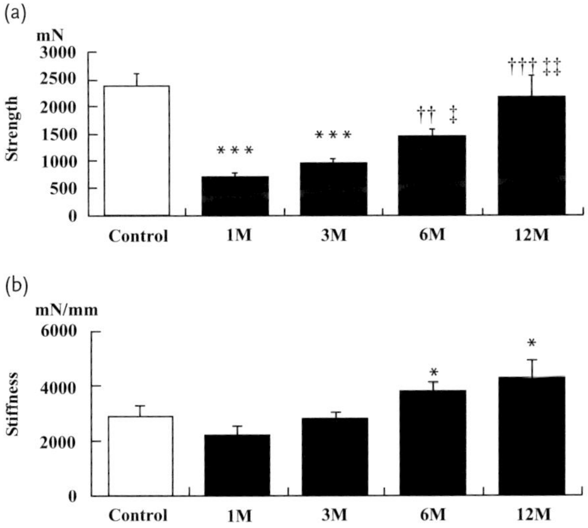

Figure 14.10 Increase in mechanical properties of TEVAs. (a and b) Tensile strength and stiffness of TEVAs. $P < 0.001$. $***P < 0.001$ vs. inferior vena cava, $^{††}P < 0.01$ vs. TEVAs at 1 month, $^{†††}P < 0.001$ vs. TEVAs at 1 month, $^{‡}P < 0.05$ vs. TEVAs at 3 months, $^{‡‡}P < 0.01$ vs. TEVAs at 3 months. (b) Stiffness/width was calculated as stiffness (mN/mm) = elastic modulus (mN/mm^2) × wall thickness (mm). $P < 0.01$. $*P < 00.05$ vs. TEVAs at 1 month. Data from dog inferior vena cava are shown as control. Data represent mean ± standard error.

size in the TCPC group. There was no evidence of aneurysm formation or calcification on cineangiography or computed tomography. All tube grafts were patent and the diameter of the tube graft increased with time (110 ± 7% of the implanted size), suggesting that these vascular structure may have the potential for growth, repair, and remodeling, and provide an important alternative to the use of prosthetic materials in the field of pediatric cardiovascular surgery.

14.7 Conclusions

It is likely that recent studies on scaffolds have focused on the novel synthesis of well-defined, biodegradable polymers, fabrication of porous structures with tailored architecture using sophisticated tools such as computer-assisted manufacture techniques, chemical modifications of scaffold surface by emulating advantageous features of the natural ECM, and genetic engineering [20–22]. Despite these continuous efforts for creating a variety of new scaffolds, the number of reports representing the results of clinical trials directly associated with tissue engineering is not increasing but rather decreasing in recent years. This tendency is similar to preclinical trials using large animal models, although it seems that small animals like rat have increasingly been used for tissue engineering studies. This trend might be potentially inevitable because the preferred motivation of scientists these days is to publish their experimental results in international journals with high impact factors, rather than to contribute to their society. Unfortunately, it is doubtful that the explosively growing life science, nanoscience, and biomedical technology will certainly encourage in warranting the clinical applications of current tissue engineering research to patients.

The engagement of biomedical industries on tissue engineering is crucial in transferring the accomplishments of tissue engineering studies to clinical applications. It is too difficult for academic people to fabricate by themselves a number of large-sized scaffolds applicable to patients. However, even if tissue engineers have developed a "promising" scaffold, companies would not be interested in manufacturing the scaffold on a large scale unless the nontoxicity of the scaffold material has already been proved by authorized procedures. It should be always kept in mind that the proved nontoxicity of biomaterials is a prerequisite in their clinical applications in preference of any other attributes of biomaterials. Recently, the regulatory organization of Japan has strictly inhibited biomedical companies to provide a surgeon with scaffolds manufactured by companies that are not yet approved, even if the project, which the surgeon attempts to apply to patients, has been approved by the IRB of his or her facility.

In this respect, close communications among tissue engineers, surgeons, biomedical manufacturers, and regulatory organizations are absolutely needed for promoting clinical trials in the tissue engineering field.

References

1 Ikada, Y. (2006) *Tissue Engineering–Fundamentals and Applications*, Academic Press, New York.
2 Gilbert, S.F. (2003) *Developmental Biology*, 7th edn, Sinauer Associates, Sunderland, MA.
3 Brockes, J.P. (1997) *Science*, **276**, 81–87.
4 Muneoka, K. and Bryant, S.V. (1984) *Dev. Biol.*, **105**, 179–187.
5 Martin, C. and Gonzalezdel Pino, J. (1998) *Clin. Orth. Relat. Res.*, **353**, 63–73.
6 Soderbeg, T., Nystrom, A., Hallmans, G., and Hulten, J. (1983) *Scad. J. Plast. Reconstr. Surg.*, **17**, 147–152.

7 Vidal, P. and Dickson, M.G. (1993) *J. Hand Surg.*, **18**, 230–233.

8 Neumann, L. (1988) *J. Trauma.*, **28**, 717–718.

9 Horton, W.A. (1990) *Growth Genet. Horm.*, **6**, 1–3.

10 Tsutsumi, S., Shimazu, A., Miyazaki, K., Pan, H., Koike, C., Yoshida, E., Takagishi, K., and Kato, Y. (2001) *Biochem. Biophys. Res. Comm.*, **288**, 413–419.

11 Tomihata, K. and Ikada, Y. (1997) *Biomaterials*, **18**, 567–575.

12 Suzuki, S., Matsuda, K., Isshiki, N., Tamada, Y., Yoshioka, K., and Ikada, Y. (1990) *Br. J. Plast. Surg.*, **43**, 47–54.

13 Suzuki, S., Matsuda, K., Nishimura, Y., Maruguchi, Y., Maruguchi, T., Ikada, Y., Morita, S., and Morota, K. (1996) *Tissue Eng.*, **2**, 267–275.

14 Wakitani, S., Imoto, K., Yamamoto, T., Saito, M., Murata, N., and Yoneda, M. (2002) *Osteoarthritis Cartilage*, **10**, 199–206.

15 Kinoshita, Y., Kobayashi, M., Fukuoka, S., Yokoya, S., and Ikada, Y. (1996) *Tissue Eng.*, **2**, 327–341.

16 Kinoshita, Y., Yokoya, S., Amagasa, T., *et al.* (2003) *Int. J. Oral Maxillofac. Surg.*, **32** (Suppl. 1), 117.

17 Shin'oka, T., Imai, Y., and Ikada, Y. (2001) *N. Engl. J. Med.*, **344**, 532–533.

18 Matsumura, G., Ishihara, Y., Miyagawa-Tomita, S., Ikada, Y., Matsuda, S., Kurosawa, H., and Shin'oka, T. (2006) *Tissue Eng.*, **12**, 3075–3083.

19 Shin'oka, T., Matsumura, G., Hibino, N., Naito, Y., Watanabe, M., Konuma, T., Sakamoto, T., Nagatsu, M., and Kurosawa, H. (2005) *J. Thorac. Cardiovasc. Surg.*, **129**, 1330–1338.

20 Martina, M. and Hutmacher, D. (2007) *Polym. Int.*, **56**, 145–157.

21 Ma, P.X. (2008) *Adv. Drug Deliv. Rev.*, **60**, 184–198.

22 Chan, G. and Mooney, D.J. (2008) *Trends Biotechnol.*, **26**, 382–392.

15
Drug Delivery Systems

Kevin M. Shakesheff

15.1
Introduction

The majority of medicines contain a polymer within their formulation. Polymers play diverse roles in the pharmacy. For example, they act as wicking and disintegration components of tablets, enteric coatings, and modifiers of release kinetics, lubricants, wetting agents, solid dispersion phases, viscosity modifiers, penetration enhancers, and more. Biodegradable polymers, which undergo chain scission as part of their function and prior to removal from the body, play a more limited role than biostable polymers in medicines. Indeed, only two classes of biodegradable polymers, poly(α-hydroxy acids) and polyanhydrides, have been used in marketed products in the United States. Other classes of biodegradable polymers, for example, polyorthoesters, having undergone decades of improvement, are now in late-stage human trials.

The very limited number of polymer types that have been developed is symptomatic of the great challenge faced in developing new biodegradable polymers for pharmaceutical applications. Additionally, the lack of new biodegradable polymers joining the above classes also reflects the ability to modify the properties of poly(α-hydroxyl acids) and polyanhydrides using copolymer chemistry to match the mechanical and degradation profiles required for many drug delivery applications. One interesting characteristic of this field of research is that so many groups have based their research on a narrow range of polymer types over a long period that a major body of literature exists on the chemistry, biological interactions, and medical application of these polymers.

Despite the slow pace of development of new biodegradable polymers in the field of drug delivery, there is a need to accelerate research into new classes. Current polymers have important weaknesses, and the requirements for biodegradable polymers that can release proteins, gene products, and cells are exposing these weaknesses.

This chapter aims to provide an overview of the current state of knowledge of poly(α-hydroxyl acids) and polyanhydrides to highlight the complexity of biological interactions of even these relatively simple polymers. The chapter then looks at

Handbook of Biodegradable Polymers: Synthesis, Characterization and Applications, First Edition. Edited by Andreas Lendlein, Adam Sisson.
© 2011 Wiley-VCH Verlag GmbH & Co. KGaA. Published 2011 by Wiley-VCH Verlag GmbH & Co. KGaA.

some examples of research into new classes of biodegradable polymers that address specific weaknesses in the current systems. The chapter does not attempt to cover the entire range of biodegradable polymers under development as drug delivery systems or to provide a complete history of the development of the poly(α-hydroxyl acids) and polyanhydrides, but the reader is referred to more comprehensive review articles throughout.

15.2
The Clinical Need for Drug Delivery Systems

Drug delivery systems modify the kinetics or the location of the escape of the drug from the medicine. Tables 15.1 and 15.2 provide generic reasons for using drug delivery systems.

For drug delivery systems containing biodegradable polymers, the major motivations for clinical use have been to deliver drugs that are required for long periods, are rapidly removed by metabolism or excretion, and are required at sites of administration that are difficult or impossible to reach with oral or conventional injection routes [1].

Cancer chemotherapy [2] and long-term replacement of human growth hormone [3] have been the major clinical foci for research on biodegradable polymer applications. Zoladex is the most successful (in terms of duration of clinical use and

Table 15.1 Use of a drug delivery system for kinetic control.

Dissolution of drug is too slow
Drug and/or formulation is physically removed from the site of action too rapidly
Metabolism or excretion of the drug is too fast
Drug is required intermittently
Administration is complex, invasive, and/or costly and therefore, dosing frequency needs to be reduced
Patient compliance (e.g., motivation to remember to take dosage) is poor and consequences of missing dosages are serious

Table 15.2 Examples of motivations to use a drug delivery system for location control.

Avoid side effects by minimizing exposure of other tissues
Concentrate drug at the site of action
Avoid rapid metabolism or excretion from the body
Accelerate drug transport across cell membranes
The route of administration is technically difficult (e.g., injection)

number of patients treated) biodegradable polymer-based formulation [4]. The primary clinical application of Zoladex LA is in the treatment of prostate cancer with the luteinizing hormone releasing hormone antagonist goserelin acetate. This drug blocks the downstream control of testosterone by the pituitary gland and thereby starves the tumor of a hormone that stimulates cancer growth. Goserelin acetate is a peptide molecule that can only be delivered by injection (it would be metabolized in the gastrointestinal track by enzymes). In addition, the drug needs to be constantly present in the blood stream for extended periods (e.g., 3 months). The polymer science underlying the release of drug from Zoladex, and the related product Lupron, is explored in Section 15.3.1.

Biodegradable polymer systems have also been employed for over a decade in the treatment of glioblastoma multiforme, an aggressive tumor within the brain [5–7]. In common with Zoladex, the systems used glioblastoma multiforme need to deliver drug over extended periods of time. Gliadel is a polyanhydride-based delivery system of the drug, 5-nitrourea, that concentrates the drug at the site of the tumor. In contrast to Zoladex, Gliadel provides local delivery of a toxic drug that must not be delivered to other sites in the body. The Gliadel product is placed at the site of the original tumor at the end of surgery to remove the primary tumor. Therefore, Gliadel achieves both temporal and spatial control of the release of a potent and toxic chemotherapy.

15.3 Poly(α-Hydroxyl Acids)

Polymers composed of lactic acid and glycolic acid dominate scientific literature on biodegradable polymers for drug delivery.

Polylactic acid:

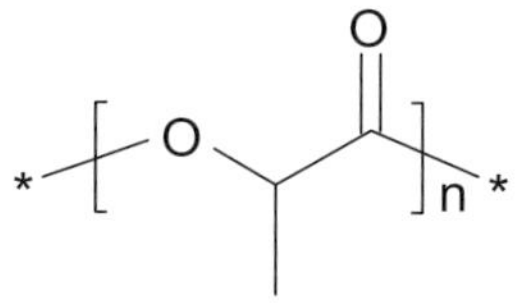

Polyglycolic acid:

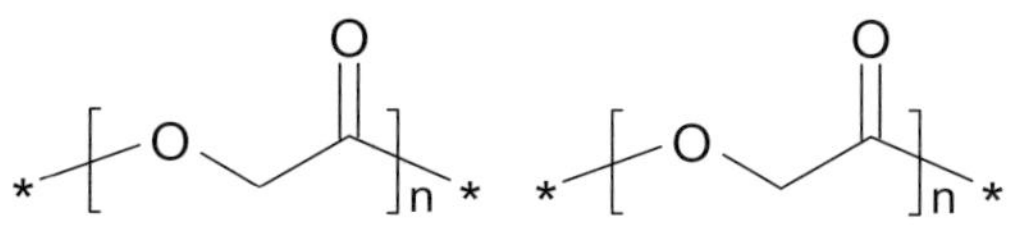

These polymers are synthesized by ring-opening polymerization of lactide and glycolide. In terms of nomenclature, the polymers are often termed polylactide, polyglycolide, and polylactide-*co*-glycolide as this reflects the monomer chemistry. However, the abbreviations PLA, PGA, and PLGA are more widely used than PL, PG, and PLG, and thus in this chapter polymer names including the acid term are used. It is very important to always specify the stereochemistry of the lactic acid component (see Section 15.3.1) as it has a profound effect on the physical and biological behavior of these polymers.

The polymers in this family have been components of biodegradable sutures and orthopedic implants for many years providing a long history of use in the human body. PLGA systems possess many attributes that make them suitable for drug delivery applications in which a slow release of a drug within a device is required [8]. Principle attributes are given below:

1) **Ability to control the kinetics of polymer degradation**

   For detailed explanation of control, see review by Anderson and Shive [6] and papers of Vert *et al.* [7–13], for example, A summary of key features of methods of control are discussed below.

2) **Numerous routes to fabrication**

   Described in Section 15.6.4.

3) **Mechanical properties**

   Sufficient compressive and tensile strength for use in applications in which the delivery system will be under compression or tension during function. For example, the polymer class is used in orthopedic implants.

4) **Widely available at medical grade**

   Synthesized to high purity and following good manufacturing practice by a number of companies across the world.

The history of use of PLGA polymers provides a number of important lessons for the development of new classes of polymers. Despite the simplicity of the polymer chemistry of PLGA polymers, the broad use of these polymers in humans and animal models has exposed significant complexity in the behavior of these polymers *in vivo*. Section 15.3.1 highlights some of the complexity and draws heavily on the excellent review of Anderson and Shive [8].

15.3.1
Controlling Degradation Rate

There are two distinct steps in the breakdown and removal of a biodegradable polymer; degradation and erosion. Degradation is the chemical breakage of bone along the polymer backbone that results in a decrease in polymer molecular weight. Erosion is the loss of mass from the delivery system due to the dissolution of the products of degradation.

The kinetics of degradation and erosion are determined by chemical and physical properties of the drug delivery system. A major attraction of the poly(α-hydroxy acids) is the ability to use the ratio of lactic acid to glycolic acid in the polymer chain to control both sets of kinetics. The methyl group on the lactic acid monomer retards hydrolysis of the neighboring ester group compared with the glycolic acid

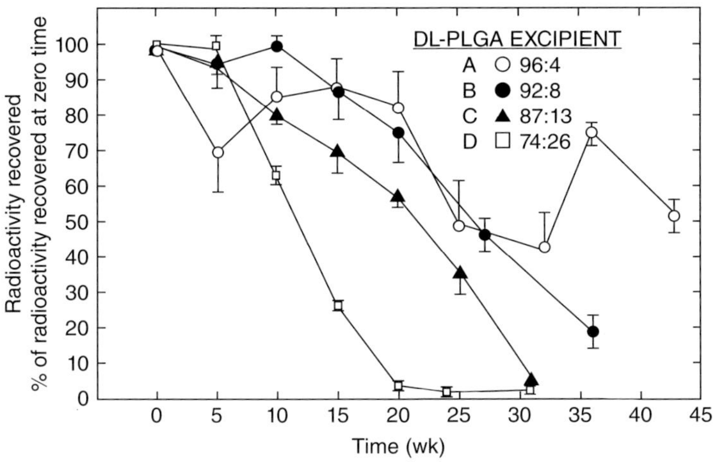

Figure 15.1 *In vivo* biodegradation of microcapsules is measured by Beck *et al.* by measuring radiolabeled P_{DL}LGA. Reproduced with permission from [10].

structure [9]. Hence, lactic acid containing homopolymers may take over one year to degrade and erode, while PGA may degrade and erode in one month. It is important to note that the exact period of time for degradation and erosion is not stated exactly because there are competing physical factors that can greatly accelerate or retard biodegradation. A study published by Beck *et al.* in 1983 demonstrates the effect of lactic acid to glycolic acid ratio on biodegradation and is reproduced in Figure 15.1 [10].

The next issue to be considered is the stereochemistry of the carbon-alpha to the ester [11, 12]. Clearly, both D and L forms of lactic acid exist, with L being the form used in nature. Lactic acid-based polymers are synthesized by the ring-opening polymerization of lactide. For drug delivery applications, both D,L-lactide and L-lactide are used. Hence, poly(D,L-lactic acid) (P_{DL}LA) and poly(L-lactic acid) (P_LLA) and both sterochemistries may be incorporated into PLGA copolymers. P_LLA and PGA are semicrystalline, while P_{DL}LA is amorphous. The degree of crystallinity affects the rate of water penetration into the drug delivery system and hence the rate of biodegradation. P_LLA may take more than 2 years to degrade *in vivo* if a semicrystalline morphology is allowed by the manufacturing route, while P_{DL}LA is removed in approximately 1 year. Li and Vert described a further complication in that the degree of crystallinity of quenched P_LLA (starting point amorphous due to quenching) and inherently amorphous P_{DL}LGA increased during degradation due to reorganization of the degradation products prior to, and delaying, erosion [13, 14].

When predicting the kinetics of degradation and erosion of PLGA polymers, it is necessary to consider the balancing contributions of polymer chemistry and crystallinity. For example, P_{DL}LGA (50:50) degrades and erodes more rapidly than PGA because the rate of penetration of water is the rate-limiting step rather than the steric hindrance to hydrolysis of the monomer structure.

A further complication in predicting and understanding the kinetics of degradation of this family of polymers is the effect of device size and architecture. Counterintuitively large device made from PLGA degrade more rapidly than microparticles in certain circumstances [15]. In addition, an important clue in the mechanism of accelerated degradation of large devices is the finding that large rods of PLGA often become hollow during degradation. These phenomena can be explained by the process of autocatalysis in which the acidic degradation products of PLGA hydrolysis accelerate local degradation. This localized catalysis is greatest within large devices due to the slow escape of the acid species. Hence, heterogeneous degradation kinetics occur across devices that have a diameter or width >300 μm.

15.4 Polyanhydrides

The second class of biodegradable polymers approved for use in humans in a drug delivery application are the polyanhydrides. The product Gliadel has been used for the treatment of brain tumors (see Section 15.2). A comprehensive review has been published by Katti *et al.* [16].

The motivation for using polyanhydrides over poly(α-hydroxy acids) is the need to restrict polymer erosion to the surface of the devise. As described in Section 15.3, the PLGA systems erode through a bulk mechanism for small particles and an autocatalytic hollowing mechanism for large rods. These mechanisms result in the encapsulated drug contacting with water for extended periods before the drug is released. Therefore, drugs that are sensitive to hydrolysis or other water-mediated instabilities could lose activity over time in the PLGA devices. A surface-eroding device would keep the drug dry prior to release. A further advantage of a surface-eroding system is the ability to control drug release kinetics via changes in surface area of the delivery system. For PLGA systems, the relationship between polymer degradation, drug release, and surface area is very difficult to predict (because bulk effects dominate and can be erratic due to physical breakup of the delivery system).

The Gliadel system is formed from a copolymer of the monomers bis(carboxyphenoxy)propane (CPP) and sebacic acid (SA). The structures of these monomers and the generic anhydride structure are shown below. Although no other polyanhydrides have been used in approved pharmaceutical products to date, the field of polyanhydride chemistry is active, and promising new structures are under investigation.

General PAA structure:

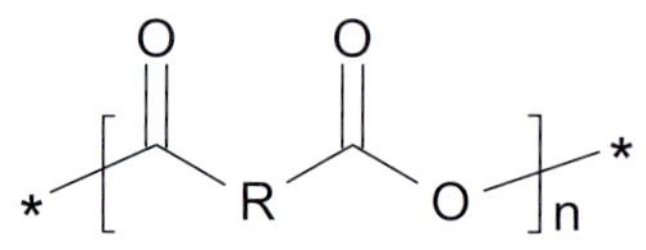

CPP:

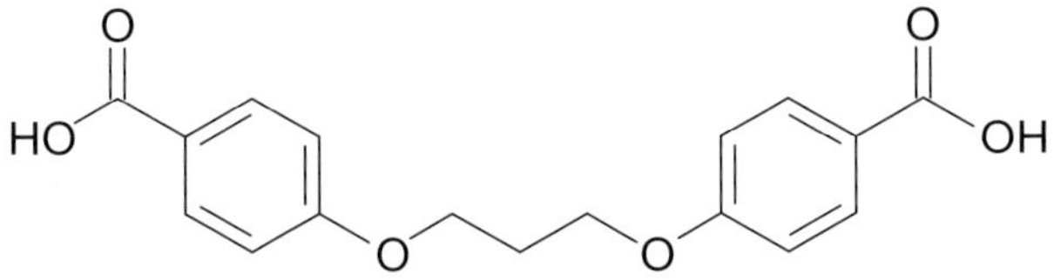

SA:

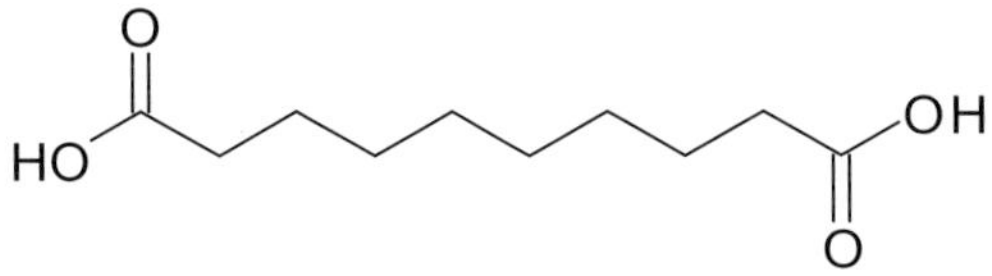

Poly(bis-(carboxypheoxy) propane-*co*-sebacic acid) (PCPPSA) is designed to achieve surface erosion and to allow biodegradation kinetics to be controlled by the ratio of the monomers. The CPP component is hydrophobic and discourages water penetration into the device. The anhydride links between monomer are very labile and break rapidly in the presence of water. Hence, water penetration is slow that polymer degradation and chain scission is limited to the surface of the device. However, CPP has low water solubility and although degradation of PCPP at the surface is quick, erosion is very slow. Hence, SA is used within the polymer structure to accelerate dissolution of degradation products. Overall, the design of these copolymers is a balance between the need to restrict water penetration and to allow erosion to occur over clinically acceptable timescales. Figure 15.2 reproduces data from the paper of Leong *et al.* that quantified degradation kinetics for the PCPPSA system [17].

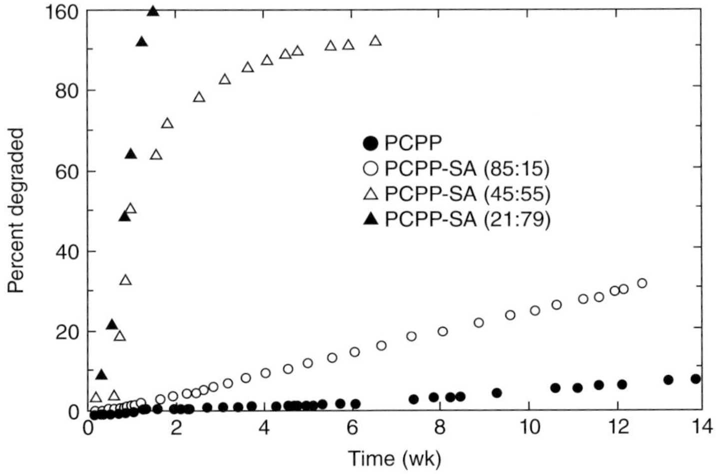

Figure 15.2 Degradation profiles of PCPPSA in 0.1 M pH 7.4 phosphate buffer at 37 °C. Reproduced with permission from [17].

15.5
Manufacturing Routes

The manufacture of drug and biodegradable polymer composites is not trivial. Both the poly(α-hydroxy acids) and polyanhydrides are water insoluble, indeed, that is an essential property that enables them to act as controlled release system. Hence, the polymer and drug phases of the delivery system will normally not share solvents that could be used to codissolve as a mobilization and mixing step in the manufacturing process. In the final product, a homogenous distribution of the drug within the polymer phase is likely to be required to generate a controlled and repeatable release rate for the drug. Therefore, it is essential that manufacturing routes achieve control of the size of the drug phases within the polymer phase and efficient dispersion of the drug phase.

A further complication in the manufacturing of these systems is the need to closely control the architecture of the finished product. For injectable formulations, a first requirement is that the drug delivery system can be expelled from the needle of syringe. For injections into the blood stream, or to sites where leakage into the blood stream will occur, the size of particle that can be used is further restricted to avoid blockage a fine capillaries in the blood system.

Emulsion-based processes are widely used to achieve the above properties in drug delivery systems. The drug can be dissolved in water and the polymer is dissolved in an organic solvent. Suspension of very small droplets of the aqueous drug solution within the organic solvent phase can be achieved within a water-in-oil (W/O) emulsion. If this W/O emulsion is suspended in a second water phase, then the droplet size of the organic solvent phase defines the maximum size of the final particle. Evaporation of the organic solvent in a stirred, open container creates solid particles containing the droplets of aqueous drug solution. Finally, sublimation of the water phase yields solid phase particles.

The above water-in-oil-in-water (W/O/W) emulsion system is widely used because it is adaptable to many polymer and drug combinations, including protein and nucleic-based drugs. However, there are numerous problems associated with the technique. In particular, the entrapment efficiency of the drug can be low as the drug can escape into the larger volume second water phase (outside of the organic solvent droplets). In addition, the formation of high surface area interfaces between the water and organic solvent phases may cause denaturing of protein drugs due to aggregation of loss of conformation.

A number of emulsion techniques have been described in the patent and scientific literature, which overcome shortcomings of the W/O/W technique. For example, Cleland *et al.* developed a human growth hormone delivery system using a novel cryogenic step in particle formation and demonstrated the importance of the manufacturing route to ensure integrity of protein drugs [18]. In addition, the manufacturing route eliminated the triphasic release profile that can hinder the use of PLGA-based microparticles. Morita *et al.* describe a useful method of creating solid dispersions of protein in polyethylene glycol (PEG) and then dispersing this composite in organic solutions of PLGA to create a solid-in-oil suspension.

This technique increases entrapment efficiency and removes any organic–water interface from the manufacturing environment [19].

An alternative to using an organic solvent to mobilize the polymer phase uses heat to melt the polymer. This has been used in the manufacture of Zoladex. The temperatures required to mobilize PLGA can be above 100 °C (depending on the composition and molecular weight) and so the technique is restricted for use with drugs that are stable at these elevated temperatures. Recently, Ghalanbor *et al.* have used hot-melt exclusion to load a protein, lysozyme, into PLGA. They demonstrated loading of up to 20% w/w of protein in the polymer with full retention of the protein enzymatic activity. The addition of PEG to the formulation eliminated the burst release of drug and drug release was controlled over a 80-day period [20].

The temperature of process of PLGA and many other polymers can be lowered to below 37 °C using CO_2 as a high-pressure processing medium. This technique relies on CO_2 depressing the glass transition temperature of amorphous polymers and lowering the viscosity of amorphous or crystalline polymer melts. High-pressure and supercritical-CO_2 processing have been described for microparticles, fibers, and highly porous scaffolds containing numerous types of protein drug [21–23].

15.6 Examples of Biodegradable Polymer Drug Delivery Systems Under Development

15.6.1 Polyketals

Polyketal-based drug delivery systems are under development for applications in which the acid degradation products from either poly(α-hydroxy acids) or polyanhydrides could cause detrimental side effects. Sy *et al.* have developed a poly(cyclohexane-1,4-diylacetone dimethylene ketal)-based delivery systems that can be used in the treatment of inflammatory diseases such as cardiac dysfunction [24].

This polyketal degrades in the presence of acid and generated neutral products. Sy *et al.* demonstrate that the encapsulation of a p38 inhibitor (SB239063) can improve the treatment of myocardial infarction.

15.6.2 Synthetic Fibrin

The biodegradable polymers discussed so far in this chapter have all used a simple water- or acid-triggered hydrolysis of a synthetic polymer backbone to lower their molecular weight and convert from water-insoluble to water-soluble forms.

A recent trend in the design of new biodegradable polymers for drug delivery has been to mimic enzymatic mechanisms of degradation used by our own bodies to remove extracellular matrix (ECM) and fibrin clots during tissue turnover or repair [25]. The need to employ this sophisticated method of controlling polymer biodegradation has been created by the demands of regenerative medicine. Within one aspect of regenerative medicine, there is a need to deliver growth factors or angiogenic factors to a localized site within the body to control tissue formation. Potent molecules such as vascular endothelial growth factor, platelet-derived growth factor, and bone morphogenetic proteins have clinical potential and applications in the formation of bone and enhancing blood vessel formation (e.g., in diabetic foot ulcers). These factors are naturally occurring within our bodies and the body has evolved methods of tightly controlling the exposure of cells to these molecules. These molecules are bound within the ECM and are exposed to cells when the cells locally degrade the ECM to reveal the next growth factor molecule. The release of the factor is, therefore, demand driven and effective dosages have been shown to be orders of lower magnitude using this mechanism as opposed to chemically driven hydrolysis of PLGA.

The approach of using matrices from fibrin or synthetic versions of fibrin have been reviewed by Lutolf *et al.* [25]. The approach to design a fully synthetic version of fibrin has been described by, for example, Kraehenbuehl *et al.* [26]. They used PEG-based hydrogels in which PEG-vinylsulfone and a four-armed PEG-OH molecule were crosslinked to form a 3D hydrogel. The gel contained the peptide Ac-GCRDGPQGIWGQDRCG-NH_2. This peptide can be cleaved by enzymes matrix metalloproteinases that are secreted by cells as they remodel fibrin or other biological matrices. Hence, the hydrogel was degraded locally by cells.

15.6.3
Nanoparticles

The importance of drug delivery system architecture was highlighted in Section 15.5. Many sites of the body that require high localized concentrations of drugs are inaccessible to any particle in micron range. Therefore, nanoparticle technologies have been used in drug delivery for many decades.

Pioneering work in this field focused on the mechanism to avoid uptake of nanoparticles within the liver. The liver has a natural function to remove potential harmful foreign particles that have been coated with plasma proteins via a process termed opsonization. Early pioneering work by Davis and Illum demonstrated that polymer nanoparticles, including PLGA, could avoid extensive liver uptake if their surfaces were engineered to present high densities of PEG [27–30].

Building on this concept, Gref *et al.* developed a nanaoparticle system using a copolymer of PLGA and PEG [31]. The particles could be formed by the one-step phase separation manufacturing step and entrapped up to 45% w/w of the drug The high density of PEG on the surface of the nanoparticles again altered biodistribution within mice. Five minutes after the administration, 66% of a dose of control particles (lacking PEG) were within the liver. This value dropped to less than 30% of the dose within the liver after 2 h for particles with a 20 kDa PEG component.

A recent study by Rothenfluh *et al.* demonstrated the ability to use nanotechnology and biological mimicry to create drug delivery systems that penetrate into articular cartilage tissue [32]. This is an especially challenging site to target drug delivery systems into because the tissue lacks a blood vessel system and possesses a very dense ECM. This team demonstrated that particles of 38 nm penetrated the cartilage structure while particles of 96 nm could not. Surface engineering of the particles with the short peptide ligand WYRGRL targeted the nanoparticles to the articular cartilages, as opposed to other tissues, to achieve a 72-fold increase in particle deposition.

15.6.4 Microfabricated Devices

One of the most ambitious drug delivery systems composed of a biodegradable polymer has been described by Grayson *et al.* [33]. This paper shows that an elegant fabrication method for an old class of polymers can generate remarkable control of drug release. The motivation for the work by Grayson *et al.* was to mimic the body's ability to release molecules in a pulsatile manner. Many hormones, for example, are required for intermittent periods and do not function if the body is constantly exposed to them due to desensitization. In addition, many vaccines require multiple injections to rechallenge the body and generate immune responses.

Using PLGA polymers and a layered microfabrication technique, it was demonstrated that pulses of drug could be achieved. The device, shown schematically in Figure 15.3a, possesses pockets that act as drug containers. The pockets are capped with a membrane of PLGA and release of the drug is restricted by the presence of the polymer. Now, the length of time it takes for the cap to be removed is dependent on the molecular weight of the PLGA. As shown in Figure 15.3b, the use of four different molecular weights of PLGA as capping materials generated four pulses of release.

A further example of using microfabrication in drug delivery is the formation of biodegradable polymer microneedles for injection without the use of a hypodermic syringe. The Prausnitz group has created a minimally invasive drug delivery system composed of an array of polymer needles in the shape of cones with a tip radius of only 2.5 μm [34]. This array can penetrate the skin to a depth of 750 μm without causing any pain. The needles are loaded with drug/polymer microparticles that release into the sublayers of the skin over periods of many days. A major advantage of this system is the ability of the patient to self-administer the drug delivery system. In comparison, many implant systems (e.g., Zoladex LA) require insertion by trained staff. The biodegradable needles may be withdrawn when the patient removes the array patch with any needles that remain within the skin layers safely degrading.

15.6.5 Polymer–Drug Conjugates

A polymer–drug conjugate is formed when a covalent bond is formed between a polymer and drug. The physical properties of the molecule become dominated by

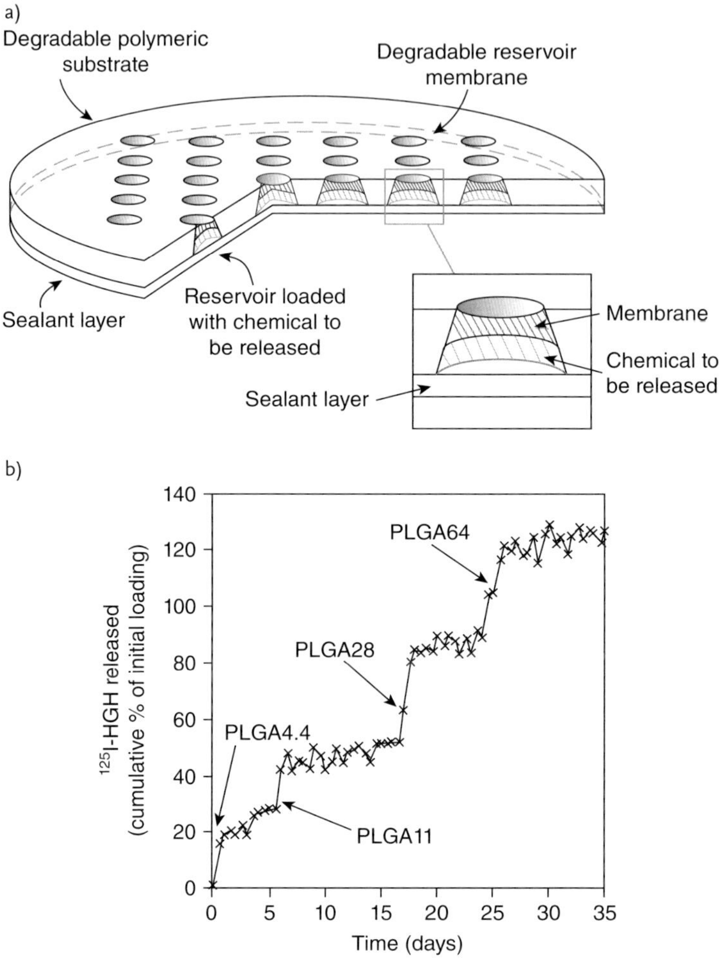

Figure 15.3 (a) Diagram of microfabricated device. The main body of the device is composed of a polymer that resists erosion until after the pockets have release their drug payload. Reservoirs or pockets of drug are capped with a membrane composed of PLGA. (b) Pulsatile release of human growth hormone achieved using PLGA of molecule weight of 4.4, 11, 28, and 64 Da. Reproduced with permission from [33].

the polymer and hence *in vivo* distribution, rate of liver excretion, and other properties that determine the time and location of drug action may be varied. Ringsdorf's initial vision for this class of polymer–drug conjugates has inspired numerous systems that have shown considerable promise in clinical trials [35, 36]. From a clinical perspective, the most important class of polymer–drug conjugates is formed using PEG. Numerous protein–PEG conjugates are used in drug therapy owning to the ability of the PEG to slow down the rate of protein metabolism and renal excretion and, hence, increase the half-life of biopharmaceutical [37].

However, PEG is not a biodegradable polymer and so we will not explore the mechanism of action of these conjugates further.

There are a number of polymer–drug conjugates that do contain biodegradable components. One key function of polymer–drug conjugates is their ability to release the drug once it has been carried into the cell by the polymer component. The high molecular mass of the polymer–drug conjugate results in an accumulation of the conjugate in certain types of tumors. This accumulation is caused by the enhanced permeability and retention effect, whereby many tumors possess leaky blood vessels that allow the conjugate to escape the blood system efficiency. In contrast, nontumor sites within the body have less-leaky blood vessels and so the drug does not enter tissues within which it would cause major side effects. Within the tumor site, the polymer–drug conjugate is taken up by cells and enters intracellular vesicles called lysosomes. The drug must escape the lysosome to have a pharmacological effect.

The Duncan group has described polymer–drug conjugates that preferentially release drug within the lysosome [38]. These lysosomotropic nanomedicinces use *N*-(2-hydroxypropyl)methacrylamide copolymer as the nondegradable backbone of the conjugate. This copolymer has been shown to be nontoxic and nonimmunogenic. The linkage between the *N*-(2-hydroxypropyl)methacrylamide copolymer and the anticancer drug is chemically broken within the lysosome when the pH falls. Hence, the drug remains as part of the conjugate until it has been successfully delivered to the target cell.

15.6.6 Responsive Polymers for Injectable Delivery

Responsive polymers undergo a phase change or gelation in result of a change in local environmental conditions. This concept has been used to great success in the development of block copolymers of PLGA–PEG–PLGA [39]. A product called ReGel is being developed for a range of drug delivery applications by exploiting the thermal gelation of this class of polymers. Gelation occurs at a temperature just below the body temperature. As a result, aqueous solutions of PLGA–PEG–PLGA are liquid at room temperature and may be injected through syringe needles into the body. Within the body, the system rapidly gels to form a delivery system that is retained at the site of administration. If a drug is included within the aqueous polymer solution, then it will be entrapped within the gel and released slowly due to retarded diffusion that accelerates as the PLGA component degrades. For example, ReGel has been used to release an anticancer agent, paclitaxel, for approximately 50 days [40].

15.6.7 Peptide-Based Drug Delivery Systems

The remarkable properties of biological molecules within living organisms have stimulated research into the replication of these properties in synthetic materials

[41]. Living systems use peptides and proteins to achieve many chemical and mechanical properties within cells. These properties can be generated in synthetic polymers built from amino acid monomers to form polymers with biodegradable amide linkages.

For example, Tirrell and coworkers have described artificial protein hydrogels with tunable erosion rates [42]. The hydrogels were formed from genetically engineered proteins and through aggregation of leucine zipper domains. The erosion rate of the hydrogel was controlled through changes in the amino acid sequence which in turn changes the network topology. This strategy generated hydrogels that are formed through physical crosslinks and possess highly predictable degradation properties.

15.7 Concluding Remarks

The poly(α-hydroxy acids) and polyanhydrides have undergone many years of research to generate the clinical products in use today. The versatility of these polymers encourages their use in a broad range of drug delivery systems in preclinical development. The weaknesses of these systems are apparent in the literature but the difficulty of replacing with new biodegradable polymer should not be underestimated. Promising new approaches are being reported based on systems that copy mechanisms of protein sequestering, thermal gelation, and cell-mediated release. Polymer–drug conjugates are being patiently optimized and clinical studies show promise in cancer chemotherapy. In addition, new fabrication techniques are opening new opportunities for established classes of biodegradable polymers.

References

1 Langer, R. (1990) New methods of drug delivery. *Science*, **249** (**4976**), 1527–1533.

2 Weinberg, B.D., Blanc, E., and Ga, J.M. (2008) Polymer implants for intratumoral drug delivery and cancer therapy. *J. Pharm. Sci.*, **97** (**5**), 1681–1702.

3 Kim, H.K., Chung, H.J., and Park, T.G. (2006) Biodegradable polymeric microspheres with "open/closed" pores for sustained release of human growth hormone. *J. Control Release*, **112** (**2**), 167–174.

4 DelMoral, P.F., Dijkman, G.A., Debruyne, F.M.J., Witjes, W.P.J., and Kolvenbag, G. (1996) Three-month depot of goserelin acetate: clinical efficacy and endocrine profile. *Urology*, **48** (**6**), 894–900.

5 Attenello, F.J., Mukherjee, D., Datoo, G., McGirt, M.J., Bohan, E., Weingart, J.D., Olivi, A., Quinones-Hinojosa, A., and Brem, H. (2008) Use of Gliadel (BCNU) wafer in the surgical treatment of malignant glioma: a 10-year institutional experience. *Ann. Surg. Oncol.*, **15** (**10**), 2887–2893.

6 Brem, S., Tyler, B., Pradilla, G., K.L., Legnani, F., Caplan, J., Brem, and H. (2007) Local delivery of temozolomide by biodegradable polymers is superior to oral administration in a rodent glioma

model. *Cancer Chemother. Pharmacol.*, **60** (**5**), 643–650.
7 Wang, P.P., J. Frazier, and H. Brem (2002) Local drug delivery to the brain. *Adv. Drug Deliv. Rev.*, **54** (**7**), 987–1013.
8 Anderson, J.M. and Shive, M.S. (1997) Biodegradation and biocompatibility of PLA and PLGA microspheres. *Adv. Drug Deliv. Rev.*, **28** (**1**), 5–24.
9 Li, S.M., Garreau, H., and Vert, M. (1990) Structure property relationships in the case of the degradation of massive poly(alpha-hydroxy acids) in aqueous media. 2. Degradation of lactide–glycolide copolymers – PLA37.5GA25 and PLA75GA25. *J. Mater. Sci. Mater. Med.*, **1** (**3**), 131–139.
10 Beck, L.R., Pope, V.Z., Flowers, C.E., Cowsar, D.R., Tice, T.R., Lewis, D.H., Dunn, R.L., Moore, A.B., and Gilley, R.M. (1983) Poly(DL-lactide-*co*-glycolide) norethisterone microcapsules – an injectable biodegradable contraceptive. *Biol. Reprod.*, **28** (**1**), 186–195.
11 Li, S.M. and Vert, M. (1994) Morphological-changes resulting from the hydrolytic degradation of stereocopolymers derived from L-lactides and DL-lactides. *Macromolecules*, **27** (**11**), 3107–3110.
12 Vert, M. (1986) Biomedical polymers from chiral lactides and functional lactones – properties and applications. *Macromol. Chem. Macromol. Symp.*, **6**, 109–122.
13 Vert, M. (2007) Polymeric biomaterials: strategies of the past vs. strategies of the future. *Progr. Polym. Sci.*, **32** (**8–9**), 755–761.
14 Vert, M., Li, S.M., Spenlehauer, G., and Guerin, P. (1992) Bioresorbability and biocompatibility of aliphatic polyesters. *J. Mater. Sci. Mater. Med.*, **3** (**6**), 432–446.
15 Grizzi, I., Garreau, H., Vert, M., and S.L. (1995) Hydrolytic degradation of devices based on poly(DL-lactic acid) size-dependence. *Biomaterials*, **16** (**4**), 305–311.
16 Katti, D.S., S. Lakshmi, R. Langer, and C.T. Laurencin (2002) Toxicity, biodegradation and elimination of polyanhydrides. *Adv. Drug Deliv. Rev.*, **54** (**7**), 933–961.
17 Leong, K.W., Brott, B.C., and Langer, R. (1985) Bioerodible polyanhydrides as drug-carrier matrices. 1. Characterization, degradation, and release characteristics. *J. Biomed. Mater. Res.*, **19** (**8**), 941–955.
18 Cleland, J.L., Johnson, O.L., Putney, S., and Jones, A.J.S. (1997) Recombinant human growth hormone poly(lactic-*co*-glycolic acid) microsphere formulation development. *Adv. Drug Deliv. Rev.*, **28** (**1**), 71–84.
19 Morita, T., Sakamura, Y., Horikiri, Y., Suzuki, T., and Yoshino, H. (2000) Protein encapsulation into biodegradable microspheres by a novel S/O/W emulsion method using poly(ethylene glycol) as a protein micronization adjuvant. *J. Control. Release*, **69** (**3**), 435–444.
20 Ghalanbor, Z., Korber, M., and Bodmeier, R. (2010) Improved lysozyme stability and release properties of poly(lactide-*co*-glycolide) implants prepared by hot-melt extrusion. *Pharm. Res.*, **27** (**2**), 371–379.
21 Gualandi, C., White, L.J., Chen, L., Gross, R.A., Shakesheff, K.M., Howdle, S.M., and Scandola, M. (2010) Scaffold for tissue engineering fabricated by non-isothermal supercritical carbon dioxide foaming of a highly crystalline polyester. *Acta Biomater.*, **6** (**1**), 130–136.
22 Davies, O.R., Lewis, A.L., Whitaker, M.J., Tai, H.Y., Shakesheff, K.M., and Howdle, S.M. (2008) Applications of supercritical CO_2 in the fabrication of polymer systems for drug delivery and tissue engineering. *Adv. Drug Deliv. Rev.*, **60** (**3**), 373–387.
23 Whitaker, M.J., Hao, J.Y., Davies, O.R., Serhatkulu, G., Stolnik-Trenkic, S., Howdle, S.M., and Shakesheff, K.M. (2005) The production of protein-loaded microparticles by supercritical fluid enhanced mixing and spraying. *J. Control. Release*, **101** (**1–3**), 85–92.
24 Sy, J.C., Seshadri, G., Yang, S.C., Brown, M., Dikalov, S., T.O., Murthy, N., Davis, and M.E. (2008) Sustained release of a p38 inhibitor from non-inflammatory microspheres inhibits cardiac dysfunction. *Nat. Mater.*, **7** (**11**), 863–869.
25 Lutolf, M.P. and J.A. Hubbell (2005) Synthetic biomaterials as instructive

extracellular microenvironments for morphogenesis in tissue engineering. *Nat. Biotechnol.*, **23** (**1**), 47–55.

26 Kraehenbuehl, T.P., Zammaretti, P., Van der Vlies, A.J., Schoenmakers, R.G., Lutolf, M.P., Jaconi, M.E., and Hubbell, J.A. (2008) Three-dimensional extracellular matrix-directed cardioprogenitor differentiation: systematic modulation of a synthetic cell-responsive PEG-hydrogel. *Biomaterials*, **29** (**18**), 2757–2766.

27 Redhead, H.M., Davis, S.S., and Illum, L. (2001) Drug delivery in poly(lactide-*co*-glycolide) nanoparticles surface modified with poloxamer 407 and poloxamine 908: *in vitro* characterisation and *in vivo* evaluation. *J. Control. Release*, **70** (**3**), 353–363.

28 Stolnik, S., Heald, C.R., Neal, J., Garnett, M.C., Davis, S.S., Illum, L., Purkis, S.C., Barlow, R.J., and Gellert, P.R. (2001) Polylactide-poly(ethylene glycol) micellar-like particles as potential drug carriers: production, colloidal properties and biological performance. *J. Drug Target.*, **9** (**5**), 361–378.

29 Riley, T., Stolnik, S., Heald, C.R., Xiong, C.D., Garnett, M.C., Illum, L., Davis, S.S., Purkiss, S.C., Barlow, R.J., and Gellert, P.R. (2001) Physicochemical evaluation of nanoparticles assembled from poly(lactic acid)-poly(ethylene glycol) (PLA-PEG) block copolymers as drug delivery vehicles. *Langmuir*, **17** (**11**), 3168–3174.

30 Dunn, S.E., Brindley, A., Davis, S.S., Davies, M.C., and Illum, L. (1994) Polystyrene-poly(ethylene glycol) (ps-peg2000) particles as model systems for site-specific drug-delivery. 2. The effect of PEG surface-density on the *in-vitro* cell-interaction and *in-vivo* biodistribution. *Pharm. Res.*, **11** (**7**), 1016–1022.

31 Gref, R., Minamitake, Y., Peracchia, M.T., Trubetskoy, V., Torchilin, V., and Langer, R. (1994) Biodegradable long-circulating polymeric nanospheres. *Science*, **263**, 1600–1603.

32 Rothenfluh, D.A., Bermudez, H., O'Neil, C.P., and Hubbell, J.A. (2008) Biofunctional polymer nanoparticles for intra-articular targeting and retention in cartilage. *Nat. Mater.*, **7** (**3**), 248–254.

33 Grayson, A.C.R., Choi, I.S., Tyler, B.M., Wang, P.P., Brem, H., Cima, M.J., and Langer, R. (2003) Multi-pulse drug delivery from a resorbable polymeric microchip device. *Nat. Mater.*, **2** (**11**), 767–772.

34 Park, J.-H., Allen, M.G., and Prausnitz, M.R. (2006) Polymer microneedles for controlled-release drug delivery. *Pharm. Res.*, **23**, 1008- 1019.

35 Ringsdorf, H. (1975) Structure and properties of pharmacologically active polymers. *J. Polym. Sci. Symp.*, **51**, 135–153.

36 Gros, L., Ringsdorf, H., and Schupp, H. (1981) Polymer antitumour agents on a molecular and on a cellular level? *Angew. Chem. Int. Ed. Engl.*, **20**, 305–325.

37 Jain, A. and Duncan, S.K. (2008) Jain, PEGylation: an approach for drug delivery. A review. *Crit. Rev. Ther. Drug Carrier Syst.*, **25** (**5**), 403–447.

38 Duncan, R. (2007) Designing polymer conjugates as lysosomotropic nanomedicines. *Biochem. Soc. Trans.*, **35**, 56–60.

39 Zentner, G.M., Rathi R., Shih C., McRea J.C., Seo M-H., Oh, H., Rhee B.G., Mestecky J., Moldoveanu Z., Morgan M., and Weitman, S. (2001) Biodegradable block copolymers for delivery of proteins and water-insoluble drugs. *J. Control. Release*, **72**, 203–215.

40 Elstad, N.L. and Fowers, K.D. (2009) OncoGel (ReGel/paclitaxel) – Clinical applications for a novel paclitaxel delivery system. *Adv. Drug Deliv. Rev.*, **61** (**10**), 785–794.

41 Langer, R. and Tirrell, D.A. (2004) Designing materials for biology and medicine. *Nature*, **428** (**6982**), 487–492.

42 Shen, W., Zhang, K.C., Kornfield, J.A., and Tirrell, D.A. (2006) Tuning the erosion rate of artificial protein hydrogels through control of network topology. *Nat. Mater.*, **5** (**2**), 153–158.

16
Oxo-biodegradable Polymers: Present Status and Future Perspectives

Emo Chiellini, Andrea Corti, Salvatore D'Antone, and David McKeen Wiles

16.1
Introduction

Synthetic and semisynthetic polymeric materials were developed for their versatility, easy processability, and durability – resistance to all forms of degradation as promoted by physical, chemical, and biological means or combinations thereof. Enhanced durability is achieved when required by including stabilizing additives (usually in combinations) and by processing under conditions that maximize the maintenance of molecular weight and functionality during fabrication and under subsequent service conditions. Macromolecular materials have been and are widely accepted because of their cost-effectiveness to provide a large variety of items that improve the comfort and quality of life in both modern industrial societies and developing countries. Moreover, the demand in the next two decades for polymeric materials is expected to increase two- to threefold primarily as a consequence of an increase in plastics consumption in developing countries, with an annual growth rate worldwide of 7–10%.

Plastics are ubiquitous because the various types that are commercially available collectively span a very wide range of useful properties. It is commonly claimed that approximately one-third of all commodity plastics are used for packaging purposes. The reason is that these materials are inexpensive, easy to fabricate, strong, tough, stretchy, have good barrier properties, and are reusable and recyclable, among other characteristics. The polyethylene (PE) shopping bag is an example of a common plastic article that is used in very large quantities because it does exactly what it is supposed to do at very low cost. It has supplanted the alternatives, for example, the brown paper bag, almost completely at checkout stations because it has overall superior properties and, most importantly, it is much less of an environmental burden to produce and transport [1, 2]. One criticism that is leveled at commodity plastics in short-lived applications, however, is that they persist too long after they are used and discarded. This is considered to lead to a serious plastic waste burden. The banning or taxing of PE shopping bags and analogous products is not the answer, however, because consumer

Handbook of Biodegradable Polymers: Synthesis, Characterization and Applications, First Edition. Edited by Andreas Lendlein, Adam Sisson.
© 2011 Wiley-VCH Verlag GmbH & Co. KGaA. Published 2011 by Wiley-VCH Verlag GmbH & Co. KGaA.

requirements need to be met and there is no acceptable substitute. Therefore, innovative technology is required.

The design, production, and consumption of polymeric materials for commodity and specialty plastic items must surely contend with all the constraints and regulations already in place or predicted to deal with the management of primary and postconsumer plastic waste. This is certain to involve the formulation of environmentally sound degradable polymers. Technologies based on the recovery of free energy content through recycling and from incineration with heat recovery will be flanked by the increasingly attractive option of environmentally degradable macromolecular materials. These latter polymers should be considered as preferred replacements for conventional commodity plastics in those product segments for which recycling is not a practical option. The strategies that are nowadays receiving a great deal of attention at both fundamental and applied levels include the design of some bio-based polymers, the introduction of hybrid polymeric formulations, and the reengineering of well-established polymers of synthetic and natural origin.

16.2
Controlled – Lifetime Plastics

During the past 20 years, science and technology have been developed for polymers that can biodegrade after being used and discarded. Everything from shopping bags to agricultural mulch films to daily landfill covers to food packaging as examples can be made to disintegrate after disposal and to yield thereby molecular fragments that are susceptible to mineralization by naturally occurring microorganisms. The carbon in these polymers is returned to the biocycle, and there are no harmful residues. These are the oxo-biodegradable polyolefins, as defined below.

According to the ASTM definition [3], a biodegradable plastic is "*a degradable plastic in which the degradation results from the action of naturally occurring microorganisms such as bacteria, fungi and algae.*" Rather more informative is the ASTM definition [4] of the environmental degradation of a plastic: "*abiotic or biotic degradation process or both that occurs in a given environment and includes photodegradation, oxidation, hydrolysis and biodegradation. Living organisms affect biotic degradation processes and abiotic degradation processes are not biological in nature.*" Two principal types of commercially viable biodegradable plastics have been developed and are finding a variety of applications in many mercantile segments and consumer products: (i) *oxo biodegradable polymers* for which degradation is the result of oxidative and cell-mediated phenomena, either simultaneously or successively and (ii) *hydro-biodegradable polymers* for which degradation is the result of hydrolytic and cell-mediated phenomena, either simultaneously or successively. Both types of biodegradable polymer feature a two-stage sequential molar mass reduction in the environment with the first stage being abiotic. Since the objective is to reduce the amount of plastic with minimum effect on the environment, the second stage is bioassimilation of the molecular fragments that are generated in the first stage. Abiotic mechanisms are generally regarded as too slow by themselves to be adequate in a variety of disposal environments.

There are several applications in which really quite rapid degradation of plastics after use is required. For example, plastics that end up in water- or sewage-treatment systems are an example of situations in which they need to lose integrity relatively rapidly so as to avoid plugging pumps, filters and the like. Hydrolytically unstable biodegradable plastics can provide an answer here. In many other uses (e.g., food packaging), however, hydrolytic instability is a disadvantage. Overall stability is required during shelf storage and use, but this should be followed by relatively rapid abiotic degradation within a specific time, depending on the disposal environment. The avoidance of the accumulation of plastic fragments requires that these be consumed through biodegradation by microorganisms in virtually all disposal environments. Effective biodegradation of such residues can be achieved when originally hydrophobic plastics acquire water-wettable (hydrophilic) surfaces and a relatively low molecular weight so that there is a significant number of molecular "ends" accessible at the surface. The science and technology of the development of commercially viable commodity plastics that can meet these criteria are the topics addressed in this chapter.

Of the current worldwide production of synthetic polymers, nearly 90% is represented by full-carbon-backbone macromolecular systems (polyvinyl and polyvinylidenics [5]), and 35% to 45% of production is for one-time use items (disposables and packaging).Therefore, it is reasonable to envisage a dramatic environmental impact attributable to the accumulation of plastic litter and other plastic waste from discarded full-carbon-backbone polymers, which are conventionally recalcitrant to physical, chemical, and biological degradation processes. The mechanism of biodegradation of full-carbon-backbone polymers requires an initial oxidation step, mediated or not by enzymes, followed by fragmentation with a substantial reduction in molecular weight. The functional fragments then become vulnerable to microorganisms present in different environments, with production (under aerobic conditions) of carbon dioxide, water and cell biomass. Figure 16.1 outlines the general features of environmentally degradable polymeric materials, which are classified as hydro-biodegradable and oxo-biodegradable. Typical examples of oxo-biodegradable polymers are PE, poly(vinyl alcohol) [6], natural rubber (poly-*cis*-1,4 isoprene) [7], and lignin, a naturally occurring structurally complex heteropolymer.

The prodegradant added to polyolefins to convert them to oxo-biodegradable status does not cause any oxidation or other degradation as long as antioxidants are present. Thus, the shelf life and use life of the plastics are maintained for a period that is controlled by the amount of antioxidant (or other stabilizing additives) present in the formulation. Once the stabilizers have been depleted, the prodegradant catalyses the oxidative degradation of the polymer, with the rate of degradation related to the concentration of prodegradant. By controlling the concentrations of these two classes of additive, one practically controls the "lifetime" of the plastic.

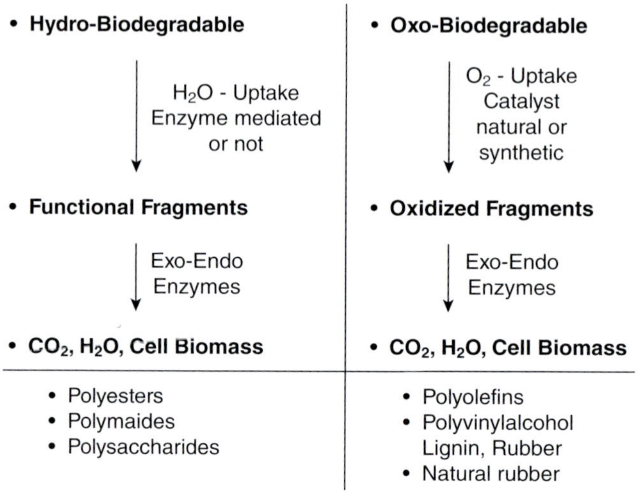

Figure 16.1 General features of the two classes of environmentally degradable polymers.

16.3 The Abiotic Oxidation of Polyolefins

The knowledge [8–10] of the thermal and photolytic peroxidation mechanisms of PE and polypropylene (PP) constitutes the basis for the development of "reengineered" polyolefins susceptible to enhanced oxidation and fragmentation, when exposed to heat or light, with the aim of overcoming the intrinsic recalcitrance of polyolefins to biodegradation. It has been established that the beginning of the sequence of reactions leading to polyolefin peroxidation is the generation of sensitizing impurities during the processing of these thermoplastics [8]. It has been recognized that carbonyl [8, 11] and hydroperoxide [8, 12, 13] groups represent the major sensitizing impurities formed during the processing of PE and of PP. At this stage, the chemical structure of the polyolefins is considered to be the most important parameter capable of influencing the oxidative degradation processes. During subsequent use and disposal steps, the oxidation of both types of resin appears to be mainly affected by structural parameters, such as the degree of polymerization, chain conformation, degree of crystallinity, and geometry [13].

In the case of PE, the poor reactivity of the nonpolar C–C and C–H bonds markedly constrains the degradation processes by radical reactions. These are generally initiated by bond-breaking processes promoted by energy input in the form of heat, UV radiation, mechanical stress, or some combination of these. Since the susceptibility of saturated bonds to scission is dependent on bond energy, the initial homolytic bond scission reactions are largely restricted to structural defects, such as branch points and double bonds. The tertiary carbon–hydrogen bonds that alternate with methylene groups along PP chains are obviously vulnerable. Following initial bond breakage, a complex series of radical reactions may lead to the total degradation of the molecules.

16.3.1
Mechanisms

The overall sequence of reactions that are the basis of PE oxidation and of polyolefins in general has been elucidated during several decades of research, producing a large number of original papers and review articles. For example, it is widely accepted that the starting point of the process is the homolytic cleavage of C–C bonds in the backbone that occurs during polymer processing as a result of shear stresses during extrusion [14]. In the presence of oxygen (as in most industrial processes), the carbon-centered radicals are converted to peroxy radicals, and then to hydroperoxides by hydrogen abstraction from nearby tertiary sites. The high reactivity of hydroperoxides subjected to heat and/or UV light promotes a further series of reactions leading to chain scission (molar mass reduction) and the formation of several different oxidized groups. In the overall peroxidation process of PE, the decomposition of hydroperoxide groups is acknowledged to be the rate-determining step [13]. This initiates a radical chain reaction which is autoaccelerating, as shown in Scheme 16.1.

Even though the aim of this chapter is not to review the large amount of literature on polyolefin degradation, a general overview of the mechanism and kinetics of PE oxidation is useful for a better understanding of the environmental fate of PE. In this connection, the basic mechanism proposed by Bolland and Gee [15]–comprising the classic steps of initiation, chain propagation, and termination–should be considered as relevant (Scheme 16.1). As already mentioned, the key intermediates in the accepted mechanism are hydroperoxides, the decomposition of which produces further free radicals and the derived oxidation products. Many of the kinetic studies of thermooxidative processes of polyolefins are for polymers in the melt, and several mechanistic studies have focused on polymers in a solution. Considering that the practical uses of polyolefins are in the solid state, numerous studies have been devoted to the investigation of the kinetics of the thermal and photooxidation of polyolefin films and sheets. Oxidation products

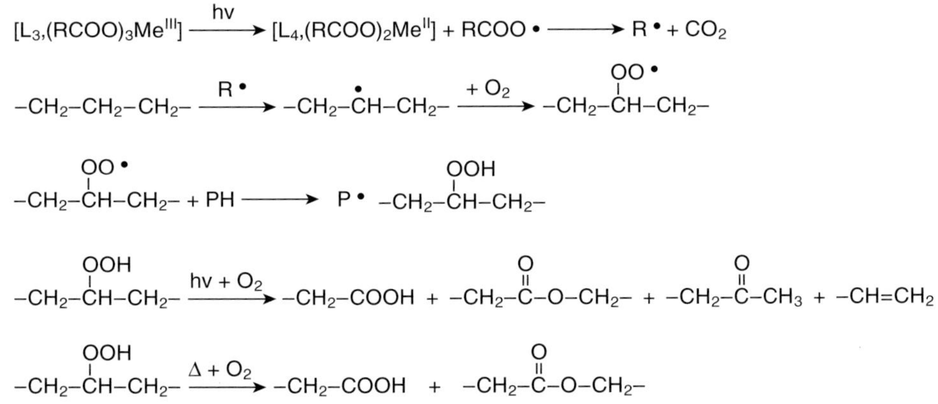

Scheme 16.1 General mechanism of peroxidation and chain cleavage in polyolefins.

have been identified and several parameters, such as oxygen pressure, temperature, and sample thickness that influence oxidation processes have been considered in the investigations.

16.3.2 Oxidation Products

Fourier transform infrared (FTIR) spectroscopy is one of the most powerful techniques to be used in studying the kinetics of PE and PP oxidation in the solid state. Significant changes can be easily monitored in various regions of the spectra of films and sheets during thermal and photooxidation. In particular, the presence of hydroperoxides is recognized from absorption bands between 3400 and 3200 cm^{-1} and absorptions in the carbonyl region – specifically between 1780 and 1700 cm^{-1} – are used to evaluate the rate and extent of the oxidative degradation of polyolefins. In addition, the absorbance variation of double-bond deformation peaks as well as the absorption from carbon–oxygen single bonds can provide valuable information on the mechanism and oxidation products involved.

When oxygen concentrations are nonlimiting, during the initiation stage, it may be assumed that all the macroradicals as they are produced (e.g., by shear stresses) are instantly oxidized to peroxy radicals which, by intra- or intermolecular abstraction, are then converted to hydroperoxides. A fairly complex series of chain reactions involving the formation/decomposition of peroxy radicals/hydroperoxides constitutes the propagation step leading to oxidation product formation and chain scissions. It has been estimated that, in oxidized solid PE, more than 80% of the oxygen-containing products are represented by carbon chains bearing ketone or carbonyl groups [13]. Briefly their formation is generally attributed to the decomposition of hydroperoxides (in the case of ketones), whereas carbonyl groups are considered to be produced by peroxide decomposition. In addition, the conversion of alkoxy macroradicals by β-scission to produce a carbonyl group and a chain-end radical can occur. It has been ascertained that there is a straightforward relationship between the number of carbonyl groups formed and the extent of chain scission. Thus quantitative FTIR analysis can be used effectively to measure the extent of the abiotic thermal degradation of PE [16].

As a consequence of the radical oxidation processes and relevant chain scissions, a fairly high number of degradation products containing functional groups have been recognized in several investigations. In particular, two different classes represented by low-medium molecular weight fractions and volatile intermediates, respectively, can be detected during kinetic studies of the thermal and photooxidation of PE. As a result, the oxidation processes of PE and in particular of low-density polyethylene (LDPE) can be monitored effectively by gravimetric analysis showing the weight increase (oxygen uptake) as a function of the thermal aging time and temperature [16]. In a case study, carried out on an LDPE sample containing a thermal prodegradant, the time profile of weight variation showed a S-shaped profile, thus accounting for the exponential accumulation of oxidized low-medium molecular weight fractions, followed by the progressive weight decrease owing to the loss of volatile intermediates. In several studies, low-medium molecular weight products containing carbonyl and hydroxyl groups have been identified [17]. It has been ascertained also that the amounts of these products account for at least 80% of the products containing ketone and carboxyl groups [13]. Carboxylic acids tend to accumulate during prolonged exposure times since other oxygen-containing products formed in the early stages of the degradation process, such as alcohols, aldehydes, and ketones, are susceptible to further oxidation to produce carboxylic acids. In classic studies [18, 19], most of the low-molecular-weight degradation products from both thermally- and photooxidized PE have been isolated and identified by solid-phase extraction coupled with gas chromatography/mass spectrometry. Accordingly, numerous semivolatile compounds have been identified [17] including alkanes, alkenes, ketones, aldehydes, alcohols, mono- and di-carboxylic acids, lactones, keto-acids, and esters. In addition, highly volatile organic products (C2–C6) have also been detected although in a very relatively small amounts. Among these, acetaldehyde represents the most important; its quantitative release profile has been monitored [13].

As discussed above, carboxylic and dicarboxylic acids have been found to be the most abundant products which are formed during both photo- and thermal oxidation. They tend to accumulate owing to their low propensity to oxidize further during prolonged aging. They presumably evolve from the oxidation of other functional groups, such as primary alcohols and aldehydes deriving from hydroperoxide decomposition followed by hydrogen abstraction or β-scission, respectively, particularly during thermal aging. Also, the photolytic cleavage of ketone groups by Norrish mechanisms can lead to the formation of carboxylic acid groups. Accordingly, dicarboxylic acids have been found to be the predominant products formed during the prolonged photoxidation of PE [20]. The presence of large amounts of carboxylic acids, as demonstrated qualitatively by FTIR spectroscopy, suggests the severe alteration of the PE matrix. Indeed, with the progressive accumulation of carboxylic groups as a function of exposure time, there is an accompanying variation in both the shape and the intensity of the absorption bands between 3800 and 3400 cm^{-1} that are associated with aliphatic carboxylic acids. Owing to their relative thermal stability (compared to photoinstability), ketone groups are considered as typically associated with the thermal degradation of PE. Other products, recognizable in the volatile and semivolatile fractions of PE oxidation products, such as keto-acids, have been identified during low-temperature thermal degradation, whereas lactones are usually generated under extreme conditions or after extensive degradation has occurred [18, 19, 21].

Since most of the oxidation products result in the first instance from the decomposition of polymer hydroperoxides, they are formed and trapped within the bulk solid, and only a small fraction can escape. In this connection, it has been noted that the estimation of the degradation products extractable with organic solvent(s) should both provide useful information about the level of oxidative degradation and help characterize the low-medium-molecular-weight oxidation products. Further information about the regioselectivity of oxidation can be obtained by analyzing the level of solvent extractable fractions and molecular weight and polydispersity of a nonextractable polymer residue. The characterization of the oxidized fraction of PE soluble in a dichlorobenzene–methanol mixture showed that it contained a large amount of oxygen-containing functional groups attached to low-molecular-weight chain fragments, whereas the nonextractable portion of the polymer contained a relatively low level of oxidation [13].

16.3.3
Prodegradant Effects

In a recent study of ours [16] the amount of extractable fraction from thermally oxidized LDPE samples containing prodegradant additives was evaluated as a function of the level of polymer matrix oxidation as assessed by the carbonyl index (CO*i*) (Figure 16.2). In particular, it was shown that the amount of acetone-extractable material is positively correlated to the level of oxidation induced by thermal treatment in an oven, thus reaching fairly high levels corresponding to 25–30% of the original sample weight. The solvent-extractable fractions, as characterized by NMR and by FTIR, were heavily oxidized and had low molecular weights (0.80–1.60 kDa). Furthermore, it was also observed that the increase in oxidation level as related to CO*i* values was matched by an increase in the quantity of oxidized plastic having reduced molecular weight.

The large amount of numerous oxidation products, as well as their relative concentrations, are accounted for by the large number of interrelating elementary reactions, and these give rise to a rather complex scheme to describe the oxidation kinetics. In spite of the major interest in these complex series of chain reactions that has attracted a great deal of attention over the past 50 years, little agreement on the kinetic models and values of specific rate constants has been achieved, from either a theoretical or an experimental basis. One of the more contentious issues is that of whether the oxidation mechanism should be considered as a homogene-

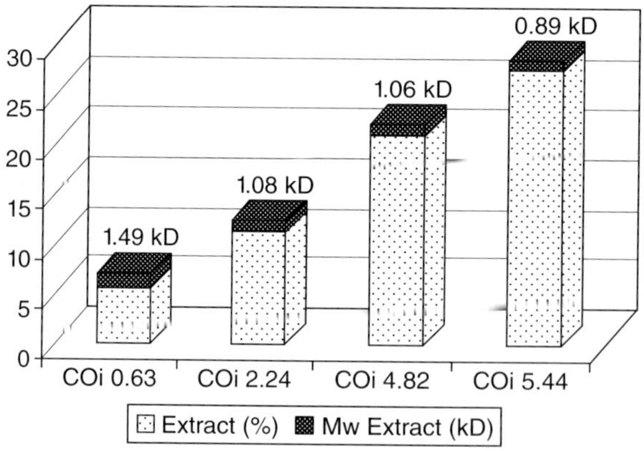

Figure 16.2 Percentage of fractions extractable with acetone and relevant molecular weight in thermally treated LDPE film at various levels of oxidation as determined by the carbonyl index (CO*i*).

ous phenomenon or as a heterogeneous process involving the spread of degradation from localized centers. This latter interpretation takes into account the semicrystalline nature of polyolefins, where amorphous regions, more susceptible to oxygen diffusion and oxidation, coexist with crystalline regions where oxidation events are hindered, largely owing to the negligible dissolved oxygen content.

Even though the aim of this chapter is not to review this topic, some basic simplified information on kinetic oxidation processes is useful to predict the ultimate oxo-biodegradation propensity of PE. In particular, the effect of physical parameters such as aging time, temperature, and radiation intensity may affect the subsequent extensive biodegradation step. Indeed, one of the most important requirements, in order to predict the oxidative behavior of polyolefin films, is the correlation between the experimental kinetics of oxidation and the chemical reactions that may occur in the bulk polymer. In several studies, therefore, it has been shown that the early stage of the thermooxidative degradation of PE, as monitored by the formation of carbonyl groups, is apparently in agreement with a typical auto-acceleration mechanism [22]. In the overall peroxidation of PE, the decomposition of hydroperoxide groups is therefore considered to be the rate-determining step [13]. Hence, additive molecules which are capable of enhancing hydroperoxide formation and decomposition to other radicals are active prodegradants since they accelerate the oxidation and cleavage of polyolefin chains.

16.4
Enhanced Oxo-biodegradation of Polyolefins

Major strategies to enhance the environmental degradation and biodegradation of polyolefins have been focused on copolymerization, blending, or grafting with functional polymers and other compounds as well as the addition of prodegradant additives. UV-absorbing carbonyl groups capable of accelerating the photooxidation process can be introduced by copolymerizing ethylene and carbon monoxide or vinyl ketones [23–26], the latter strategy being the technical process used in the production of Ecolyte polyolefins [25]. Another strategy to improve the environmental degradability of PE and PP films is the addition of prodegradant additives during processing [27]. It has been suggested, in fact, that this latter alternative may provide a more efficient control of the degradation rate, thus making the shelf life and use life of the polyolefins compatible with a very wide range of applications and disposal environments [14]. Most of the prodegradant additives used commercially are organic complexes of transition metals, those capable of yielding two metal ions differing in oxidation number by one unit. Several polymer-soluble metal carboxylates and acetylacetonates of Co^{3+}, Fe^{3+}, and Mn^{2+} are very effective photoprodegradants for polyolefins, capable of initiating the degradation process through metal salt photolysis to give the reduced form of the metal ion and a free radical. The anion radical promotes a fast hydrogen abstraction from the polymer with the subsequent formation of hydroperoxide. Afterward, the general radical oxidation reactions of polyolefins proceed with enhancement by the usual redox

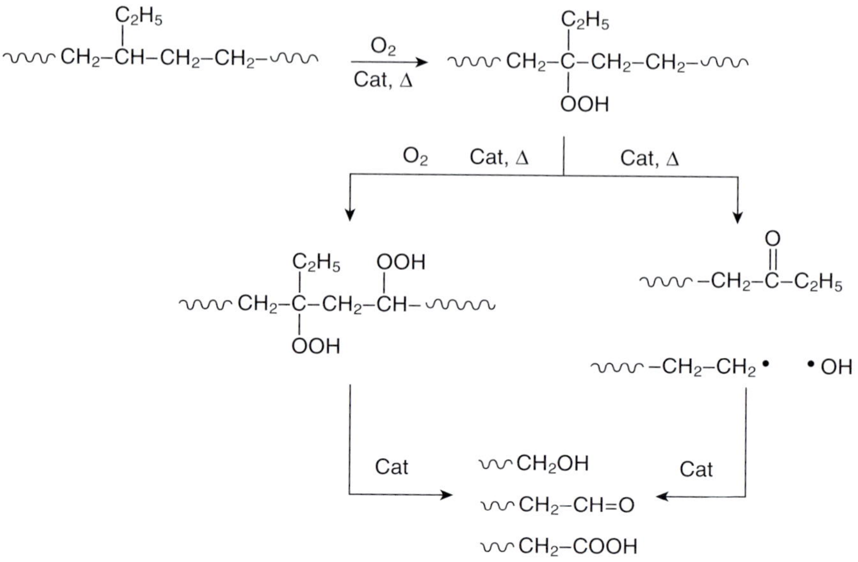

Scheme 16.2 Radical chain reactions in polyolefins as promoted by transition metal ion prodegradants.

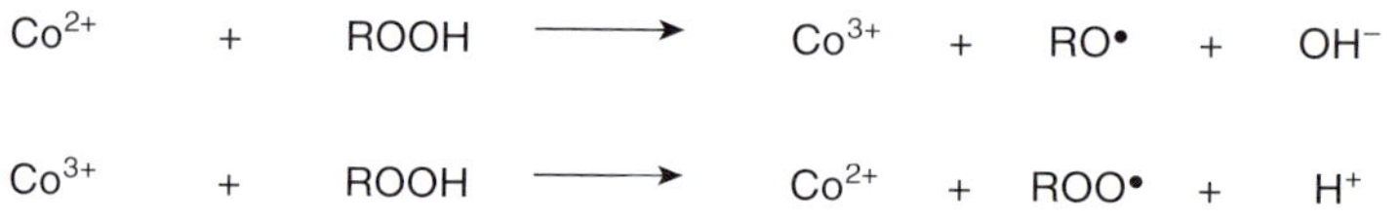

Scheme 16.3 Hydroperoxides decomposition as mediated by transition metal ions.

reactions between hydroperoxides and metal ions. Alternatively, prodegradant additives can induce the peroxidation process in polyolefins by absorbing energy as heat. In this case also, the activity derives from transition metal ions typically added to the final products in the form of stearates or acetylacetonates. The most commonly used cations are Mn^{2+} [28] and Co^{2+} [29]. Instead of Fe^{3+} complexes which are significant in photooxidation processes, Mn^{2+} and Co^{2+} are used to accelerate the radical reactions in polyolefin oxidation through the decomposition of hydroperoxides and peroxides induced by heat absorption [30] (Schemes 16.2 and 16.3). It must be emphasized that the transition metal salts catalytically induce the rapid decomposition of polyolefin hydroperoxides; very small amounts only are required to speed up the peroxidation of polyolefins by several orders of magnitude.

Another type of "photosenstizer" prodegradant for PE is Fe^{3+} dithiocarbamates or dithiophosphates. These compounds initially act as antioxidants by decomposing hydroperoxides by an ionic mechanism [31, 32] after which the ligands are destroyed and the free transition metal ions perform as a prodegradant. In this way, in these compounds both antioxidant and prodegradant functionalities co-exist. This characteristic has been used to finely control the lifetime before the photooxidation of PE commences. This is the basis of the Scott–Gilead technology [12] which led to the production and commercialization of controlled-lifetime photodegradable mulching films and analogous products, under the trade name Plastor. Nowadays, several agricultural plastic items based on oxo-biodegradable PE contain prodegradants comprising transition metal ions with organic ligands. They have been developed by several companies for sale as master batches (for blending on conventional equipment with normal resins) under different trade marks (TDPA, EPI Environmental Plastics Inc.), or end products (Envirocare, CIBA Specialty Chemicals).

Several studies have reported the significant reduction of molecular weight after thermal and photodegradation of PE samples containing prodegradants [20, 33] as well as the extraction, isolation and identification of oxidation products, including carboxylic acids, ketones, esters, and low-molecular-weight hydrocarbons [34, 35]. The overall rate and extent of the abiotic peroxidation of polyolefins are related to structural parameters such as chain defects and branching. The latter give rise to facile oxidation due to the susceptibility to hydrogen abstraction from tertiary carbon atoms through a vicinal hydrogen-bonded intermediate which can obviously be extensive in poly-α-olefins such as PP. The role of vicinal hydroperoxides is of particular importance in carbon chain cleavage and it leads to the release of small molecules carboxylic acids, alcohols, and ketones [20] even though random chain scission is considered to be the predominant process initially. Taking into account these considerations, it has been shown repeatedly that the decreasing order of susceptibility of polyolefins to peroxidation is: iPP > LDPE > LLDPE > HDPE [36, 37].

The ultimate environmental fate of "degradable" polyethylenes has to be recognized as the results of the combined action of abiotic factors and microorganisms. This has suggested the definition of "oxo-biodegradable" materials in keeping with the processes of biodegradation of lignin and natural rubber [11], for example, since the evaluation of the extent and rate of peroxidation of these materials represents a powerful tool for the prediction of their biodegradation. Several studies have therefore been carried out with the aim of determining the mechanisms of the photo- and thermal oxidation of polyolefins containing prodegradant additives. Nevertheless, most of these studies have been carried out under strictly controlled laboratory conditions, including accelerated conditions that cannot be considered as representative of natural environments. In fact, only a few investigations have been performed by assessing the synergistic effects of temperature, humidity, and sunlight exposure that are collectively involved in outdoor exposures. In the context of the general mechanism of the radical oxidation of PE (Scheme 16.1), the following parameters can be monitored during abiotic degradation testing of LDPE and LLDPE blown films: (i) weight variation; (ii) CO*i*; (iii) surface wettability; (iv) molecular weight changes; and (v) fractionation by solvent extraction. In particular, gravimetric analysis can be used effectively to understand the weight changes as a consequence of the oxygen uptake during the early stages of oxidation as well as the weight loss due to volatilization of low molar mass fragments after prolonged thermal and photodegradation.

Another powerful tool for the qualitative and quantitative evaluation of the oxidation processes is the CO*i* as determined by FTIR spectroscopy. It has been reported repeatedly that most of the degradation intermediates from PE peroxidation contain carbonyl groups so that their concentration, as determined by CO*i* measurements, can be used to monitor the progress of degradation [22]. These determinations are usually carried out on test films by recording the ratio of the optical density of the carbonyl absorption bands in the range 1780–1700 cm^{-1} to the optical density of the band at 1463 cm^{-1} (CH_2 in-plane vibration – scissoring peak). In addition, FTIR analyses provide information on the presence and formation over time of oxidation products with absorption maxima at 1712 cm^{-1} (carboxylic acids), 1723 cm^{-1} (ketones), 1730 cm^{-1} (aldehydes), and 1780 cm^{-1} (lactones) [35].

16.4.1 Biodegradation of Polyolefin Oxidation Products

The determination of the wettability of film surfaces by contact angle measurements may provide useful information about the increasing polarity of film surfaces as a consequence of oxidation and the formation of functional groups. This information is also useful in order to predict the propensity for microbial attack on PE films, by recognizing that one of the reasons suggested to explain the intrinsic recalcitrance of PE to biodegradation is the hydrophobic character that hinders the adhesion and interaction of microbial cells. In terms of potential ultimate biodegradation (i.e., conversion to CO_2 and H_2O, mineralization), the assessment of molecular weight changes is of fundamental importance as well. Indeed, it has been suggested from a theoretical point of view that since PE is a nominally straight-chain hydrocarbon, it should be metabolized according to the biochemical pathway for linear alkanes. On the other hand, it has been established that there is a molecular weight upper limit for the utilization of *n*-alkanes as a carbon source by microorganisms. Haines and Alexander established that linear hydrocarbons with more than 44 carbon atoms (tetratetracontane) cannot be metabolized by soil microorganisms [38]. Recently, this dimensional limit has been extended to 0.72 kDa corresponding to 60 carbon atoms in a study using single bacterial strains [39]. In any case, these limits are thought to be related to the bacterial metabolism of *n*-alkanes that need the accessibility to methyl chain ends by extracellular oxidizing enzymes to start the biodegradation process. Thus, the rate and eventually the ultimate extent of biodegradation of solid *n*-alkanes is strongly affected by the availability of $-CH_3$ chain ends susceptible to enzymatic oxidation. It follows that the chain ends present at the surface of solid *n*-alkane decrease with an increase in molecular weight with extremely low values in the case of high-molecular-weight PE.

Finally, other information about the relationship between the levels of oxidation reached during the abiotic stage of degradation of "degradable" PE – the molecular weight reduction as well as the potential to be biodegraded in the environment – can be effectively obtained by the fractionation of pre-aged specimens using solvent extraction. This procedure may also provide, especially if carried out by using solvents with different polarities, further information on the relative amounts of different classes (e.g., carboxylates, alkanes, etc.) of degradation products deriving from peroxidation and cleavage of PE chains.

16.4.2 Standard Tests

Further to the evaluation of the abiotic oxidation of "degradable" PE, the final step to be investigated in order to envisage the ultimate environmental fate of these materials is the estimation of the extent of biodegradation under different conditions. The requirement of two steps, abiotic and biotic, in the degradation mechanism of oxo-biodegradable plastic materials has recently led to the preparation and approval of ASTM D6954-04 "Standard guide for exposing and testing plastics that degrade in the environment by a combination of oxidation and biodegradation" [4]. This standard provides a framework to assess and compare the degree of degradation attainable under controlled thermal and photooxidation tests as well as the degree of biodegradation and ecological impacts in defined environments after abiotic degradation. Evaluations in ASTM D6954-04 are divided into three levels relevant to: (i) accelerated aging in standard tests for both thermal- and photooxidations and determination of the degree of abiotic degradation (Tier 1); (ii) measuring biodegradation (Tier 2); (iii) assessing the ecological impact after these processes (Tier 3). In order to implement Tier 1, the standard suggests the use of test conditions for thermal or photooxidation likely to occur in application and disposal environments for which the test material is designed. In other words, accelerated oxidation should be carried out at temperatures and humidity ranges typical of application and disposal conditions. Test materials resulting from the accelerated oxidation tests are therefore exposed to appropriate use or disposal environments (soil, landfill, compost) in standard respirometric (biometric) tests in order to assess the rate and the degree of biodegradation (Tier 2). Finally, any residues of the materials under test, deriving from both the abiotic oxidation stage and the biodegradation tests must be submitted to ecotoxicity tests to demonstrate their ultimate environmental compatibility (Tier 3).

As a case study, the oxo-biodegradation behavior of LDPE blown film containing proprietary prodegradant[1)] additives has been reported. In accordance with the general scheme of oxo-biodegradation, the study has been divided into two stages: (i) Tier 1 represented by the abiotic pre-treatment and structural characterization of the sample, and (ii) Tier 2 in which the ultimate biodegradation of the oxidized LDPE sample has been evaluated under different environmental conditions. Finally, the relationship between the degree of oxidation achieved during the abiotic oxidizing step and the propensity to biodegradation has been established. As repeatedly reported, the general mechanism of thermal or photooxidation is

1) Various patents assigned to EPI-Environmental Products Inc.–Vancouver, Canada.

Table 16.1 Structural changes recordable during the oxidation process of oxo-biodegradable polyolefins.

Parameters to be monitored during the preaging of LDPE		
Parameter	**Meaning**	
Weight variation	Increase: oxygen uptake	Decrease: loss of volatile intermediates
Carbonyl index by FTIR	Formation of cleavable oxidized groups in the main chain	Preliminary characterization of the oxidized functional groups
Wettability	Formation of an oxidized polar group at the surface	Increase: raise of the propensity to microbial colonization
Molecular weight	Evaluation of the degree of chain scissions	Decrease: raise of the propensity to microbial attack
Fractionation by solvent extraction	Estimation of the level of degradation	Characterization of oxidized intermediates

best described as an autocatalytic radical chain reaction leading to the oxidation and scission of polymer molecules, with the concomitant formation of oxidized low-molecular-weight fragments.

The parameters reported in Table 16.1 have been monitored during the preaging step carried out at 70 °C in an air convection oven. After an induction period, an appreciable weight increase according to a sigmoidal profile, attributable to the uptake of oxygen, is observed. Subsequently, as a result of prolonged treatment time, the sample weight starts to decrease owing to the loss of volatile (i.e., low molecular weight) degradation products. Since the carbonyl groups usually account for most of the oxidation products of the thermooxidative degradation of PE, the concentration of carbonyl groups on oxidation products, as determined from the CO*i*, can be used in monitoring the progress of degradation [40]. In line with the recorded weight increase, the CO*i* also shows a sigmoidal increasing shape. In addition, FTIR spectroscopy shows progressive broadening in the 1700–1780 cm^{-1} range for carbonyl groups with overlapping bands corresponding to carboxylic acids (1712 cm^{-1}), ketones (1723 cm^{-1}), aldehydes (1730 cm^{-1}), and esters (1740 cm^{-1}) during the aging period, thus indicating the formation of different oxidation products as aging progresses, as reported previously [41].

The increase in surface wettability during the oxidation of PE film containing prodegradant is an important indicator of the loss in hydrophobic character of this plastic. It has been shown that a few days of thermal treatment at 70 °C is sufficient to cause a dramatic decay in the contact angle of the LDPE film surface as hydrophobicity decreases. In addition, as a result of thermal oxidation, bulk density increases and film disintegration is observed, by floating the test film in tap water.

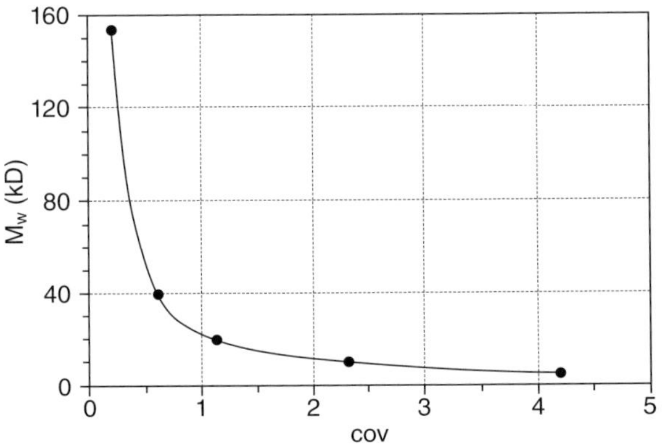

Figure 16.3 Molecular weight versus carbonyl index (CO*i*) relationship in LDPE film sample thermally treated in air in oven at 70 °C.

The disintegration of film samples occurred within 28 days, with film debris tending to sink to the bottom of the vessel. In parallel with the advancing oxidation process as monitored by CO*i*, weight changes and wettability increase during thermal degradation; a dramatic decrease in the molecular weight of the test sample has been recorded after a few days of oven aging. The progressive shift toward lower molecular weights as the CO*i* increases with the aging time can be observed by HT-GPC analysis. The relationship between the M_w and CO*i* can be expressed by a mono-exponential trend (Figure 16.3). Accordingly, CO*i* values may be used in order to predict the M_w decrease as a function of the level of oxidation. Moreover, the recorded trend is in agreement with a statistical chain scission mechanism, as suggested for the thermal- and photodegradation of polyolefins [13].

The feasibility to separate oxidized LDPE films into high- and low-molecular-weight fractions using a relatively simple extraction procedure with acetone has also been demonstrated [16]. In particular, the level of oxidation, as related to the carbonyl index, illustrates the increase in the amount of the solvent extractable fraction in parallel with a significant decrease in molecular weight. Accordingly, from heavily oxidized test samples, more that 25% by weight of acetone extracts can be obtained, thus also showing a very low *Mw* (0.85–1.05 kDa) (Figure 16.2). These data once more confirm that LDPE containing prodegradant additives can be effectively oxidized and massively degraded to low molar mass fractions which, owing to their wettability and polar functionality, become vulnerable to microorganisms.

16.4.3 Biometric Measurements

The second stage in the assessment of the environmental fate of "degradable" polyolefins is the evaluation of the ultimate biodegradation of them under different test conditions aimed at reproducing disposal or accidental littering environments.

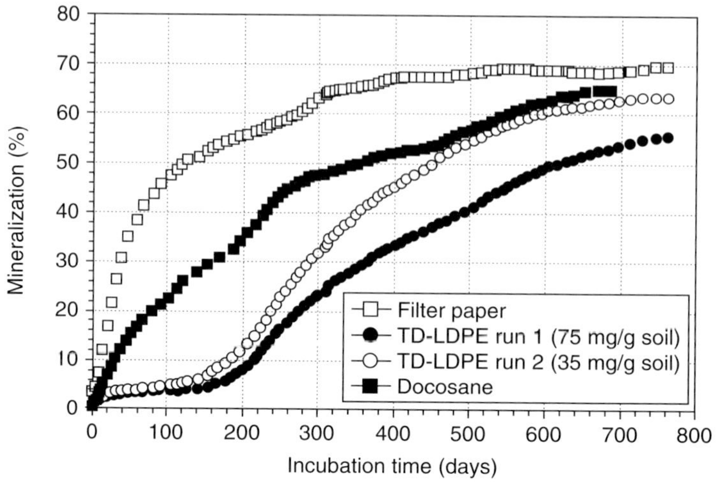

Figure 16.4 Mineralization profiles of thermally oxidized LDPE films and cellulose in a soil burial respirometric test.

In this connection, by using the materials retrieved from different abiotic degradation tests, several biometric respirometric tests have been carried out. The aim was to assess the susceptibility to mineralization under test conditions representative of soil, compost, and river water environments of LDPE that had reached different degrees of oxidation during the abiotic degradation stage.

Highly reproducible results have been obtained from several biodegradation tests carried out in soil burial biometer flasks aimed at assessing the biodegradation behavior of thermally oxidized LDPE film samples [42]. In all cases, the mineralization in soil of thermally oxidized samples does not show appreciable lag phases but it tends to a first plateau at about 5–7% mineralization in a few weeks (Figure 16.4). After that, a prolonged stasis (4–6 months) in the microbial conversion to CO_2 of the carbon in the samples has been observed repeatedly before further and markedly exponential increases in the biodegradation rate. Fairly high (55–65%) degrees of mineralization are observed after 18–24 months of incubation at room temperature. This two-step biodegradation behavior of thermally oxidized LDPE samples has been observed also in mature compost biodegradation tests. Therefore, in contrast to previous studies [41, 43] showing only limited and slow conversion to CO_2 of UV-irradiated LDPE samples, samples with no preaging and additive-free LDPE samples in natural soils, very large degrees of mineralization have been recorded although these were obtained over a relatively long time frame ranging between 22 and 30 months [42].

The first exponential phase, occurring during the first days of incubation in the biodegradation of thermally oxidized LDPE in soil, could be attributed to the fast assimilation by soil microorganisms of low-molecular-weight oxidized intermediates whose formation on the film surfaces has been demonstrated by the increased wettability observed during the abiotic stage of degradation. The ready biodegradability of these compounds has been suggested previously because they disappear once oxidized samples are incubated in the presence of hydrocarbon-degrading microorganisms such as *Arthrobacter paraffineus* [20, 40, 44, 45]. Additional support for this hypothesis is obtained from the FTIR characterization of the LDPE films exposed for a few months to soil microorganisms during biometric tests. Indeed, a significant reduction of the absorbance in the carbonyl region with respect to the values recorded at the beginning of the test has been observed repeatedly. In contrast, the number of double bonds in the carbon–carbon polymer chains was found to increase during the soil burial experiments, with a corresponding dramatic change in the fingerprint region of the IR spectra of the LDPE samples. These observations suggest that, during the soil burial tests, preoxidized LDPE samples undergo an ongoing degradation process, mediated by both abiotic and biological factors, which leads to the formation of large amounts of oxidized molecular fragments capable of being assimilated as carbon sources by soil microorganisms. This might explain the two-step biodegradation behavior that has been observed repeatedly for preoxidized PE films in soil.

The effect of different levels of oxidation as reached during the preaging (e.g., thermal degradation) step on the biodegradation propensity has also been evaluated in respirometric tests carried out in an aqueous medium in the presence of river water microbial populations. The complex biodegradation profile characterized by the presence of alternating plateau and exponential phases has been observed in this case also, which suggests that this behavior can be considered as typical of the biodegradation of oxidized LDPE. A straightforward relationship between the level of oxidation as determined by CO*i* and the biodegradation behavior of thermally oxidized LDPE samples has been observed also in biodegradation tests carried out in an aqueous medium [46]. In this experiment, heavily oxidized fractions of thermally degraded LDPE films as well as LDPE films having medium and high CO*i* values were supplied as the sole energy and carbon source in a mineral salt medium to a microbial consortium obtained from a river water sample. During incubation for 140 days at room temperature, degrees of mineralization ranging between 10% and 50% were recorded in the case of thermally oxidized LDPE samples having CO*i* values between 4.6 and 20.7, respectively (Figure 16.5). Negligible mineralization was observed in the case of lightly oxidized LDPE with a CO*i* value of only 2.3. These data also suggest, therefore, that readily biodegradable oxidized LDPE samples can be obtained, depending on the level of oxidation reached during the abiotic pretreatment.

16.5 Processability and Recovery of Oxo-biodegradable Polyolefins

The use of activating additives of the types described above does not affect the processing characteristics of conventional polyolefin resins. These are "run" on the usual equipment at normal speeds. The products are indistinguishable from

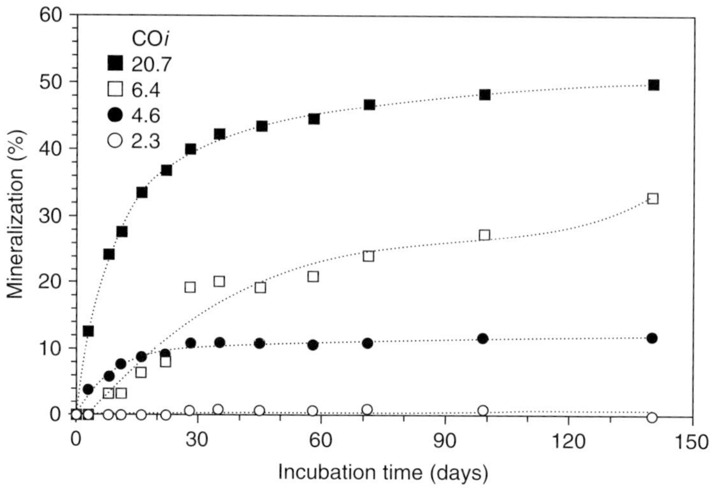

Figure 16.5 Mineralization profiles of thermally oxidized LDPE materials having different levels of oxidation in aqueous respirometric tests.

the same products made without the prodegradant additives. The oxo-biodegradable technology adds yet another desirable characteristic to the long list of useful properties for which the polyolefins are well known. This technology also provides environmental benefits at nominal extra cost that consumers would like to support but are usually reluctant to pay much of a premium for.

In many countries there are formal as well as informal programs for recycling postconsumer plastics. The recycling of used plastics can be a significant challenge, [47] but it is an important part of striving for a sustainable society. It is significant therefore to note that EPI's TDPA–PE materials, in spite of being oxo-biodegradable, can be recycled with regular PE recycling operations. This is because the prodegradants involved are not just simple oxidizing agents. The former do not affect the properties of the plastic until something else, for example, heat or UV light, initiates oxidative degradation, and this will not occur until all the antioxidants are consumed.

16.6
Concluding Remarks

For a number of decades, the polyolefins have been among the most useful and versatile materials. This is because they have a wide variety of desirable properties and are relatively inexpensive. In order to provide for years of reliable service, especially outdoors, it has been necessary to determine the kinetics and mechanisms by which polyolefins lose their useful properties over time, that is, to elucidate the details of oxidative degradation. On the basis of this fundamental information, highly effective stabilizers (e.g., radical scavengers, peroxide decomposers, UV absorbers) have been developed; they are used extensively in a wide variety of commercial formulations.

More recently, the demand for polyolefin products having a shorter lifetime has arisen, primarily in single-use packaging applications but also for a variety of agricultural products and in hygiene applications. As a result, oxo-biodegradable polyolefins have been invented and developed. These "clever" products are based on the following principles. (i) The actual requirement is for polyolefins having controlled lifetimes, that is, shelf life/use life combinations that can be varied between a few months and several years, depending on the formulation. (ii) In order to achieve such controlled lifetimes, it is required to enhance the rate of oxidative degradation – after the polyolefin articles have been used and discarded – by several orders of magnitude. This cannot be done simply by adding an oxidizing agent or by omitting the addition of stabilizers. It is being done by adding transition metal/fatty acid salts in catalytic quantities to conventional polyolefin resins prior to product fabrication. These salts catalyze the decomposition of hydroperoxide groups attached to the polymer molecules, but only after stabilizing additives in the resins have been depleted. (iii) Polyolefins are resistant to biodegradation by naturally occurring microorganisms, but their degradation products are biodegradable. (iv) The combination of abiotic oxidation and biodegradation provides for the required shelf life/use life values and sufficiently rapid bioassimilation to avoid the buildup of discarded plastics in a variety of environments. The oxo-biodegradation of suitable polyolefin formulations when buried in soil occurs at rates which permit the retention and use of as-produced biomass; (v) no toxic byproducts are produced in either the abiotic or subsequent biotic degradation of oxo-biodegradable polyolefins.

In the immediate future, it is expected that oxo-biodegradable polyolefins will become available for even more eco-compatible products and applications.

References

1 Guillet, J.E. (1995) Plastics and the environment, in *Degradable Polymers: Principles and Applications*, (eds G. Scott and D. Gilead), Chapman & Hall, London, pp. 216–246.

2 Wiles, D.M. (2005) *Oxo-biodegradable Polyolefins in Packaging*, in *Biodegradable Polymers for Industrial Applications*, CRC Press, FL, USA, pp. 437–450. Chapter 16.

3 ASTM (2009) D6400-04 Standard specification for compostale plastics. http://www.astm.org./Standards (accessed April 1, 2009).

4 ASTM (2009) D6954-04 Standard guide for exposing and testing plastics that degrade in the environment by a combination of oxidation and biodegradation. http://www.astm.org./Standards (accessed 1 April 2009).

5 Scott, G. and Wiles, D.M. (2001) *Biomacromolecules*, **2**, 615–622.

6 Sakai, K., Hamada, N., and Watanabe, Y. (1986) *Agric. Biol. Chem.*, **50**, 989–996.

7 Karsten, R. and Steinbüchel, A. (2005) *Appl. Environ. Microbiol.*, **71**, 2803–2812.

8 Scott, G. (1993) *Atmospheric Oxidation and Antioxidants*, Elsevier, The Netherlands.

9 Billingham, N.C.,and Calvert, P.D. (1983) The degradation and stabilisation of

polyolefins –An introduction, in *Degradation and Stabilization of Polyolefins* (ed. N.S. Allen) Applied Science Publishers, London, UK, pp. 1–28.
10 Guillet, J.E. and Norrish, R.G.W. (1954) *Nature*, **173**, 625–627.
11 Scott, G. (2000) *Polym. Degrad. Stab.*, **68**, 1–7.
12 Gilead, D. and Scott, G. (1982) *Developments in Polymer Stabilisation-5*, Applied Science Publishers, London.
13 Iring, M. and Tüdos, F. (1990) *Prog. Polym. Sci.*, **15**, 217–262.
14 Wiles, D.M. and Scott, G. (2006) *Polym. Degrad. Stab.*, **91**, 1581–1592.
15 Bolland, J.L. and Gee, G. (1946) *Trans. Faraday Soc.*, **42**, 236–243.
16 Chiellini, E., Corti, A., D'Antone, S., and Baciu, R. (2006) *Polym. Degrad. Stab.*, **91**, 2739–2747.
17 Hakkarainen, M. and Albertsson, A.-C. (2004) *Adv. Polym. Sci.*, **169**, 177–199.
18 Hakkarainen, M., Albertsson, A.-C., and Karlsson, S. (1997) *J. Environ. Polym. Degrad.*, **5**, 67–73.
19 Hakkarainen, M., Albertsson, A.-C., and Karlsson, S. (1997) *J. Appl. Polym. Sci.*, **66**, 959–967.
20 Albertsson, A.-C., Barenstedt, C., Karlsson, S., and Lindberg, T. (1995) *Polymer*, **36**, 3075–3083.
21 Karlsson, S. and Albertsson, A.-C. (1998) *Polym. Eng. Sci.*, **38**, 1251–1253.
22 Gugumus, F. (1996) *Polym. Degrad. Stab.*, **52**, 131–144.
23 Heskins, M. and Guillet, J.E. (1968) *Macromolecules*, **1**, 97–98.
24 Harlan, G. and Kmiec, C. (1995) Ethylene-carbon monoxide copolymers, in *Degradable Polymers: Principles and Applications* (eds G. Scott and D. Gilead), Chapman & Hall, London, UK, pp. 151–168. chap 8.
25 Guillet, J.E. (1973) US Patent 3 753 952.
26 Guillet, J.E. (1973) Polymers with controlled life times, in *Polymers and Ecological Problems* (ed. J.E. Guillet), Plenum, New York, USA, pp. 1–25.
27 Scott, G. (1994) Environmental biodegradation of hydrocarbon polymers: initiation and control, in *Biodegradable Plastics and Polymers* (eds K. Doi and R. Fukuda), Elsevier, Amsterdam, pp. 79–91.
28 Jakubowicz, I. (2003) *Polym. Degrad. Stab.*, **80**, 39–43.
29 Weiland, M., Daro, A., and David, C. (1995) *Polym. Degrad. Stab.*, **48**, 275–289.
30 Koutny, M., Lemaire, J., and Delort, A.-M. (2006) *Chemosphere*, **64**, 1243–1252.
31 Al-Malaika, S., Chakraborty, B., and Scott, G. (1983) *Dev. Polym. Stab.*, **6**, 73–80.
32 Al-Malaika, S. and Scott, G. (1983) *Polym. Degrad. Stab.*, **5**, 415–424.
33 Albertsson, A.-C., Erlandsson, B., Hakkarainen, M., and Karlsson, S. (1998) *J. Polym. Environ.*, **6**, 187–195.
34 Albertsson, A.-C., Barenstedt, C., and Karlsson, S. (1993) *Acta Polym.*, **45**, 97–103.
35 Khabbaz, F. and Albertsson, A.-C. (2000) *Biomacromolecules*, **1**, 665–673.
36 Iring, M., Foldes, E., Barabas, K., Kelen, T., and Tüdos, F. (1986) *Polym. Degrad. Stab.*, **14**, 319–332.
37 Winslow, F.H. (1977) *Pure Appl. Chem.*, **49**, 495–502.
38 Haines, J.R. and Alexander, M. (1974) *Appl. Environ. Microbiol.*, **28**, 1084–1085.
39 Heat, D.J., Lewis, C.A., and Rowland, S.J. (1997) *Org. Geochem.*, **26**, 769–785.
40 Albertsson, A.-C., Andersson, S.O., and Karlsson, S. (1987) *Polym. Degrad. Stab.*, **18**, 73–87.
41 Albertsson, A.-C. and Karlsson, S. (1988) *J. Appl. Polym. Sci.*, **35**, 1289–1302.
42 Chiellini, E., Corti, A., and Swift, G. (2003) *Polym. Degrad. Stab.*, **81**, 341–351.
43 Ohtake, Y., Kobayashi, T., Asabe, H., and Muratami, N. (1998) *Polym. Degrad. Stab.*, **60**, 79–84.
44 Volke-Sepulveda, T., Saucedo-Castaneda, G., Manzur-Guzman, A., Limon-Gonzalez, M., and Trejo-Quintero, G. (1999) *J. Appl. Polym. Sci.*, **73**, 1435–1440.
45 Bonhomme, S., Cuer, A., Delort, A.-M., Lemaire, J., Sancelme, M., and Scott, G. (2003) *Polym. Degrad. Stab.*, **81**, 441–452.
46 Chiellini, E., Corti, A., and D'Antone, S. (2007) *Polym. Degrad Stab.*, **92**, 1378–1383.
47 Karlsson, S. (2004) Recycled polyolefins. Material properties and means for quality determination, in *Long-Term Properties of Polyolefins* (ed. A.-C. Albertsson), Springer, Berlin, pp. 210–229.

Index

Handbook of Biodegradable Polymers: Synthesis, Characterization and Applications, First Edition. Edited by Andreas Lendlein, Adam Sisson.
© 2011 Wiley-VCH Verlag GmbH & Co. KGaA. Published 2011 by Wiley-VCH Verlag GmbH & Co. KGaA.